Chemoinformatics of Natural Products

Also of interest

Chemoinformatics of Natural Products
Volume 2: Advanced Concepts and Applications
Ntie-Kang (Ed.), 2020
ISBN 978-3-11-066888-9, e-ISBN 978-3-11-066889-6

Computational Sciences
Ramasami (Ed.), 2017
ISBN 978-3-11-046536-5, e-ISBN 978-3-11-046721-5

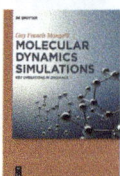

Molecular Dynamics Simulations.
Key Operations in GROMACS
Mongelli, 2021
ISBN 978-3-11-052605-9, e-ISBN 978-3-11-052689-9

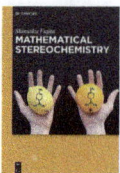

Mathematical Stereochemistry
Fujita, 2015
ISBN 978-3-11-037197-0, e-ISBN 978-3-11-036669-3

Physical Sciences Reviews.
e-ISSN 2365-659X

Chemoinformatics of Natural Products

Volume 1: Fundamental Concepts

Edited by
Fidele Ntie-Kang

DE GRUYTER

Editor
Dr. Fidele Ntie-Kang
Senior Scientist (AG Sippl)
Institute for Pharmacy
Martin-Luther-Universität Halle-Wittenberg
Kurt-Mothes-Str. 3
06120 Halle (Saale), Germany
fidele.ntie-kang@pharmazie.uni-halle.de

And

Department of Chemistry
Faculty of Science
University of Buea
P. O. Box 63, Buea
South West Region, Cameroon
fidele.ntie-kang@ubuea.cm

And

Department of Informatics and Chemistry
University of Chemistry and Technology
Prague
Technická 5 166 28 Prague 6
Dejvice
Czech Republic
ntiekanf@vscht.cz

ISBN 978-3-11-057933-8
e-ISBN (PDF) 978-3-11-057935-2
e-ISBN (EPUB) 978-3-11-057945-1

Library of Congress Control Number: 2019951926

Bibliographic information published by the Deutsche Nationalbibliothek
The Deutsche Nationalbibliothek lists this publication in the Deutsche Nationalbibliografie; detailed
bibliographic data are available on the Internet at http://dnb.dnb.de.

© 2020 Walter de Gruyter GmbH, Berlin/Boston
Cover image: iPandastudio / iStock / Getty Images Plus
Typesetting: Integra Software Services Pvt. Ltd.
Printing and binding: CPI books GmbH, Leck

www.degruyter.com

To Prof. Joseph T. Mbafor for his contributions to the advancement of natural products chemistry in Cameroon.

About the Editor

Dr. Fidele Ntie-Kang studied Chemistry at the University of Douala in Cameroon between 1999 and 2004, leading to BSc and MSc degrees. His Ph.D. work in Physical Sciences at the Centre for Atomic Molecular Physics and Quantum Optics (CEPAMOQ) was based on the computer-aided design of anti-tubercular agents. He later pursued a "Habilitation" in medicinal chemistry at Martin-Luther University, Halle-Wittenberg, Germany, after having carried out a period of extensive research on chemoinformatics applications for natural products drug discovery from African medicinal plants under Prof. Wolfgang Sippl. He is experienced in molecular modeling and has been involved in the design and management of databases of natural products from African flora for virtual screening. Fidele has formerly worked as a Scientific Manager/Senior Instructor at the Chemical and Bioactivity Information Centre (CBIC), hosted at the Chemistry Department of the University of Buea, Cameroon. Dr. Ntie-Kang currently teaches physical chemistry, computational chemistry and medicinal chemistry at the undergraduate and postgraduate levels in the University of Buea, Cameroon and in several summer schools, since 2008. Fidele has been awarded several fellowships and distinctions, including a doctoral fellowship from the German Academic Exchange Services (DAAD), Our Common Future Fellowship by the VolksWagen Foundation (Germany), UNIDO Fellowship in ICS-UNIDO Trieste (Italy), Commonwealth Professional Fellowship in Leeds (UK), Georg Forster Postdoctoral Fellowship by the Alexander von Humboldt Foundation (Germany), and ChemJets fellowship from the Czech Ministry of Youth, Sports and Education (Czech Republic). He is and has been a member of several scientific organizations, including the African-German Network of Excellence in Science (AGNES), the African Scientific Institute (ASI), the German Chemical Society, the African Network for Drug Diagnostics and Innovation (ANDI) and the Cameroon Academy of Young Scientists (CAYS). Fidele has published above 70 journal articles and book chapters, with over 1200 citations. He is a reviewer for several funding organizations, several high impact factor journals, editorial board member of several journals, and member of the scientific committee of international conferences, as well as a guest co-editor of special issues focused on natural products in the journals *Molecules* (MDPI) and *Frontiers in Pharmacology*.

https://doi.org/10.1515/9783110579352-202

Contents

Part I: Foundational Chemoinformatics Concepts for Natural Product-based Drug Discovery

Mohd Athar, Alfred N. Sona, Boris D. Bekono and Fidele Ntie-Kang

Fidele Ntie-Kang, Kennedy D. Nyongbela, Godfred A. Ayimele and Suhaib Shekfeh

Part II: Chemoinformatics Tools and Methods for Natural Product Structure Elucidation

Aurélien F. A. Moumbock, Fidele Ntie-Kang, Sergi H. Akone, Jianyu Li, Mingjie
Gao, Kiran K. Telukunta and Stefan Günther

Marilia Valli, Helena Mannochio Russo, Alan Cesar Pilon, Meri Emili Ferreira
Pinto, Nathalia B. Dias, Rafael Teixeira Freire, Ian Castro-Gamboa and
Vanderlan da Silva Bolzani

Marilia Valli, Helena Mannochio Russo, Alan Cesar Pilon, Meri Emili Ferreira
Pinto, Nathalia B. Dias, Rafael Teixeira Freire, Ian Castro-Gamboa
and Vanderlan da Silva Bolzani

Gabin T. M. Bitchagno and Serge A. F. Tanemossu

Part III: Chemoinformatics Tools and Methods for Lead Compound Discovery and Development

Eleni Koulouridi, Marilia Valli, Fidele Ntie-Kang and Vanderlan da Silva Bolzani

Rita C. Guedes and Tiago Rodrigues

Ricardo Bruno Hernández-Alvarado, Abraham Madariaga-Mazón and Karina
Martinez-Mayorga

Part IV: **Case Studies**

Samuel Egieyeh, Sarel F. Malan and Alan Christoffels

Samuel Egieyeh, Sarel F. Malan and Alan Christoffels
**16   Cheminformatics techniques in antimalarial drug discovery and
       development from natural products 2: Molecular scaffold and machine
       learning approaches —— 397**

Eleni Koulouridi, Sergi Herve Akone, Marilia Valli, Vanderlan da Silva Bolzani,
Fernanda I. Saldívar-González, Angélica Pilón-Jiménez, José L. Medina-Franco,
Berhanu M. Abegaz, Hanok Kinfe, Aurélien F. A. Moumbock, Mohd Athar, Gabin
T.M. Bitchagno, Serge A.T. Fobofou, David Newman, Rita C. Guedes, Tiago
Rodrigues, Maruca Annalisa, Bagetta Donatella, Lupia Antonio, Ricardo
B. Hernández-Alvarado, Abraham Madariaga-Mazón, Karina Martinez-
Mayorga, Stefan Günther and Fidele Ntie-Kang
**Glossary of terms used in chemoinformatics of natural products: fundamental
principles —— 417**

List of contributing authors

Berhanu M. Abegaz
Department of Chemistry
University of Johannesburg
Auckland Park Campus PO Box 524
Auckland Park 2006
Johannesburg,
Republic of South Africa
abegazm@uj.ac.za
Chapter 1

Sergi H. Akone
Faculty of Science
Department of Chemistry
University of Douala
P. O. Box 24157
Douala, Cameroon
And
Institute for Pharmaceutical Biology
and Biotechnology
Heinrich-Heine-University Düsseldorf
Universitätsstrasse 1
Geb. 26.23
40225 Düsseldorf
Germany
Chapter 6

Stefano Alcaro
Department of Health Sciences
University "Magna Græcia" of Catanzaro
Viale Europa
88100 Catanzaro
Italy
And
"Magna Græcia" University of Catanzaro
Net4Science Academic Spin-Off
"S. Venuta" Catanzaro
Italy
Chapter 12, 13

Francesca Alessandra Ambrosio
Department of Health Sciences
University "Magna Græcia" of Catanzaro
Viale Europa
88100 Catanzaro
Italy
And
"Magna Græcia" University of Catanzaro
Net4Science Academic Spin-Off
"S. Venuta" Catanzaro
Italy
Chapter 12, 13

Anna Artese
Department of Health Sciences
University "Magna Græcia" of Catanzaro
Viale Europa
88100 Catanzaro
Italy
And
"Magna Græcia" University of Catanzaro
Net4Science Academic Spin-Off
"S. Venuta" Catanzaro
Italy
Chapter 12, 13

Mohd Athar
School of Chemical Sciences
Central University of Gujarat
Sector 30
Gandhinagar 382030 Gujarat
India
mathar93@gmail.com
https://orcid.org/0000-0001-6337-1026
Chapter 2

Godfred A. Ayimele
Pharmacochemistry Research Group
Department of Chemistry
University of Buea
P. O. Box 63 Buea
Buea, Cameroon
Chapter 3

Donatella Bagetta
Department of Health Sciences
University "Magna Græcia" of Catanzaro
Viale Europa
88100 Catanzaro

https://doi.org/10.1515/9783110579352-203

Italy
And
"Magna Græcia" University of Catanzaro
Net4Science Academic Spin-Off
"S. Venuta" Catanzaro
Italy
Chapter 12, 13

Boris Davy Bekono
Department of Physics
Ecole Normale Supérieure
University of Yaoundé I
P.O. Box 47
Yaoundé, Cameroon
Chapter 2

Gabin T. M. Bitchagno
Institute for Organic Chemistry
Johannes Gutenberg Universitat Mainz
Mainz
Germany
And
Department of Chemistry
University of Dschang
Dschang, Cameroon
Chapter 9

Ian Castro-Gamboa
Nuclei of Bioassays
Biosynthesis and Ecophysiology of Natural
Products (NuBBE)
Department of organic Chemistry
Institute of Chemistry
São Paulo State University (UNESP)
Av. Prof. Francisco Degni 55
Araraquara, SP
Brazil
Chapter 7, 8

Raffaella Catalano
Department of Health Sciences
University "Magna Græcia" of Catanzaro
Viale Europa
88100 Catanzaro
Italy
And
"Magna Græcia" University of Catanzaro

Net4Science Academic Spin-Off
"S. Venuta" Catanzaro
Italy
Chapter 12, 13

Alan Christoffels
University of the Western Cape
School of Pharmacy
7535 Bellville
South Africa
Chapter 15, 16

Giosuè Costa
Department of Health Sciences
University "Magna Græcia" of Catanzaro
Viale Europa
88100 Catanzaro
Italy
Chapter 12

Vanderlan da Silva Bolzani
Nuclei of Bioassays
Biosynthesis and Ecophysiology of Natural
Products (NuBBE)
Department of organic Chemistry
Institute of Chemistry
São Paulo State University (UNESP)
Av. Prof. Francisco Degni 55
Araraquara, SP
Brazil
bolzaniv@iq.unesp.br
Chapter 7, 8, 10

Nathalia B. Dias
Scientific and Technological Bioresource
Nucleus (BIOREN)
Universidad de La Frontera (UFRO)
Francisco Salazar 01145
Temuco
Chile
Chapter 7, 8

Samuel Egieyeh
University of the Western Cape
South African Medical Research Council
Bioinformatics Unit
South African National Bioinformatics
Institute

7535 Bellville Cape Town
South Africa
And
University of the Western Cape
School of Pharmacy
7535 Bellville
South Africa
segieyeh@uwc.ac.za
Chapter 15, 16

Rafael Teixeira Freire
Medway Metabonomics Research
Group
University of Greenwich (UoG)
Chatham Maritime
Kent
United Kingdom
Chapter 7, 8

Mingjie Gao
Albert-Ludwigs-Universität Freiburg
Pharmaceutical Bioinformatics
Institute of Pharmaceutical Sciences
Hermann-Herder-Str. 9
79104 Freiburg
Germany
Chapter 6

Rita C. Guedes
Research Institute for Medicines (iMed.
Ulisboa)
Faculdade de Farmácia da Universidade
de Lisboa
Av. Prof. Gama Pinto
1649-003 Lisboa
Portugal
Chapter 11

Stefan Günther
Albert-Ludwigs-Universität Freiburg
Pharmaceutical Bioinformatics
Institute of Pharmaceutical Sciences
Hermann-Herder-Str. 9
79104 Freiburg
Germany
Chapter 6

Ricardo Bruno Hernández-Alvarado
Instituto de Química
Universidad Nacional Autónoma de México
Avenida Universidad 3000
Mexico City
Mexico
Chapter 14

Henok H. Kinfe
Department of Chemistry
University of Johannesburg
Auckland Park Campus PO Box 524
Auckland Park 2006
Johannesburg,
Republic of South Africa
hhkinfe@uj.ac.za
Chapter 1

Eleni Koulouridi
National and Kapodistrian University of Athens
Athens
Greece
elenikoulouridi@gmail.com
Chapter 10

Jianyu Li
Albert-Ludwigs-Universität Freiburg
Pharmaceutical Bioinformatics
Institute of Pharmaceutical Sciences
Hermann-Herder-Str. 9
79104 Freiburg
Germany
Chapter 6

Antonio Lupia
Department of Health Sciences
University "Magna Græcia" of Catanzaro
Viale Europa
88100 Catanzaro
Italy
And
"Magna Græcia" University of Catanzaro
Net4Science Academic Spin-Off
"S. Venuta" Catanzaro
Italy
Chapter 12, 13

Abraham Madariaga-Mazón
Instituto de Química
Universidad Nacional Autónoma de México
Avenida Universidad 3000
Mexico City
Mexico
Chapter 14

Sarel F. Malan
University of the Western Cape
School of Pharmacy
7535 Bellville
South Africa
Chapter 15, 16

Karina Martinez-Mayorga
Instituto de Química
Universidad Nacional Autónoma de México
Avenida Universidad 3000
Mexico City
Mexico
kmtzm@unam.mx
Chapter 14

Annalisa Maruca
Department of Health Sciences
University "Magna Græcia" of Catanzaro
Viale Europa
88100 Catanzaro
Italy
And
"Magna Græcia" University of Catanzaro
Net4Science Academic Spin-Off
"S. Venuta" Catanzaro
Italy
maruca@unicz.it
Chapter 12, 13

José L. Medina-Franco
Department of Pharmacy
School of Chemistry
Universidad Nacional Autónoma de Mexico
Av. Universidad 3000
Mexico City 04510
Mexico
jose.medina.franco@gmail.com
Chapter 4

Federica Moraca
Department of Health Sciences
University "Magna Græcia" of Catanzaro
Viale Europa
88100 Catanzaro
Italy
And
Department of Chemical Sciences
University of Napoli Federico II
Via Cinthia 4
I-80126 Napoli
Italy
And
"Magna Græcia" University of Catanzaro
Net4Science Academic Spin-Off
"S. Venuta" Catanzaro
Italy
Chapter 12, 13

Aurélien F. A. Moumbock
Albert-Ludwigs-Universität Freiburg
Pharmaceutical Bioinformatics
Institute of Pharmaceutical Sciences
Hermann-Herder-Str. 9
79104 Freiburg
Germany
aurelien.moumbock@pharmazie.uni-frei-
burg.de
Chapter 6

David J. Newman
Newman Consulting llc
Wayne
PA 19087
USA
djnewman664@verizon.net
Chapter 5

Fidele Ntie-Kang
Pharmacochemistry Research Group
Department of Chemistry
University of Buea
P. O. Box 63 Buea
Buea, Cameroon
And
Department of Pharmaceutical Chemistry
Martin-Luther University Halle-Wittenberg

Wolfgang-Langenbeck Str. 4
06120 Halle (Saale)
Germany
And
Department of Informatics and Chemistry
University of Chemistry and Technology
Prague
Technická 5 166 28 Prague 6
Dejvice
Czech Republic
fidele.ntie-kang@pharmazie.uni-halle.de
Chapter 2, 3, 6, 10

Kennedy D. Nyongbela
Pharmacochemistry Research Group
Department of Chemistry
University of Buea
P. O. Box 63 Buea
Buea, Cameroon
Chapter 3

Alan Cesar Pilon
Nuclei of Bioassays
Biosynthesis and Ecophysiology of Natural
Products (NuBBE)
Department of organic Chemistry
Institute of Chemistry
São Paulo State University (UNESP)
Av. Prof. Francisco Degni 55
Araraquara, SP
Brazil
And
Nucleus of Research in Natural Products
and Synthetics
Department of Physics and Chemistry
Faculty of Pharmaceutical
Sciences of Ribeirão Preto
São Paulo University
Ribeirão Preto
Brazil
Chapter 7, 8

B. Angélica Pilón-Jiménez
Department of Pharmacy
School of Chemistry
Universidad Nacional Autónoma de Mexico
Av. Universidad 3000

Mexico City 04510
Mexico
Chapter 4

Meri Emili Ferreira Pinto
Nuclei of Bioassays
Biosynthesis and Ecophysiology of Natural
Products (NuBBE)
Department of organic Chemistry
Institute of Chemistry
São Paulo State University (UNESP)
Av. Prof. Francisco Degni 55
Araraquara, SP
Brazil
Chapter 7, 8

Roberta Rocca
Department of Health Sciences
University "Magna Græcia" of Catanzaro
Viale Europa
88100 Catanzaro
Italy
And
Department of Experimental and Clinical
Medicine
Magna Graecia University and Traslational
Medicinal Oncology Unit
Salvatore Venuta University Campus
Catanzaro
Italy
And
"Magna Græcia" University of Catanzaro
Net4Science Academic Spin-Off
"S. Venuta" Catanzaro
Italy
Chapter 12, 13

Tiago Rodrigues
Instituto de Medicina Molecular
Faculdade de Medicina
Universidade de Lisboa
Av. Prof. Egas Moniz
1649-028 Lisboa
Portugal
tiago.rodrigues@medicina.ulisboa.pt
Chapter 11

Isabella Romeo
Department of Health Sciences
University "Magna Græcia" of Catanzaro
Viale Europa
88100 Catanzaro
Italy
And
"Magna Græcia" University of Catanzaro
Net4Science Academic Spin-Off
"S. Venuta" Catanzaro
Italy
Chapter 12, 13

Helena Mannochio Russo
Nuclei of Bioassays
Biosynthesis and Ecophysiology of Natural
Products (NuBBE)
Department of organic Chemistry
Institute of Chemistry
São Paulo State University (UNESP)
Av. Prof. Francisco Degni 55
Araraquara, SP
Brazil
Chapter 7, 8

Fernanda I. Saldívar-González
Department of Pharmacy
School of Chemistry
Universidad Nacional Autónoma de Mexico
Av. Universidad 3000
Mexico City 04510
Mexico
fer.saldivarg@gmail.com
Chapter 4

Suhaib Shekfeh
Medicinal Chemistry
MTS
Wolframstr. 3
86161 Augsburg
Germany
Chapter 3

Alfred Ndeme Sona
Department of Chemistry
University of Buea
P. O. Box 63 Buea

Buea, Cameroon
Chapter 2

Carmine Talarico
Department of Health Sciences
University "Magna Græcia" of Catanzaro
Viale Europa
88100 Catanzaro
Italy
And
"Magna Græcia" University of Catanzaro
Net4Science Academic Spin-Off
"S. Venuta" Catanzaro
Italy
Chapter 12, 13

Serge Alain Fobofou Tanemossu
Department of Biological Chemistry and
Molecular Pharmacology
Harvard Medical School
Boston, MA
USA
And
Institute of Pharmacy
Martin-Luther University Halle-Wittenberg
Halle Saale
Germany
And
Department of Bioorganic Chemistry
Leibniz Institute of Plant Biochemistry
Halle Saale
Germany
SergeAlain.FobofouTanemossu@ipb-halle.de
Chapter 9

Kiran K. Telukunta
Albert-Ludwigs-Universität Freiburg
Pharmaceutical Bioinformatics
Institute of Pharmaceutical Sciences
Hermann-Herder-Str. 9
79104 Freiburg
Germany
And
IT International Solar Energy Society eV
Wiesentalstr 50
Freiburg
Baden-Württemberg

Germany
Chapter 6

Marilia Valli
Laboratory of Medicinal and Computational
Chemistry (LQMC)
Centre for Research and Innovation in
Biodiversity and Drug
Discovery
Physics Institute of São Carlos
University of São Paulo (USP)
Avenida João Dagnone, nº 1100
São Carlos, SP
Brazil

And
Nuclei of Bioassays
Biosynthesis and Ecophysiology of Natural
Products (NuBBE)
Department of organic Chemistry
Institute of Chemistry
São Paulo State University (UNESP)
Av. Prof. Francisco Degni 55
Araraquara, SP
Brazil
mariliava@ifsc.usp.br
Chapter 7, 8, 10

Fidele Ntie-Kang

Editorial: Chemoinformatics at the service of natural product discovery

The term "*chemoinformatics*" (written interchangeably as "*cheminformatics*", but also known as "*chemical informatics*" or "*molecular informatics*") refers to a broad array of computer-based techniques and applications to solve chemistry problems. Generally speaking, *chemical informatics* helps chemists investigate new problems, organize and analyze scientific data in order to develop novel compounds, materials, and processes through the application of *information technology* (IT). *Chemical informatics* is an evolving field with many facets. Major aspects include:

- Information acquisition: methods for generating and collecting data empirically (experimentation) or theoretically (molecular simulation).
- Information management: storage and retrieval of information.
- Information use: data analysis, correlation, and application to problems in the chemical and biochemical sciences.

Chemoinformatics involves the application of IT to chemical data and includes topics such as the construction of chemical databases, combinatorial library design, structure-activity relationships (SAR), structure-based drug design (SBDD), text searching of structures and substructure searching of databases and patents. *Chemoinformatics* is closely related to *chemical computation* (e. g. quantum mechanics, statistical methods like regression, neural networks, quantitative structure-activity relationships (QSAR), graph theory, DNA computing, etc.), *molecular modeling* (e. g. 3D structure generation, 3D pharmacophore searching, docking, etc.) and *biopharmaceutical computation* (e. g. computer-aided drug design, combinatorial chemistry, protein and enzyme structure, membrane structure and ADME-related research). In the latter, the computer is used to analyze the interactions between the drug and the receptor site and to design molecules with an optimal fit. Once drug targets are developed, libraries of compounds are screened for activity with one or more relevant assays using high throughput screening (HTS). Virtual (computer-based) screening (VS) is the equivalent of HTS which helps to drastically reduce the number of compounds to be tested in assays. Basically, *chemoinformatics* for drug discovery (a branch of computer-based drug discovery) is a discipline that involves the mixing of information resources to transform data into information and information into knowledge, for the intended purpose of making decisions faster in the arena of drug lead identification and optimization.

Broad based *chemoinformatics* textbooks exist [1–3] and many chemoinformatics tools and methods have been dedicated to drug discovery [4]. However, no existing *chemoinformatics* textbook is exclusively dedicated to the investigation of natural

https://doi.org/10.1515/9783110579352-001

products (NPs) or secondary metabolites (SMs), a source of lead compounds of major significance for drug discovery [5]. This discipline has recently undergone serious advancement, leading to the development of various methods, software tools, databases and webservers that connect researchers involved in various cross disciplines (e. g. *bioinformatics, biophysics, biochemistry, computational biology, biostatistics and biopharmaceuticals*) [6]. This book has been written with inputs from several experts in the field, working in different continents and from diverse backgrounds. The focus has been on the application of *chemical informatics* methods to NP discovery and NP-based drug discovery. The book is divided into four parts; topics dealing with the fundamental chemoinformatics concepts for NP-based drug discovery, chemoinformatics tools and methods for NP structure elucidation, chemoinformatics tools and methods for lead compound discovery and development, the fourth and last part dealing with case studies.

The first volume of "Chemoinformatics of Natural Products" is focused on laying down the fundamental concepts, which any non initiated reader should grasp before further exploring the subject in the second volume of this series.

Part I of this volume examines the foundational chemoinformatics concepts for natural product-based drug discovery and begins with a description of the various structural classes of NPs, their structural diversity, bioactivity, and ecological functions [7]. This is followed by two chapters dealing with the concepts of "drug-likeness", "lead-likeness" and "NP-likeness" [8, 9]. Afterwards, a chapter is presented which deals with an investigation of the chemical space of SMs [10]: The first part then concludes with a survey of NPs already in the market or currently going through the pipeline towards approval as drugs [11].

In Part II, which comprises four chapters, the focus is on chemoinformatics tools and methods for NP structure elucidation, focusing on mass spectrometry (MS) and nuclear magnetic resonance (NMR) tools. This section begins with a descriptions of tools, software and methods for natural product fragment and mass spectral analysis [12], followed by a description of the computational background methods implemented in structure elucidation [13–15].

Part III is dedicated to chemoinformatics methods and tools proper for drug discovery, i. e. lead compound discovery and development. This section is vital for identifying if a SM could be a potential drug, as well as verifying its likelihood to portray adverse effects towards its development into a drug, e. g. if it is toxic. This part begins with a description of database resources for lead compound discovery *via* VS [16], drug target prediction using bio- and chemoinformatics approaches like pharmacophore screening, molecular docking, similarity-based searches, and statistical learning or machine intelligence [17]. Two further chapters describe VS methods and tools with several applications, successful VS campaigns that have led from nature to the identification of compounds with multi-target activity, along with natural compounds active on nucleic acids [18, 19]. Also given is a vivid presentation of molecular dynamics, metadynamics, and the use of 3D dynamic pharmacores (dynophores), i.e.

the application of molecular dynamics in pharmacore modeling for the identification of lead compounds from natural sources. This part of the book ends with methods for predicting if a SM could be toxic [20].

Part IV is dedicated to case studies. This section provides an analysis of SMs having anti-malarial activities using diverse chemoinformatics approaches, including the use of molecular descriptors, molecular diversity analysis, quantitative-structure activity/property relationships, and machine learning [21, 22].

Another essential feature of the book series is the inclusion of a pseudo chapter at the end, which has been dedicated to basic definitions of terms used in the book. This serves as a "Glossary of Terms Used in Chemoinformatics of Natural Products" and could really enhance reading, as a non-initiated reader has the chance to constantly look through the list to find new terms and get familiar with them through reading.

Acknowledgements: FNK acknowledges a return fellowship and an equipment subsidy from the Alexander von Humboldt Foundation, Germany. Financial support for this work is acknowledged from a ChemJets fellowship from the Ministry of Education, Youth and Sports of the Czech Republic awarded to FNK.

References

[1]   Leach AR, Gillet VJ. An introduction to chemoinformatics. Kluwer, 2003. ISBN 1-4020-1347-7.
[2]   Gasteiger J, Engel T, editors. Chemoinformatics: a textbook. Weinheim: Wiley-VCH, 2003. ISBN 3-527-30681-1.
[3]   Gasteiger J, editor. Handbook of chemoinformatics: from data to knowledge, vol. 4, Weinheim: Wiley-VCH, 2003. ISBN 3-527-30680-3.
[4]   Brown FK. Chemoinformatics, what it is and how does it impact drug discovery. Annu Rep Med Chem. 1998;33:375–84.
[5]   Newman DJ, Cragg GM. Natural products as sources of new drugs from 1981 to 2014. J Nat Prod. 2016;79:629–61.
[6]   Glen R. Developing tools and standards in molecular informatics. Chem Comm. 2002;23:2745–7.
[7]   Abegaz BM, Kinfe H. Secondary metabolites, their structural diversity, bioactivity, and ecological functions: an overview. Phys Sci Rev. 2018. DOI: 10.1515/psr-2018-0100.
[8]   Athar M, Sona AN, Bekono BD, Ntie-Kang F. Fundamental physical and chemical concepts behind "drug-likeness" and "natural product-likeness". Phys Sci Rev. 2018. DOI: 10.1515/psr-2018-0101.
[9]   Ntie-Kang F, Nyongbela KD, Ayimele GA, Shekfeh S. "Drug-likeness" properties of natural compounds. Phys Sci Rev. 2018. DOI: 10.1515/psr-2018-0169.
[10]   Saldívar-González FI, Pilón-Jiménez BA, Medina-Franco JL. Chemical space of naturally occurring compounds. Phys Sci Rev. 2018. DOI:10.1515/psr-2018-0103.
[11]   Newman DJ. From natural products to drugs. Phys Sci Rev. 2018. DOI: 10.1515/psr-2018-0111.
[12]   Moumbock AF, Ntie-Kang F, Akone HS, Li J, Gao M, Telukunta KK, et al. An overview of tools, software and methods for natural product fragment and mass spectral analysis. Phys Sci Rev. 2019. DOI: 10.1515/psr-2018-0126.
[13]   Valli M, Russo HM, Pilon AC, Pinto ME, Dias NB, Freire FT, et al. Computational methods for NMR and MS for structure elucidation I: software for basic NMR. Phys Sci Rev. 2018. DOI:10.1515/psr-2018-0108.

[14] Valli M, Russo HM, Pilon AC, Pinto ME, Dias NB, Freire FT, et al. Computational methods for NMR and MS for structure elucidation II: database resources and advanced methods. Phys Sci Rev. 2018. DOI:10.1515/psr-2018-0167.

[15] Bitchagno GT, Tanemossu, SAF. Computational methods for NMR and MS for structure eluci-dation III: more advanced approaches. Phys Sci Rev. 2018. DOI:10.1515/psr-2018-0109.

[16] Koulouridi E, Valli M, Ntie-Kang F, Bolzani VS. A primer on natural product-based virtual screening. Phys Sci Rev. 2018. DOI: 10.1515/psr-2018-0105.

[17] Guedes R, Rodrigues T. Drug target prediction using chem- and bioinformatics. Phys Sci Rev. 2018. DOI: 10.1515/psr-2018-0112.

[18] Maruca A, Ambrosio FA, Lupia A, Romeo I, Rocca R, Moraca F, et al. Computer-based techniques for lead identification and optimization I: basics. Phys Sci Rev. 2018. DOI: 10.1515/psr-2018-0113.

[19] Lupia A, Moraca F, Bagetta D, Maruca A, Ambrosio FA, Rocca R, et al. Computer-based techni-ques for lead identification and optimization II: advanced search methods. Phys Sci Rev. 2018. DOI: 10.1515/psr-2018-0114.

[20] Hernández-Alvarado RB, Madariaga-Mazón A, Noriega-Colima KO, Osnaya-Hernández A, Martinez-Mayorga K. Prediction of toxicity of secondary metabolites. Phys Sci Rev. 2019. DOI: 10.1515/psr-2019-0107.

[21] Egieyeh S, Malan SF, Christoffels A. Cheminformatics techniques in antimalarial drug discovery and development from natural products, part 1: basic concepts. Phys Sci Rev. 2018. DOI: 10.1515/psr-2018-0130.

[22] Egieyeh S, Malan SF, Christoffels A. Cheminformatics techniques in antimalarial drug discovery and development from natural products, part 2: machine learning approaches. Phys Sci Rev. 2018. DOI: 10.1515/psr-2018-0029.

Part I: **Foundational Chemoinformatics Concepts for Natural Product-based Drug Discovery**

Berhanu M. Abegaz and Henok H. Kinfe

1 Secondary metabolites, their structural diversity, bioactivity, and ecological functions: An overview

Abstract: Natural products are also called secondary metabolites to distinguish them from the primary metabolites, i.e. those natural compounds like glucose, amino acids, etc. that are present in every living cell and are used and required in the essential life processes of cells. Natural products are classified according to their metabolic building blocks into alkaloids, fatty acids, polyketides, phenyl propanoids and aromatic polyketides, and terpenoids. The structural diversity of natural products is explored using the scaffold approach focusing on the characteristic carbon frameworks. Aside from discussing specific alkaloids that are either pharmacologically (e.g. boldine, berberine, galantamine, etc.) or historically (caffeine, atropine, lobeline, etc.) important alkaloids, a single chart is presented which shows the typical scaffolds of the most important subclasses of alkaloids. How certain classes of natural products are formed in nature from simple biochemical 'building blocks' are shown using graphical schemes. This has been done for a typical tetra-ketide (6-methylsalicylic acid) from acetyl coenzyme A, or in general to all the major subclasses of terpenes. An important aspect of understanding the structural diversity of natural products is to recognize how some compounds can be visualized as key intermediates for enzyme mediated transformation to several other related structures. This is seen in the case of how arachidonic acid can transform into prostaglandins, or geranyl diphosphate to various monoterpenes, or squalene epoxide to various pentacyclic triterpenes, or cholesterol transforming to sex hormones, bile acids and the cardioactive cardenolides and bufadienolides. These are presented in carefully designed schemes and charts that are appropriately placed in the relevant sections of the narrative texts. The ecological functions and pharmacological properties of natural products are also presented showing wherever possible how the chemical scaffolds have led to developing drugs as well as commercial products like sweeteners.

Keywords: Natural products, secondary metabolites, alkaloids, fatty acids, prostaglandins, macrolides, phenyl propanoids, terpenoids, artemisinin, gossypol, isolongifolenone, phorbol, stevioside, cafestol, kahweol, steroids

This article has previously been published in the journal *Physical Sciences Reviews*. Please cite as:
Abegaz, B.M., Kinfe, H.H. Secondary metabolites, their structural diversity, bioactivity, and ecological functions: An overview. *Physical Sciences Reviews* [Online] **2019** DOI: 10.1515/psr-2018-0100

https://doi.org/10.1515/9783110579352-002

1.1 Introduction

1.1.1 Definitions

"Natural Products" is an imprecise term used to describe substances isolated from plants, microorganisms, insects, mammals, *etc*. Let us take each of the two words and discuss the issues that arise.

"Natural" is used to indicate that they are not synthetic. But, natural compounds such as those found in all living cells are **not included**. Like DNA, or the 20 amino acids that are involved in the natural synthesis of proteins, or the usual sugars like glucose, fructose, mannose, *etc*. that are essential for metabolic (building-up) as well as catabolic (breakdown) processes in the living cell. Such compounds are essential for the function and survival of the cell, be it that of animals, plants, or even microorganisms. The processes leading to these important and ubiquitous substances are called "primary metabolism".

The second word "Product" is used in a "chemical sense" to show that they are outputs of chemical reactions in the cell and by, necessity, mediated by enzymes. But "products" has broader meaning. "Compounds" is more precise than "natural products". Natural products are those compounds that are not needed in the fundamental life processes of living cells and so are not ubiquitous. The process of their biochemical synthesis is referred to as "secondary metabolism". Therefore, "Secondary metabolites" is more appropriate than "natural products", but, it might sound more technical for the non-scientific community and for this reason, both are used interchangeably.

1.1.2 The roles of secondary metabolites

Secondary metabolites do have roles for the plant or animal that has developed (evolved) them for whatever purposes. Insects have pheromones which are secondary metabolites that allow them to communicate with members of their species. Some are used to attract sexual partners (sex pheromones); others, to raise alarm in time of danger (alarm pheromones); and still others are produced to aggregate when group efforts are needed for action (aggregating pheromones). Likewise, the toxic, bitter, sweet, colorful, *etc*., constituents of plants have various important purposes such as to attract insects (pollinators), feeding deterrent, and protection to name a few. Some secondary metabolites such as flavonoids occur in many species, genera and even families, while others have restricted taxonomical distribution. Some metabolites are so rare that they are only found in a single taxon. For example, the cyanogenic pyridone alkaloid ricinine **1** (Figure 1.1) has been found only in the castor plant, *Ricinus communis* (Euphorbiaceae).

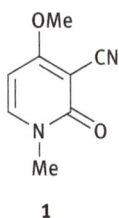

1

Figure 1.1: Pyridone alkaloid, ricinine, isolated from the castor plant *Ricinus communis*.

1.1.3 Classification of secondary metabolites

The most reasonable way to classify secondary metabolites is by grouping them according to biosynthesis pathways. Just as DNA is made from nucleotides with a unique assembly of four nitrogen bases (adenine, thymine, guanine and cytosine), polysaccharides are made from simple pentose and hexose sugars; just as proteins are made from amino acids, secondary metabolites are also made from simple building blocks, which vary for the respective classes of metabolites. For example, simple acetate in the form of a thioester, Coenzyme A, is the starting material for the biosynthesis of fats, prostaglandins, and a wide class of natural compounds known as polyketides. The *ca.* 30,000 naturally occurring alkaloids are largely derived from just a few amino acids: lysine, ornithine, histidine, *etc.* Phenyl propanoids are derived from the two aromatic amino acids: phenyl alanine and tyrosine. Mevalonate and methyleryhthritol are used to make one of the largest families of secondary metabolites called terpenoids that include steroids. The discussion in this chapter follows this biosynthesis-based classification. The respective building blocks will be introduced as we discuss the various classes of secondary metabolites in this chapter.

1.1.4 Structural complexity of secondary metabolites

The simplest secondary metabolite may very well be α-fluoroacetic acid **2** (Figure 1.2) first isolated from a South African plant *Dichapetalum cymosum* (Dichapetalaceae), commonly known as gifblaar (Afrikaans) or poison leaf, by Marais in 1944 [1]. This publication is on record reporting the *first* natural occurrence of a compound with a covalently bonded fluorine atom. α-Fluoroacetic acid is highly toxic and has been the cause of death of many animals and humans. Very complex natural products are also known with the dimeric iboga alkaloid **3** and ixoratannin A2 **4** serving as typical examples (Figure 1.2). The alkaloid **3**, for example, has no less than ten stereogenic centers and thus 2^{10} (1024) isomeric possibilities. It is a wonder that nature's enzyme systems have the sophistication and selectivity to make just one of them! Advances in NMR spectroscopy and X-ray crystallography have made it possible to elucidate such complex structures.

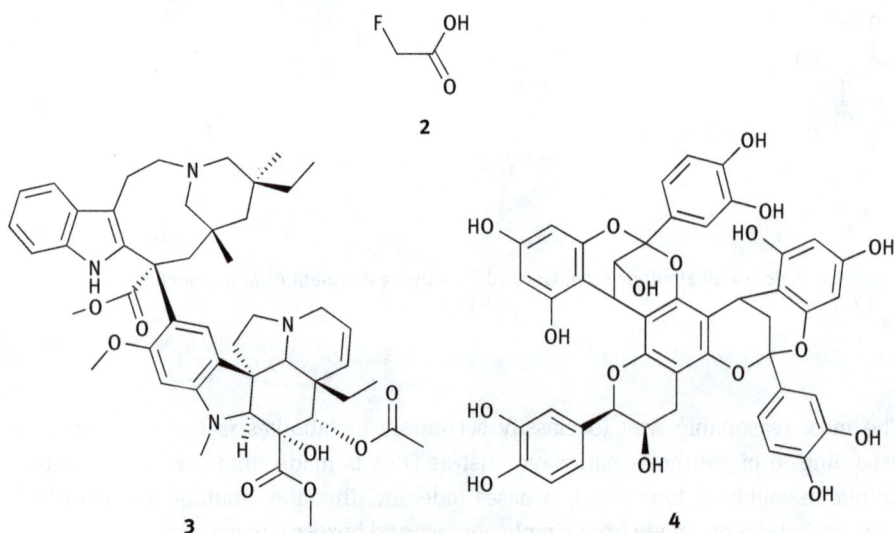

Figure 1.2: Examples of metabolites of different level of structural complexity.

1.1.5 Weaning and re-emergence of interest in secondary metabolites

Secondary metabolites have served as important sources of drugs in human as well as veterinary health. It is possible to give just a few examples of drugs derived from natural products without whose contributions it would be difficult to imagine what loss of life the world would have encountered. The natural product arsenals used against malaria (quinine and derivatives, artemisinin, *etc.*), the β-lactam antibiotics (the various penicillins) and the tetracyclines (TCs) against infectious diseases are just a few of many examples. The most popular anticholesterolemic drug (Lovastatin) for the control of the lifestyle-associated disease, hypercholesterolemia, has annual sales in tens of billions of dollars and has been a blockbuster and lucrative drug of the pharmaceutical industry.

Despite the foregoing success stories, there have been weaning and re-emerging interests during the last and current centuries. Critical issues for the loss of interest in drug discovery from natural sources have been partly economic as well as political (arising from the ways of implementing the convention on biodiversity and Nagoya Protocol by various governments) but by and large technical. The isolation and characterization procedures of secondary metabolites are rather tedious and even then, often lead to quantities that are too small for full evaluation of their drug potential. Another reason for loss of interest of drug companies has been the very low hit rate when random testing was employed. Many bioassay techniques have been challenged by interfering bulk substances like chlorophyll, tannins and polyphenols for which some solutions have been proposed [2, 3]. These have worked in some cases while others have been disappointing and frustrating when further fractionation of active extracts led to diminished activities presumably due to the loss of synergistic actions of constituents which fall into different fractions during the purification process. A resurgence in interest appeared with the development of multiple and high throughput screening procedures. This increased the rate of assays tremendously but did not significantly improve the hit-rates in relation to the investment that went into it. Hyphenated techniques and dereplication approaches did increase the hit rates and led to the discovery of novel molecules. Nevertheless, loss of interest in natural products surfaced again with the advent of combinatorial chemistry in the 1990s which did not deliver the expectations that existed in the beginning. Interest in natural products has resurfaced again in the twenty-first century with new genomic and metabolomic approaches [4] and with advances in computational methods. An organizational approach using the structural classification of natural products as organizing principle for charting the known chemical space explored by nature has been proposed [5]. "*In-silico*" screening including scaffold analysis [6, 7], topological consideration of the surfaces of the active site of enzymes, modelling of protein–protein interactions [8] and predicting how "metabolite like" natural products may prevent those interactions are opening new possibilities for developing the drug candidate libraries [9].

In the following sections, the important classes of secondary metabolites will be presented starting with alkaloids and their major sub-classes in Section 1.2. This will be followed by a succinct introduction on fatty acids, where reference is made to arachidonic acid as a bridge to link to the next section on prostaglandins, and the section concludes with a brief description on macrolides in Section 1.3. Mono- and poly-aromatics which include phenolics, quinones and chromanones as well as phenyl propanoids together with flavonoids that are formed by mixed biosynthetic processes will be introduced in Sections 1.4 and 1.5, respectively. The Section 1.6 and last sections are devoted to the largest class of natural compounds, namely terpenoids. Thereafter, the section discusses some representative members of the terpenes. Much care is taken to avoid discussion of the well-known examples like Taxol, vincristine, digitonin, *etc* since they are commonly described in standard text books that focuses on natural products. For others like podophylotoxin, betulinic acid and artemisinin, the discussion focuses on recent findings and new applications. We have also made reference to the work of computational chemists and chemoinformatic experts to enhance the chapter's relevance to the intended readers of the book. We have curtailed biosynthetic and synthetic details and inserted instead structure scaffolds in order to raise the interest of the target readers.

1.2 Alkaloids

Most alkaloids, nitrogen containing secondary metabolites, are derived from any of the seven amino acids, *viz.*, ornithine **5**, lysine **6**, tyrosine **7**, anthranilic acid **8**, picolinic acid **9**, histidine **10** and tryptophan **11** (Figure 1.3). The nitrogen of the

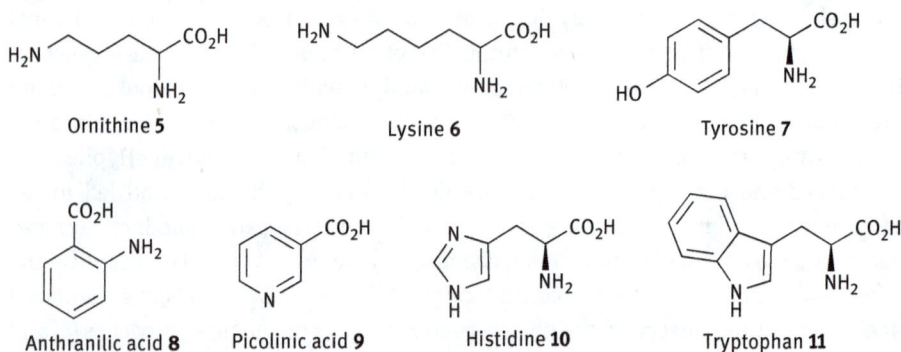

Ornithine **5**   Lysine **6**   Tyrosine **7**

Anthranilic acid **8**   Picolinic acid **9**   Histidine **10**   Tryptophan **11**

Figure 1.3: Amino acids that give rise to various alkaloids.

amino acid is retained often in a heterocyclic form. Some metabolic processes involving transamination reactions may introduce a nitrogen atom to a non-amino acid derived compound to give rise to a class of compounds which are referred to as "pseudoalkaloids" to denote the non-amino acid origin of the nitrogen contained in them. Steroidal alkaloids and xanthine alkaloids for instance, obtain their nitrogen in this manner.

Although the number of known alkaloids is estimated to be just fewer than 30,000, only 53 alkaloids have reached commercial exploitation by the pharmaceutical industry [10]. Computational chemists have reviewed these structures and applied predictive rules such as the Lipinski's rule of five[1] and concluded that 6000–7000 alkaloids have the potential for development and at least 600 of these alkaloids ought to have potential drug applications [10]. This may suggest that the full potential of these substances is still not realized and that the contributions of alkaloids as drug candidates are currently underexplored.

Owing to their large number and structural diversity, there are many ways of classifying alkaloids into sub-classes. The most common classification system is based on the basic heterocyclic chemical entity that appears in the structure of the alkaloid (Figure 1.4).

Below are some examples of the sub-classes of alkaloids that have either historical or current applications in the pharmaceutical industry.

---

1 The rule was developed to evaluate the drug likeness or determine if a chemical compound with a certain pharmacological or biological activity has properties that would make it a likely orally active drug in humans. Lipinski's rule states that, in general, an orally active drug has no more than one violation of the following criteria: (i) No more than 5 hydrogen bond donors (the total number of nitrogen–hydrogen and oxygen–hydrogen bonds); (ii) No more than 10 hydrogen bond acceptors (all nitrogen or oxygen atoms); (iii) A molecular mass less than 500 Daltons; and (iv) An octanol-water partition coefficient [11] log P not greater than 5. Note that there are four rules and the 5 in the rule of five is due to the numbers in the rules being multiples of 5.

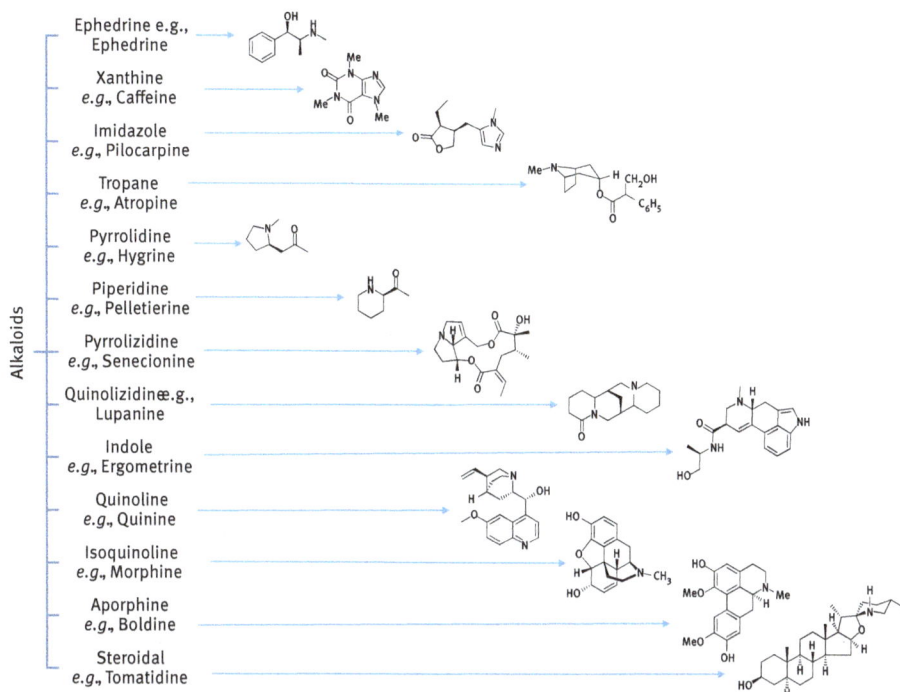

Figure 1.4: Classification of alkaloids.

1.2.1 Tropane alkaloids

The simplest alkaloids are the tropane alkaloids which contain a pyrrolidine moiety (derived from ornithine) with a second piperidine ring formed by condensation and cyclization of a three-carbon unit coming from two acetate units. Examples include:

Hyoscyamine (12, 13) and **atropine (14)** (Figure 1.5): these tropane alkaloids are found in several genera of the Solanaceae, including *Atropa, Datura, etc.* Hyoscyamine has *levo-* and *dextro*-forms arising from the chiral center α to the carbonyl group. The bicyclic ring has a chiral center but exists in the *meso-* form since it is symmetrical and hence does not lead to any optical isomerism. Although tropane nucleus contains two asymmetric carbon atoms at C-1 and C-5, it is optically inactive due to intramolecular compensation. However, the α-hydroxy group at C-3 is usually esterified with chiral tropic acid which is responsible of the optical activity of hyoscyamine.

The racemic (*R/S*) form of hyoscyamine is known as atropine (**14**). Historically, the plant source for the two enantiomers of hyoscyamine is *Atropa belladonna* (Solanaceae). The name belladonna is derived from the alleged use of a preparation from the plant by Italian women to dilate their pupils and look very attractive, and is

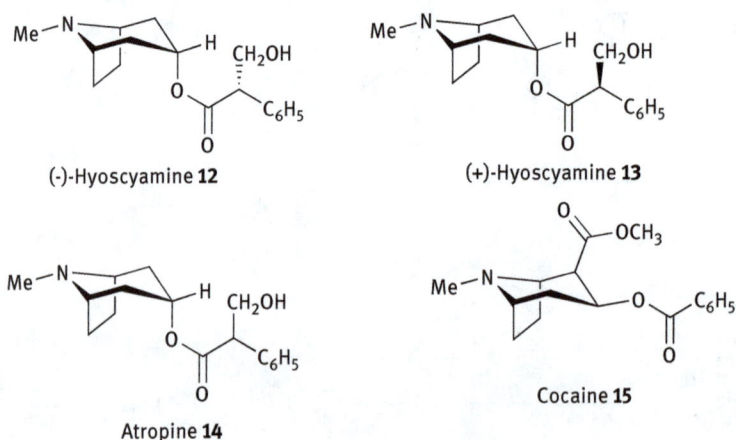

(-)-Hyoscyamine 12

(+)-Hyoscyamine 13

Atropine 14

Cocaine 15

Figure 1.5: Examples of commercially produced tropane alkaloids.

still in use by some modern fashion women to enhance their visual appeal [12]. Atropine was first synthesized by Willstätter [13]. Clinically it used to be an agent to reduce salivation and bronchial secretions prior to major surgeries. It is also used to dilate the pupil prior to eye examination and is prescribed but it is contraindicated to be used with people suffering from intraocular pressure (i.e., glaucoma patients). As an anticholinergic agent it is used to overcome the effects of exposure to organophosphorus pesticides. Atropine is medically used as antispasmodic, anti-Parkinson, cycloplegic drug and is sold under the names: Abdominol ®, Espasmo®, Protecor®, Tonaton®, etc.

Cocaine (15) has been in use in South America as far back as 3000 BC. Cocaine was first extracted from coca leaves in 1859 by German chemist Niemann. Austrian psychoanalyst Freud is reported to have been a habitual user of the substance and even published in 1884 an article in which he claimed the "benefits" of cocaine, and called it a "magical" substance. In 1886 Pemberton, a US pharmacist, and founder of Coca-Cola Company produced his beverage which included extracts of coca leaves as an ingredient. The popularity of the drink skyrocketed due to the euphoric and energizing effects of the drink by the turn of the century. As the dangers associated with the intake of cocaine became known, Coca-Cola Company succumbed to public pressure and removed the cocaine from the soft drink in 1903. Cocaine is an illegal drug, a powerful, addictive stimulant and is classified as a controlled substance. Possession of it can lead to serious penalties in many countries. It causes vasoconstriction and thus can be used clinically to reduce bleeding at the point of incision and as a topical anesthetic. Pharmacologically, cocaine is a parasympathomimetic; it acts as adrenergic stimulant by noradrenaline reuptake blockade. Its use as a local anesthetic is now entirely confined to ophthalmic, ear, nose and throat surgery. However, its use is strictly in highly regulated clinical settings.

1.2.2 Piperidine alkaloids

Piperidine alkaloids are derived from lysine, the higher homolog of ornithine and examples include the following:

Pelletierine (**16**), lobeline (**17**) and sparteine (**18**) (Figure 1.6). Pelletierine (**16**) is a piperidine alkaloid present in the bark of *Punica granatum* (Punicaceae) together with *N*-methylpelletierine and pseudopelletierine (**19**). It is used as a taenifuge to expel ringworms and nematodes.

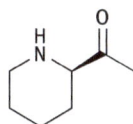

Pelletierine **16**

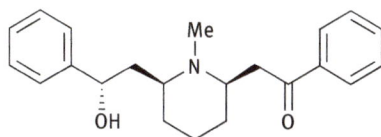

Lobeline **17**

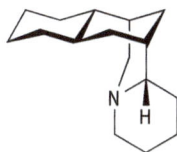

(-)-sparteine **18**

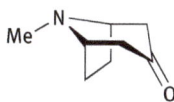

Pseudopelletierine **19**

Figure 1.6: Examples of piperidine alkaloids.

Lobeline (**17**) is the main alkaloid of *Lobelia inflata* (Campanulaceae), a plant native to northern North America. The plant is known to contain as many as 20 piperidine alkaloids and lobeline (**17**) is the major and the most biologically active alkaloid of the plant. The Penobscot Indians of America smoke the dried leaves of *L. inflata* as a substitute for tobacco. Lobeline (**17**) is currently the subject of renewed interest for the treatment of drug abuse and neurological disorders and may very well present a new class of promising therapeutic agents acting on the central nervous system. An excellent review on the history, chemistry and biology of alkaloids from *L. inflata* is published by Felpin and Leberton [14] for further reference.

1.2.3 Xanthine alkaloids

Purine itself does not occur in nature, but numerous derivatives are biologically significant. The pharmaceutically important bases of this group are methylated derivatives of the oxidized form of purine (**20**): 2, 6-dioxopurine which are usually designated as xanthines.

This is a small group of alkaloids containing the xanthine core structure (Figure 1.7). Three representatives, caffeine (**21**), theophylline (**22**) and theobromine (**23**) will be discussed.

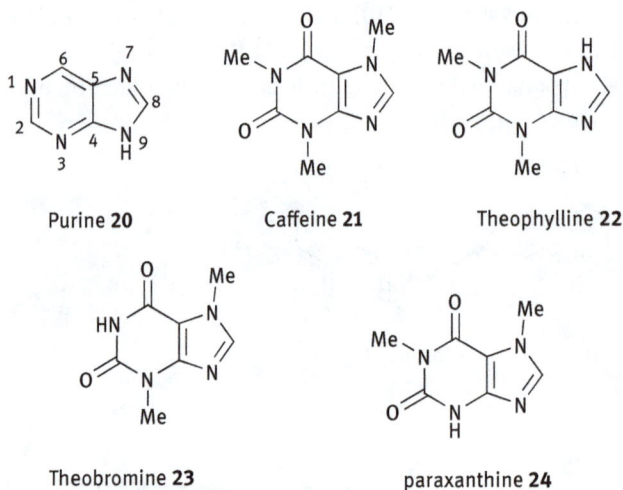

Figure 1.7: Examples of xanthine alkaloids.

Caffeine (21), theophylline (22) and theobromine (23): Caffeine is a stimulant found principally in coffee, tea and the cola nut, but it is known to occur in the seeds, nuts and leaves of as many as 60 plants. Although coffee is traded among all nations of the world it is produced only by a few countries in South America, Africa and Asia. The origin of coffee, *Coffea arabica* (Rubiaceae), is believed to be in the Kaffa region of Ethiopia, the only place in the world where it is found in the wild. Coffee is consumed primarily for two reasons: to overcome drowsiness and be alert as well as to improve physical performance. Caffeine has an additional use in medicine to assist infants with respiration. Caffeine has a striking resemblance to adenine and guanine, which are components of DNA. Caffeine has a remarkable solubility in both water (hot) as well as non-polar media and hence the ease with which it crosses the lipid barrier to get to the brain. Its stimulant properties are a result of its ability to reversibly block the action of adenosine by binding on its receptors. Coffee is also regarded as a diuretic, but this is probably due to the theobromine which is a metabolic product of caffeine.

Caffeine is metabolized in the liver by cytochrome P450 oxidase into three demethylated derivatives namely: paraxanthine (84 %), theobromine (12 %) and small amounts of theophylline (4 %). Paraxanthine (**24**), after formation in the liver, can produce increased lipolysis in the blood plasma. Theobromine (**23**) dilates blood vessels and increases the urine volume. Theophylline (**22**) can relax the bronchial smooth muscles but is not produced in sufficient quantities to be of therapeutic value during asthmatic attack.

Theophylline (**22**) occurs naturally in tea (*Camellia sinensis*) and cacao (*Theobroma cacao*). The stimulant effect of tea is partly attributed to the presence of theophylline. Clinical tests have indicated that theophylline improves the sense of smell of anosmia patients effectively [15]. It possesses broncho-dilating and anti-inflammatory effect and hence has been widely used clinically for the treatment of asthma and Chronic Obstructive Pulmonary Disease [16].

Despite their close relationships, the three xanthine alkaloids (**21–23**) were first isolated from different plants. The first to be discovered was caffeine by its isolation from coffee by Runge in 1819. Voskresensky first discovered theobromine (**23**) from extracts of cacao beans in 1841 while Kossel reported the first extraction of theophylline (**22**) from tea leaves [17]. The structural proof and total synthesis of all three was done after 1888 by Fischer who obtained the Nobel Prize 1902 for his work on these and other alkaloids.

Theobromine (**23**) is also found as a component of the cacao tree (*Theobroma cacao*, Malvaceae) and in the naturally caffeine-free plant commonly known as cacao tea (*Camellia ptilophylla*, Theaceae) as well as in *Camellia irrawadiensis* [18]. It is believed that the cacao tree produces the slightly bitter theobromine as a component of its chemical defensive mechanism. It also renders chocolate and cacao their typical bitter taste. Although theobromine is not toxic to humans, it is highly toxic and sometimes lethal to animals, especially domestic animals, if consumed [19]. It is, therefore, not wise to let dogs feed on chocolate as the theobromine present in the chocolate can kill them.

1.2.4 Benzyl isoquinoline alkaloids

There are many medicinally important alkaloids (Figure 1.8) that contain the benzyl-isoquinoline structural unit and examples include:

Papaverine (30): is a benzyl isoquinoline alkaloid isolated from the opium plant that has quite different pharmacological and structural features from the other opium alkaloids. It was first isolated in 1848 by Merck, the founder of the Merck pharmaceutical company (now merged with Sigma-Adrich). It is a direct-acting smooth muscle relaxant of the coronary, cerebral, pulmonary and peripheral vasodilation presumed to be *via* inhibition of the cyclic nucleotide phosphodiesterase and is mainly used for the treatment of erectile dysfunction [20]. It has also been reported to exhibit antiviral potency against respiratory syncytial virus, cytomegalovirus, measles and HIV [21].

Morphine (28): is the major alkaloid found naturally in opium poppy (*Papaver somniferum*, Papaveraceae). It is a narcotic analgesic mainly used in conditions requiring analgesic effect without the loss of consciousness. It is a highly addictive

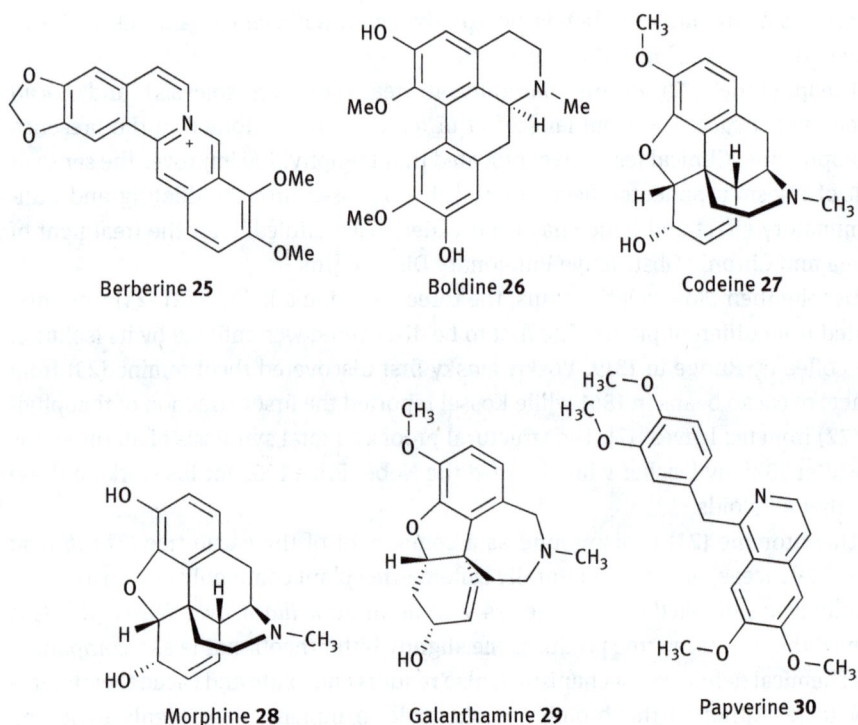

Figure 1.8: Examples of benzyl isoquinoline alkaloids.

drug and thus its use is highly controlled. Morphine was first isolated by Sertürner in 1804 from the opium poppy and in 1826 it was unveiled as the first commercial pure natural product to enter the drug market. Gates and Tschudi proved the structure of morphine by chemical synthesis [22]. Besides its use as an analgesic, it has been used as a precursor for the synthesis of codeine (*o*-methylated morphine) and the more potent and most commonly abused narcotic heroin (diacetyl morphine).

Codeine (**27**): is a mild sedative alkaloid isolated from the opium poppy, *Papaver somniferum* var. album (Papaveraceae). It is the most effective, safe and widely used analgesic, antitussive as well as anti-diarrheal drug and is ranked as one of the most commonly used drugs according to WHO. It was first isolated in 1832 by Robiquet. Considering the high worldwide demand and its presence in the opium poppy as a minor alkaloid, codeine is synthesized *via* O-methylation of the major alkaloid morphine (**28**) as well as *via* conversion of thebaine [23] which is the most abundant alkaloid isolated from *Papaver bracteatum* [24]. Although the use of codeine is regulated in most countries, codeine is sold over-the-counter in South Africa.

Boldine (**26**): is the major alkaloid found in the leaves and bark of a Chilean tree "boldo" (*Peumus boldus*, Monimiaceae). It belongs to the benzyl isoquinoline-derived family and accounts for 12–19 % and 6 % of the total alkaloid content of boldo leaves and bark, respectively. The plant contains more than 17 alkaloids, of which boldine is the major constituent. Lately, it has been isolated along with other three related alkaloids (norboldine, reticuline and linderegatine) from the roots of Linderaaggregate, which is found in China, Japan and Southeast-Asia [25]. Owing to its ability to scavenge highly reactive free radicals, it is regarded as one of the potent natural antioxidants [26]. Moreover, it is one of the Pan-African Natural Product Library that underwent virtual screening and was correlated with experimental data to be identified as a novel HIV-1 inhibitor [27].

Berberine (**25**): is a non-basic ammonium salt found as the major quaternary alkaloid in the stem, bark, root and rhizome of plants belonging to the genera *Berberis* (Berberidaceae), *Alpinia* (Zingiberaceae) and others. It is one of the few natural products that has undergone numerous clinical and preclinical tests. It possesses a wide spectrum of biological activities that include analgesic, antiarrhythmic, anti-bacterial, anticancer, antidepressant, anti-diarrheal, anti-inflammatory, antihypertensive, antihyperglycemic, antioxidant, anxiolytic, hypolipidemic activities as well as neuroprotective and memory enhancement properties [28]. Owing to its strong yellow colour, it has been used to dye wool, silk and leather. The chemical structure of berberine (**25**) was elucidated by Perkin and Robinson in 1910 [29] and Kametani and co-workers reported its total synthesis in 1969 [30].

Galantamine (**29**): is a naturally occurring alkaloid first isolated in 1947 from *Galanthus nivalis* (common snowdrop). It was later isolated from other *Galanthus* species (Amaryllidaceae). It gained reputation when it was approved for the treatment of Alzheimer's disease in 2000. Although there is yet no intervention to deal with the root cause of the disease, galantamine offers a modest symptomatic treatment [31]. It acts *via* a reversible inhibition of acetylcholinesterase and the complementary allosteric modulation of the nicotinic acetylcholinesterase receptors. Unlike physostigmine which inhibits both acetylcholinesterase and butyryl cholinesterase, galantamine is selective towards the inhibition of acetylcholinesterase. Furthermore, it has been reported to act as a potential antidote for organophosphate poisoning [32]. Barton and Kobayashi reported the first synthesis of galantamine [33]. In 1999, 1 Kg of galantamine was costing *ca*. $50,000, while currently 100 tablets of 4 mg cost *ca*. $65, which is still out of reach for many. Although several syntheses protocols have been reported, the commercial source of this important drug is extraction from the bulbs of daffodils.

1.3 Fatty acids

Fatty acids are straight chain carboxylic acids with <u>even</u> number of carbon atoms and chain length varying between 4 and 28 carbons. Fatty acids are biosynthesized from acetyl-CoA through an enzyme mediated decarboxylative condensation reaction with malonyl-CoA followed by the successive reduction, dehydration and hydrogenation of the resulting β-ketobutanoyl derivative. A repeat of such process leads to a product that grows by two carbons at a time. Hence the even chain length of the natural fatty acids. Examples of the saturated series of fatty acids are: butyric **31** (butanoic – C_4), caproic **32** (hexanoic – C_6), lauric **33** (dodecanoic – C_{12}), myristic **34** (tetradecanoic – C_{14}), palmitic **35** (hexadecanoic – C_{16}) and arachidic **36** (Eicosanoic – C_{20}) and are shown in Figure 1.9.

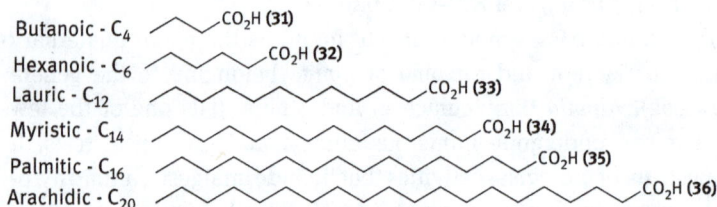

Butanoic - C_4

Hexanoic - C_6

Lauric - C_{12}

Myristic - C_{14}

Palmitic - C_{16}

Arachidic - C_{20}

Figure 1.9: Saturated natural fatty acids.

Unsaturated fatty acids are also known with comparable chain length as the saturated ones but containing from one up to six double bonds generally encountered in a conjugated manner. The geometry of these double bonds is "*cis*" and this gives the fatty acids a curved conformation. Some examples of unsaturated vegetable fatty acids include $\propto$-linoleic (**37**) (18:3, 9*c*,12*c*,15*c*), arachidonic (**38**) (20:4, 5*c*,8*c*,11*c*,14*c*), and eicosapentaenoic (**39**) (20:5, 5*c*,8*c*,11*c*,14*c*,17*c*) acids (Figure 1.10).

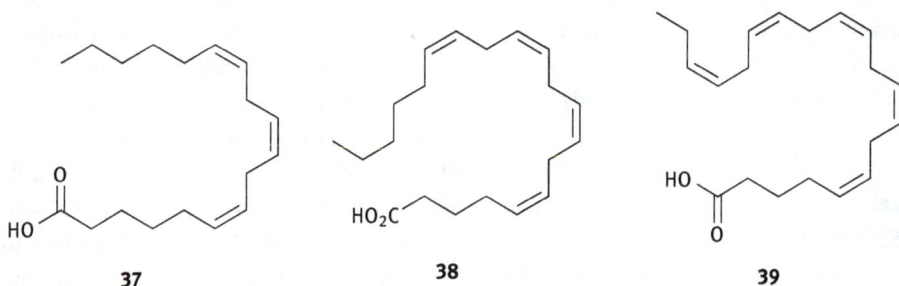

37

38

39

Figure 1.10: Poly-unsaturated natural fatty acids.

Omega fatty acids ($\acute{\omega}$-3, $\acute{\omega}$-6, $\acute{\omega}$-9) are a class of essential fatty acids found in fish oils, especially from salmon and other cold-water fish. They are claimed to lower the levels of cholesterol and low-density lipoproteins in the blood. The two main $\acute{\omega}$-3 fatty acids are eicosapentaenoic acid (**39**) and docosahexaenoic acid. The 3 in omega-3 is to show the position of the double bond on the third carbon from the methyl end of the chain (likewise, ώ-6 and ώ-9).

1.3.1 Prostaglandins – metabolites of arachidonic acid

Prostaglandins are present in very small quantities in the various tissues of the human body and perform important functions. As illustrated in Figure 1.11, they are derived from arachidonic acid (**38**). Prostaglandins were first isolated from sheep semen in the 1940s by von Euler. S. Bergstrom, a Swedish scientist, subsequently worked on prostaglandins and isolated sufficient quantities [34] to research on their properties. He shared the Nobel Prize in 1982 with B. I. Samuelsson and J. R. Vane for the isolation, identification and analysis of prostaglandins. There are well over 200 prostaglandins identified so far. Examples include prostaglandin E_2 (**41**) (PGE$_2$), prostaglandin D_2 (**42**) (PGD$_2$), prostaglandin I_2 (**43**) (PGI$_2$) and prostaglandin $F_{2\alpha}$ (PGF$_{2\alpha}$).

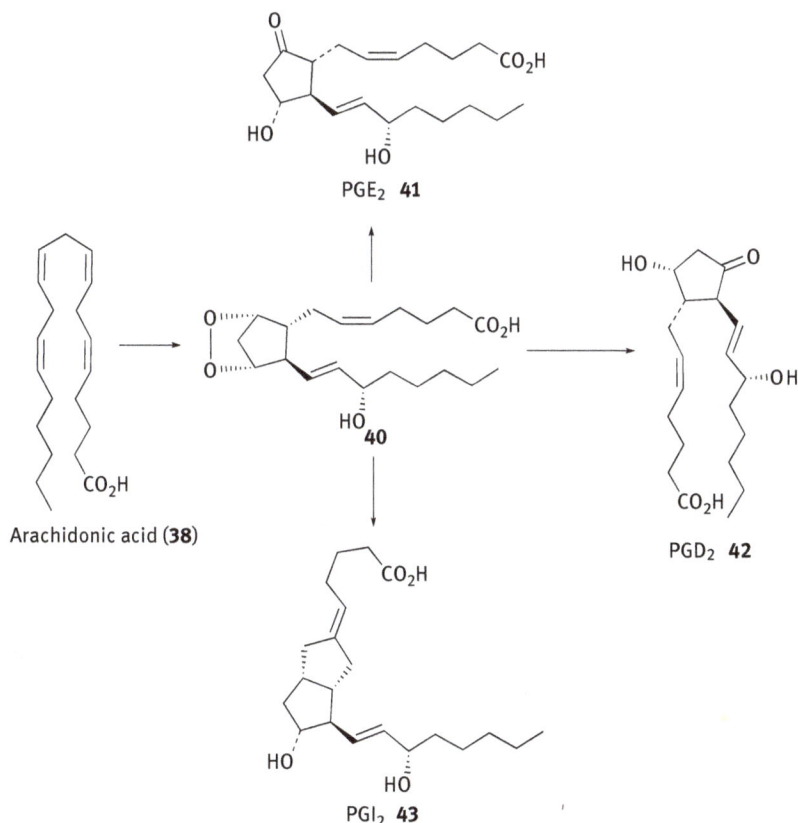

Figure 1.11: Biotransformation of arachidonic acid (**38**) to prostaglandins.

PGE$_2$ (**41**), also known as dinoprostone is used clinically to induce labor or for termination of pregnancy. It is also given to babies with congenital heart defects until surgery can be carried out.

Relatively large amounts of PGD$_2$ (**42**) are found only in the brain and in mast cells of mammals. Research has shown a clear connection between this substance and the development of allergic diseases such as asthma as well as the inhibition of hair growth. Topical application of PGD$_2$ (**42**) results in the inhibition of the growth of hair.

PGI$_2$ (**43**) also called prostacyclin is used to treat pulmonary artery hypertension. Besides being an inhibitor of platelet activation, it is also an effective vasodilator.

1.3.2 Macrolides

Macrolides are a group of macrocyclic natural products that have given rise to many important antibiotics such as erythromycins (**44**), avermictin (**45**) and azithromycin (**46**) (Figure 1.12). The word macrolide was initially coined by Woodward for medium ring sizes as in erythromycin A, cethromycin, azithromycin, spiramycin, tylosin, *etc*, but now even larger rings as big as 60 are known (see quinolidomicin A1 **47**) that belong to the group. Most of the antibiotics have been obtained from microorganisms (Streptomyces) but plants [35], lichens [36] as well as insects do produce macrolides. A homologous series of simple macrolides containing up to 36 membered rings (e.g., 36-hexatriacontanolide) have been isolated from the heads of the termite defense soldiers of *Amitermes neotenous* (Termitidae) [37]. The largest ring so far recorded is in quinlidoycin A1 isolated from the methanol extract of the cultured mycelium of an Actinomycete *Micromonospora* sp. by Japanese workers. This macrolide was found to be highly cytotoxic against P388 murine leukemia cells (IC$_{50}$ 8 nM) [38].

1.4 Polyketides

Polyketides occur widely in plants and microorganisms. These metabolites are referred to as tetra-, hexa-, octaketides, *etc.*, by considering the number of acetate groups that the substance is derived from. For example, the simple aromatic compound 6-methylsalicylic acid (**48**) is a tetraketide formed by a series of enzyme mediated reactions within the cell as shown in Figure 1.13. The enzyme systems are generally referred to as polyketide synthases. The process begins with thio-Claisen condensations of malonyl-CoA (and/or its derivatives) with acetyl Coenzyme A (AcetylSCoA).[2] The intermediate **I** could rearrange to the folded geometry to undergo the cascade of transformations to

---

[2] In the case of the macrolides the thio-Claisen condensation occurs between propionyl-CoA and methylmalonyl-CoA [39] p. 69.

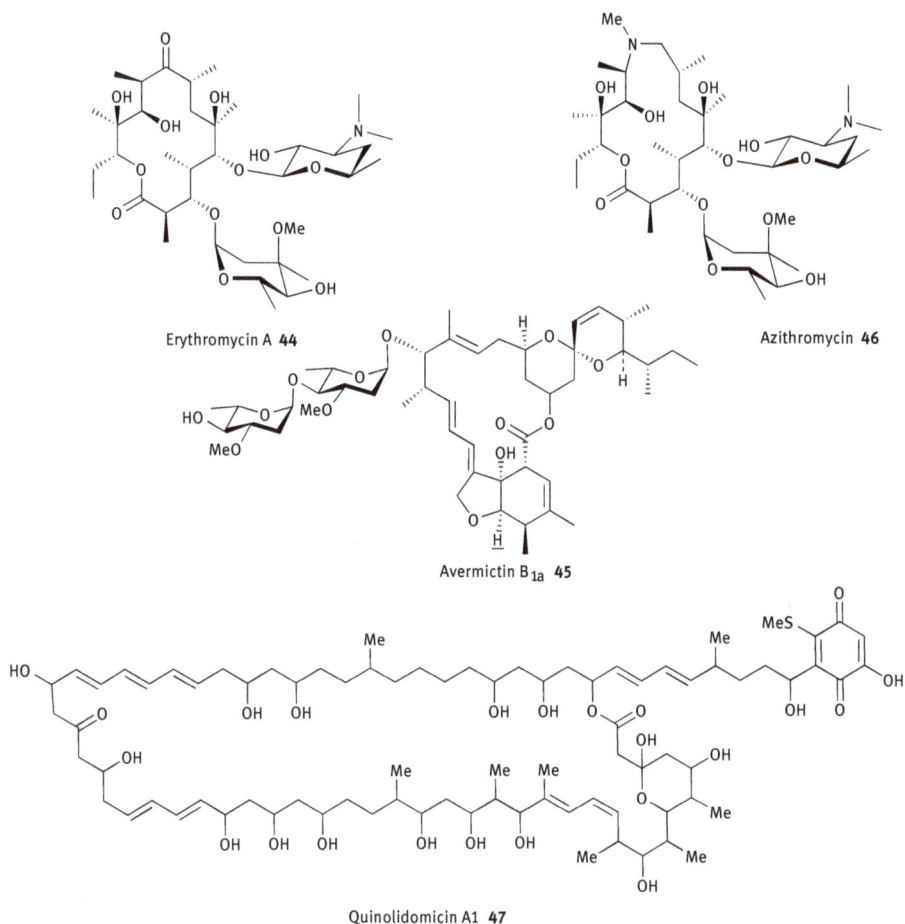

Erythromycin A **44**

Azithromycin **46**

Avermictin B$_{1a}$ **45**

Quinolidomicin A1 **47**

Figure 1.12: Examples of macrolides.

provide tetraketide **48**. A reduction of the keto group at position C-5 to an alcohol followed by elimination to form a double bond and aldol-like cyclization of the alpha carbon of the acid with the carbonyl at position C-7 followed by dehydration will form the six-membered ring, which will finally aromatize by enolization of the keto group at position C-3 to give 6-methylsalicylic acid (**48**).

Higher polyketides are formed from corresponding homologues following appropriate folding and transformations. The 5,7-dihydroxy-2-methylchromone (**49**) is a pentaketide, the aglycones of aloenin (**50**) and aloesin (**51**) are hexaketides while that of barbaloin (**52**) is an octaketide metabolite (Figure 1.14) [40]. The later three (**50, 51** and **52**) have been isolated as their glucose derivatives and are called glycosides. The compounds without the sugar moieties are called aglycones.

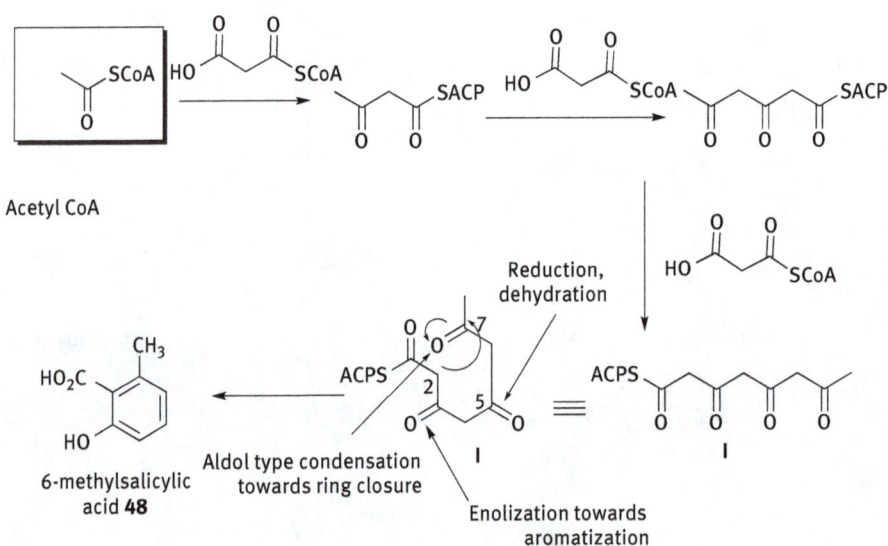

Figure 1.13: Biosynthesis of tetraketide 6-methylsalicylic acid (48).

Anthraquinones, which are octaketides are an interesting class of secondary metabolites and have been the subject of investigation for a long time. They are coloured substances and hence the extraction and separation of these compounds can be easily followed visually during the analytical and preparative procedures of thin layer and column chromatography. They are rather flat structures and therefore stereochemical considerations become important only in cases such as floribundone (53) [41] and knipholone (54) which give rise to atropisomers [42] (Figure 1.14). The chirality of these compounds arises due to the lack of axial symmetry along the axis joining the two aryl moieties. Conformationally, one moiety is along the plane of the paper while the other is orthogonal to it.

Quinones and anthraquinones are known to inhibit cytochrome P450. *In silico* studies to find effective inhibitors initially led to the identification of emodin **55** as a lead compound. Further docking studies and quantitative structure–activity relationship using the emodin (**55**) scaffold led to potential inhibitors of which the chloroamine derivative **56** was found to be up to 30 times more effective inhibitor [43].

Pravastatin (**57**), another example of a polyketide, is one of the most successful drugs in the market. It is a nonaketide and is a member of an important class of compounds known as statins. Statins are lipid-lowering drugs which reduce the level of cholesterol in the blood. When used in combination with diet, exercise and managed body mass it leads to effective lowering of cholesterol and helps to prevent cardiovascular disease. The mechanism of action is believed to be by competitively inhibiting β-hydroxy-β-methylglutaryl-**CoA** (HMG-CoA) reductase, the enzyme that

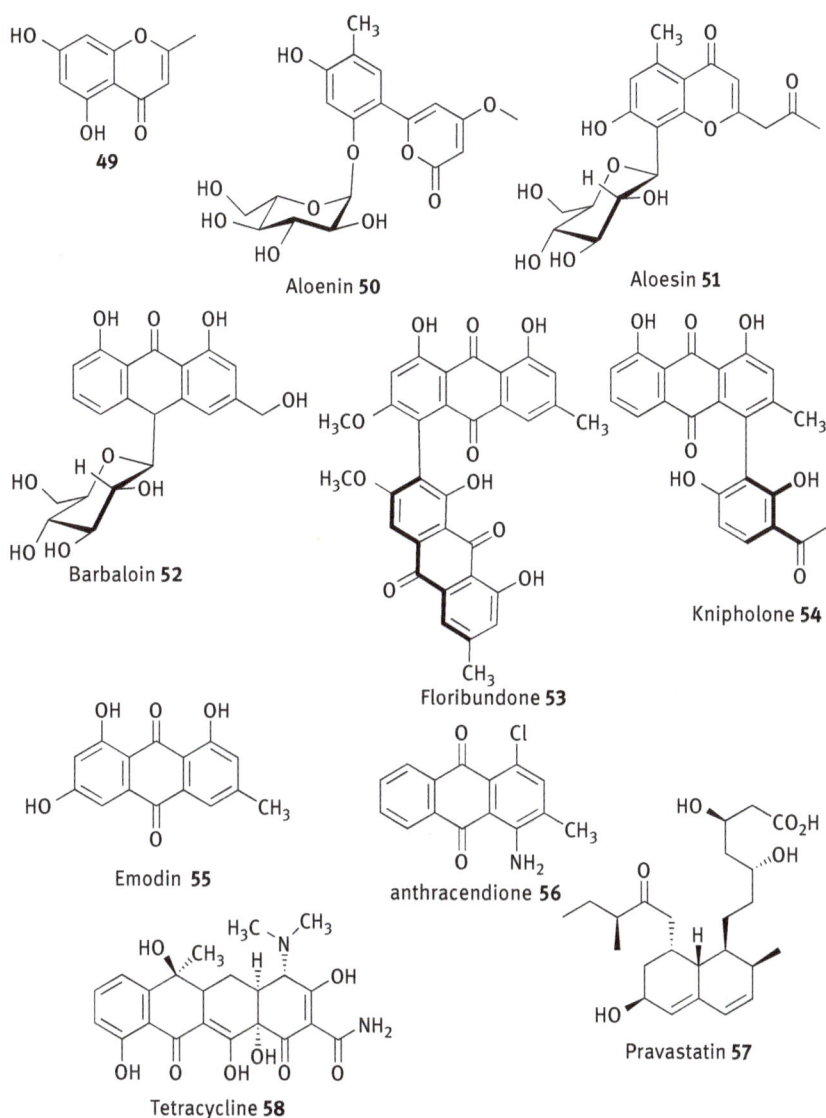

Figure 1.14: Examples of polyketides.

catalyzes the early rate-limiting step in cholesterol biosynthesis for conversion of HMG-CoA to mevalonate (MEV).

Tetracycline (TC) **58** is a broad-spectrum polyketide antibiotic produced by the *Streptomyces* genus of Actinobacteria. It has been used widely to treat respiratory infections such as pneumonia, acne, genital, urinary and skin infections. Its use (and abuse) over many decades has led to the development of resistant bacteria.

Tetracycline residues are now a serious concern and more effort is needed to find ways of effectively removing it from biological systems in the environment. A recent report addresses concerns of danger to human health from TC residues from the environment entering the body through the food chain and describes fluorescence spectroscopy and molecular docking methods under simulated physiological conditions to study the functional protein, bovine serum albumin (BSA) as a target of TC-induced toxicity. The aim is to examine BSA interactions with TC, anhydrotetracycline (ATC) and epitetracycline [44].

1.5 Phenyl propanoids and aromatic polyketides

1.5.1 Phenyl propanoids

Many natural products are made up of an Aryl-C-C-C scaffold or this unit may be a key component of the overall molecular architecture of a larger compound. These phenyl propanoids are derived from the aromatic amino acids: phenyl alanine **59** and tyrosine **7**. The compounds **60–67** are important examples (Figure 1.15).

Four examples of natural phenylpropanoid compounds with more complex structures are: rosmarinic acid **68**, syringin **69** and the lignans: pinoresinol **70** and podophyllotoxin **71** (Figure 1.15). Rosmarinic acid **68** and podophyllotoxin **71** are discussed below in detail.

Rosmarinic acid **68** is the ester of caffeic acid and 3,4-dihydroxyphenyllactic acid first isolated from *Rosmarinus officinalis* (Lamiaceae) in 1958 by Italian chemists. It is present in high amounts in the rosemary plant and in Perilla oil. Rosmarinic acid is reported to possess antioxidant, anti-inflammatory and antimicrobial activities. Its antioxidant property is greater than vitamin E. It is also used for preserving sea food in Japan. A 2003 review presents a comprehensive report on rosmarinic acid including its discovery, distribution in the plant kingdom, chemical and biosynthesis, biological properties and the state of knowledge for its production using plant *in vitro* cultures [45]. Several synthetic strategies are reported since then including a high-yielding synthesis using bioengineered *Escherichia* coli [46].

Podophyllotoxin **71** has been known for 140 years since it was first isolated in 1880 by Podwyssotzki from the slow growing North American plant *Podophyllum peltatum* (Berberidaceae). It has been used for the treatment of venereal warts and serves as a synthetic precursor for the chemotherapeutic drugs: etoposide **72** and teniposide **73** (Figure 1.16). Although the mechanism of action of podophyllotoxin as anticancer is based on its prevention of polymerization and assembly of tubulin into microtubules in a dose-dependent fashion these semisymmetric derivatives act by different mechanism of action, by inhibition of topoisomerase II (an enzyme that aids in DNA unwinding) for the treatment of testicular and lung cancer, leukemia, lymphoma and Kaposi's sarcoma.

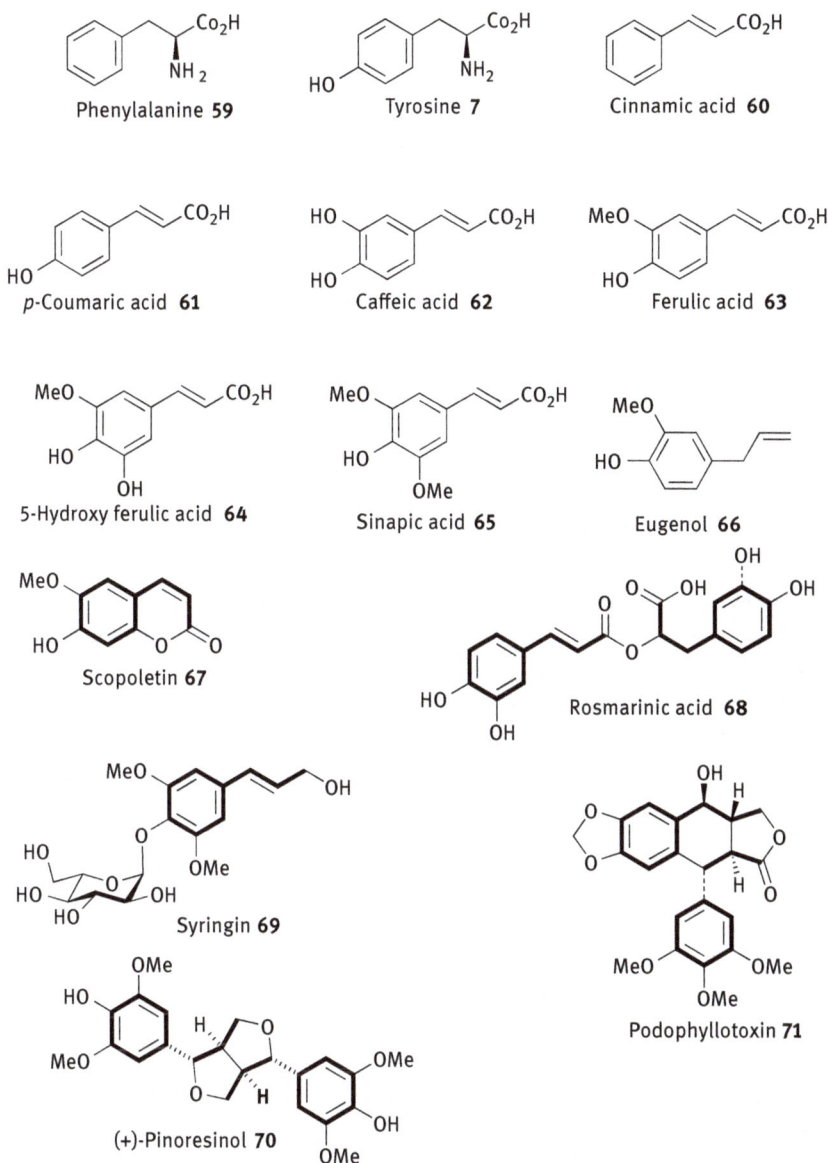

Figure 1.15: Examples of phenyl propanoids (the solid bonds represent the phenyl propane fragments).

The perceived future scarcity of podophyllotoxin as well as the desire to produce superior anticancer agents, has spurred activities in two areas. One is the search for new natural sources and as many as 38 plants have been identified [47] of which only the Himalayan mayapple (*Podophyllum hexandrum*) has served as a viable source for

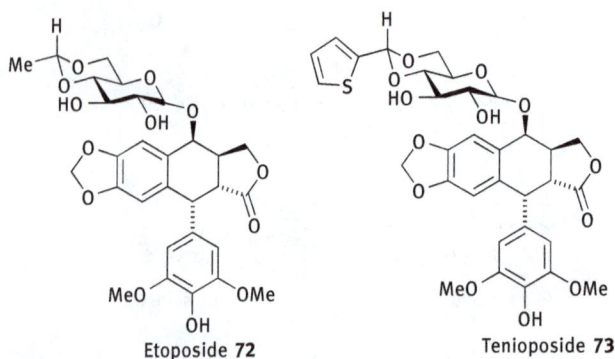

Figure 1.16: Podophylotoxin based drugs.

some time. It is important to note that podophyllotoxin is synthesized by the plant only when it is under stress such as being wounded and thus tedious manual wounding must be applied to the plant before harvest. Secondly, there have been no less than 30 total syntheses reported during the last 50 years with newer and more efficient processes being revealed [48–51], though none of them have been developed for its commercial production.

To solve the mismatch in the demand and supply of podophylotoxin a synthetic biology facility has been developed involving the engineering of the fast-growing tobacco plant *Nicotiana benthamiana* (Solanaceae) as a bioreactor to produce (-)-4'-desmethyl-epipodophyllotoxin [52], which is a direct precursor of etoposide **72**.

1.5.2 Aromatic polyketides-flavonoids

Flavonoids constitute a class of natural products which are widely distributed in plants. They are subdivided into small groups such as chalcones **74**, flavans **75**, flavanols **76**, flavanone **77**, flavone **78**, isoflavone **79**, anthocyanidins **80**, flavan-3-ol **81**, *etc*. In each group there are hundreds of compounds which differ in the number of OH groups and relative positions of the flavan scaffold and if they are methoxylated or not. The names of each subgroup and examples of compounds belonging to each are shown in Figure 1.17.

The structural complexity of flavonoids is not confined to just oxygenation alone. Glycoside formation through each of these hydroxyl groups (*O*-glycosides), and less frequently through carbon (*C*-glycosides), are possible. Furthermore, the positions *ortho*- and/or *para*- to each of the -OH carbons can be alkylated and this leads to, say, prenylated flavonoids which can undergo further elaboration including cyclization. Some examples of such derivatives are given below in Figure 1.18. Rotenone **82** is a natural insecticide and fish poison that is no longer used as of 2007 due to

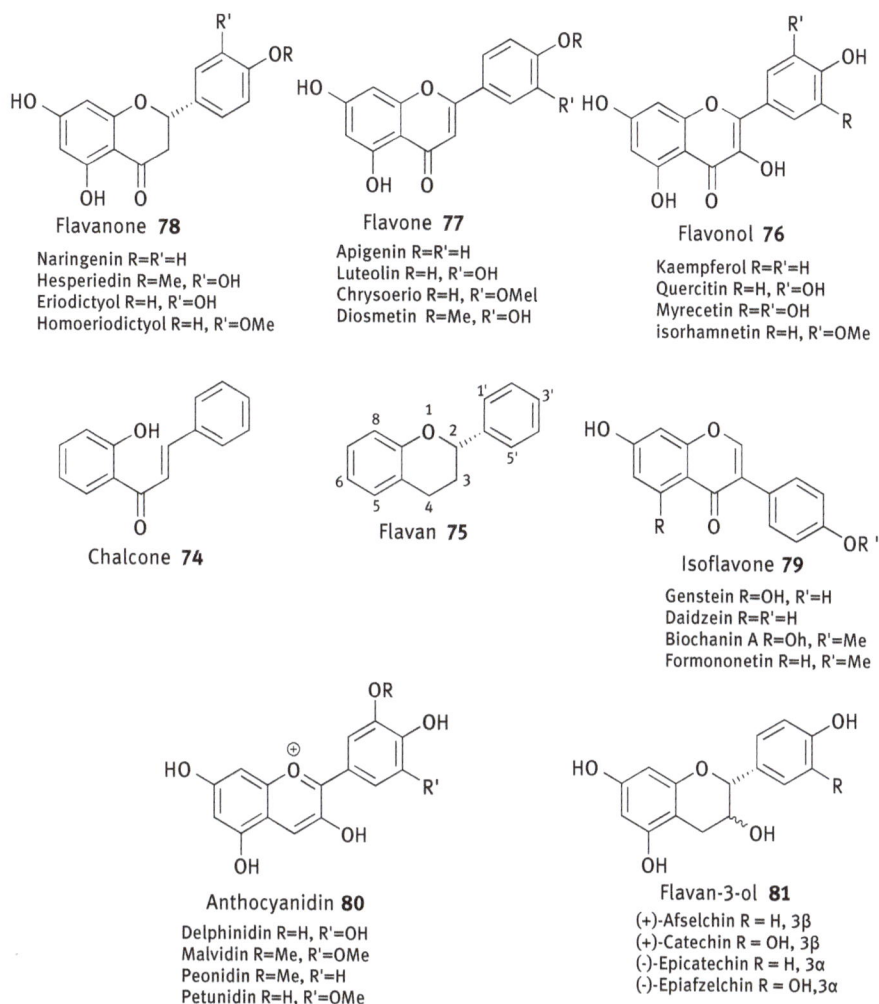

Figure 1.17: Different sub-groups of flavonoids and the chalcone precursor.

environmental concerns. Prorepensin **83** was reported from a Cameroonian medicinal plant whose total synthesis was also reported by Korean workers [53].

Flavonoids have received great attention during the last two decades because of their possible protective role against a number of age-related disorders. Fruits and vegetables contain very high levels of flavonoids besides other important constituents like fibers, vitamins (A, C and E) and β-carotene. Apples, chocolate and grapes contain fairly high levels of flavan-3-ols. Catechins occur in green tea but the amount is reduced significantly in black tea. The infusion of the former is rich in (-)-epigallo-catechin and (-)-epigallocatechin-3-O-gallate. Flavanols – in particular kaempferol,

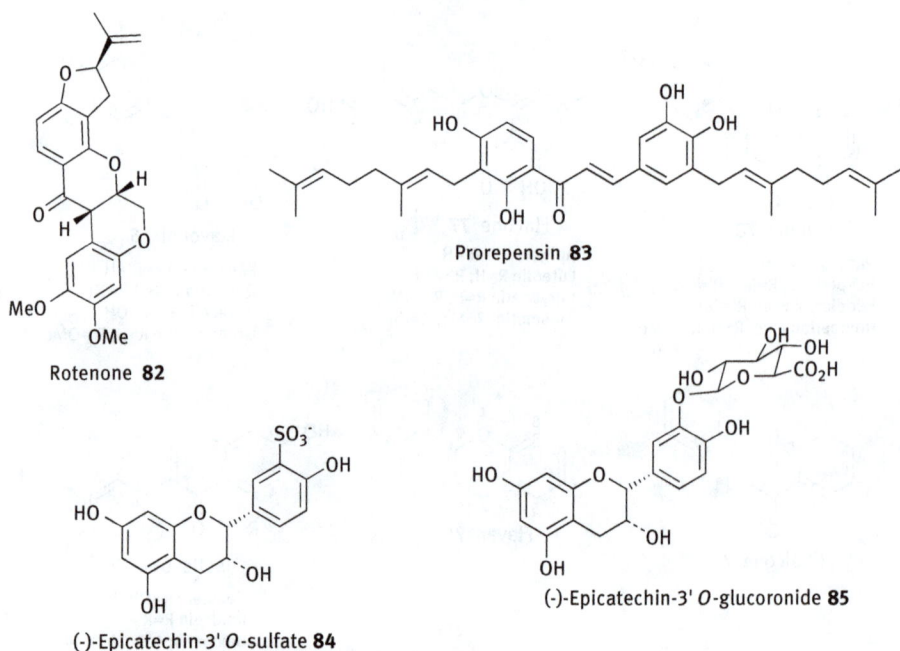

Rotenone **82**

Prorepensin **83**

(-)-Epicatechin-3' *O*-sulfate **84**

(-)-Epicatechin-3' *O*-glucoronide **85**

Figure 1.18: Metabolic products of flavonoids.

quercetin, isorhamnetin and myricetin – are ubiquitous in edible plants such as broccoli, tomatoes, leeks, black grapes, blueberries, blackcurrants, apples and apricots. Onion bulbs are major sources of dietary flavonols, and as many as 25 different flavonols have been isolated from them. It should be noted that humans do not biosynthesize flavonoids. Recent dietary studies have addressed the issue of their metabolism, and bioavailability in humans [54]. Flavonoids are modified and conjugated into a variety of derivatives such as conjugation with glucuronic acid, sulfate and methyl groups. Examples of identified human metabolites include epicatechin-3'-O-sulphate **84**, epicatechin-3'-O-glucoronide **85** (Figure 1.18), *etc*.

1.6 Terpenoids

Terpenoids are a class of natural products that arise from the isopentenyl-dimethylallyl pyrophosphate (IPP ⇋ DMAPP) building block as shown in Figure 1.19. There are two established biosynthetic routes to the IPP/DMAPP system, namely the MEV pathway and the 2-*C*-methyl-*D*-erythritol-4-phosphate pathway. The detailed steps in the biosynthesis and presentation of the enzymes and cofactors involved are beyond the intended purposes of this chapter, and hence the interested reader is referred to standard texts on the subject [39].

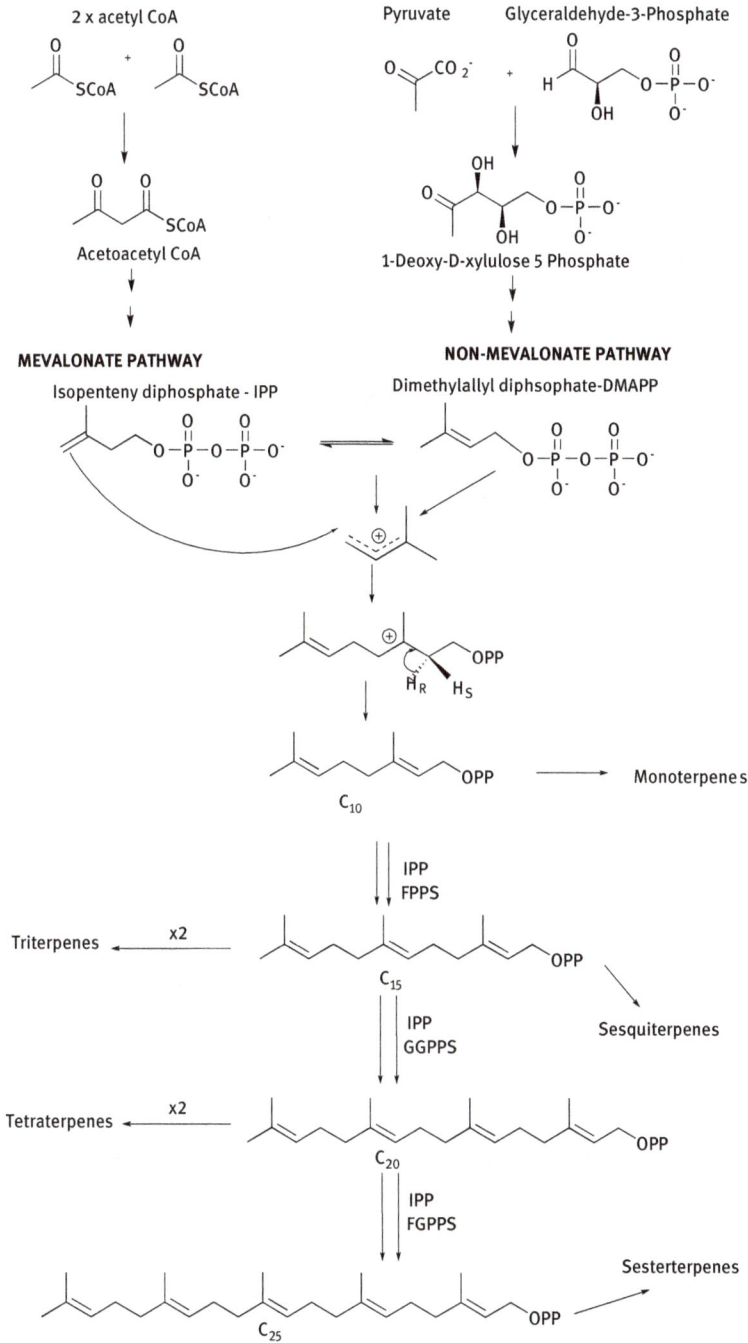

Figure 1.19: Biosynthesis of terpenoids.

1.6.1 Monoterpenes

Monoterpenes **87–93** are a class of terpenoids that are made up of ten carbon units. Many of the flavour and fragrance compounds like menthol **87**, citral **92**, geraniol **91**, pinenes **90**, limonene **88**, carvone **93**, *etc.* (Figure 1.20) are examples of monoterpenes. Besides providing pleasant smell they also have medicinal applications. Limonene is used as an antianxiety and an antidepressant agent. α-Pinene is an anti-inflammatory, bronchodilator, antibacterial and also helps in enhancing memory. Geraniol acts as a mosquito repellant.

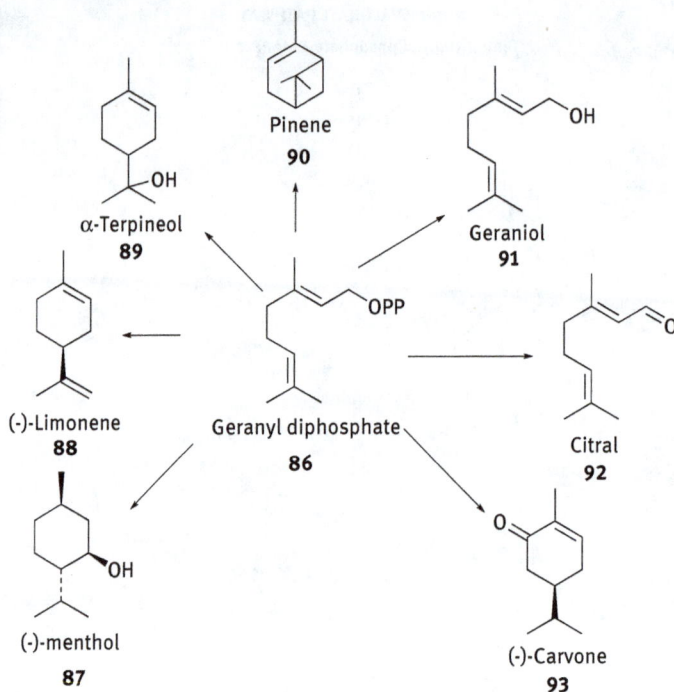

Figure 1.20: Examples of monoterpenes.

Monoterpenes (*S*)-ipsdienol **94** and (*S*)-ipsenol **95** (Figure 1.21) are the aggregating pheromones of the *Ips* spp. bark beetles. The precise ratio of these compounds is

Figure 1.21: The structures of (*S*)-ipsdienol 94 and (*S*)-ipsenol 95.

regulated by the enzymes that make them and is a key factor in determining their semiochemical function.

1.6.2 Sesquiterpenes

Sesquiterpenes may formally be considered as higher terpene homolog of the monoterpenes formed by addition of one isopentenyl group to give a class of compounds with 15 carbon atoms. Examples include: costunolide **96**, dehydrocostuslactone **97**, artemisinin **98**, the dimeric sesquiterpene gossypol **99** and isolongifolenone **100** which are di-, tri-, pentacyclic and dimeric-, polycyclic aromatic structures (Figure 1.22).

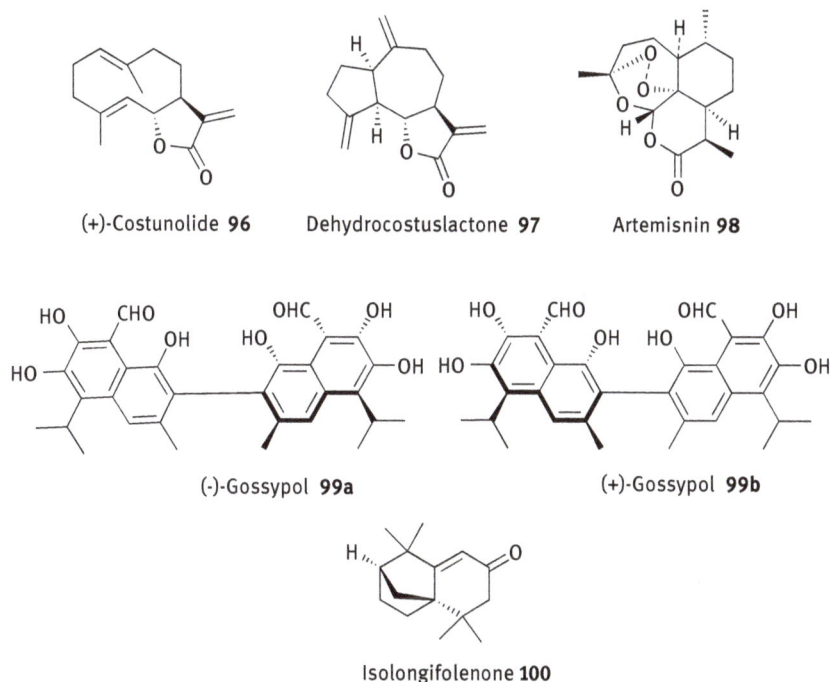

(+)-Costunolide **96**   Dehydrocostuslactone **97**   Artemisnin **98**

(-)-Gossypol **99a**   (+)-Gossypol **99b**

Isolongifolenone **100**

Figure 1.22: Examples of sesquiterpene derivatives.

Costunolide **96** and dehydrocostuslactone **97** co-occur in many plants and are important constituents of traditional medicines of many countries such as china, Japan and Ethiopia. The important plant sources of these co-occurring substances are mokko (*Saussurea lappa*, Asteraceae), a plant popularly known as "costus", *Laurus nobilis* (Lauraceae) [55], the roots of *Aucklandia lappa* (Asteraceae) and *Echinops kebericho* (Asteraceae) [56]. The literature lists a wide range of biological activities that include: anti-inflammatory, anticancer, antiviral, antimicrobial, antifungal, antioxidant, antidiabetic, antiulcer and anthelmintic activities. Reported mechanisms of action include: causing cell cycle arrest, apoptosis and differentiation, promoting the aggregation of microtubule protein, inhibiting the activity of telomerase, inhibiting metastasis and invasion, reversing multidrug resistance as well as restraining angiogenesis. Many of these claims have been assessed in a recent review [57] which confirmed that co-treatment with both sesquiterpenes showed synergistic anti-cancer effect yet stressing the need for further studies to fully establish the therapeutic benefits.

1.6.2.1 Artemisinin

Artemisinin **98** is a sesquiterpene first isolated from the Chinese plant *Artemisia annua* (Asteraceae). It is the principal component of the Artemisia-combination therapy (ACT) of the WHO prioritized treatment for *Plasmodium falciparum*-caused malaria. Youyou Tu a pharmacologist from the China Academy of Traditional Chinese Medicine, shared the 2015 Nobel Prize for physiology and medicine with two other scientists Satoshi Omura and William C. Campbell. Tu was recognized for her work on artemisinin, while the other two had worked on the macrolide avermictins (**45**) which have potent anthelmintic, insecticidal and acaricidal properties. These recognitions have sent a strong message to the importance of these secondary metabolites but, generally, to the practitioners of the field of phytomedicine and medicinal chemistry based on traditional practices. A lot has been written on artemisinin and hence we will focus on more recent aspects.

It is now realized that artemisinin has many other potential benefits other than curing malaria. Recent studies have revealed that artemisinin and its derivatives may be used against cancer, human cytomegalovirus infections and other viral as well as schistosomiasis [58].

The importance of ACT and the need for it among the developing world has underscored the need to find non-profit motivated, environmentally friendly processes for the industrial scale production of artemisinin. In this regard it is very encouraging to notice two major advances. The first is the possibility of employing synthetic biology in combination with metabolic engineering of *Saccharomyces cerevisiae* to achieve the conversion of acetyl-CoA to amorphadiene and using plasmids the latter was transformed to artemisinic acid [59]. The second important development has been the chemical conversion of artemisinic acid **101/102** to artemisinin. An environmentally friendly photochemical oxidation process using CO_2 [60] as a solvent was reported which was later developed for an industrial scale manufacturing as shown in Figure 1.23 [61].

The success of the large-scale production and availability of artemisinin will have a tremendous impact in the fight against malaria.

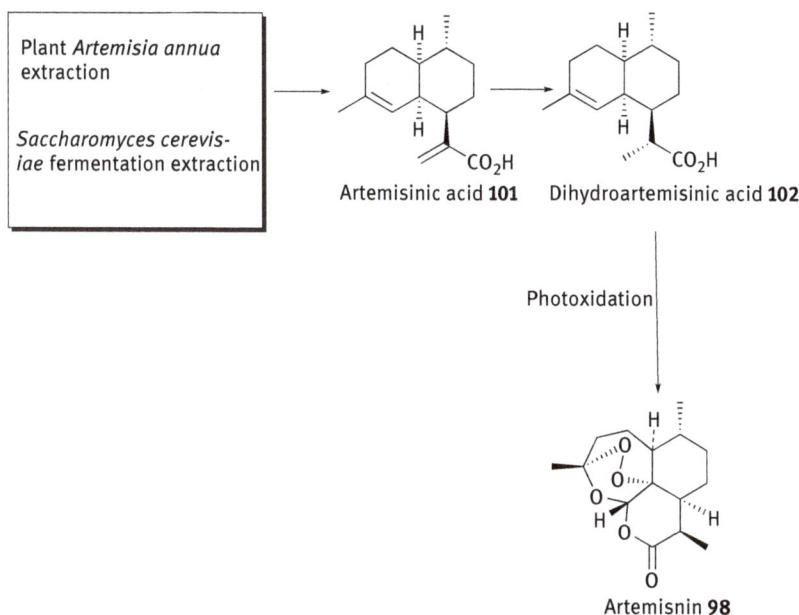

Figure 1.23: Industrial scale synthesis of artemisinin.

1.6.2.2 Gossypol

Gossypol **99** is a natural compound found in the seeds, stem and leaves of cultivated as well as wild cotton plants (*Gossypium hirsutum*, Malvaceae). Extremely low-level of fertility in communities which consumed high-level of cottonseed oil in their diet in China led to the belief that gossypol may be the suspect agent. Studies later revealed that gossypol was a male contraceptive. Large scale studies on humans were conducted but the toxicity observed led to the abandonment of the project. The main manifestations were, besides the infertility, potassium depletion.

Gossypol is a sesquiterpene dimer formed within the plant by oxidative coupling of two C-15 fragments through the mediation of an enzyme, therefore, leading to a non-racemic mixture of products, the optical activity arising due to the stereogenic axis joining the two fragments. Gossypol, thus, exists as two enantiomers. The (-)-enantiomer is toxic to animals and it is this isomer that has the antifertility property in male humans.

The ratio of the non-toxic (+) to the toxic (-) isomer varies from 95:5 in *Gossipium hirsutum*, to a near racemic (44:56) in *G. barbadense* [62].

1.6.2.3 Isolongifolenone

Isolongefolene is a sesquiterpene isolated from the stems and leaves of *Humiria balsamifera* (Humiriaceae)[104]). It has potent anti-tyrosinase activity and toxicity towards breast cancer cells. Isolongifolenone also repels blood-feeding arthropods and was found to be superior to the synthetic chemical repellent *N*,*N*-diethyl-3-methyl benzamide in repelling ticks and deterring feeding mosquitoes [63, 64] both of which are known spreaders of diseases such as malaria, West Nile virus and Lyme disease. This aroused interest to find ways of making this interesting compound and a facile and efficient synthesis from isolongifolene was achieved using *t*-butyl hydroperoxide and chromium hexacarbonyl in acetonitrile/benzene in 98 % yield and 93 % purity (Figure 1.24) [65].

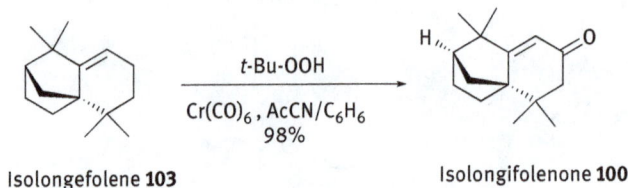

Isolongefolene **103**   Isolongifolenone **100**

Figure 1.24: Synthesis of isolongifolenone 100.

1.6.3 Diterpenes

The biosynthesis of diterpenes involves the adding of the 5-carbon isopentenyl diphosphate (IPP) unit to a farnesyl derivative to give geranylgeranyl diphosphate unit which then undergoes various transformations to give a linear, bicyclic, tricyclic, *etc.* metabolites. Formally the diterpenes have 20 carbons which may then undergo oxidation to alcohol or acid functional groups to form esters, glycosides and other derivatives. Examples of diterpenes include the highly toxic phorbol **104**, the commercial sweetener stevioside **105**, the well-known cancer drug paclitaxel (Taxol) or the coffee diterpenes cafestol **106** and kahweol **107** (Figure 1.25). As Taxol is widely known and discussed in most text books, this chapter will focus on other diterpenes that are rarely covered in other treatise.

1.6.3.1 Phorbol

Phorbol **104** was first obtained in 1934 by hydrolysis of the oil of *Croton tiglium* (Euphorbiaceae). It belongs to the tiglane class of diterpenes which are typical products of the Euphorbiaceae and Thymelaeaceae families of plants. The structure consists of a rigid tetracyclic scaffold of a 5-membered ring (A), a 7-membered ring (B), a 6-membered ring (C) and a 3-membered ring (D). It has eight stereogenic centers, six of which are on the six-membered ring C. There were many proposed structures, but the correct structure was proposed and confirmed by X-ray in 1967 [66].

The total synthesis of phorbol **104** has been a challenge for nearly 80 years because of the many stereogenic centers and the highly substituted nature of the

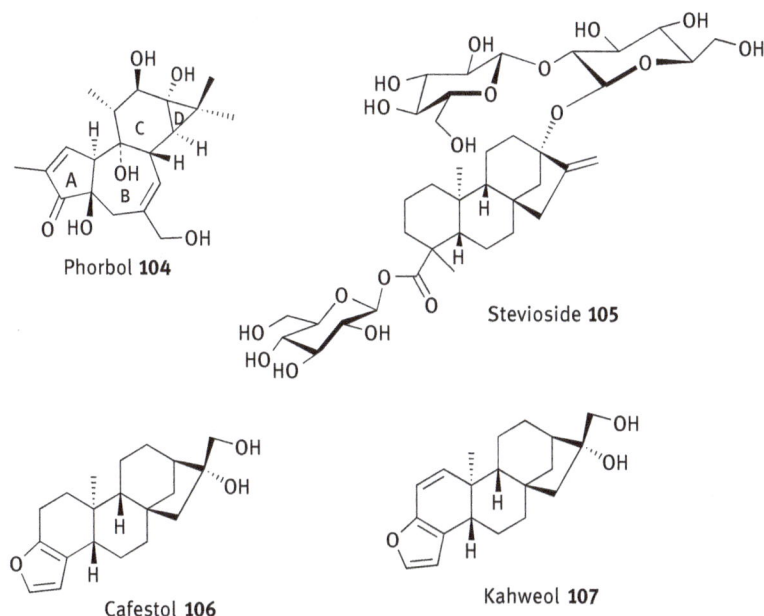

Figure 1.25: Examples of diterpenes.

molecule. Two total synthesis have been achieved recently by two groups [67, 68], in 19 steps starting from the monoterpene carene-3. This is a huge improvement since earlier reports involved more than twice reaction steps [69].

Phorbol **104** occurs in the form of various ester derivatives which are tumour promoters and thus are harmful substances. The mechanism of their action has been reasonably well explored. They activate the protein kinase isozyme which depends on optimal lipophilicities of fatty acid side chains to play the important role in cellular signal transduction. Tetradecanoyl phorbol acetate has been shown to strongly activate the protein kinase isozyme promoting active development of tumor even at 20 nM concentration. The phorbol ester phorbol-12-myristate-13-acetate is used in carcinogenesis research.

1.6.3.2 Stevioside

Stevioside **105** is a natural, non-calorific sweetener commercialized and widely used in Brazil and Japan. It was first isolated by French chemists, Bridel and Lavielle in 1931 [70], from a South American plant *Stevia rebaudiana* (Asteraceae). The leaves contain 3–10 % of the sweet substance which is 250–300 times sweeter than sucrose. Rebaudioside A is also another sweetener occurring in the same plant. It is even sweeter than stevioside (300–400 × *cf* sucrose) but is present in small amounts of 2–4 %.[3] The aglycone of stevioside is a diterpene having the ent-kaurane skeleton with a double bond at C-16, an alcohol at C-13 and a carboxylic acid at position C-4 (i.e., *ent*-13-hydroxykaur-16-en-19-oic acid). The sweet taste of the compound resides in the total structure of the bidesmosidic glycoside, that is, the aglycone with two glucose units in a glycosidic linkage to the C-13 OH group and the carboxyl group esterified with glucose.

Rebaudioside A differs from stevioside by having an additional glucose unit attached to the disaccharide at C-13. Two undesirable features of stevioside are its poor solubility and bitter aftertaste. Rebaudioside A scores highly in this regard but its quantity in the mixture is limited. In Dec 2008, US Food and Drug Administration granted approval to stevia containing a minimum of 95 % rebaudioside-A as general-purpose sweetener for use in foods and beverages.

Taking note of the above [73] Singla and Jaitak have utilized the computational method to investigate the binding mode of the two glycosides against the G-protein hTAS2R4 receptor and have shown how both would bind at the receptor site but the larger molecule would fail to trigger the bitter taste signalling due to the presence of the additional glucose moiety at C-13. These same authors developed an enzymatic transglycosylation using β-1,3-glucanase from *Irpex lacteus* (Steccherinaceae) to obtain rebaudioside A from stevioside.

An important aspect of relatively abundant secondary metabolites is the possibility of using them for chemical transformation to more useful substances. This is exemplified by the recent conversion of stevioside **105** to (–)-tripterifordin **108** and (–)-neotripterifordin **109** (Figure 1.24), two potent inhibitors of HIV replication [74] as shown in Figure 1.26.

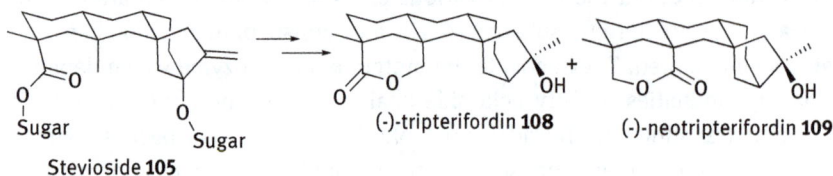

Figure 1.26: Multistep conversion of stevioside 105 to tripterifordin 108 and neotripterifordin 109.

1.6.3.3 Cafestol and kahweol

Roasted coffee contains, besides caffeine, phenolic acids, melanoidins, *N*-methylpyridinium, polyphenols such as chlorogenic acid, quinic and caffeic acids, and

---

3 Vleggaar and coworkers [71] have reported an alkaloid, Monatin, from a South African plant *Schlerochitin ilicifolius*. Monatin is probably one of the sweetest natural substance which is estimated to be 3000 times sweeter than sucrose [72].

diterpenes kahweol and cafestol. The amount of these diterpenes in brewed coffee is dependent on the mode of preparing the drink. The highest amounts (*ca.* 5–7 mg for each diterpene) are found in Scandinavian-style boiled coffee and Turkish-style coffee [75]. This report reveals the negligible levels in instant and drip-filtered brews and minor amounts in espresso coffee. In view of the high level of coffee consumption globally, there have been numerous studies to find out if these diterpenes confer any health benefit. It is to be remembered that the public has been given different recommendations about the risks and benefits of coffee consumption https://www. aarp.org/health/healthy-living/info-10-2013/coffee-for-health.html. Caffeine itself is not tolerated by certain individuals and may even be lethal, but fortunately the lethal doses are not easily reachable since one may have to consume nearly 200 cups a day. But there have been many studies on the other constituents, particularly the diterpenes cafestol **106** and kahweol **107**. A recent review [76] evaluates epidemiological findings linking coffee consumption to potential health benefits through prevention of several chronic and degenerative diseases such as cancer, cardiovascular disorders, diabetes and Parkinson's disease. There are strong indications that both cafestol [77] and kahweol [78] induce apoptosis to leukemia cells [79]. The mechanism of anti-inflammatory action of cafestol is probably through the inhibition of both PGE2 production and the mRNA expression of COX-2 [80].

Our discussion about cafestol **106** and kahweol **107** has so far been about roasted and brewed coffee. The structures of both diterpenes contain a furan ring and this may lead to the speculation that the furan moiety may be formed during the roasting and brewing processes and if this is so, then these compounds would be artefacts. However, this concern appears to be ruled out by the data available on green coffee beans.

Relatively less chemical work is reported in the literature about green coffee. Fatty acid esters (C_{18} – stearate, oleate and linoleate, and C_{16} – palmitate) of cafestol and kahweol occur in the green coffee which probably hydrolyses during the roasting and brewing process. Some decomposition does take place during the roasting process and one of the identified semi-degraded products is *seco*-kahweol. Surprisingly considerable amounts of these free diterpenes occur in the green coffee, which means that they are true secondary metabolites and not artefacts. This leads to the following important realization.

There are about **25** major species within the genus *Coffea* but there are only **two** main species that are cultivated for commercial coffee consumption; *Coffea arabica*, and *Coffea canephora* (generally referred to as Robusta). Cafestol **106** and kahweol **107** have not been reported from other species other than the various species of coffee. The two diterpenes occur in many wild species [81] and *Coffea arabica* with smaller quantities noted in *C. canephora*. This gives these two diterpenes a recognition as true markers of coffee, even more so than caffeine which has been reported to occur in over 60 plants belonging to different families (Rubiaceae, Theaceae, Malvaceae, *etc.*).

The structure of cafestol **106** was fully elucidated in 1959 by Djerassi et al. [82] and that of kahweol **107** was deduced by reducing the double bond to the former [83]. Two total syntheses have been reported for cafestol **106**, one by E. J. Corey [84], and a recent one by Zhu which allows the formation of the key tricyclic intermediate on gram quantities [85]. The synthesis of cafestol **106** was achieved in 20 steps from the easily available vinyl iodide **109** as Shown in Figure 1.27.

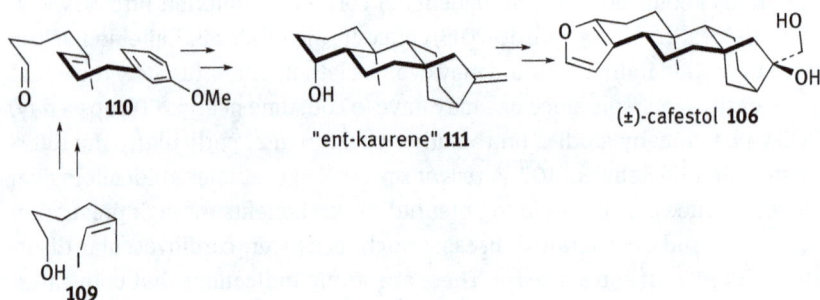

Figure 1.27: Synthesis of cafestol **106** developed by Zhu.

1.6.4 Triterpenes

The basic triterpene skeleton has 30 carbons with the simplest polyene hydrocarbon, squalene **112** (Figure 1.28), found in shark oil, amaranth seed, rice bran, wheat germ and olives. The number of triterpenes known so far is estimated to be more than 20,000 with over 200 skeletons mostly composed of tetra-, or pentacyclic systems.

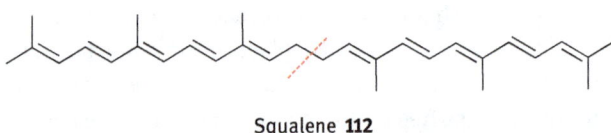

Squalene **112**

Figure 1.28: Squalene formed by coupling of two farnesyl (C_{15}) units.

The triterpenes are classified based on the number of rings into pentacyclic and tetracyclic triterpenes. Among them various subgroups such as gammaceranes **113**, hopanes **114**, lupanes **115**, oleananes **116** and ursanes **117** which differ in the E-ring as shown in Figure 1.29.

The transformation of the *deca*-olefinic hydrocarbon, squalene **112** to a pentacyclic structure is one of the fascinating aspects of triterpene chemistry and Figure 1.30 gives a glimpse of the complex enzymatic multistep biosynthesis of how nature begins with

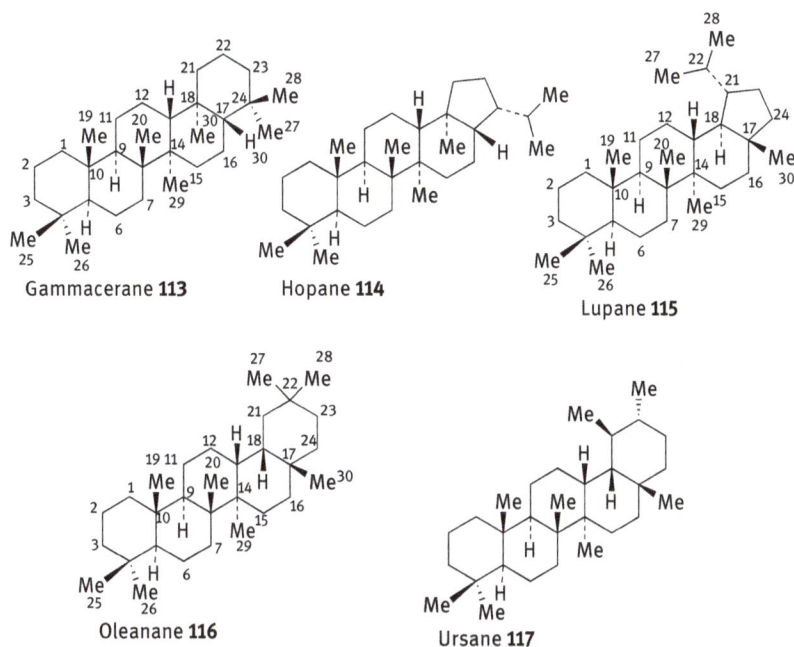

Figure 1.29: Examples of triterpenes.

an epoxide of one of the end-double bonds of squalene **118** to make the pentacyclic triterpene, lupeol **119**.

Triterpenes have been shown to have anticancer, antiviral, antifungal, anti-inflammatory and antibacterial properties. Many triterpenes show cytotoxic properties against tumor cells with only low or minimal activity toward normal cells. This differentiation between cancerous and normal cells is very critical in cancer-drug development. The literature on the various biological properties of triterpenes is quite extensive and hence we will only provide information only on lupane class of triterpenes, and principally, on betulinc acid **121** [86, 87], (Figure 1.31).

Betulin **120** was first isolated in 1788 by the German-Russian chemist, T. Lovitz who obtained it by sublimation from birch bark [88]. Scientific interest in betulinic acid **121** made a sharp rise in 1995, (over 200 years after its discovery) when it was demonstrated that it selectively caused the apoptosis of malignant melanoma cells without affecting normal cells. Research work has since intensified in many fronts: (1) evaluating the biological activity and it is realized that it also has significant antiviral, antiretroviral, antiplasmodial and anti-inflammatory properties; (2) identifying biological sources. *Betula* bark, bark of *Platanus acerifolia* (Malvaceae), *Vochysia divergens*, *Ficus pandurate* (Moraceae), and the leaves of *Vitex negundo* (Lamaceae) and *Pterospermum heterophyllum* (Malvaceae) were identified as potential sources. Of

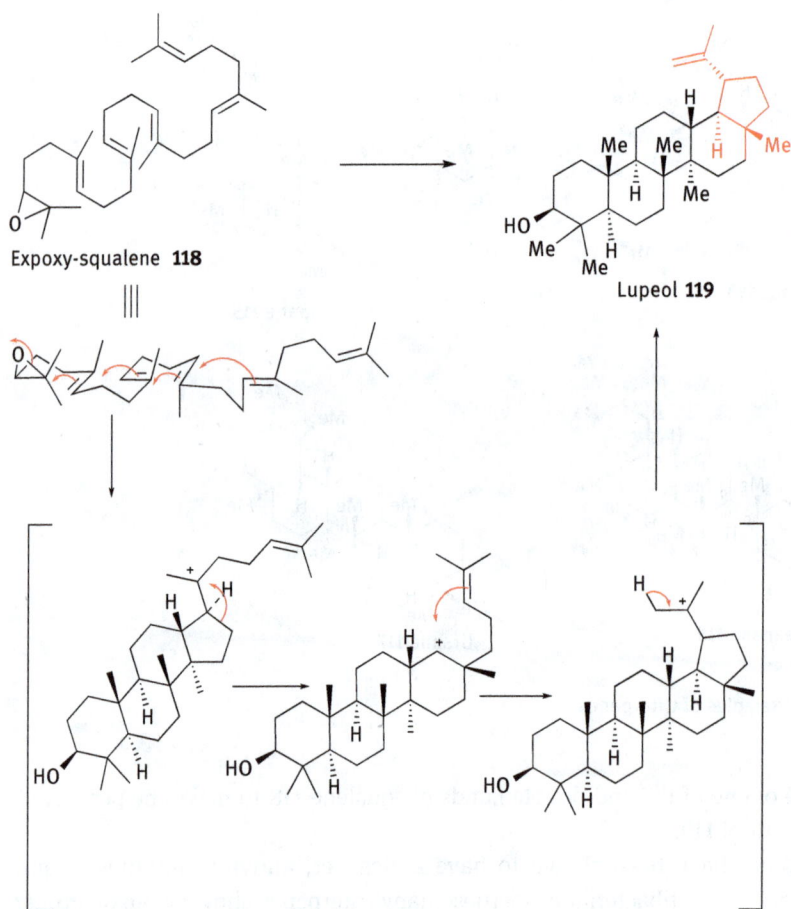

Figure 1.30: Biosynthetic transformation of squalene to lupeol.

these the bark of *Betula alba* (Betulaceaeae) contains 10–18 g of betulin/100 g of dry birch bark and *Plantanus acerifolia* bark as a rich source of betulinic acid (2.4–3.3 g/100 g of dry bark) [89]; (3) finding ways of converting betulin to betulinic acid. The Jones oxidation of betulin **120** to betulinic acid **121** also oxidizes the C-3 OH to a ketone. NaBH$_4$ reduction gives a mixture of epimers which are difficult to separate. This problem was addressed by carrying out the oxidation using CrO$_3$/SiO$_2$ which gave the aldehyde (betulinal) in reasonable yield. Conversion of the latter to betulinic acid with permanganate was quantitative [90]; (4) further studies to realize the thereapeutic potential of betulinic acid; and (5) using the betulinic acid scaffold to prepare various derivatives that can be tuned effectively to particular diseases.

There have been at least 12 reviews, including one patent review since 2005 on various aspects of research on betulinic acid [91–94]. A snapshot summary of

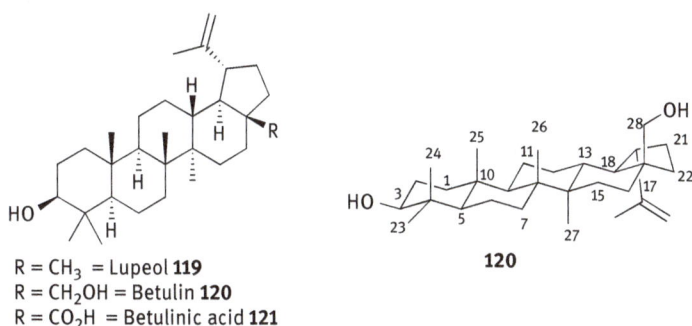

R = CH₃ = Lupeol **119**
R = CH₂OH = Betulin **120**
R = CO₂H = Betulinic acid **121**

Figure 1.31: Structures of lupeol 119, botulin 120 and betulinic acid 121.

these reports is as follows: (1) the efficacy of betulinic acid go beyond malignant melanoma cells into most solid types of tumour from different regions of the body; (2) betulinic acid exerts its effects directly on the mitochondrion triggering the apoptosis of cancerous cells; (3) it regulates the cell cycle and the angiogenic pathway *via* specific protein transcription factors – cyclin D1 and epidermal growth factor receptor; (4) inhibition of the signal transducer and activator of transcription 3 and nuclear factor -κB signaling pathways; and (5) it prevents the invasion and metastasis of malignancies *via* epithelial-mesenchymal transition and inhibition of topoisomerase I.

The betulinic acid scaffold has also allowed the preparation of derivatives taking into consideration the reactivity sites at positions 3, 20 and 28. Figure 1.32 summarizes some of the type of derivatives that have been made.

1.6.5 Steroids

Steroids are tetracyclic structures which are derived from triterpenes but contain mostly 27 (occasionally, 28 or 29) instead of 30 carbons. The three carbons that are removed during the metabolic process are the methyl group at C-14 and the two methyl groups at C-4. The basic steroid structure, the numbering of the ring carbons and the designation of the four rings (A, B, C and D) are as shown in **122** (Figure 1.33). A typical steroid is cholesterol **123** and the numbering of the angular methyl groups and the 8-carbon chain at C-7 is as shown. Sterols are found in the membranes of plants, animals and microorganisms and are termed phytosterols, zoosterols and mycosterols, respectively. The widely found mammalian sterol is cholesterol **123**. The most important and widely available plant sterols are sitosterol **124** (C₂₈), stigmasterol **125** (C₂₉) and campesterol **126** (C₂₈). The most abundant mycosterol in fungi, for example in edible mushrooms, is ergosterol **127**. These sterols play, in their respective hosts, essential roles in membrane function, regulating its fluidity, plasma membrane biogenesis and

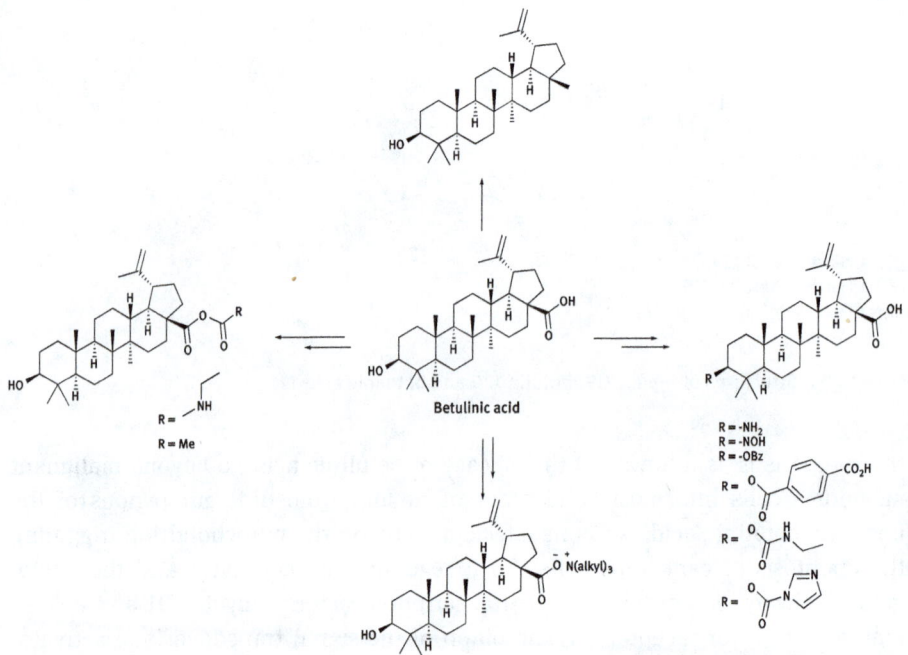

Figure 1.32: Transformation of betulinic acid 121 to various derivatives.

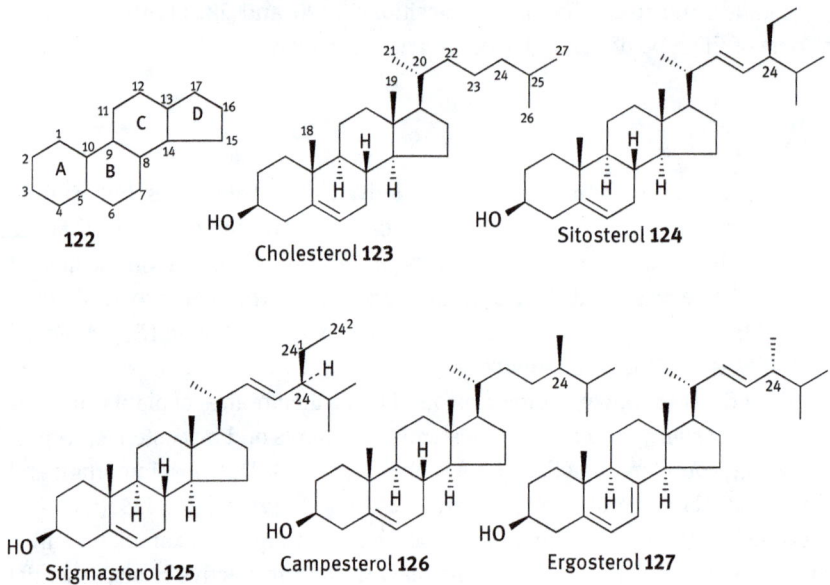

Figure 1.33: Examples of steroids.

permeability. It should be noted that cholesterol is not found in plants, and ergosterol is found in fungi only.

Phytosterols have an assumed hypocholesterolemic effect because of their structural similarity to cholesterol. They, therefore, can combine with cell receptors in the human intestine, and compete with cholesterol in the absorption process. They may also have other benefits since they have been shown to possess a variety of other properties *in vivo* such as antiproliferative, anticancer, anti-inflammatory, antidiabetic and immunomodulatory effects [95].

Further metabolites **128–138** of cholesterol are formed through modification of the side chain on the D-ring (Figure 1.34) leading to bile acids, corticosteroids, mammalian sex hormones, cardenolides, bufadienolides, sterols, steroidal saponins, *etc.*

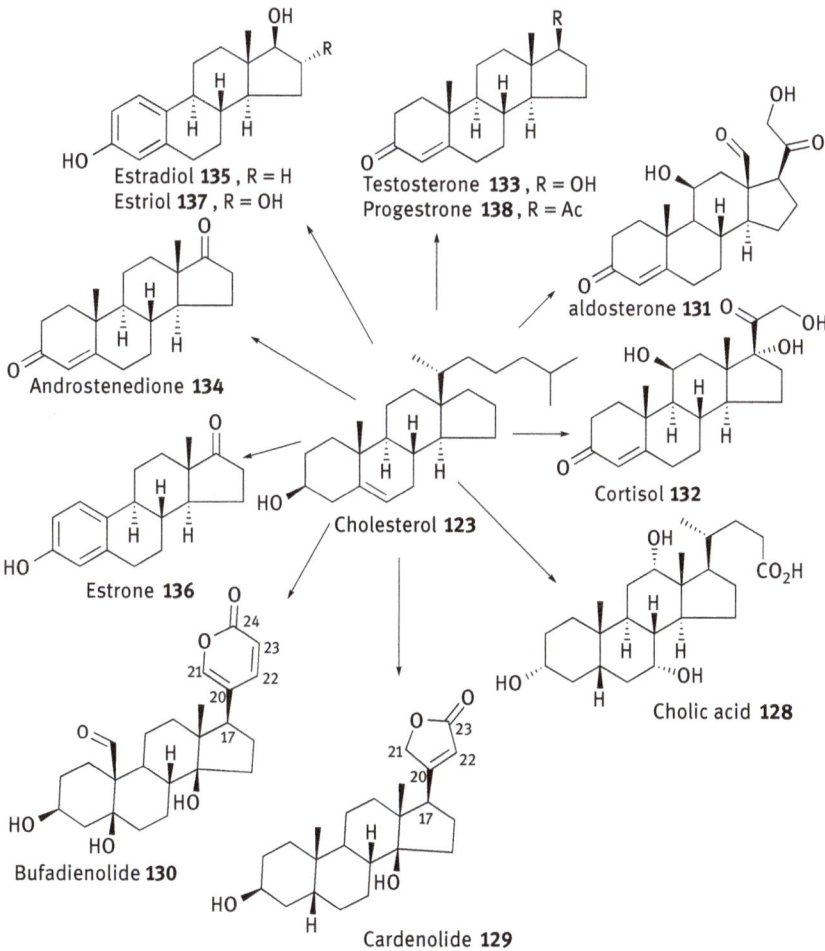

Figure 1.34: Bioderivatization of cholesterol.

Bile acids – We will discuss only one of these acids, cholic acid **128**. The transformation of cholesterol to cholic acid takes place in the liver and proceeds not only by degrading the side chain to a four-carbon unit but also changing the fusion of the A/B ring from *trans* to *cis*. The OH group at C-3 and two introduced OH groups at C-7 and C-12 have the alpha configuration. Cholic acid **128** is generally stored in the gall bladder, but after conjugation at the carboxyl end with the smallest amino acid, glycine, flows to the gut to do the amazing job of emulsifying the food as part of the digestion and absorption process follows. The changes in conformation described above and the polar OH groups orientation makes the molecule easily adaptable for micelle formation to perform detergent like function. Note that the conversion of cholic acid to sodium glycolate gives a water-soluble salt at physiological pH.

Corticosteroid and mammalian sex hormones: hormones synthesized from cholesterol are referred to as steroid hormones. They are grouped into two, namely, adrenocortical hormones and sex hormones. They have a C_{21} pregnane skeleton with the sidechain at C-17 of cholesterol degraded to a α-hydroxy-keto fragment. In ring-A the 3-OH is converted to a α,β-unsaturated-3-keto functionality. Aldosterone **131** and cortisol **132** are adrenocortical hormones produced by the adrenal gland, found next to each kidney. The function of aldosterone is to control the electrolyte balance by enhancing the rate of reabsorption of sodium ions and increasing the excretion of potassium ions. Water retention in tissues is influenced by the level of sodium ions; as a result, aldosterone promotes water retention. Cortisol regulates the synthesis of glucose from fatty acids and amino acids and the deposition of glycogen in the liver. Cortisol also inhibits the inflammatory response of tissue to injury or stress. For this reason, steroid drugs are used in the treatment of severe skin allergies and autoimmune diseases such as rheumatoid arthritis.

The sex hormones include progesterone, the estrogens, and testosterone and their derivatives. The natural steroidal sex hormones are made by the ovaries or testes (gonads) by adrenal glands, or by conversion from other sex steroids in other tissues such as liver or fat. Androgens, estrogens and progestrogens are the main classes of sex steroids. The so called male hormones (androgens) are testosterone **133**, dihydrotestosterone, androstenediol and androstenedione **134**. For the female hormones (estrogens) the main ones are estradiol **135**, estrone **136**, estriol **137** and progesterone **138**. Note that all the sex hormones occur in both males and females, but in different amounts. Progesterone is the most important and only naturally occurring human progestogen.

Cardenolides and bufadienolides: Cardenolides are C_{23} steroids consisting of an unsaturated five-membered lactone ring attached to a steroid nucleus at C-17 and a sugar moiety at C-3. The bufadienolides differ by having a six-membered ring lactone on the C_{24} steroid scaffold as shown in Fig. 37. Biosynthetically, they are made from cholesterol through intermediates, 20α-hydroxycholesterol → pregnenolone → progesterone → cardenolide **129**/bufadienolide **130**).

The historical beginning of cardenolides and bufadienolides as cardiac glycosides may be traced back to 1785 when W. Withering meticulously documented the
efficacy of the leaves of the common foxglove plant (*Digitlis purprea*
(Plantaginaceae)) (http://www.historyofscience.com/articles/jmnorman-william-
withering.php). The classic story has been their outstanding roles in the therapy of
congestive heart failures. But in very recent times they have become important
subjects for investigation as anticancer agents as they can induce apoptosis and
inhibit the growth of cancer cell lines and it is this latter aspect that is the focus in this
chapter. The first report demonstrated the exclusive involvement of ClC-3 Cl –
channel in bufadienolide-induced apoptotic cell death [96]. Therefore, cardenolides
could be a new therapeutic agent that targets to inhibit cancer cell survival due to
altering the distribution of Na^+/K^+-ATPase in cancer versus normal cells [97].

Bufadienolides are known to occur as skin secretions of toads from the genus
Bufo, as well as in the snake *Rhabdophis tigrinus*, which sequesters bufadienolides
from toads for defensive purposes [98] and in the arthropod insects (fire fly) [99].
They also occur in plants in six families: Crassulaceae (Kalanchoe), Hyacinthaceae,
Iridaceae, Melianthaceae, Ranunculaceae and Santalaceae. The cardenolides occur
more widely, but principally in the Apocynaceae (e. g. *Calotropis, Thevetia, Nerium,
Strophanthus*) [97–100] Asparagaceae (*Convallaria*), and Scrophulariaceae
(*Digitalis*).

It has been noted that as many as 200 cardenolides have been isolated from the
Apocynaceae family and only 25 % of them were active anticancer agents. A recent
review [100] analysed the active and inactive cardenolides isolated from *Calotropis*
species and drew conclusions on structure and activity correlations. Another group
has also conducted molecular docking to study the structure-activity relationship of
bufadienolides derivatives from *Kalanchoe pinnata* (Crasullaceae) to the inhibition of
Na^+/K^+-ATPase on bufadienolides isolated from *Kalanchoe* species [11]. The conclusions were similar in that the predicted binding energies were correlated with
experimental data. A similar conclusion was also reached for cardenolides which
were studied by the Chan group mentioned above. Calactin **139** (Figure 1.35) isolated

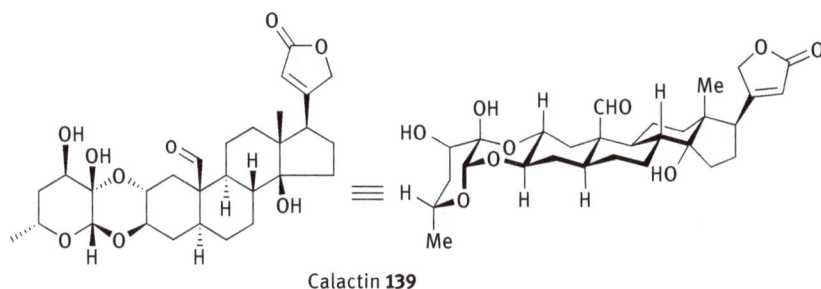

Calactin **139**

Figure 1.35: The structure of calactin.

from Egyptian *C. procera* exhibited cytotoxic activities with IC_{50} values of 0.036 μM and 0.083 μM against A-549 and Hela cells, respectively, but did not show antimicrobial activity [101].

Saponins are glycosides of triterpenes or steroids. They are called saponins because they have soap-like properties and can act as detergents. Saponins hemolyze red blood cells. The non-sugar portion of saponins may be triterpenes or steroids. Ginseng, from *Panax ginseng* (Araliaceae), which is widely used to provide general wellbeing and reduce stress, is a saponin composed of the triterpene dammarane and the sugars rhamnose and glucose. It was the parasitologist Aklilu Lemma who observed dead snails along the banks of a river in Adwa (Northern Ethiopia) where villagers regularly use the soapberry plant (*Phytolacca dodecandra*, Phytolacaceae) for washing clothes that made him investigate the molluscicidal properties and eventually developed ways of controlling bilharzia infections. This villagers' self-help approach for public health use was eventually approved. Furthermore, a patent was drawn for the possible use of the same plant for controlling zebra mussels which were a serious nuisance clogging the cooling systems of nuclear reactors in the Great Lakes region of the US. The molluscicidal properties of *P. dodecandra* have been attributed to the glycosides of oleanolic acid **140** (Figure 1.36) which occurs in the fruits of the plant [102]. Unlike the triterpenoid saponins, the sapogenins of steroidal saponins have 27 carbons in which the side chain of the cholesterol has been modified to a spiroketal or a hemiketal functional group. An example of a steroidal saponin is dioscin (**141**), diosgenin bis-α-*L*-rhamnopyranosyl-(1→2, 1→4)-β-*D*-glucopyranoside found in many varieties of *Discorea* (Figure 1.36).

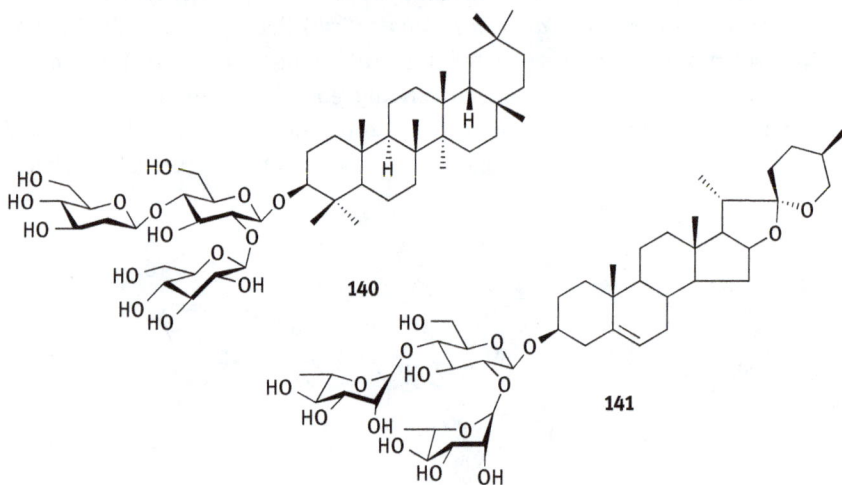

Figure 1.36: Examples of saponins.

Funding: The authors gratefully acknowledge the University of Johannesburg and the National Research Foundation (NRF) of South Africa for funding the project and to Prof. S. A. Khalid and Dr Tarekegn G. Yesus for reading the manuscript and making useful comments and improvements.

References

[1]   Marais J, Du Toit P. Monofluoroacetic acid, the toxic principle of "gifblaar", *Dichapetalum cymosum* (Hook) Engl. Onderstepoort. J Vet Sci Anim Ind. 1944;20:67–73.

[2]   Wall ME, Wani MC, Brown DM, Fullas F, Olwaldi JB, Josephson FF, et al. Effect of tannins on screening of plant extracts for enzyme inhibitory activity and techniques for their removal. Phytomedicine. 1996;3:281–5.

[3]   Picker P, Vogl S, McKinnon R, Mihaly-Bison J, Binder M, Atanasov AG, et al. Plant extracts in cell-based anti-inflammatory assays—pitfalls and considerations related to removal of activity masking bulk components. Phytochem Lett. 2014;10.

[4]   Harvey A, Edrada-Ebel RQ, Quinn RJ. The reemergence of natural products for drug discovery in the genomics era. Nat Rev Drug Discov. 2015;1–19.

[5]   Koch MA, Schuffenhauer A, Scheck M, Wetzel S, Casaulta M, Odermatt A, et al. Charting biologically relevant chemical space: a structural classification of natural products (SCONP). Proc Natl Acad Sci (PNAS). 2005;102:17272–7.

[6]   Khanna V, Ranganathan S. Structural diversity of biologically interesting datasets: a scaffold analysis approach. J Cheminform. 2011;3:1–14.

[7]   Schäfer T, Kriege N, Humbeck L, Klein K, Koch O, Mutzel P. Scaffold Hunter: a comprehensive visual analytics framework for drug discovery. J Cheminform. 2017;9:1–18.

[8]   Rodrigues T, Reker D, Schneider P, Schneider G. Counting on natural products for drug design. Nat Chem. 2016;8:531–541.

[9]   Skinnider MA, Dejong CA, Franczak BC, McNicholas PD, Magarvey NA. Comparative analysis of chemical similarity methods for modular natural products with a hypothetical structure enumeration algorithm. J Cheminform. 2017;9:1–15.

[10]   Amirkia V, Heinrich M. Alkaloids as drug leads – a predictive structural and biodiversity-based analysis. Phytochem Lett. 2014;10

[11]   Yusuf M, Firdaus AR, Supratman U. Computational study of bufadienolides from Indonesia's *Kalanchoe pinnata* as Na+/K+-ATPase inhibitor for anticancer agent. J Young Pharmacists. 2017;9:475–9.

[12]   Al B. The source-synthesis- history and use of atropine. J Acad Emergency Med. 2014;13:2–3.

[13]   Willstätter R. Umwandlung von tropidin in tropin. Berichte der Deutschen Chemischen Gesellschaft, 1901;34:3163–5.

[14]   Felpin F-X, Leberton J. History, chemistry and biology of alkaloids from *Lobelia inflata*. Tetrahedron. 2004;60:10127–53.

[15]   Henkin R, Velicu I, Schmidt L. An open-label controlled trial of theophylline for treatment of patients with hyposmia. Am J Med Sci. 2009;337:396–406.

[16]   Barnes PJ. Theophylline. Pharmaceuticals. 2010;3:725–47.

[17]   Kossel A. Über eine neue base aus dem pflanzenreich. Ber Dtsch Chem Ges. 1888;21:2164–7.

[18]   Hiroshi A, Hiromi K. Biosynthesis of purine alkaloids in *Camellia* plants. Plant Cell Physiol. 1987;28:535–9.

[19]   Alexander J, Benford D, Cockburn A, Cravedi J-P, Dogliotti E. Theobromine as undesirable substances in animal feed: scientific opinion of the panel on contaminants in the food chain. Eur Food Saf Authority. 2008;725:1–66.

[20] Abusnina A, Lugnier C. Therapeutic potentials of natural compounds acting on cyclic nucleotide phosphodiesterase families. Cell Signal. 2017;39:55–65.

[21] Turano A, Scura G, Caruso A, Bonfanti C, Luzzati R, Bassetti D, et al. Inhibitory effect of papaverine on HIV replication *in vitro*. AIDS Res Hum Retroviruses. 1989;5:183–92.

[22] Gates M, Tschudi G. The synthesis of morphine. J Am Chem Society. 1956;78:1380–4.

[23] Barber RB, Rapoport H. Conversion of thebaine to codeine. J Med Chem. 1976;19:1175–80.

[24] Fairbairn JW, Hakim F. *Papaver bracteaturn* Lindl.-a new plant source of opiates. J Pharm Pharmac. 1973;25:353–8.

[25] Han Z, Zheng Y, Chen N, Luan L, Zhou C, Gan L, et al. Simultaneous determination of four alkaloids in *Lindera aggregata* by ultra-high-pressure liquid chromatography–tandem mass spectrometry. J Chromatogr. 2008;1212:76–81.

[26] O'Brien P, Carrasco-Pozo C, Speisky H. Boldine and its antioxidant or health-promoting properties. Chem Biol Interact. 2006;159:1–17.

[27] Tietjen I, Ntie-Kang F, Mwimanzi P, Onguéné PA, Scull MA, Idowu TO, et al. Screening of the Pan-African Natural Product Library identifies ixoratannin A-2 and boldine as novel HIV-1 inhibitors. PLoS One. 2015;10(4):1–19.

[28] Kumar A, Ekavali KC, Mukherjee M, Pottabathini R, Dhull DK. Current knowledge and pharmacological profile of berberine: an update. Eur J Pharmacol. 2015;761:288–97.

[29] Perkin WH, Robinson R. Strychnine, berberine, and allied alkaloids. J Chem Society, Trans. 1910;97:305–23.

[30] Kametani T, Noguchi I, Saito K, Kaneda S. Studies on the syntheses of heterocyclic compounds. Part CCCII. Alternative total syntheses of nandinine, canadine, and berberine iodide. J Chem Society. C. 1969:2036–8.

[31] Janssen B, Schäfer B. Galantamine. ChemTexts. 2017;3:1–21.

[32] Indu TH, Raja D, Manjunathai D, Ponnusankar S. Can galantamine act as an antidote for organophosphate poisoning? A review. Indian J Pharm Sci. 2016;78:428–35.

[33] Barton DHR, Kirby GW. 153. Phenol oxidation and biosynthesis. Part V. The synthesis of galantamine. J Chem Soc (Resumed). 1962;806– DOI: 10.1039/JR9620000806.

[34] Kresge N, Simoni RD, Hill RL. The Prostaglandins, Sune Bergström and Bengt Samuelsson. J Bio Chem. 2016;28:e9–e11.

[35] Abegaz B, Atnafu G, Duddeck H, Snatzke G. Macrocyclic pyrrolizidine alkaloids of *Crotalaria rosenii*. Tetrahedron. 1987;43:3263–8.

[36] Dubost C, Marko IE, Ryckmans T. A concise total synthesis of the lichen macrolide (+)-aspicilin. J Am Chem Soc. 2006;8:5137–40.

[37] Prestwich GD, Collins MS. Macrocyclic lactones as the defense substances of termite genus *Armitermes*. Tetrahedron Lett. 1981;22:4587–90.

[38] Hayakawa Y, Shin-Ya K, Furihata K, Seto H. Structure of a vovel 60-membered macrolide, quinolidomicin A. J Am Chem Soc. 1993;115:3014–5.

[39] Dewick PM. Medicinal natural products: a biosynthetic approach. Chichester: John Wiley & Sons Ltd, 2009.

[40] Mizuuchi Y, Shi SP, Wanibuchi K, Noguchi H, Abe I. Novel type III polyketide synthases from *Aloe arborescens*. FEBS J. 2009;276:2391–401.

[41] Alemayehu A. Bianthraquinones and a spermidine alkaloid from *Cassia floribunda*. Phytochemistry. 1988;27:3255–8.

[42] Bringmann G, Menche D. First atropo-enantioselective total synthesis of the axially chiral phenylanthraquinone and 6'-O-methylknipholone. Angew Chem Int Ed. 2001;40:1687–90.

[43] Sridhar J, Liu J, Foroozesh M, Stevens CL. Inhibition of cytochrome P450 enzymes by quinones and anthraquinones. Chem Res Toxicol. 2011;25:357365.

[44] Tong X, Mao M, Xie J, Zhang K, Xu D. Insights into the interactions between tetracycline, its degradation products and bovine serum albumin. SpringerPlus. 2016 5:. DOI: 10.1186/s40064-016-2349-4.

[45] Petersena M, Simmonds MS. Rosmarinic acid. Phytochemistry. 2003;62:121–5.

[46] Jiang J, Bi H, Zhuang Y, Liu S, Ma Y. Engineered synthesis of rosmarinic acid in *Escherichia coli* resulting production of a new intermediate, caffeoyl-phenyllactate. Biotechnol Lett. 2016;38:81–8.

[47] Ardalani H, Avan A, Ghayour-Mobarhan M. Podophyllotoxin: a novel potential natural anticancer agent. Avicenna J. Phytomedicine. 2017;7:285–94.

[48] Hajra S, Garai S, Hazra S. Catalytic enantioselective synthesis of (–)-podophyllotoxin. Org Lett. 2017;19:6530–3.

[49] Ting CP, Maimone TJ. C-H bond arylation in the synthesis of aryltetralin lignans: a short total synthesis of podophyllotoxin. Angew Chem Int Ed. 2017;53:3115–9.

[50] Liu Y, Zhu S, Gu K, Guo Z, Huang X, Wang M, et al. GSH-activated NIR fluorescent prodrug for podophyllotoxin delivery. ACS Applied Materials & Interfaces. 2017 8 25;9:29496–504. DOI: 10.1021/acsami.7b07091.

[51] Vishnuvardhan MVPS, Reddy VS, Chandrasekhar K, Lakshma VN, Sayeed IB, Alarifi A, et al. Click chemistry-assisted synthesis of triazolo linked podophyllotoxin conjugates as tubulin polymerization inhibitors. MedChemComm. 2017;8:1817–23. DOI: 10.1039/c7md00273d

[52] Service RF. Genetic engineering turns a common plant into a cancer fighter. Science. 2015.

[53] Jung EM, Lee YR. First total synthesis of prorepensin with a bis-geranylated chalcone. Bull Korean Chem Soc. 2009;30:2563–6.

[54] Calani L, Dall'Asta M, Bruni R, Del Rio D, Flavonoid occurrence, bioavailability, metabolism, and protective effects in humans: focus on flavan-3-ols and flavonols. In: Annalisa Romani, Vincenzo Lattanzio, and Stéphane Quideau, editors. Recent Advances in Polyphenol Research, Chapter 8, vol 4, 1st ed. John Wiley & Sons, Ltd., 2014:239–79.

[55] Ferrari B, Castilho P, Tomi F, Rodrigues AI, Costa MDC, Casanova J. Direct identification and quantitative determination of costunolide and dehydrocostuslactone in the fixed oil of *Laurus novocanariensis* by 13C-NMR spectroscopy. Phytochemical Anal. 2005;16:104–7.

[56] Abegaz B, Tadesse M, Majinda R. Distribution of sesquiterpene lactones and polyacetylenic thiophenes in *Echinops*. Biochem Syst Ecol. 1991;19:323–8.

[57] Lin X, Peng X, Su C. Potential anti-cancer activities and mechanisms of costunolide and dehydrocostuslactone. Int J Mollecular Sci. 2015;16:10888–906.

[58] Efferth T, Zacchino S, Georgiev MI, Liu L, Wagner H, Panossian A. Nobel Prize for artemisinin brings phytotherapy into the spotlight. Nature. 2015;22:A1–A3.

[59] Paddon CJ, Keasling JD. Semi-synthetic artemisinin: a model for the use of synthetic biology in pharmaceutical development. Nat Rev Microbiol. 2014;12:355–67.

[60] Amara Z, Bellamy JFB, Horvath R, Miller SJ, Beeby A, Burgard A, et al. Applying green chemistry to the photochemical route to artemisinin. Nature Chemistry. 2015 5 11;7:489–95. DOI: 10.1038/NCHEM.2261

[61] Burgard A, Gieshoff T, Peschl A, Hörstermann, D, Keleschovsky C, Villa R, et al. Optimisation of the photochemical oxidation step in the industrial synthesis of artemsinin. Chem Eng J. 2016;294:83–96.

[62] Stipanovic R, Puckhaber LS, Bell AA, Percival AE, Jacobs J. Occurrence of (+)- and (-)-gossypol in wild species of cotton and in *Gossypium hirsutum* Var. Marie-galante (Watt) Hutchinson. J Agric Food Chem. 2005;53:6266–71.

[63] Zhang A, Jerome A, Klun JA, Wang S, Carrol JF, Debbouns M. Isolongifolenone: a novel sesquiterpene repellent of ticks and mosquitoes. J Med Entomol. 2009;46:100–06.

[64] Klun JA, Kramer M, Zhang SA, Wang S, Debboun M. A quantitative *in vitro* assay for chemical mosquito-deterrent activity without human blood cells. J Am Mosq Control Assoc. 2008;24:508–12.

[65] Wang S, Zhang A. Facile and efficient synthesis of isolongifolenone. Org Prep Proced: New J Org Synth. 2009;40:405–10.

[66] Hoppe W, Brandl F, Strell I, Rohrl M, Gassmann I, Hecker E, et al. X-ray structure analysis of neophorbol. Angew Chem Int Ed. 1967;6:809–10.

[67] Kawamura S, Chu H, Felding J, Baran PS. Nineteen-step total synthesis of (+)-phorbol. Nature. 2016;532:91–3.

[68] Keisuke N. 19 step total synthesis of phorbol using a concept of innovative strategy "two-phase synthesis". J Synth Org Chem Jpn. 2017;75:257–8.

[69] Lee K, Cha JK. Formal Synthesis of (+)-Phorbol. J Am Chem Soc. 2001;123: 5590–1.

[70] Bridel M, Lavielle R. Sur le principe sucré des feuilles de Kaâ-hê-é (*Stevia rebaundiana* B). CR De Acad Sci. 1931;192:1123–5.

[71] Vleggaar R, Ackerrnan LG, Steyn PS. Structure elucidation of monatin, a high-intensity sweetener isolated from the plant *Schlerochiton ilicifolius*. Journal Chem Soc, Perkin 1. 1992;3095–8.

[72] Amino Y, Kawahara S, Mori K, Hirasawa K, Sakata H, Kashiwagia T. Preparation and characterization of four stereoisomers of monatin. Chem Pharm Bull. 2016;64:1161–71.

[73] Singla R, Jaitak V. Synthesis of rebaudioside A from stevioside and their interaction model with hTAS2R4 bitter taste receptor. Phytochemistry. 2016;125:106–11.

[74] Kobayashi S, Shibukawa K, Hamada Y, Kuruma T, Kawabata A. Syntheses of (–)-tripterifordin and (–)-neotripterifordin from stevioside. J Org Chem. 2018;83:1606–13.

[75] Gross G, Jaccaud G, Huggett AC. Analysis of the content of the diterpenes cafestol and kahweol in coffee brews. Food Chem Toxicol. 1997;35:547–54.

[76] Ludwig IA, Clifford MN, Lean ME, Ashiharad H. Coffee: biochemistry and potential impact on health. Food Funct. 2014;5:1695–717.

[77] Limaa CS, Spindola DG, Bechara A, Garcia DM, Palmeira-dos-Santos C, Peixoto-da-Silvaa J, et al. Cafestol, a diterpene molecule found in coffee, induces leukemia cell death. Biomed Pharmacother. 2017;92:1045–54.

[78] Oh JH, Lee JT, Yang ES, Chang J-S, Lee DS, Kim SH, et al. The coffee diterpene kahweol induces apoptosis in human leukemia U937 cells through down-regulation of Akt phosphorylation and activation of JNK. Apoptosis. 2009;14:1378–86.

[79] Lee K-A, Chae J-I, Shim J-H. Natural diterpenes from coffee, cafestol and kahweol induce apoptosis through regulation of specificity protein 1 expression in human malignant pleural mesothelioma. J Biomed Sci. 2012;19:2–10.

[80] Shen T, Lee J, Lee E, Kim SH, Kim W, Cho JY. Cafestol, a coffee-specific diterpene, is a novel extracellular signal regulated kinase inhibitor with AP-1-targeted inhibition of prostaglandin E2 production in Lipopolysaccharide-activated Macrophages. Biol Pharm Bull. 2010;33.

[81] De Roos B, van der Weg G, Urgert R, van de Bovenkamp P, Charrier A, Katan MB. Levels of cafestol, kahweol, and related diterpenoids in wild species of the coffee plant *Coffea*. J Agric Food Chem. 1997;45:3065–309.

[82] Djerassi C, Cais M, Mitscher L. Terpenoids XXXVII. The structure of the pentacyclic diterpene cafestol. On the absolute configuration of diterpenes and alkaloids of the Phyllocladene group. J Am Chem Soc. 1959;81:2386–98.

[83] Kaufmann HP, Gupta AK. Terpene als bestandteile des unverseifbaren von fetten, IV.†‡ Zur konstitution des kahweols, II. Chem Ber. 1964;97:2652.

[84] Corey EJ, Wess G, Xiang YB, Singh AK. Stereospecific total synthesis of (±)-cafestol. J Am Chem Soc. 1987;109:4717–18.

[85] Zhu L, Luo J, Hong R. Total synthesis of (±)-cafestol: a late-stage construction of the furan ring inspired by a biosynthesis strategy. Org Lett. 2014;16:2162–5.

[86] Chudzik M, Korzonek-Szlacheta IK. Triterpenes as potentially cytotoxic compounds. Molecules. 2015;20:1610–25.

[87] Ríos J. Effects of triterpenes on the immune system. J Ethnopharmacol. 2010;128:1–14.

[88] Kuznetsova SA, Skvortsova GP, Maliara N, Skurydina ES, Veselova OF. Extraction of betulin from birch bark and study of its physico-chemical and pharmacological properties. Russian J Bioorg Chem. 2014;40:742–7.

[89] Sebastian Jäger S, Trojan H, Kopp T, Laszczyk MN, Scheffler A. Pentacyclic triterpene distribution in various plants – rich sources for a new group of multi-potent plant extracts. Molecules. 2009;14:2016–31.

[90] Pichette A, Liu H, Roy C, Tanguay S, Simard F, Lavoie S. Selective oxidation of betulin for the preparation of betulinic acid, an antitumoral compound. Synth Commun. 2004;34:3925–37.

[91] Rios JL, Manez S. New pharmacological opportunities for betulinic acid. Planta Med. 2018;84:8–19.

[92] Ali-Seyed M, Jantan I, Vijayaraghavan K, Nasir S, Bukhari A. Betulinic acid: recent advances in chemical mechanisms of a promising anticancer therapy. Chem Biol Drug Des. 2016;87: 517–36.

[93] Zhang X, Hu J, Chen Y. Betulinic acid and the pharmacological effects of tumor suppression (Review). Mol Med Rep. 2016;14:4489–95.

[94] Csuk R. Betulinic acid and its derivatives: a patent review (2008 – 2013). Expert Opin Ther Pat. 2014;24:1–11.

[95] Correa RC, Peralta RM, Bracht A, Ferreira IC. The emerging use of mycosterols in food industry along with the current trend of extended use of bioactive phytosterols. Trends Food Sci Technol. 2017;67:19–35.

[96] Liu J. Discovery of bufadienolides as a novel class of ClC-3 chloride channel activators with antitumor activities. J Medicinal Chemistry. 2013 7 10;56:5734–43. DOI: 10.1021/jm400881m.

[97] Wen S, Chen Y, Lu Y, Wang Y, Ding L, Jiang M. Cardenolides from the Apocynaceae family and their anticancer activity. Fitoterapia. 2016;112:74–84.

[98] Hutchinson D, Savitzky AH, Burghardt GM, Nguyen C, Meinwald J, Schroeder FC. Chemical defense of an Asian snake reflects local availability of toxic prey and hatchling diet. J Zool. 2013;289:270–8.

[99] Krenn L, Kopp B. Bufadienolides from animal and plant sources. Phytochemistry. 1998;48: 1–29.

[100] Chan EW, Sweidan NI, Wong SK, Chan HT. Cytotoxic cardenolides from *Calotropis* species: a short review. Records Nat Prod. 2017;11:334–44.

[101] Mohamed NH, Liu M, Abdel-Mageed WM, Alwahibi LH, Dai H, Ismail MA, et al. Cytotoxic cardenolides from the latex of *Calotropis procera*. Bioorg Med Chem Lett. 2015;25:4615–20.

[102] Thiilborg ST, Christensen SB, Cornett C, Olsen CE, Lemmich E. Molluscicidal saponins from a Zimbabwean strain of *Phytolacca dodecandra*. Phytochemistry. 1994;36:753–9.

[103] Bartlet GR. Biology of free and combined adenine; distribution and metabolism. Transfusion. 1977;17:339–50.

[104] Da Silva L, Alves VL, Mendonça LVH, Conserva LM, da Rocha, EMM, Andrade EHA, et al. Chemical constituents and preliminary antimalarial activity of *Humiria balsamifera*. Pharm Biol. 2004;42:94–7.

Mohd Athar, Alfred N. Sona, Boris D. Bekono and Fidele Ntie-Kang

2 Fundamental physical and chemical concepts behind "drug-likeness" and "natural product-likeness"

Abstract: The discovery of a drug is known to be quite cumbersome, both in terms of the microscopic fundamental research behind it and the industrial scale manufacturing process. A major concern in drug discovery is the acceleration of the process and cost reduction. The fact that clinical trials cannot be accelerated, therefore, emphasizes the need to accelerate the strategies for identifying lead compounds at an early stage. We, herein, focus on the definition of what would be regarded as a "drug-like" molecule and a "lead-like" one. In particular, "drug-likeness" is referred to as *resemblance* to existing drugs, whereas "lead-likeness" is characterized by the similarity with structural and physicochemical properties of a "lead"compound, i.e. a reference compound or a starting point for further drug development. It is now well known that a huge proportion of the drug discovery is inspired or derived from natural products (NPs), which have larger complexity as well as size when compared with synthetic compounds. Therefore, similar definitions of "drug-likeness" and "lead-likeness" cannot be applied for the NP-likeness. Rather, there is the dire need to define and explain NP-likeness in regard to chemical structure. An attempt has been made here to give an overview of the general concepts associated with NP discovery, and to provide the foundational basis for defining a molecule as a "drug", a "lead" or a "natural compound."

Keywords: fragment analysis, drug discovery, natural products, natural product score

2.1 Introduction

2.1.1 What makes drugs "drug-like"?

Despite all the endeavors of the pharmaceutical industry, around 9 out of every 10 drug candidates fail to obtain approval, resulting in the high cost of drug development [1]. This is due to problems associated with pharmacokinetics and pharmacodynamics for which drug-likeness or lead-likeness suggest ways to improve the failure rate. For a

---

This article has previously been published in the journal *Physical Sciences Reviews*. Please cite as: Athar, M., Sona, A.N., Bekono B.D., Ntie-Kang, F. Fundamental physical and chemical concepts behind "drug-likeness" and "natural product-likeness". *Physical Sciences Reviews* [Online] **2019** DOI: 10.1515/psr-2018-0101

https://doi.org/10.1515/9783110579352-003

substance to be considered a drug, otherwise for it to exhibit "drug-like" properties, it must have access to the site of action in the body after administration, where it should carry out its physiological or pharmacological function. Meanwhile, a drug is expected to have a certain range of physicochemical properties to have access to the body cells and be able to carry out the aforementioned functions, while remaining safe to the host organism. One would, therefore, admit that "drug-likeness" is a qualitative concept of drug design that can be generally defined as a balance between molecular and structural features used to determine how drug-like a substance is with respect to bioavailability. It helps to predict the "drug-like" ability of compound from its chemical structure and relies on the principle of molecular similarity with known drugs. In other words, "drug-like" properties or "drug-likeness" would appeal to the intrinsic properties of a chemical compound that are needed to be optimized by the medicinal chemist in order to achieve the desired optimal pharmacological properties. Studies have shown that such properties confer greater propensity to become a successful drug.

Broadly speaking, "drug-like" properties fall under three broad classes (see Figure 2.1):
- structural properties (hydrogen bond forming moieties, lipophilicity, molecular weight (MW), polar surface area),
- physico-chemical (solubility, permeability, stability)
- biochemical properties (metabolism, protein and tissue binding, transport modality) and pharmacokinetics/toxicity (clearance, half-life, bioavailability).

2.1.2 Structural features affect ADMET

It must be noted that, among these descriptive properties, structural properties can be most easily tailored and used as filters for screening compound-libraries. It is also reported that such properties or descriptors will, in turn, affect the bioavailability features characterized by absorption, distribution, metabolism, and excretion (ADME). Besides, drug-likeness addresses other pharmaceutical properties, for example, solubility, chemical stability, and distribution profile.

2.1.3 Christopher Lipinski, the father of "drug-likeness"

Christopher Lipinski first conceptualized "drug-likeness" as "the acceptable ADME/tox properties of the compound that make them survive through the completion of human Phase 1 trials" [2]. Later, it was found that structural properties of molecules affect not only the physicochemical properties but also the solubility, permeability, and drug absorption. As an outcome from such appraisals, a notion was proposed that many molecular properties are of interest in drug discovery. Nowadays, such drug-likeness parameters (Figure 2.1) can play an important role in the transformation of a lead molecule to a marketable drug.

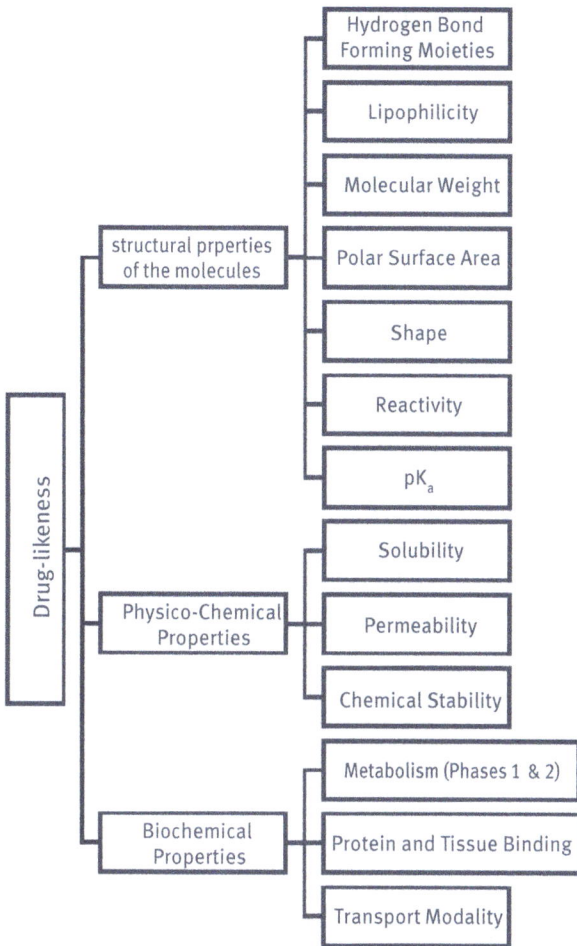

Figure 2.1: An overview of the elements of drug-likeness.

2.1.4 Bioavailability *versus* "drug-likeness"

In general, biological availability or bioavailability is the ability of a molecule to have access to the cells by crossing the cell membranes. Thus, bioavailability is an essential feature of all drugs, since without it the whole purpose of designing and administering a drug molecule into the body is futile. Optimal bioavailability is a crucial paradigm for the pharmacological design of a potent drug molecule, apart from its tight binding properties with the target [3]. Quite often, the molecules eradicated during the preclinical/clinical stages and sometimes even after market launch are due to low bioavailability, high toxicity, poor pharmacokinetics, or drug-drug interactions. Owing to limited time and the need for expensive resources to carry out *in vivo* studies, the early prioritization of the compounds using computational predictions has become a top priority in rational drug discovery. Compounds with undesirable properties, especially poor ADMET profiles can, therefore, be quickly expelled from further development. Suitable pharmacokinetic properties are vital for a drug molecule to be easily transported from the site of entry to the site of action. Such attributes ensure the presence of the optimal concentration of the drug molecule in order to have optimal pharmacodynamic effects (binding to a drug target).

2.1.5 Molecular properties used for predicting bioavailability and simple filters

Various molecular descriptors encode a biologically informative picture of a pharmaceutically relevant molecule. As a result of combining such predictive descriptors, various rules have been put forth to describe the drug-likeness of a molecule. Such approaches for drug-likeness evaluation include simple rules/filters, which are based on simple molecular properties derived from chemical structures. These include the "Rule of five" (Ro5), the "Rule of three" (Ro3), Veber rules, and Ghose rules or filters [2, 4, 5]. Besides, advanced pattern recognition algorithms (similarity metrics) and sophisticated machine learning methods have been developed. With time, the implications of these computational methods have proven that such rules are not void of shortcomings. Some of them include conformation dependency and the ability to only encode for the static nature of the structure and physicochemical attributes. However, in reality, the molecules in question are not frozen, meaning that they have a highly dynamic behavior and occupy greater property space than those defined by simple metrics like the Ro5, Ro3, etc.

2.1.5.1 Rule of five

According to Lipinski's Ro5 [2], a molecule is likely orally bioavailable (hence "drug-like") if it doesn't violate more than two of the rules: MW ≤ 500 Daltons (Da), the calculated logarithm of the n-octanol/water partition coefficient (clogP) < 5 units, number of hydrogen-bond acceptors (HBAs) ≤ 10 (roughly defined as the number of O atoms and N atoms) and number of hydrogen-bond donors (HBDs) ≤ 5 (roughly defined as the number of OHs and NHs). An additional "rule" for the number of rotatable bonds (NRB ≤ 10) is often included. This collection of simple molecular filters derived its appellation "Rule of 5" from the fact that each of the above numbers is a multiple of 5. The definitions of HBA and HBD, however, still remain ambiguous in many software tools (more on this in the next chapter).

2.1.5.2 Rule of three

This is a set of guidelines describing desirable physicochemical properties for molecules in fragment-based drug design (FBDD) screening collections [6], i.e. used to define a molecule that would be defined as "fragment-like", i.e. MW < 300, the cLogP

≤ 3, HBD ≤ 3 and HBA ≤ 3. The latter two criteria have not been widely adopted, but are also ambiguous in how donors and acceptors are defined. An additional criterion of NRB ≤ 3 is sometimes included. Again, the appellation "rule of 3" comes from the cutoff numbers being multiples of 3.

2.1.5.3 Veber's rules

These are a set of more stringent "rules" or cutoffs set as filters for predicting if a molecule would exhibit good oral bioavailability [5]. This ignores the MW criterion and focuses on the NRB (≤10) and the polar surface area (PSA ≤ 140 Å^2).

2.1.5.4 Ghose filters

This is an improved version of Lipinski's Ro5 for predicting drug-likeness [7], i.e. −0.4 < clog P < +5.6, 180 ≤ MW ≤ 480, molar refractivity (MR) in the range 40 ≤ MR ≤ 130, Number of atoms (N) in the range 20 ≤ N ≤ 70 (including HBAs and HBDs).

2.1.5.5 Ganesan's rules (beyond the classical "rule of five")

These are a set of new cutoffs of known descriptors for peptide drug design, which go beyond the Ro5 (bRo5) since peptides occupy quite a different chemical space from small molecule drugs from which Lipinski's Ro5 was initially derived. These rules are based on an extensive analysis of known peptide drugs and clinical candidates having peptide features [8]. The descriptors used include MW, total polar surface area (TPSA), NRB, clogP, HBA, HBD, and the fraction of sp^3-hybridized carbon atoms (Fsp3). New cut-offs for drug-like peptides were set, which go well beyond Lipinski's criteria.

2.1.6 Property filters for ADME/tox predictions as rough estimates for "drug-likeness"

Recent developments in the field of drug-likeness include protein target-based pharmaceutical filters, the development of newer descriptors like molecular sensitivity (which takes how many physicochemical properties vary as a function of the flexibility of a molecule), the quantitative estimate of drug-likeness (QED), etc [9]. Several software tools have been developed, which compute molecular properties (descriptors) of a new molecule, based on previously computed quantitative structure-activity relationships (QSAR) and compare these with those of known drugs, as a measure of predicting how drug-like a new molecule could be. Table 2.1 shows a number of such selected descriptors related to ADMET along with the estimated range of values for known drugs (as predicted by the QikProp software, commercialized by Schrodinger). This is based on the basic principle that for a drug to be acceptable as orally absorbed, it should easily be transported to the site of action, be chemically stable in the presence of metabolizing enzymes, have a low binding affinity to HERG K$^+$ channels (which affect the rhythm of the heart) and easily eliminated from body (ADMET).

Table 2.1: Selected computed ADMET-related descriptors and their recommended ranges for 95% of known drugs [10].

Descriptor	Definition	Required range
S_{mol}	the total solvent-accessible molecular surface, in Å^2 (probe radius 1.4 Å)	300 to 1000 Å^2
$S_{mol,hfob}$	the hydrophobic portion of the solvent-accessible molecular surface, in Å^2 (probe radius 1.4 Å)	0 to 750 Å^2
V_{mol}	the total volume of molecule enclosed by the solvent-accessible molecular surface, in Å^3 (probe radius 1.4 Å)	500 to 2000 Å^3
$\log S_{wat}$	the logarithm of aqueous solubility	−6.0 to 0.5
$\log K_{HSA}$	the logarithm of predicted binding constant to human serum albumin	1.5 to 1.2
$\log_{B/B}$	the logarithm of predicted blood/brain barrier partition coefficient	−3.0 to 1.0
BIP_{caco-2}	the predicted apparent Caco-2 cell membrane permeability, in nm s^{-1} (in Boehringer–Ingelheim scale	<5 low, >100 high
MDCK	the predicted apparent Madin–Darby canine kidney cell permeability in nm s^{-1}	<25 poor, >500 great
Ind_{coh}	the index of cohesion interaction in solids, calculated from the number of hydrogen-bond acceptors (HBA), donors (HBD) and the surface area accessible to the solvent, SASA (Smol) by the relation $Ind_{coh} = HBA \times \sqrt{HBD/S_{mol}}$	0.0 to 0.05
Glob	the globularity descriptor, $Glob = (4\pi r^2)/S_{mol}$, where r is the radius of the sphere whose volume is equal to the molecular volume	0.75 to 0.95
QP_{polrz}	the predicted polarizability	13.0 to 70.0
$\log HERG$	the predicted IC_{50} value for blockage of HERG K^+ channels	concern<−5
$\log K_p$	the predicted skin permeability	−8.0 to −1.0
#metab	the number of likely metabolic reactions	1 to 8

2.1.7 The distinction between "drug" and "leads"

The distinction between drug-likeness and lead-likeness could be perceived by comparing the property distributions of the natural product (NP) libraries shown in Table 2.2 (the properties of "lead-like" compounds are discussed in the next chapter). A promising lead compound may, therefore, be defined as one which combines potency with an attractive ADMET profile, permitting it to be developed into a marketable drug. As such, compounds with uninteresting predicted ADMET profiles may be completely dismissed from the list of potential drug candidates early enough (even if these prove to be highly potent), thus saving cost. As an example, there were only 239 (out of 610 drug-like) and 326 (out of 925 drug-like) lead-like molecules in AfroDb and StreptomeDB NP libraries, respectively [11, 12]. Other attributes of both the libraries were varied such as clogP values, e.g. 2.44 for the complete library (as

Table 2.2: Summary of average pharmacokinetic property distributions of the total StreptomeDB and AfroDb libraries, in comparison with the various subsets [11, 12].

Library	AfroDb				StreptomeDB			
Property	Total	Drug-like	Lead-like	Fragment-like	Total	Drug-like	Lead-like	Fragment-like
[a]Library size	1008	610	239	51	2444	925	326	127
[b]% compl.	48	73	75	76	577	459	207	40
[c]MW (Da)	406	328	266	219	485	262	230	151
[d]clogP	3.99	2.99	2.44	1.89	1.30	1.19	1.61	0.96
[e]HBA	5.76	4.89	3.91	3.39	1169	5.27	3.96	2.99
[f]HBD	1.67	1.25	0.87	0.60	347	1.92	1.46	1.07
[g]NRB	6.30	4.24	3.43	1.40	1146	4.78	3.68	1.28
[h]LogB/B	−0.-90	−0.63	−0.57	−0.29	−2.25	−1.01	−0.72	−0.31
[i]BIP$_{caco-2}$ (nms^{-1})	1516	1663	2032	1983	522	734	986	1275
[j]S$_{mol}$ (Å 2)	674	568	492	424	748	490	460	343
[k]Smol,hfob (Å^2)	415	312	235	139	417	213	179	112
[l]V$_{mol}$ (Å^3)	1265	1030	860	712	1426	840	768	536
[m]LogS$_{wat}$ (S in mol L^{-1})	−5.11	−3.88	−3.11	−2.50	−3.20	−2.42	−2.50	−1.14
[n]LogK$_{HSA}$	0.59	0.21	0.20	−0.20	−0.48	−0.37	−0.29	−0.63

[a]Number of total compounds in library; [b]Number of compounds with #star = 0; [c]Molar weight (range for 95 % of drugs: 130–725 Da); [d]Logarithm of partitioning coefficient between n-octanol and water phases (range for 95 % of drugs: −2 to 6); [e]Number of hydrogen bonds accepted by the molecule (range for 95 % of drugs: 2–20); [f]Number of hydrogen bonds donated by the molecule (range for 95 % of drugs: 0–6).; [g]Number of rotatable bonds (range for 95 % of drugs: 0–15); [h]Logarithm of predicted blood/brain barrier partition coefficient (range for 95 % of drugs: −3.0 to 1.0); [i]Predicted apparent Caco-2 cell membrane permeability in Boehringer–Ingelheim scale, in nm/s (range for 95 % of drugs: < 5 low, > 100 high); [j]Total solvent-accessible molecular surface, in Å^2 (probe radius 1.4 Å) (range for 95 % of drugs: 300–1000 Å^2); [k]Hydrophobic portion of the solvent-accessible molecular surface, in Å^2 (probe radius 1.4 Å) (range for 95 % of drugs: 0–750 (Å^2); [l]Total volume of molecule enclosed by solvent-accessible molecular surface, in Å^3 (probe radius 1.4 Å) (range for 95 % of drugs: 500–2000 Å^3); [m]Logarithm of aqueous solubility (range for 95 % of drugs: −6.0 to 0.5); [n]Logarithm of predicted binding constant to human serum albumin (range for 95 % of drugs: −1.5 to 1.2).

compared with a value of 2.99 for the drug-like subset) and 1.61 for the complete library (as compared with a value of 1.19 for drug-like for the drug-like subset) for AfroDb and StreptomeDB, respectively. Similarly, other descriptor values for the fragment-like libraries were significantly varied as can be perceived from Table 2.2.

2.1.8 Rationale of this work

In this work, the discussion is focused on the origin of the concept of "drug-likeness" and NP-likeness, as well as their implications in drug development processes, since NPs have increasing been the source of many lead compounds [13]. Our focus will remain on drug-likeness, lead-likeness, and NP-likeness, while the drug-likeness of NPs will be the subject of the following chapter. In order to comprehend the biological complexity and to re-define the chemical space by NP fragments, the concept of fragment-likeness is also discussed. It is to be noted that about one-third of all approved small molecule drugs have been inspired by NPs [13]. As a result, NPs have recently become the center of attraction of the pharmaceutical industry which renders them a promising and reliable source of new bioactive molecules. Therefore, proposed rules for small molecules cannot be blindly applied on NPs. Rather, one needs to understand NP-likeness and its distinction from drug-likeness.

2.2 Re-canvassing drug-likeness "rule of thumbs" for natural products

It is popularly believed that NPs fell from the scope of the rule of thumbs (see Section 2.1.5), which explains why to adhere to better drug discovery and new opportunities, it is usually suitable to ignore the known rule of thumbs or better still create a strategic distance from them. There are three principal reasons for this:
- these standards were built with a way toward creating orally bioavailable drugs
- carrier-mediated active transporters in the cell membrane easily facilitate the absorption of several molecules that could fail the "drug-likeness" test as defined by simple rule of the thumbs, typically NPs [14], which were excluded from such a definition
- these principles depend on the review investigation of existing drugs and neither give any information of the target scope involved in the drug action nor the physicochemical space of undiscovered drugs

It is important to recognize that the vast majority of known chemical libraries are based on existing target knowledge and available drugs. Such libraries of drug-like molecules have proven ineffective against a variety of challenging targets, such as protein–protein interactions (PPI), nucleic acid complexes, and antibacterial modalities. It is considered that such compounds are characterized by underrepresented scaffolds [15]. Leveraging these well-established structural classes (by NPs) has been viewed as a means to identify new ligands having desirable properties for subsequent drug development. Figure 2.2 shows that a significant proportion of newly drugs year by year from 1981 to 2014 were NPs with unmodified structures of NP-derived. For all the reasons mentioned above, there is a dire need to understand the biological complexity and in what way this can be fulfilled by the broad chemical space offered by the NPs.

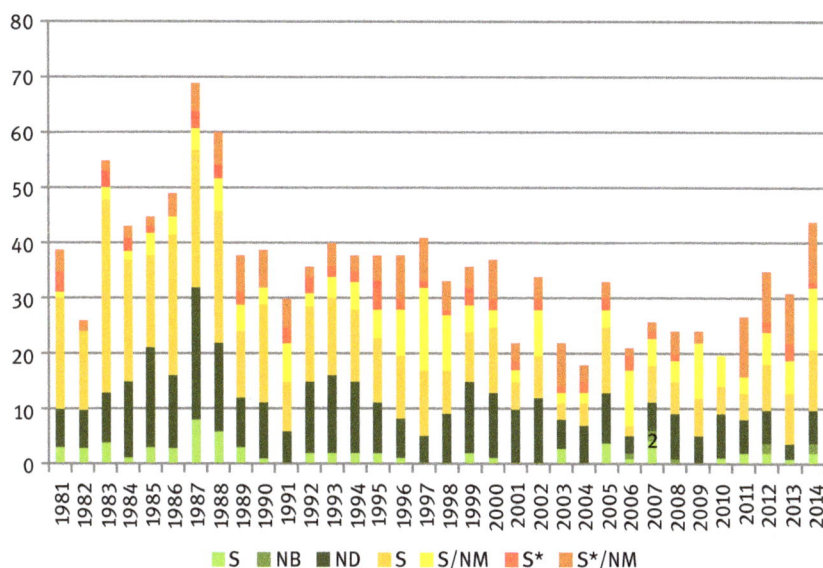

Figure 2.2: Approved drugs by source/year. "N": Natural product, unmodifed in structure, though might be semi- or totally synthetic "NB": Natural product "botanical drug" (in general these have been recently approved) "ND": Derived from a natural product and is usually a semisynthetic modification "S": Totally synthetic drug, often found by random screening/modification of an existing agent "S*": Made by total synthesis, but the pharmacophore is/was from a natural product "V": Vaccine [13]. Figure reproduced by permission.

2.3 The need to increase the chemical space and target space for addressing biological complexity

Present drug space is being poorly served from the challenging biological targets as well as chemical space offered by the therapeutics (chemical fragments). However, recent years have witnessed an amazing extension of the "drug space" that is significant for treating new ailments. In recent times, the chemical space relevant to drug discovery is extended to incorporate proteins, peptides, and ribonucleic acids. This can be clearly indicated by a case study reported by Overington, which states that current drugs only address 207 protein targets of the human genome [16]. It was further pointed out that half of all drugs are centered on just four classes of protein targets, i.e. rhodopsin-like G-protein coupled receptors, atomic receptors, and voltage-and ligand-gated particle channels. By utilizing existing medication like particles, this study estimated that just ≈10–14 % of the proteins encoded in the human genome are "druggable" [17]. Since the biological organization of cells is quite complex in nature and not currently fully comprehended, e.g. the action of kinases; it has earlier been thought that such drug targets are non-tractable due to multiple active sites. Besides, PPI, for example, were previously considered un-druggable [18].

An effective methodology to fill these gaps is by investigating the zones of NP mimics that can be subsequently applied to filter and screen compounds. All kind of thanks and credit ought to have been given to NPs that has made it possible to unveil such targets. The example and details are given in the next section. Navigation and mapping of chemical space relevant for drug discovery can be achieved by NPs. The methodologies render novel chances to combine NP-inspired compound comprising NP fragments and useful rings. Furthermore, NP target fishing enables us to access the regions of drug space and target space that are less investigated, and subsequently help to astound the impediments in the utilization of NPs in drug discovery, for example, the absence of openness and synthetic feasibility.

2.4 Why are natural products distinct from synthetic drugs?

2.4.1 They have a wider chemical space and increased diversity

Efforts are in the pipeline to identify lead molecule with the appropriate drug space to deliver answers to the challenging biological targets. In this regards, NP can effectively regulate targets and compound libraries that are characterized by the underrepresented scaffolds and limited chemical space [15].

Shoichet and colleagues have shown that 83% of NP and 20% of metabolite scaffolds (with ≤11 heavy atoms) are missing from industrially accessible compounds [19]. Subsequently, libraries having explicit, underrepresented scaffolds may address complex target by giving new pharmacophores and binding geometries. This can be explained on the basis of large chemical space [20], which can be defined as the complete set of possible combinations of small molecules (10^{30}–10^{200} structures). Further, this number depends on the parameter utilized as explained by the authors [21]. Just a small fraction of this chemical space (10^6 molecules) can be possibly tested in a virtual high throughput screening. Nevertheless, in spite of the overwhelming numbers, it appears to be likely that just a segment of chemical space is pertinent to drug discovery. The response to this can be effectively answered by NPs and their related functionalities.

Recent studies highlighted that about 34% of approved small new chemical entities (NCEs) were either NPs or their semi-synthetic derivatives, as reported in the period of 1981–2006, [17, 22]. Besides, more than 100 compounds derived from NP were shown to be present in a variety of remedies as therapeutics in clinical trials [23, 24]. NP compounds are biogenic and the inclusion of these compounds in screening libraries can lead to better results [19]. In addition, NPs are more similar to metabolites and according to an investigation of hit rates from HTS campaigns in Novartis, NPs represent as the most diverse class among synthetic and combinatorial molecules

[25]. In another study, Schneider and Grabowski analyzed the diversity of NP databases that emphasize the scaffolds used in combinatorial chemistry [26]. The Waldmann Group has proposed ways to exploit the intersection between the space of natural and biological products [27, 28]. Later on, in a quest to investigate chemical space, these authors developed a tool named "Structural Classification of Natural Products (SCONP)," which sorts out scaffolds in a hierarchical tree fashion. These scaffolds can be used to design a set of molecules around a biologically relevant scaffold. A similar concept has recently been utilized to develop a related tool called "Scaffold Hunter", which helps to navigate chemical and biological spaces interactively and provide new synthetic directions based on the scaffold hierarchy [29, 30].

Previously, Sauer *et al.* evaluated the chemical space of NP based on their 3D molecular shapes (Figure 2.3) *via* Principal Moment of Inertia (PMI) and Plane of Best

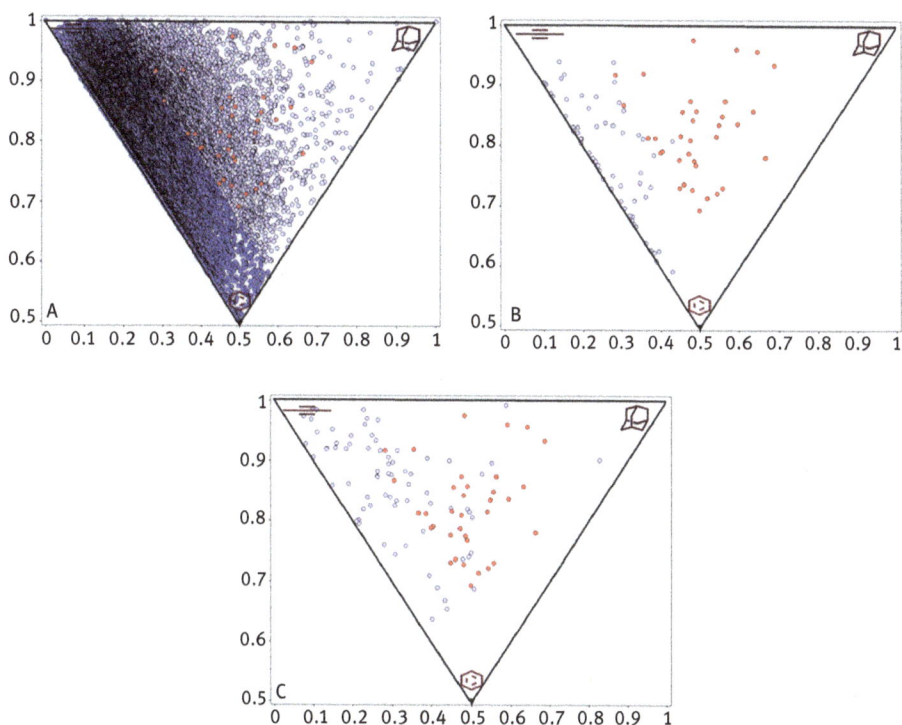

Figure 2.3: Distribution of NP fragments based on molecular shape. Each corner of the triangular plot represents certain molecular shape, Acetylene or rod-like shaped (top left corner), disc or benzene-like (at the bottom corner) and adamantine or spherical-like (at top right). (A) PMI plot with 35 3D fragments (red) and 18,534 fragments from the ZINC database (gray). (B) PMI plot based on heavy atom counts similarity of 62 best-matched ZINC fragments (gray) and 35 3D fragments (C) PMI plot based on best match physiochemical descriptors 76 ZINC fragments with 35 3D fragments and 76 of best-matched fragments from ZINC based on physiochemical descriptors [31]. Figure reproduced by permission.

Fit (PBF) [31]. As illustrated in these triangular plots of standardized PMI ratios, rod, disk, and sphere shaped molecules, i.e. acetylene, benzene, and adamantane have been shown at the top left, bottom corner and top right of the triangles, respectively. In Figure 2.3A, the darker blue region accounts for 75 % of 18,534 fragments from the ZINC database, suggesting that the region is dominated by acetylene or benzene-like shape molecules. Firth *et al.* described PBF as a quantitative measure of how far the atoms in the minimized 3D structure of a molecule are from the plane that best fits through the molecule [32].

The usefulness of NP can be further highlighted by the discovery of NPs kinase inhibitors in the 80s and 90s as key signal transducers. Earlier, such drug targets were believed to be untractable due to their remarkable homology with mostly multiple binding sites and high ATP concentration in the cytoplasm [33]. Another success story in NPs regime is their ability to deal with PPI, which again had been considered to be difficult and drug-free. It is generally accepted that PPI modulation offers a great deal of promise for disease treatment [34–38]. NP-like analogs of rapamycin and antibiotics acting by inhibiting or modulating PPI are already successful therapeutics. Since small molecules have the advantage of resembling drugs, scientists continue to study the nature of PPI and find ways to modulate them with small molecules. This has led to an increasing number of "clues" in the literature that suggest ways to move forward. Several recent papers summarize PPI strategy's success and identify some common themes for effective "protein mimicry" [39, 40]. A report by Wells and McLendon analyzes recent examples of small molecular blockers of PPI and argues to dispel some misconceptions about drug discovery possibilities in this area [18].

In another study, principal component analysis (PCA) was performed using structural and physicochemical characteristics of 40 top-selling drugs and 60 NPs. It was found that the top-selling drugs cluster largely in one region of the plot, and the drug-like libraries overlap with this region (Figure 2.4). NPs, span a much broader range of chemical space and in PCA few outliers were NPs and their derivatives. Moreover, NPs displayed higher polarity/decreased hydrophobicity, higher MWs (to left on the *x*-axis) and more stereochemical features with fewer aromatic rings (to bottom on the *y*-axis) compared to synthetic drugs and drug-like libraries [15].

2.4.2 Unique chemical properties of natural products at the atomic and molecular levels

Many research groups have compared NPs with drug molecules and compound libraries [41–43]. The characteristics of NPs have been compared with drugs derived from combinatorial synthesis and the following conclusions were presented to define NPs:
–   an increase in rigidity
–   a predominance of fused, bridged, and spiro ring systems

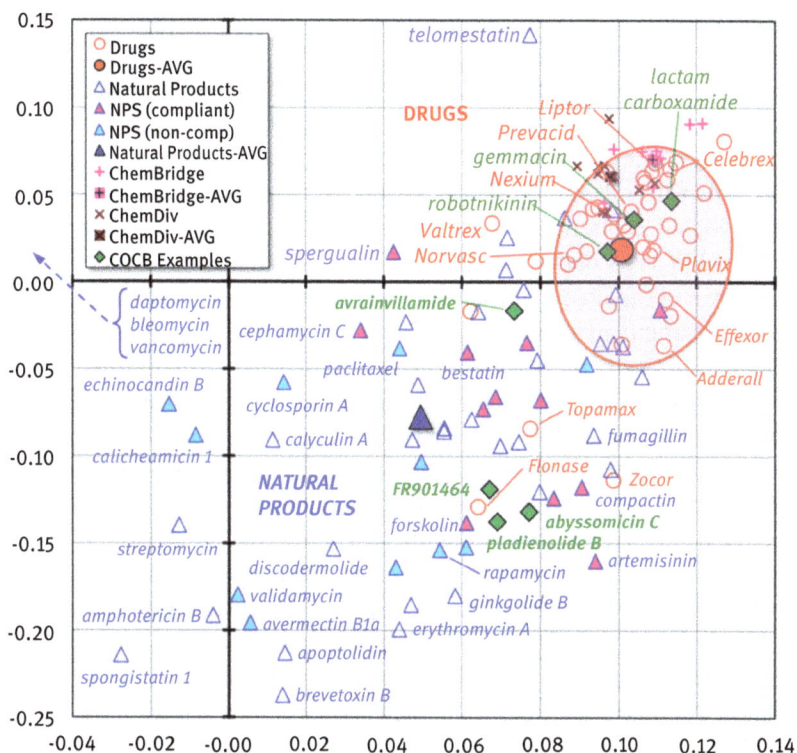

Figure 2.4: PCA analysis of 20 structural and physicochemical characteristics of 40 top-selling drugs (red circles), 60 NPs (blue triangles), including Ganesan's rule-of-five compliant (pink filled) and non-compliant (blue filled) subsets, and 20 compounds from commercial drug-like libraries (ChemBridge, pink plusses; Chem Div, maroon crosses). This study illustrates the narrow focus of existing drugs and drug-like libraries in chemical space in contrast to NPs [15]. Figure reproduced by permission.

- a lower percentage of aromatic atoms
- a larger number of chiral centers
- a higher count of carbon–oxygen bonds
- a lower count of carbon–nitrogen bonds
- an increase in diversity of scaffolds

NPs are structurally distinct [44] and hold a large number of sp^3 hybridized carbons, spiro, bridged connected systems and high structurally diverse stereogenic centers [45]. Earlier, structural rigidity had favorably correlated with successful clinical candidate translations, a property which depends on the NRB descriptor [46]. Substantial evidence in regard to these factors has been reported from the bioavailability studies in the rat model [5].

2.4.3 Greater biocompatibility

NPs are biologically pertinent chemical entities that are produced by living organisms and their origin is attributed to complex metabolic pathways which give special benefits to host organism [46, 47]. Since NPs are produced by proteins interacting with other proteins within living cells, they would be considered as naturally validated scaffolds [48][49]. Of particular interest for the discovery of drugs, NPs are also sometimes referred to as secondary metabolites, i.e. metabolites not directly necessary for the survival of the host. They are typically produced by organisms such as bacteria, plants or various marine invertebrates. Their principal role usually being serving for or participating in processes described as "chemical warfare" that protect the parent organisms against pathogens and predators. NPs are said to have been optimized for optimum interactions with biological macromolecules in a very long natural selection process. NPs, therefore, represent an excellent source of validated substructures for new drug design [50]. For these many reasons, it is common practice to include "metabolite-like" scaffolds in a screening library. The class of NPs referred to as secondary metabolites involved in defense or signaling are particularly important in drug discovery because they are said to have been optimized during evolution to interact effectively with biological receptors [51].

Therefore, NPs are an excellent source of validated substructures for the design of new bioactive molecules because of their increased biocompatibility [44]. Dobson *et al.* compared drugs and library compounds to endogenous metabolites in a related work and found that drugs are much more similar to metabolites than a library of synthesized compounds [52]. Owing to the fact that NPs strongly encode the biogenic character, it is expected to increase the number of molecules around scaffolds that resemble under-represented biogenic motifs in a screening collection.

2.4.4 Natural product-likeness in terms of natural product derived fragments

NPs differ from small molecules in comprising a variety of biological fragments. For this reason, FBDD is often considered as a complementary strategy in drug discovery. This tactic is helpful as it utilizes molecules with a low MW (fragment-like moieties) as starting points (scaffolds) in addition to exploring libraries of organic fragments [53–56]. Therefore, it allows the exploitation of larger areas of chemical space than in classical drug discovery. Although, fragment-likeness subjugates biocompatibility it does not usually represent the true bioactivity of the compounds [57]. In practice, NP-derived fragments differ considerably from synthetic compounds in structural features, design and obtaining a library component [58–62]. The NP-based fragments are generally different from synthesized fragments with a focus on heteroatom distribution, the number of rings/aromatic rings, the degree of ring fusion and degree of 3D distribution of atoms in space [26, 28, 42, 63–65]. The addition of NP fragments increases the chances of improving bioactivity, making this approach successful and less structurally complex in achieving the desired biological selectivity. The Dictionary of Natural Products (DNP) database contains 210,213 NPs with 134,102 unique fragments (of MW = 100 to 300 g mol^{-1}). Many of approved drugs contain these fragments, which can be applied in the computational pseudo-retrosynthetic scheme (RECAP). All such fragments can be used as starting points in *de novo* drug design, that previously revolutionized FBDD [66]. A similar method has been utilized in the development of three drugs; ipilimumab (monoclonal antibody), vemurafenib (anti-melanoma) and dabrafenib (BRAF kinase inhibitor) [67].

Over *et al.* classified NPs into the representative NP-fragment libraries containing 2000 cluster centers [63]. Wide collections of NP databases have recently been developed, meaning that their implication toward the development of bioactive compounds can be achieved in three ways:

– filtering the fragment from an NP library and initiating the chemical modification for designing new compounds
– generating fragments by *in silico* methods from compound collections obtained synthetically or commercially
– cleaving the high MW compounds and generating diverse libraries by chemical modification.

2.5 Quantifying "natural product-likeness"

Different cheminformatic strategies could be useful in assessing NPs, and the after-effects of such examinations would provide useful insights into the drug discovery process. To gauge NP-likeness, it is essential to recognize similar structural space covered by known NPs. Specifically, NP-resemblance of a compound is determined as the comparability of the molecular structure space secured by known NPs. It fills in as a criterion in the selection of lead compounds and in tailoring novel drugs. Thus, scoring of NP-likeness requires the molecules to be pre-processed so as to dismiss small disconnected fragments which fall outside the scope of NPs. Further, scoring of a query molecule for the NP-likeness is based on signatures of NP and synthetic molecular information that remains separately indexed in training datasets. In particular, the frequency of the NP fragments is calculated based on their atomic signatures that search up for the recurrence of atom pieces.

Mathematically, NP-likeness score is a Bayesian measure that considers the assurance of closeness with NP, as can be depicted in the following equation [44]:

$$Fragment_i = \log\left[\frac{NP_i}{SM_i} \cdot \frac{SM_t}{NP_t}\right] \tag{2.1}$$

$$Score_N = \sum_{i=0}^{N} Fragment_i \qquad (2.2)$$

$$NP - likeness\ score = \frac{Score_N}{N} \qquad (2.3)$$

where $Fragment_i$ is the contribution of a single fragment, NP_i is the total number of NPs in the training dataset containing $Fragment_i$, SM_i is the total number of synthetic molecules in the training dataset containing $Fragment_i$, SM_t is the sum total of synthetic molecules in the dataset and NP_t is the sum total of NPs in dataset. In the second step, the contribution of the individual fragments is added up to give the total score, $Score_N$. This summed score is then normalized by the number of atoms in the molecule (N) under examination to give the final NP-likeness score. This score takes different properties of NPs into account, e.g. the number of sp^3 carbons, as initially depicted by Lovering et al. [46]. The use of this score has earlier been found to be crucial in distinguishing between NPs and non-NPs. Furthermore, the use of this method does not require the 3D structures of compounds. This calculation can be implemented in the open-source cheminformatics toolbox RDKit (http://www.rdkit.org) and is currently being used as one of the filters to screen libraries and metabolites in computer assisted structure elucidation (CASE) in drug discovery campaigns (https://omictools.com/np-like ness-tool).

2.6 Natural product space in chemical space

A recent survey by Harvey et al. involved an analysis of 22,724,825 commercially available compounds in the ZINC database [68] (which is represented as a large brown circle in Figure 2.5a) in comparison with ~160,000 NPs from the DNP (which is represented by the small green circle in approximate scale on the same figure). In the study, biologically relevant compounds which could be defined as "hits" against any biological target are shown on the (b) part of Figure 2.5.

The ZINC database (~22 million represented by the yellow square) was originally designed for virtual screening (VS) campaigns, i.e. available in ready-to-dock three-dimensional formats. It must be noted that most VS campaigns begin with the ZINC library by removing compounds that clearly violate Lipinski's Ro5, leading to a reasonably sized library (often referred to as the "drug-like" ZINC library, represented by the blue square in Figure 2.5b) or by only focusing of compounds that fulfill the criteria for "lead-like" compounds as defined by Oprea et al. [69, 70] (represented by the purple square in Figure 2.5b). For a more in-depth definition of "lead-likeness", the reader is advised to consult the next chapter in this book.

a Relative number of products

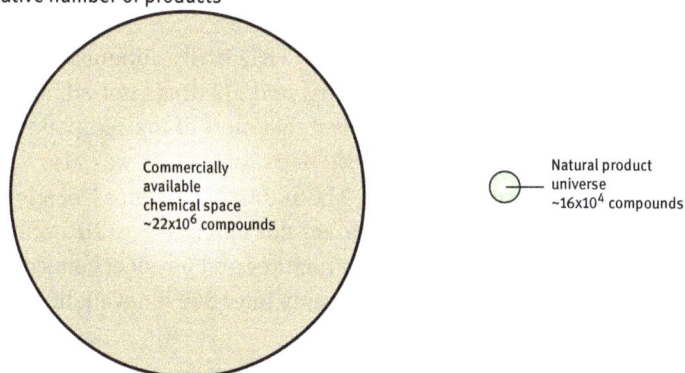

b Choosing biologically relevant chemical space

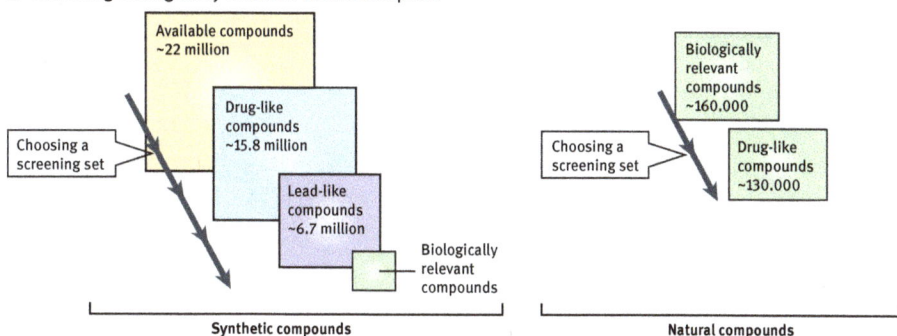

Figure 2.5: An illustration to demonstrate that biologically relevant chemical space is wider in NPs than by synthetic compounds [67]. Figure reproduced by permission.

Since "lead-likeness" filters are a lot more stringent, "lead-like" screening libraries from ZINC would have a reduced molecular mass and lipophilicity, implying that typical sizes of physical screening libraries would range from ~100,000 compounds to just a few million compounds, e.g. the "drug-like" subset of the ZINC database by the time Harvey et al. conducted the study contained 15,798,630 unique compounds, when compared with the "lead-like" subset containing 6,687,370 compounds [71]. In a typical VS study, the selection of a specific screening subset which should be enriched with biologically relevant compounds for a given drug target, as represented by the small green square in Figure 2.5b, would require considerable effort and expert knowledge. Although previous studies have shown that ~85 % of a certain NP library collection did not violate Lipinski's Ro5 and were drug-like [72], protein fold topology has revealed that NPs do occupy biologically relevant chemical space [73, 74]. A chapter dedicated to the chemical space of NPs is included in this book [75].

2.7 Drug-like and non drug-like natural products

Another study [76] involving 1,064 compounds, i.e. 852 orally administered drugs (administered to achieve systemic concentrations) and 212 drugs not administered orally to achieve systemic concentrations showed that most of the compounds that exhibit 2 or more Ro5 violations were NPs or NP derivatives. This was also true for non-oral drugs (see Table 2.2 of reference [76]). The author of reference [76] admitted, like Christopher Lipinski, that the Ro5 criteria are not applicable to NPs and their derivatives. However, looking at the chemical structures and physicochemical properties of the NPs shown on Table 2.3, we could simply infer that bioavailability could

Table 2.3: Comparison of the physico-chemical properties of two marketed NP drugs using known "drug-likeness" filters.

Chemical structure		
Compound name	Crofelemer	Ingenol mebutate (or ingenol-3-angelate)
Biological origin	*Croton lechleri* (South American tree)	*Euphorbia peplus* (a plant commonly known as milkweed)
Trade name	Mytesi® (or Fulyzaq®)	Picato®
For treatment of	Diarrhea associated with anti-HIV drugs, acute diarrhea in children, and patients with irritable bowel syndrome.	Anticancer, reactivation of latent HIV and tattoo removal
Administration route	Oral (tablets)	Topical (gel)
Bioavailability	0 (Little or no absorption from the gut)	Below detection level
DrugBank entry	https://www.drugbank.ca/drugs/DB04941	https://www.drugbank.ca/drugs/DB05013
MW	1347.65 g mol^{-1} (average)	430.53 g mol^{-1} (average)
logP (ChemAxon)	2.14	2.51
HBD	19	3
HBA	29	5
NRB	42	4
TPSA	547.52 Å^2	104.06 Å^2

(continued)

Table 2.3 (continued)

MR	$349.9 \text{ m}^3 \text{ mol}^{-1}$	$117.86 \text{ m}^3 \text{ mol}^{-1}$
Ro5 violations	3 of 4	0 of 4
Ghose filter violations	3 or 4	0 of 4
Veber's rule violations	2 of 2	0 of 2
Formula	$(C_{15}O_{6,7}H_{12})_n$	$C_{25}H_{34}O_6$
CAS Number	148465-45-6	75567-37-2
PubChem ID	17397714	6918670

only be a requirement if a drug is administered orally (if the target is located within the cells). This would be the case of the drug Ingenol mebulate whose desired effect and mode of administration are topical (on the surface of the skin). On the contrary, the anti-diarrhea drug Crofelemer which acts within the gut would not, therefore, require to be absorbed into the intestinal walls. Besides, it must be noted that only 51 % of all FDA-approved drugs had been shown to comply with the famous Ro5 [77]. As a recall, we would emphasize, from other analyses, that ~60 % of compounds from the DNP and 85 % of a natural product library (NPL) [72] do not violate the Ro5.

2.8 Do orally administered drugs in the market violate Lipinski's "rules"?

Apart from peptides (earlier discussed) which openly violate Lipinski's Ro5 and which have a new set of rules for predicting bioavailability (bRo5), several well-marketed drugs have shown that an open-minded drug discovery campaign should look well beyond simple "rules" if the desired goals must be achieved. A chapter, focused on NP and NP-derived drugs, has been included in this book [78], a lot of them violating the Ro5. We provide two examples here.

2.8.1 Talaprevir

Vertex's hepatitis C drug Telaprevir is a known orally available drug that almost violates most of, if not all, Lipinski's rules (Table 2.4), including the variant rules of rotatable bonds, and PSA.

2.8.2 Lipitor

In terms of general violation of the known rules by orally available drugs, let us consider the fact that Lipitor, Pfizer's best-selling drug and major player in their

Table 2.4: Examples of marketed drugs that clearly violate the Ro5 or other property filters.

Chemical structure		
Compound name	Telaprevir (VX-950)	Lipitor (or Atorvastatin)
Marketers	Vertex Pharmaceuticals and Johnson & Johnson.	Pfizer
Trade name	Incivek® (or Incivo®)	Sortis® (or Atorvalan®)
For treatment of	hepatitis C	lower cholesterol and triglycerides in the blood.
Administration route	Oral	Oral (tablets)
Bioavailability	0	
DrugBank entry	https://www.drugbank.ca/drugs/DB05521	https://www.drugbank.ca/drugs/DB01076
MW	679.85 g mol^{-1} (average)	558.64 g mol^{-1} (average)
logP (ChemAxon)	2.58	5.39
HBD	4	4
HBA	8	5
NRB	14	12
TPSA	179.56 Å^2	111.79 Å^2
MR	180.04 m^3 mol^{-1}	158.20 m^3 mol^{-1}
Ro5 violations	1 of 4	2 of 4
Ghose filter violations	3 of 4	3 of 4
Veber's rule violations	2 of 2	1 of 2
Formula	$C_{36}H_{53}N_7O_6$	$C_{33}H_{35}FN_2O_5$
CAS Number	402,957-28-2	134,523-00-5
PubChem ID	3,010,818	60,823

success, violates two of the four basic Lipinski rules (MW and clogP), comes close to violating a third (HBD) and also violates the variant rule of NRB. Had Pfizer chosen to follow the rules set by their own scientist (Lipinski worked at Pfizer when he published one of the most cited papers in medicinal chemistry), then this company would never have had the molecule that made is so rich today. Besides, Bruce Roth discovered Lipitor prior to the publication of Lipinski's famous paper.

2.9 Conclusions

Oral bioavailability is a major paradigm in the design of orally administered drugs. In looking for molecules to be administered as drugs, it is logical to presuppose that such a molecule must fit into the category of known drugs, i.e. be "drug-like" or fit into known molecular property filters identifiable with a majority of known drugs. Nowadays, simple molecular property filters (Lipinski's Ro5 being one of the key players) often guide medicinal chemists to tailor lead compounds. Omeprazole (Figure 2.6) is an example of a Lipinski-compliant drug used to treat several ailments, including peptic ulcer disease. However, if one went by such rules, then Paclitaxel (marketed under the brand name Taxol, Figure 2.6), one of the most important plant-based NPs used as an anticancer drug for several decades, would have probably been left out of the "prepared" supposed Lipinski-compliant VS library, just based on such filters (MW = 853.91 g mol^{-1}, HBA = 10, NRB = 14, TPSA = 221.29 Å^2, MR = 218.29 m^3 mol^{-1}). The good thing is that this drug is only administered intravenously. This implies that once all the drugs which are not administered orally are removed from known "drug-like" chemical space, a large proportion of the remaining compounds would simply not abide by such rules. This hypothesis could be verified by simply taking a look at the top 200 selling drugs. Those drugs violating such property filters would be mostly consisting of biologics or injectable drugs, even many orally administered compounds would be outliers from the "rules." This might imply that the general decline in drug discovery productivity in terms of newly launched drugs over the last three decades could be partly attributed to the introduction and general acceptance of these "rules" for drug discovery in the late 1990s after the publication of Lipinski's Ro5 [79]. Lipinski's Ro5 should however not be dictated, but be taken as a retrospective analysis of what properties will maximize the chance (or probability) of oral bioavailability. It is important to note that it was originally intended as a set of guidelines. Even Dr. Lipinski often advices to use it with caution, as a guideline rather than a rule. It must be noted that the original intention of the Ro5 was to drive synthetic chemists toward compounds which have better biophysical properties and are thus better orally active drug candidates. While this may be seen as a rather simplistic view, it shouldn't be ignored, as there would be no point taking a molecule

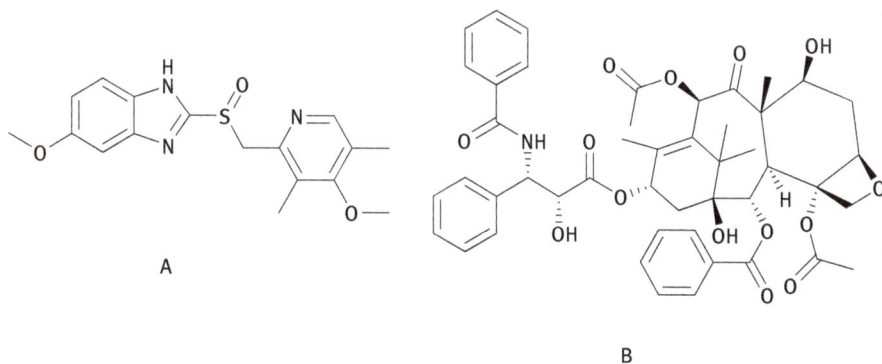

Figure 2.6: Chemical structures of two popular drugs; (A) Omeprazole and (B) Taxol, a NP marketed for cancer treatment.

forward and wasting time and resources, if it only turns out not be absorbed or pass across cell membranes. Regarding NPs, Lipinski explicitly excluded this class of compounds from his "rules" [80], the primary reasons being that secondary metabolites have supposedly evolved to be bioactive, and because they often utilize transmembrane transporters rather than passive diffusion to enter cells.

Acknowledgements: MA acknowledges the generous support from the Department of Science and Technology (DST), Government of India in the form of SRF-INSPIRE fellowship (IF150167) and Central University of Gujarat, India. FNK acknowledges a return fellowship and an equipment subsidy from the Alexander von Humboldt Foundation, Germany. BDB thanks the African-German Network of Excellence in Science (AGNES) for granting a Mobility Grant in 2017, generously sponsored by German Federal Ministry of Education and Research and the Alexander von Humboldt Foundation, Germany. Financial support for this work is acknowledged from a ChemJets fellowship from the Ministry of Education, Youth and Sports of the Czech Republic awarded to FNK.

List of abbreviations

Abbreviation	Full meaning
ADME	absorption, distribution, metabolism, and excretion
ADME/tox	absorption, distribution, metabolism, excretion and toxicity
bRo5	beyond the Ro5
clogP	the calculated logarithm of the n-octanol/water partition coefficient
CASE	computer assisted structure elucidation
DNP	Dictionary of Natural Products
FBDD	Fragment-based drug design
Fsp^3	sp^3-hybridized carbon atoms
HBA	number of hydrogen-bond acceptors
HBD	number of hydrogen-bond donors
MR	molar refractivity
MW	molecular weight
NCEs	new chemical entities
NP	natural product
NRB	number of rotatable bonds
PBF	the plane of Best Fit
PCA	principal component analysis
PMI	Principal Moment of Inertia
PPI	protein–protein interactions
PSA	polar surface area
QED	quantitative estimate of drug-likeness
QSAR	quantitative structure–activity relationships
Ro3	"Rule of three"
Ro5	"Rule of five"
SCONP	Structural Classification of Natural Products
TPSA	total polar surface area

References

[1] CSDD-Tufts Center for the Study of Drug Development. CNS drugs take 20 % longer to develop and to improve vs. non CNS drugs. Tufts CSDD impact reports, 2018. Available at: https://csdd.tufts.edu/tuftscsddreports/. Assessed: 26 Dec 2018.

[2] Lipinski CA. Lead-and drug-like compounds: the rule-of-five revolution. Drug Discov Today. 2004;1:337–41.

[3] Athar M, Lone MY, Jha PC. First protein drug target's appraisal of lead-likeness descriptors to unfold the intervening chemical space. J Mol Graph Model. 2017;72:272–82.

[4] Ghose AK, Herbertz T, Hudkins RL, Dorsey BD, Mallamo JP. Knowledge-based, central nervous system (CNS) lead selection and lead optimization for CNS drug discovery. ACS Chem Neurosci. 2011;3:50–68.

[5] Veber DF, Johnson SR, Cheng H-Y, Smith BR, Ward KW, Kopple KD. Molecular properties that influence the oral bioavailability of drug candidates. J Med Chem. 2002;45:2615–23.

[6] Congreve M, Carr R, Murray C, Jhoti H. A 'Rule of Three' for fragment-based lead discovery. Drug Discov Today. 2003;8:876–7.

[7] Ghose AK, Viswanadhan VN, Wendoloski JJ. A knowledge-based approach in designing combinatorial or medicinal chemistry libraries for drug discovery. 1. A qualitative and quantitative characterization of known drug databases. J Comb Chem. 1999;1:55–68.

[8] Santos GB, Ganesan A, Emery FS. Oral administration of peptide-based drugs: beyond Lipinski's rule. ChemMedChem. 2016;11:2245–51.

[9] Bickerton GR, Paolini GV, Besnard J, Muresan S, Hopkins AL. Quantifying the chemical beauty of drugs. Nat Chem. 2012;4:90.

[10] Ntie-Kang F, Lifongo LL, Judson PN, Sippl W, Efange SM. How "drug-like" are naturally occurring anti-cancer compounds? J Mol Model. 2014;20:2069.

[11] Ntie-Kang F. An in silico evaluation of the ADMET profile of the StreptomeDB database. SpringerPlus. 2013;2:353.

[12] Ntie-Kang F, Zofou D, Babiaka SB, Meudom R, Scharfe M, Lifongo LL, et al. AfroDb: a select highly potent and diverse natural product library from African medicinal plants. PLoS One. 2013;8:e78085.

[13] Newman DJ, Cragg GM. Natural products as sources of new drugs from 1981 to 2014. J Nat Prod. 2016;79:629–61. (add 3 refs each till the end).

[14] Dobson PD, Kell DB. Carrier-mediated cellular uptake of pharmaceutical drugs: an exception or the rule? Nat Rev Drug Discov. 2008;7:205.

[15] Bauer RA, Wurst JM, Tan DS. Expanding the range of 'druggable'targets with natural product-based libraries: an academic perspective. Curr Opin Chem Biol. 2010;14:308–14.

[16] Overington JP, Al-Lazikani B, Hopkins AL. How many drug targets are there? Nat Rev Drug Discov. 2006;5:993.

[17] Hopkins AL, Groom CR. The druggable genome. Nat Rev Drug Discov. 2002;1:727.

[18] Wells JA, McClendon CL. Reaching for high-hanging fruit in drug discovery at protein–protein interfaces. Nature. 2007;450:1001.

[19] Hert J, Irwin JJ, Laggner C, Keiser MJ, Shoichet BK. Quantifying biogenic bias in screening libraries. Nat Chem Biol. 2009;5:479.

[20] Dobson CM. Chemical space and biology. Nature. 2004;432:824.

[21] Bohacek RS, McMartin C, Guida WC. The art and practice of structure-based drug design: a molecular modeling perspective. Med Res Rev. 1996;16:3–50.

[22] Newman DJ, Cragg GM. Natural products as sources of new drugs over the last 25 years. J Nat Prod. 2007;70:461–77.

[23] Butler MS. Natural products to drugs: natural product-derived compounds in clinical trials. Nat Prod Rep. 2008;25:475–516.

[24] Al H. Natural products in drug discovery. Drug Discov Today. 2008;13:894–901.

[25] Sukuru SC, Jenkins JL, Beckwith RE, Scheiber J, Bender A, Mikhailov D, et al. Plate-based diversity selection based on empirical HTS data to enhance the number of hits and their chemical diversity. J Biomolecul Screen. 2009;14:690–9.

[26] Grabowski K, Schneider G. Properties and architecture of drugs and natural products revisited. Curr Chem Biol. 2007;1:115–27.

[27] Koch MA, Wittenberg L-O, Basu S, Jeyaraj DA, Gourzoulidou E, Reinecke K, et al. Compound library development guided by protein structure similarity clustering and natural product structure. Proc Nat Acad Sci USA. 2004;101:16721–6.

[28] Koch MA, Schuffenhauer A, Scheck M, Wetzel S, Casaulta M, Odermatt A, et al. Charting biologically relevant chemical space: a structural classification of natural products (SCONP). Proc Nat Acad Sci USA. 2005;102:17272–7.

[29] Wetzel S, Klein K, Renner S, Rauh D, Oprea TI, Mutzel P, et al. Interactive exploration of chemical space with Scaffold Hunter. Nat Chem Biol. 2009;5:581.

[30] Renner S, Van Otterlo WA, Seoane MD, Möcklinghoff S, Hofmann B, Wetzel S, et al. Bioactivity-guided mapping and navigation of chemical space. Nat Chem Biol. 2009;5:585.

[31] Hung AW, Ramek A, Wang Y, Kaya T, Wilson JA, Clemons PA, et al. Route to three-dimensional fragments using diversity-oriented synthesis. Proc Nat Acad Sci USA. 2011;108:6799–804.

[32] Firth NC, Brown N, Blagg J. Plane of best fit: a novel method to characterize the three-dimensionality of molecules. J Chem Inf Model. 2012;52:2516–25.

[33] Liao JJ. Molecular recognition of protein kinase binding pockets for design of potent and selective kinase inhibitors. J Med Chem. 2007;50:409–24.

[34] Zinzalla G, Thurston DE. Targeting protein–protein interactions for therapeutic intervention: a challenge for the future. Future Med Chem. 2009;1:65–93.

[35] Murray JK, Gellman SH. Targeting protein–protein interactions: lessons from p53/MDM2. Biopolymers. 2007;88:657–86.

[36] Verdine GL, Walensky LD. The challenge of drugging undruggable targets in cancer: lessons learned from targeting BCL-2 family members. Clin Cancer Res. 2007;13:7264–70.

[37] Berg T. Small-molecule inhibitors of protein–protein interactions. In: Martin Zacharias, editor, Protein-protein complexes: analysis, modeling and drug design. London, UK: World Scientific, 2010:318–39.

[38] Wilson AJ. Inhibition of protein–protein interactions using designed molecules. Chem Soc Rev. 2009;38:3289–300.

[39] Fry DC. Drug-like inhibitors of protein-protein interactions: a structural examination of effective protein mimicry. Curr Protein Pept Sci. 2008;9:240–7.

[40] Keskin O, Gursoy A, Ma B, Nussinov R. Principles of protein– protein interactions: What are the preferred ways for proteins to interact? Chem Rev. 2008;108:1225–44.

[41] Singh N, Guha R, Giulianotti MA, Pinilla C, Houghten RA, Medina-Franco JL. Chemoinformatic analysis of combinatorial libraries, drugs, natural products, and molecular libraries small molecule repository. J Chem Inf Model. 2009;49:1010–24.

[42] Feher M, Schmidt JM. Property distributions: differences between drugs, natural products, and molecules from combinatorial chemistry. J Chem Inf Comput Sci. 2003;43:218–27.

[43] Shelat AA, Guy RK. Scaffold composition and biological relevance of screening libraries. Nat Chem Biol. 2007;3:442.

[44] Ertl P, Roggo S, Schuffenhauer A. Natural product-likeness score and its application for prioritization of compound libraries. J Chem Inf Model. 2008;48:68–74.

[45] Silva DG, Emery FS. Strategies towards expansion of chemical space of natural product-based compounds to enable drug discovery. Braz J Pharm Sci. 2018;54:e01004.

[46] Lovering F, Bikker J, Humblet C. Escape from flatland: increasing saturation as an approach to improving clinical success. J Med Chem. 2009;52:6752–6.

[47] Firn RD, Jones CG. Natural products–a simple model to explain chemical diversity. Nat Prod Rep. 2003;20:382–91.

[48] Maplestone RA, Stone MJ, Williams DH. The evolutionary role of secondary metabolites—a review. Gene. 1992;115:151–7.

[49] Balamurugan R, Dekker FJ, Waldmann H. Design of compound libraries based on natural product scaffolds and protein structure similarity clustering (PSSC). Mol Biosyst. 2005;1:36–45.

[50] Haustedt L, Mang C, Siems K, Schiewe H. Rational approaches to natural-product-based drug design. Curr Opin Drug Discov Devel. 2006;9:445–62.

[51] Jayaseelan KV, Moreno P, Truszkowski A, Ertl P, Steinbeck C. Natural product-likeness score revisited: an open-source, open-data implementation. BMC Bioinformics. 2012;13:106.

[52] Dobson PD, Patel Y, Kell DB. 'Metabolite-likeness' as a criterion in the design and selection of pharmaceutical drug libraries. Drug Discov Today. 2009;14:31–40.

[53] Baker M. Fragment-based lead discovery grows up. Nat Rev Drug Discov. 2013;12:5–7.

[54] Chen H, Zhou X, Wang A, Zheng Y, Gao Y, Zhou J. Evolutions in fragment-based drug design: the deconstruction–reconstruction approach. Drug Discov Today. 2015;20:105–13.

[55] Joseph-McCarthy D, Campbell AJ, Kern G, Moustakas D. Fragment-based lead discovery and design. J Chem Inf Model. 2014;54:693–704.

[56] Murray CW, Rees DC. Opportunity knocks: organic chemistry for fragment-based drug discovery (FBDD). Angew Chem Int Ed Engl. 2016;55:488–92.

[57] Scott DE, Coyne AG, Hudson SA, Abell C. Fragment-based approaches in drug discovery and chemical biology. Biochemistry. 2012;51:4990–5003.

[58] Genis D, Kirpichenok M, Kombarov R. A minimalist fragment approach for the design of natural-product-like synthetic scaffolds. Drug Discov Today. 2012;17:1170–4.

[59] Austin MJ, Hearnshaw SJ, Mitchenall LA, McDermott PJ, Howell LA, Maxwell A, et al. A natural product inspired fragment-based approach towards the development of novel anti-bacterial agents. MedChemComm. 2016;7:1387–91.

[60] Pascolutti M, Campitelli M, Nguyen B, Pham N, Gorse A-D, Quinn RJ. Capturing nature's diversity. PLoS One. 2015;10:e0120942.

[61] Prescher H, Koch G, Schuhmann T, Ertl P, Bussenault A, Glick M, et al. Construction of a 3D-shaped, natural product like fragment library by fragmentation and diversification of natural products. Bioorg Med Chem. 2017;25:921–5.

[62] Rodrigues T, Reker D, Schneider P, Schneider G. Counting on natural products for drug design. Nat Chem. 2016;8:531.

[63] Over B, Wetzel S, Grütter C, Nakai Y, Renner S, Rauh D, et al. Natural-product-derived fragments for fragment-based ligand discovery. Nat Chem. 2013;5:21.

[64] Zaid H, Raiyn J, Nasser A, Saad B, Rayan A. Physicochemical properties of natural based products versus synthetic chemicals. Open Nutraceuticals J. 2010;3:194–202.

[65] Wetzel S, Schuffenhauer A, Roggo S, Ertl P, Waldmann H. Cheminformatic analysis of natural products and their chemical space. CHIMIA Int J Chem. 2007;61:355–60.

[66] Lewell XQ, Judd DB, Watson SP, Hann MM. Recap retrosynthetic combinatorial analysis procedure: a powerful new technique for identifying privileged molecular fragments with useful applications in combinatorial chemistry. J Chem Inf Comput Sci. 1998;38:511–22.

[67] Flaherty KT, Yasothan U, Kirkpatrick P. Vemurafenib. Nat Rev Drug Discov. 2011;10:811–12.

[68] Irwin JJ, Sterling T, Mysinger MM, Bolstad ES, Coleman RG. ZINC: a free tool to discover chemistry for biology. J Chem Inform Model. 2012;52:1757–68.

[69] Hann MM, Oprea T. Pursuing the leadlikeness concept in pharmaceutical research. Curr Opin Chem Biol. 2004;8:255–63.

[70] Oprea TI, Davis AM, Teague SJ, Leeson PD. Is there a difference between leads and drugs? a historical perspective. J Chem Inf Comput Sci. 2001;41:1308–15.

[71] Harvey AL, Edrada-Ebel R, Quinn RJ. The re-emergence of natural products for drug discovery in the genomics era. Nat Rev Drug Discov. 2015;14:111–29.

[72] Quinn RJ, Carroll AR, Pham NB, Baron P, Palframan ME, Suraweera L, et al. Developing a drug-like natural product library. J Nat Prod. 2008;71:464–8.

[73] McArdle BM, Campitelli MR, Quinn RJ. A common protein fold topology shared by flavonoid biosynthetic enzymes and therapeutic targets. J Nat Prod. 2006;69:14–7.

[74] Kellenberger E, Hofmann A, Quinn RJ. Similar interactions of natural products with biosynthetic enzymes and therapeutic targets could explain why nature produces such a large proportion of existing drugs. Nat Prod Rep. 2011;8:1483–92.

[75] Saldívar-González FI, Pilón-Jiménez BA, Medina-Franco JL. Chemical space of naturally occurring compounds. Phys Sci Rev. 2018. doi:10.1515/psr-2018-0103.

[76] Benet LZ, Hosey CM, Ursu O, Oprea TI. BDDCS, the rule of 5 and drugability. Adv Drug Deliv Rev. 2016;101:89–98.

[77] Zhang MQ, Wilkinson B. Drug discovery beyond the 'rule-of-five'. Curr Opin Biotechnol. 2007;18:478–88.

[78] Newman DJ. From natural products to drugs. Phys Sci Rev. 2018. doi:10.1515/psr-2018-0111.

[79] Lipinski CA, Lombardo F, Dominy BW, Feeney PJ. Experimental and computational approaches to estimate solubility and permeability in drug discovery and development settings. Adv Drug Delivery Rev. 1997;23:3–25.

[80] Lipinski CA, Lombardo F, Dominy BW, Feeney PJ. Experimental and computational approaches to estimate solubility and permeability in drug discovery and development settings. Adv Drug Deliv Rev. 2001;46:3–26.

Fidele Ntie-Kang, Kennedy D. Nyongbela, Godfred A. Ayimele
and Suhaib Shekfeh

3 "Drug-likeness" properties of natural compounds

Abstract: Our previous work was focused on the fundamental physical and chemical concepts behind "drug-likeness" and "natural product (NP)-likeness". Herein, we discuss further details on the concepts of "drug-likeness", "lead-likeness" and "NP-likeness". The discussion will first focus on NPs as drugs, then a discussion of previous studies in which the complexities of the scaffolds and chemical space of naturally occurring compounds have been compared with synthetic, semisynthetic compounds and the Food and Drug Administration-approved drugs. This is followed by guiding principles for designing "drug-like" natural product libraries for lead compound discovery purposes. In addition, we present a tool for measuring "NP-likeness" of compounds and a brief presentation of machine-learning approaches. A binary quantitative structure–activity relationship for classifying drugs from nondrugs and natural compounds from nonnatural ones is also described. While the studies add to the plethora of recently published works on the "drug-likeness" of NPs, it no doubt increases our understanding of the physicochemical properties that make NPs fall within the ranges associated with "drug-like" molecules.

Keywords: cheminformatics, drugs, drug-likeness, drug discovery, natural products

3.1 Introduction

In a previous work, focused on the definition and classification of natural products (NPs), NPs were defined as substances isolated from plants, micro-organisms, insects, mammals, etc. [1]. Since the term NPs is not inclusive of products of primary metabolism or those found in all living cells, e. g. proteins, nucleic acids, carbohydrates and compounds that are substrates for biological transporters, the definition of NPs is rather restricted to products of secondary metabolism. Thus, NPs could also be referred to as secondary metabolites. Consequently, the terms NP, secondary metabolite or naturally occurring metabolite or compound (NOC) will be used interchangeably throughout the text. NPs and NP mimics comprise an important source for human therapeutics and are estimated to compass a significant market share among the approved drugs [2]. Throughout the human history, nature's chemical library has been proven to be a rich resource for many biologically active medicinal

This article has previously been published in the journal *Physical Sciences Reviews*. Please cite as: Ntie-Kang, F., Nyongbela, K.D., Ayimele, G.A., Shekfeh, S. "Drug-likeness" Properties of Natural Compounds. *Physical Sciences Reviews* [Online] **2019** DOI: 10.1515/psr-2018-0169

https://doi.org/10.1515/9783110579352-004

leads and drugs. However, major changes in trends of the drug discovery programs have occurred during the last four decades. Thus, drug discovery programs started to focus on target-based methods; after the emergence of *in vitro* assays and the development of large combinatorial chemical libraries. With the increasing need to have access to large libraries or chemical library collections for screening, which was not clearly possible with classical NP extraction and purification methods, these last changes enabled the shift from the NP-based discovery programs to the high-throughput screening (HTS) technology as the main strategy for target-based drug discovery programs [3].

Following the availability of huge chemical screening libraries from combinatorial synthesis and valuable biological activities data collected from HTS, it became clear that medicinal chemists needed some criteria to distinguish between biologically active compounds and drugs. Christopher Lipinski was able to analyze a wealth of data that had been accumulated from the HTS and failed drug discovery programs that had been stored in the World Drug Index (WDI) during the 1980s and 1990s. He then suggested that the concept of "drug-likeness" was linked to oral bioavailability, hence to the famous "rule of 5" (RO5). Oral bioavailability is the ability of a drug to be administered orally in an efficient manner, a concept often linked to the absorption, distribution, metabolism, excretion and toxicity (ADME/T) of drug molecules. This is the ability of the drug to cross the intestinal walls, go through general blood circulation, reach it's intended target site and eventually stay at the target site in sufficient time to carry out its pharmacological function, then be eliminated efficiently so as not to accumulate into amounts that are unsafe (toxic) to the body. The RO5 comprises a simple set of four physical–chemical property ranges, which give the biologically active compound higher probability to be orally bioavailable and promising favorable pharmacokinetic (ADME/T) profile [4]. This rule includes

1. molecular weight (MW) less than 500 Da,
2. computed logarithm of on *n*-octanol/water partition coefficient (clogP) less than 5,
3. number of hydrogen bond acceptors (HBA, defined as the number of N and O atoms) less than 10 and
4. number of hydrogen bond donors (HBD, defined as the sum of OH and NH groups) less than 5.

The RO5 derives its appellation from the fact that these numbers are all multiples of 5. A further proposal by Oprea and colleagues suggested more stringent rules to identify what is called lead-likeness [5, 6]. Compounds to be classified as "leads" were by definition

1. less complex compounds with less chemical features,
2. display good biological activity with good ADMET profile and
3. amenable for chemical optimization to improve the biological activity or enhance the pharmacokinetic properties.

Hence, Oprea and coworkers were able to distinguish the lead-like chemical space from the drug-like space by stating the following lead-likeness conditions, otherwise known as Oprea lead-likeness filters:

1. MW: maximum 450
2. clogP: between −3.5 and +4.5
3. HBA: maximum 8
4. HBD: maximum 5

The statistical analysis of three compound classes, NPs, molecules from combinatorial synthesis and drug molecules, displayed significant differences between the combinatorial synthetic libraries and NP libraries [7]. Elsewhere, Saldívar-González et al. discussed some major compound databases of NPs and cheminformatics strategies that have been used to characterize the chemical space of NPs. The authors analyzed NPs from different sources and their relationships with other compounds are also discussed using novel chemical descriptors and data mining approaches that are emerging to characterize the chemical space of naturally occurring compounds [8]. In this work, our discussion will first focus on NPs as drugs, then a discussion of previous studies comparing NPs and drugs, a brief discussion on NPs, principles of designing NP libraries, the concept of "NP-likeness" and finally tools and methods used for the prediction of "drug-likeness" and NP-likeness.

3.2 Natural products as drugs

3.2.1 The proportion of natural products in catalogs of drugs

NPs play important roles in drug discovery, providing scaffolds as starting points for hit/lead discovery [9, 10]. Several known drugs, e. g. the anticancer compounds (**1–5**, Figure 3.1), are from natural sources [11]. We must note that NPs continue to play roles as drugs [2], as biological probes and as study targets for synthetic and analytical chemists [12]. About half of all approved drugs between 1981 and 2010 were shown to be NP-based [13]. The aforementioned study also showed that, of all approved drugs; NPs constituted 6 % (unaltered), 26 % (NP derivatives), 32 % (NP mimics) or from NP pharmacophores, 73 % of small molecule antibacterial and 50 % of anticancer drugs (including taxol, vinblastine, vincristine, topotecan, etc.), Table 3.1. This implies that if structural features provided by nature are successfully incorporated into synthetic drugs (SDs), this would increase the chemical diversity available for small-molecule drug discovery [2]. However, the reasons for the decline of interest by the pharmaceutical industry during the last two decades include the time factor involved in the search for NP lead compounds to the labor intensiveness of the whole process [14]. This has now been rendered much easier within industrial settings by streamlined screening procedures and enhanced organism sourcing mechanisms [15].

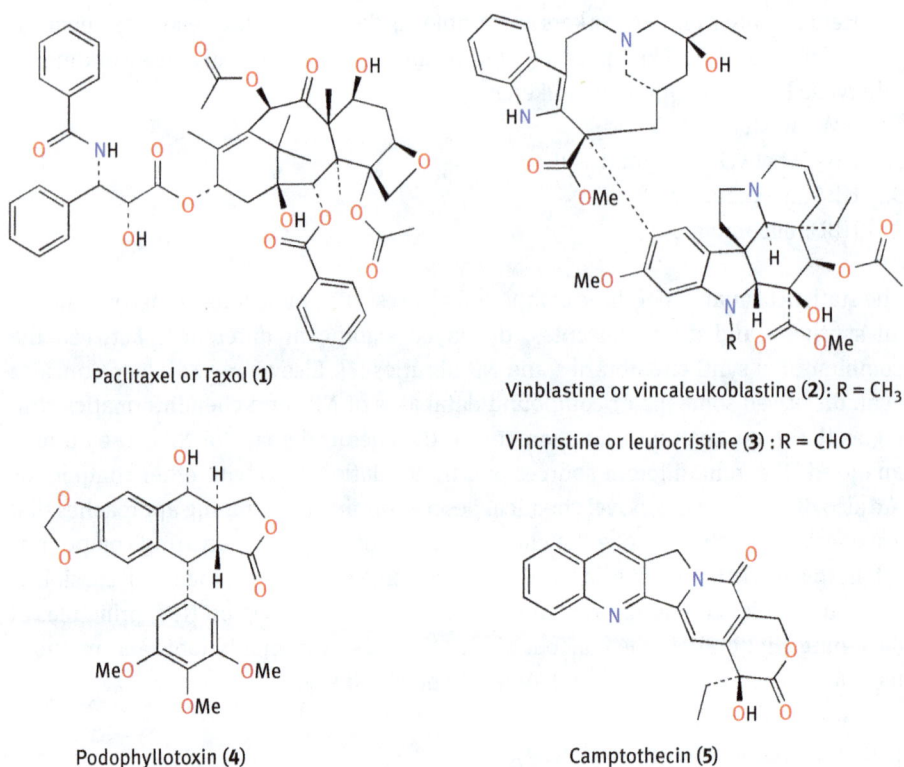

Paclitaxel or Taxol (**1**)

Vinblastine or vincaleukoblastine (**2**): R = CH₃

Vincristine or leurocristine (**3**) : R = CHO

Podophyllotoxin (**4**)

Camptothecin (**5**)

Figure 3.1: Selected naturally occurring NP anticancer drug leads.

3.2.2 The future of natural product drug discovery

Despite their evolving role in drug discovery [16, 17], a recent chemoinformatic study involving a dataset of all published microbial and marine-derived compounds since the 1940s (comprising 40,229 NPs) showed that most NPs being published today bear close similarity to previously published structures, with a plateau being observed since the mid-1990s [18], Figure 3.2. The authors observed a general trend that the rate of discovery of new NPs had flattened out since the 1990s, structures with novel scaffolds had become scarce (Figure 3.3). In the mentioned study, two compounds were considered to be dissimilar by taking a Tanimoto cutoff of $T_c < 0.4$. This study had, thus, suggested that the range of scaffolds readily accessible from nature is limited, i. e. scientists were close to having described all of the chemical space covered by NPs, even though appreciable numbers of NPs with no structural precedents continue to be discovered.

A reproduction of the same study using another dataset of 32,380 NPs showed the same trend [19]. However, a similar analysis on a dataset of randomly selected

Table 3.1: Comparative summary between natural products and synthetic drugs.

Property	Natural products	Synthetic drugs
Samples	Limited quantities (time consuming due to cumbersome extraction processes)	Readily available from combinatorial libraries
Drug-likeness	Weaker bioavailability (generally poorer ADME/T profiles)	More bioavailable
Chemistry	Complex scaffolds, more stereogenic centers	Less O-atoms, less aromatics, etc.
Proportion of marketed drugs	−6% (unaltered) −26% (NP derivatives) −32% (NP mimics) or from NP pharmacophores −73% of small molecule antibacterial −50% of anticancer drugs (taxol, vinblastine, vincristine, and topotecan, etc.)	

Data derived from Ref. [2].

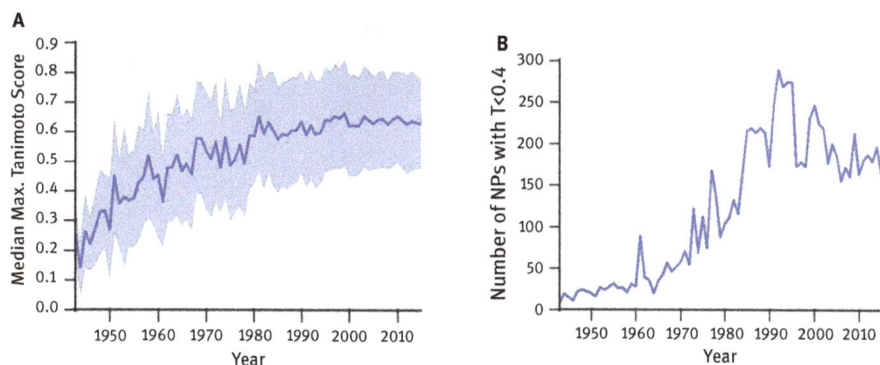

Figure 3.2: Presentation of structural diversity (a) by plotting the median maximum Tanimoto scores as a function of time. The median average deviation shown as shaded blue region. (b) By plotting the absolute number of low similarity compounds ($T_c < 0.4$) per year [18]. Figures reproduced by permission.

compounds from the ZINC database having overall lower structural similarity led the authors of the latter study further to prove that such trends may be a feature of any growing database of chemical structures, rather than reflecting trends specific to NP discovery. Besides, a Kolmogorov–Smirnov test conducted on the dataset of 40,229 NPs, with $P = 6.2 \times 10^{-14}$, showed that since 1990, the rate of structurally novel compound discovery has dramatically outpaced random expectation, with

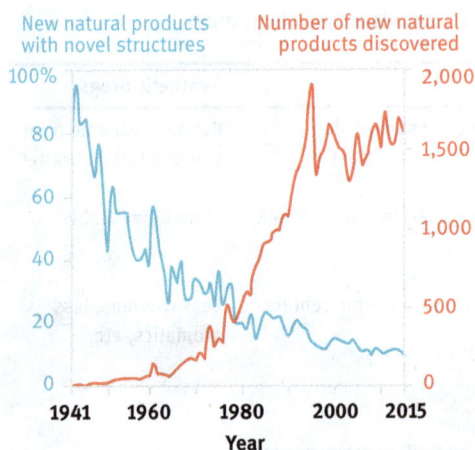

Figure 3.3: Presentation of structural diversity by plotting the number of compounds published per year and rate of novel compounds isolated as a percentage of total natural products isolated [18]. Figure reproduced by permission.

$P = 7.6 \times 10^{-11}$[19]. This implies that NPs discovered within the last three decades have been characterized by unprecedented chemical diversity, suggesting that the dream of continuously discovering new chemical structures from nature remains positive.

3.3 Natural *versus* synthetic drugs

3.3.1 The uniqueness and potential of natural products for drug discovery

NPs are unique, when compared with SDs in that they often contain more complex scaffolds and chiral centers, with more O-atoms and aromatic groups [20], Table 3.1. In addition, a study involving a comparison of SDs *versus* NPs showed that drugs derived from NP-based structures display greater chemical diversity and occupy wider regions of chemical space [2]. This is because drugs which are synthesized based on NP pharmacophores often exhibit lower hydrophobicity and greater stereo-chemical content when compared with drugs which are completely of synthetic origin. NPs are mostly more potent with higher binding affinities to a specific biological receptor. Consequently, their biological activities are often more selective than synthetic compounds. A property distribution of three investigated datasets consisting of 3,287 NPs, 10,968 drug molecules and 13,506 randomly selected combi-natorially derived lead candidates, respectively, led to the analysis of the number of chiral centers, rotatable bonds, aromatic rings, complex ring systems, degrees of saturation, as well as the ratios of different heteroatoms (O, N, etc.) [7]. This study showed that the main structural differences between NPs and combinatorially derived libraries arise from properties introduced during the synthetic process in order to render combinatorial synthesis more efficient. Moreover, it was shown that, since drug molecules originate from both natural and synthetic sources, they occupy a joint area of chemical space spread between NPs and combinatorially derived compounds.

Although NPs are often said not to satisfy all criteria of the RO5, a large proportion of NP libraries provide very good leads for drug development. For example, 60 % of the 126,140 unique compounds in the Dictionary of Natural Products (DNP) were found as "drug-like", complying with the RO5 [21]. Moreover, other investigations revealed that only 10 % of analyzed NP libraries violated two or more of Lipinski's RO5 [17, 22]. Attempts to quantify biosynthetic bias in screening libraries showed that 83 % (12,977) of core ring scaffolds present in NPs were missing in the combinatorial databases [23, 24], and the inclusion of these missing NP fragments inside the screening libraries would improve the hit rates [23]. In order to bring the drug-like space of the synthesized chemicals closer to the properties of NPs, a new measure called NP-likeness score was proposed by Ertl et al. (Section 3.7) [25].

3.3.2 The complexity and diversity of natural product scaffolds

NPs are, generally, compounds with large, diverse and structurally complex scaffolds (see previous chapter [1]), due to their often complex biosynthesis processes. The NPs contained in the DNP were previously classified according to their origins using a classification tree approach [26], with the aim of analyzing systems of rings that are typical according to the source. The high selectivity of NPs is attributed to their higher degree of complexity, higher number of stereogenic centers, more polar functional groups and different ratios of atom types, e. g. N, O, S and halogens [7]. This study sheds the light on the remarkable diversity of NPs occupying different regions of the chemical space with distinct ranges of the physicochemical properties. The complexity and diversity of NPs have been illustrated by the use of the tool ChemGPS-NP [27], which was designed for handling the chemical diversity encountered in NP research, in contrast to previously designed chemical global positioning system (ChemGPS) [28], which focused on the much more restricted drug-like chemical space. The uniqueness of the ChemGPS-NP tool, as contrasted to ChemGPS, is that a better representation of biologically relevant chemical space is achieved by including complex structural examples from the creative chemistry of naturally occurring bioactive molecules. Rules for plotting the chemical space maps include aspects of size, shape, lipophilicity, polarity, polarizability, flexibility, rigidity and hydrogen bond capacity. In ChemGPS-NP, the chemical space map coordinates are t-scores derived from principal component analysis (PCA) [29]. This is achieved through a carefully selected subset of 35 descriptors that evaluate rules on a total set of 1,779 chosen satellite and core structures [27]. In Figure 3.4, we illustrate the complexity of NPs by the diversities of the three most important principal component values or t-scores (t_1, t_2 and t_3) [27].

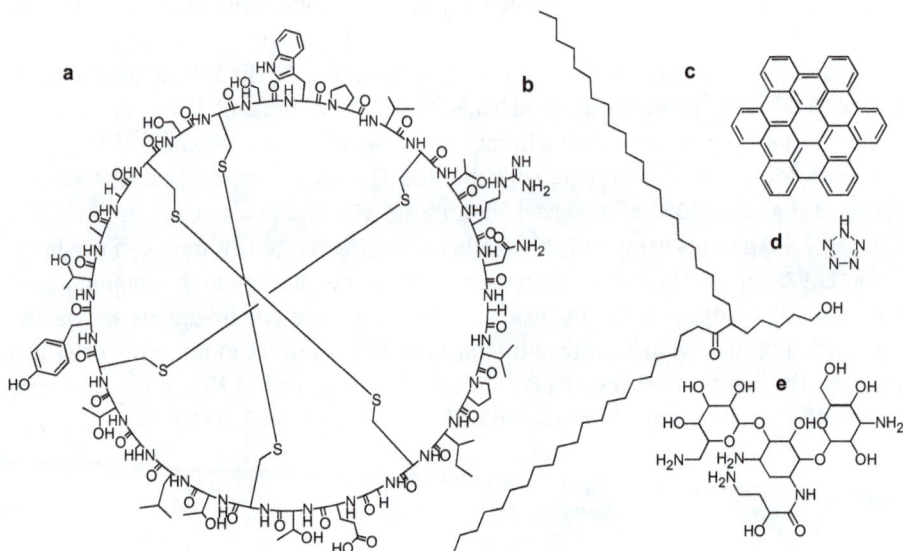

Figure 3.4: Five selected structures and their ChemGPS-NP coordinates for the three most important principal components (a) varv F: $t_1 = 46.4$, $t_2 = -10.2$, $t_3 = -5.84$; (b) 23-(5-hydroxypentyl)-22-pentatetracontanone: $t_1 = 4.83$, $t_2 = -4.52$, $t_3 = 7.58$; (c) hexabenzocoronene: $t_1 = 4.33$, $t_2 = 12.0$, $t_3 = 4.79$; (d) pentazole: $t_1 = -4.82$, $t_2 = -2.32$, $t_3 = -2.98$; (e) amikacin: $t_1 = 5.01$, $t_2 = -3.85$, $t_3 = -6.89$ [27]. Figure reproduced by permission.

3.4 Navigating the natural product chemical space

Several investigations of the three-dimensional (3D) chemical space, occupied by compounds of synthetic and natural origins, using PCA have been published [2, 7, 27, 30–35]. It was generally observed that, when compared with the Food and Drug Administration (FDA)-approved drugs and SDs, the distribution of NPs in chemical space cover regions that lack representation in synthetic medicinal chemistry compounds (Figure 3.5), thus showing that NPs have a much wider coverage of chemical space. In the following subparagraphs, we examine a few case studies in more detail.

3.4.1 The Universal Natural Products Database *versus* FDA-approved drugs

Figure 3.4(a) shows an example of the visualization of the chemical space of NPs according to the origin of the compounds from the Universal Natural Products

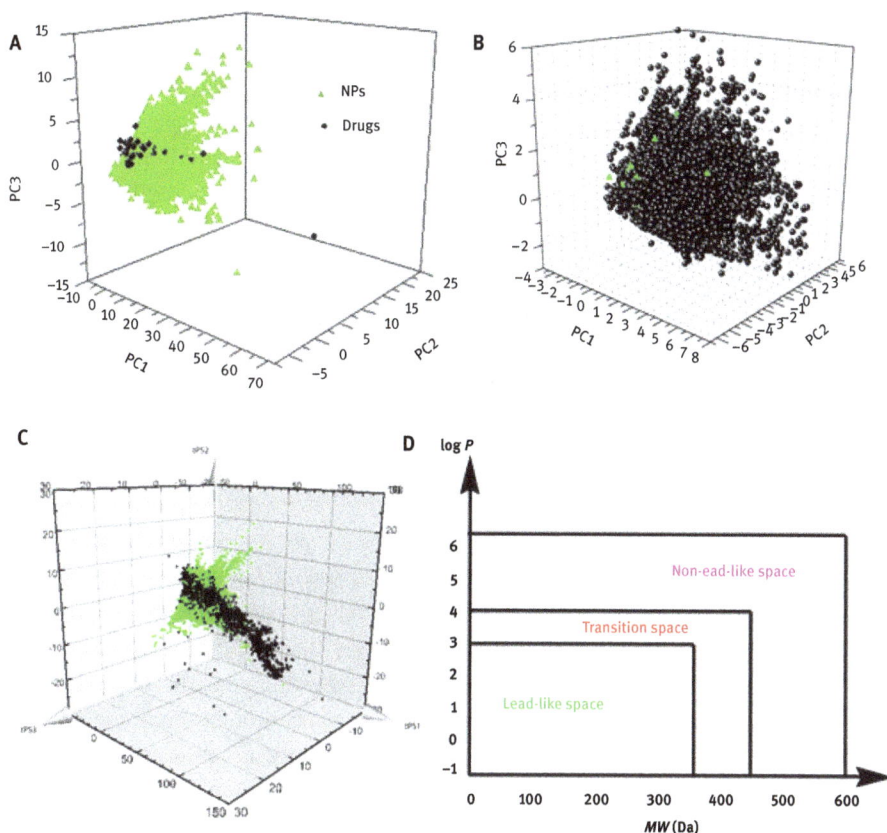

Figure 3.5: The distribution of biologically relevant chemical space of NPs, when compared with SDs: (a) PCA analysis of NPs in the Universal Natural Products Database (UNPD) and FDA-approved drugs. The green triangles and black dots represent natural products and FDA-approved drugs, respectively [30]. Material reproduced from data originally published under a Creative Commons (CC BY) License. (b) PCA analysis of NPs contained in medicinal plants and 25 FDA-approved drugs for the treatment of type II diabetes mellitus (T2DM). The black dots and green triangles represent natural products and FDA-approved drugs, respectively [32]. Material reproduced from data originally published under a Creative Commons (CC BY) License. (c) Predicted score (tPS) plots of NPs (in green) and bioactive medicinal chemistry compounds from the World of Molecular Bioactivity (WOMBAT) database (in black) [31]. (d) Property space representation for lead-like molecules of some selected chemical libraries [24, 35]. Figures reproduced by permission.

Database (UNPD), shown in green, when compared with a dataset of drugs approved by the FDA, USA, shown in black [30]. In this study, Gu and colleagues collected a total 197,201 NPs, by including data structures from the Reaxys database [36], the Chinese Natural Product Database [37], the Traditional Chinese Medicines Database [38] and the Chinese traditional medicinal herbs database [39]. The authors then used PCA to explore their chemical space, by superposing with that of FDA-approved drugs. This study showed a much large portion of overlap between NPs and FDA-approved drugs in the chemical space, indicating that the investigated NPs had a large quantity of potential lead compounds not yet approved by the FDA, thus NPs have a vast chemical diversity when compared with known drugs. Besides, the authors explored the network properties of NP-target networks and found that their polypharmacology was greatly enriched to those compounds with large degree and high betweenness centrality. Although a vast number of the NPs included in the UNPD had no biological activities, by docking all the derived 3D structures toward the 332 target proteins of the FDA-approved drugs, it was shown, based on a docking score-weighted prediction model, that NPs have good drug-like properties and could interact with multiple cellular target proteins.

3.4.2 Antidiabetic medicinal plant-based bioactive natural products *versus* known antidiabetic drugs

In this study, the authors developed a docking score-weighted prediction model based on drug-target network in order to evaluate the efficacy of medicinal plants for the treatment of type II diabetes mellitus (T2DM). The docking dataset was composed of >208,000 medicinal plant-based NPs retrieved from the UNPD *versus* drugs from DrugBank [40], which were FDA-approved for T2DM treatment. The both datasets were docked against X-ray or NMR structures for each protein from the protein databank (PDB) [41] which was related to T2DM pathogenesis, based on information of these proteins from KEGG Pathway database [42] and DrugBank. The binding free energy-based docking score (pK_i) was used to evaluate the affinity between each compound and each protein and compared with the experimental binding affinities of each of the FDA-approved T2DM drugs against their respective target proteins. It could be inferred that most of the NPs would be drug-like. Besides, the wide distribution of the investigated NPs in chemical space (Figure 3.5b) showed that there would be vast structural and functional diversity. Moreover, the large overlap between NPs and the 25 FDA-approved small-molecule drugs for T2DM demonstrated that the NPs contained in the medicinal plants had a hopeful prospect for drug discovery for T2DM.

3.4.3 Analysis of the World of Molecular Bioactivity dataset against the Dictionary of Natural Products

In an attempt to prove NPs to be a rich source of novel compound classes and new drugs, researchers from the working group of Tudor Oprea used the chemical space navigation tool ChemGPS-NP to evaluate the chemical space occupancy by NPs (from the DNP) and bioactive medicinal chemistry compounds from the World of Molecular Bioactivity (WOMBAT) database. Euclidean distances D_{pq} between points $P = (p_1, p_2, , ..., p_n)$ and $Q = (q_1, q_2, , ..., q_n)$ in Euclidean n-space, computed using

ChemGPS-NP scores of the compounds, based on computed molecular descriptor, were determined using the following formula:

$$D_{pq} = \sqrt{\sum_{i=1}^{n} (p_i - q_i)^2} = \sqrt{(p_1 - q_1)^2 + (p_2 - q_2)^2 + \ldots(p_n - q_n)^2} \qquad (3.1)$$

It was observed that the two datasets differed in coverage of chemical space (Figure 3.5c). Besides, several "lead-like" NPs were found to cover regions of chemical space not present in WOMBAT. The authors also used property-based similarity calculations to identify NP neighbors of approved drugs and showed from this method that several of the NPs exhibited the same activities as their drug neighbors in WOMBAT. It could be concluded that NPs could be identified via this method as useful lead compounds for drug discovery in searching for novel leads with unique properties. From Figure 3.5(c), it could be clearly seen that NPs cover parts of chemical space not represented in medicinal chemistry compound space, showing that these areas of chemical space are yet to be investigated and could be of interest in drug discovery.

3.5 The design of "drug-like" natural product libraries and implications for drug discovery

3.5.1 Strategy for designing a library with focused properties

Classical NP drug discovery is only able to undertake drug-likeness analysis after the compounds are isolated and their structures elucidated. However, there are success stories using approaches that address front-loading of both extracts and subsequent fractions with desired physicochemical properties prior to screening for drug discovery [24, 35]. If NPs are often referred to as "sources of inspiration", it simply implies that "lead-like" libraries could be designed, starting from NP scaffolds, with many examples available in the literature [24, 35, 43–46]. However, w hen a NP is used as the guiding structure for the creation of "NP-like" libraries, controlling certain molecular descriptors (e. g. MW, clogP, etc.) during the synthetic process is of major importance for the generation of "lead-like" libraries [24, 35]. This simply means preparing a RO5-compliant library can ensure the timely development of NP lead compounds at a reasonable rate. The reader is invited to carefully read reference [24] for a summary of what to take most seriously when preparing a NP drug-like library.

NPs are known for containing fused medium-sized rings (Figure 3.6). In an attempt to mimic such NPs, Ventosa-Andrés et al. synthesized several molecular scaffolds containing medium-sized fused heterocycles using amino acids [46]. This is because amino acids are known to be useful building blocks used in natural reservoirs as well as chemistry laboratories to create structural diversity. The authors employed a traditional Merrifield solid-phase peptide synthesis, and cyclization was carried out through acid-mediated tandem endocyclic N-acyliminium ion formation. The last steps were nucleophilic addition with internal nucleophiles. These led to seven-, eight- and nine-membered ringed molecular scaffolds with newly generated stereogenic centers in most cases, using a variety of heteroatoms contained in the bicycles, e. g. N, O and S. The details of the synthetic strategy are beyond this discussion. The reader is invited to consult the original paper for further details [46].

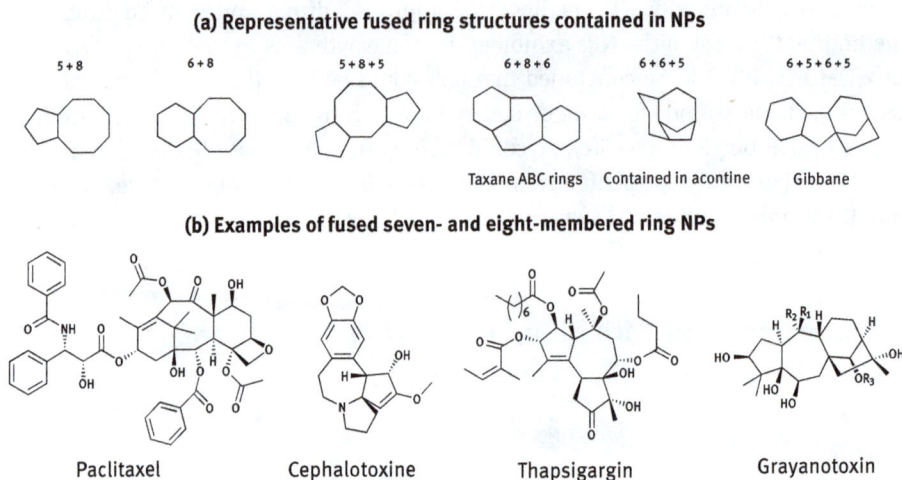

(a) Representative fused ring structures contained in NPs

5 + 8   6 + 8   5 + 8 + 5   6 + 8 + 6   6 + 6 + 5   6 + 5 + 6 + 5

Taxane ABC rings   Contained in acontine   Gibbane

(b) Examples of fused seven- and eight-membered ring NPs

Paclitaxel   Cephalotoxine   Thapsigargin   Grayanotoxin

Figure 3.6: Fused structures and examples of natural products containing fused medium-sized rings [46]. Figure adapted from the original publication.

3.5.2 "Drug-likeness" prediction on available electronic natural product libraries for chemoinformatics analysis

An entire chapter on NP databases and datasets for virtual lead discovery is available in this collection [47]. Researchers within the Kirchmair group have also provided a recent analysis of available NP virtual and physical (vendor and academic) compound libraries which are highly useful for lead compound discovery [48, 49]. These include 25 virtual and 31 physical NP datasets employable for chemoinformatics projects, e. g. chemical space exploration, fragment-based design, NP mimicking and virtual screening. For each library, the authors provide detailed information on the extent of available structural information and the overlap between the different datasets. From the analysis, it was observed that at least 10 % of known NPs belong to the readily purchasable space (including small-sized NPs for fragment-based design and macrocycles) and that with the renewed interest in NPs as lead structure, many more NPs and NP derivatives are being made available through on-demand sourcing, extraction and synthesis services.

3.5.2.1 Virtual libraries

Chen et al. recently described a large number of NP libraries, most of which can be freely accessed for chemoinformatics purposes toward lead compound discovery [48], further characterizing the chemical space thereof [49]. Most of these libraries were curated by academic groups based on literature information.

3.5.2.2 Physical collections, vendor libraries and their drug target space

This collection was done by Chen et al., by keying in the data vendor catalogs from compound suppliers and collections from academic groups [48]. With the goal of characterizing the chemical space of extent of coverage of chemical space by known and readily obtainable NPs and by individual NP databases, the authors compiled comprehensive datasets of known and readily obtainable NPs from 18 virtual databases (including the DNPs), 9 physical libraries and the PDB [49]. After removing all sugars and sugar-like moieties, which are not of interest in drug discovery projects, the authors were able to show that the readily obtainable NPs are highly diverse and populate regions of chemical space that are of high relevance to drug discovery. In some cases, substantial differences in the coverage of NP classes and chemical space by the individual databases are observed, while >2,000 NPs were found to be co-crystallized with at least one biomacromolecule in an X-ray crystal structure within the PDB.

3.6 Computational methods for estimating drug-likeness and ADME/T

It has been regrettably observed that many drugs often fail to enter the market due to poor pharmacokinetic (ADME/T) profiles [50]. This has necessitated the inclusion of pharmacokinetic considerations at earlier stages of drug discovery programs [51, 52]. However, due to the high cost of such experiments, the use of computer-based methods is often sufficient at early stages of lead discovery to save time and cost [53–55]. It requires, for example, less than 1 min to screen 20,000 molecules in an *in silico* (computer-based) model, when compared with 20 weeks in the "wet" laboratory to do the same exercise [51]. *In silico* modeling of drug-likeness often employs standard filters that have been established using the accumulated ADME/T data in the late 1990s. Thus, many pharmaceutical companies now prefer computational models that, in some cases, are replacing the "wet" screens [51]. This has spurred up the development of several theoretical methods and software programs for ADMET prediction [56–59], even though some of the predictions could be disappointing [60]. Most software tools currently used for ADMET prediction make use of statistical models like quantitative structure–activity relationships (QSAR) modeling [60, 61] or knowledge-based methods [62–64]. A promising lead compound may, therefore, be defined as one which combines an interesting biological activity against a drug target (potency) with an attractive ADMET profile. This saves time and cost by discarding compounds with uninteresting predicted ADMET profiles from the list of potential drug candidates early enough, even if these prove to be highly potent. Otherwise, the DMPK properties are "fine-tuned" in order to improve their chances of making it to clinical trials [65]. Machine learning has now become very useful in the ADME/T profiling and drug-likeness prediction of compounds aimed at drug discovery [66].

3.7 Computational methods for estimating natural product-likeness

3.7.1 The natural product-likeness score

The concept of "NP-likeness" has been around for over a decade [25]. It simply connotes the similarity of a molecule to the structure space covered by NPs and is a useful criterion in screening compound libraries and in designing new lead compounds.

Ertl et al. used a Bayesian measure which allows for the determination of how molecules are similar to the structural space covered by NPs. The NP-likeness score is an efficient approach to separate NPs from synthetic molecules. This score is very useful in virtual screening, prioritization of compound libraries toward NP-likeness and the design of building blocks for the synthesis of "NP-like" libraries [25]. The NP-likeness score NP_{score} in eq. (3.2) ranges from –5 to 5 and is computed for a whole molecule, as a sum of contributions of M fragments, f_i (considered to be independent of each other, eq. (3.3)), in the molecule, normalized relative to the molecular size:

$$NP_{score} = \sum_{i=0}^{M} f_i \tag{3.2}$$

$$f_i = \log\left(\frac{A_i}{B_i} \cdot \frac{B_{tot}}{A_{tot}}\right) \tag{3.3}$$

where A_i is the number of NPs which contain fragment i, B_i is the number of synthetic molecules which contain fragment i, A_{tot} is the total number of NPs and B_{tot} is the total number of synthetic molecules in the training set.

3.7.2 Implementations of the natural product-likeness score

The NP-likeness measure has now been implemented in several open-source, open-data tools, e. g. in the Taverna 2.2 workflow [67], which is available under the Creative

Commons Attribution-Share Alike 3.0 Unported License [68]. It is also available for download as an executable stand-alone java package under Academic Free License [69]. This scoring system can be used as a filter for metabolites in computer-assisted structure elucidation or to select natural-product-like molecules from molecular libraries for the use as leads in drug discovery. A distribution of the scores for the training (synthetic molecules and NPs) and the test datasets have been shown in Figure 3.7. In Figure 3.7, a comparison is carried out between NPs, SDs, known drugs and compounds from the Human Metabolome Database (HMDB).

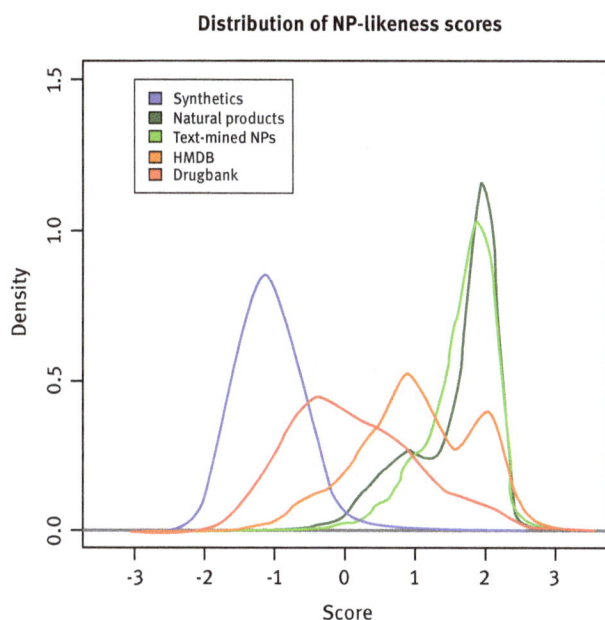

Distribution of NP-likeness scores

- Synthetics
- Natural products
- Text-mined NPs
- HMDB
- Drugbank

Figure 3.7: Distribution of NP-likeness score for the training (synthetic molecules and natural products) and the test datasets [67]. Material reproduced from data originally published under a Creative Commons (CC BY) License.

3.8 Machine-learning methods to classify drugs from nondrugs

Attempts to define molecules likely to be drugs have been limited to simple numerical rules related to computed physicochemical parameters, based on the RO5, which was derived following a statistical analysis of known drugs, e. g. 70 % of the "drug-like" compounds are known to have 0–2 HBDs, 2–9 HBAs, 2–8 rotatable bonds and 1–4 rings. Although such models are quite simple to implement and very fast to compute, by simply putting off molecules that fail two or more of the criteria, more sophisticated computational models of "drug-likeness" have been developed using machine-learning techniques (e. g. neural networks or decision trees). Machine-learning models begin with a training set of compounds with divergent properties, e. g. drugs and nondrugs. A number molecular descriptors are computed for each dataset. The model is then developed using the training set and its computed descriptors. Using a dataset of drugs from the WDI and a set of compounds from Available Chemicals Directory (ACD) with no known activities (regarded as nondrugs), and using a set of Ghose–Crippen atom type count descriptors, Sadowski and Kubinyi developed neural network model with 92 input nodes, 5 hidden nodes and 1 output node to predict "drug-likeness" [70]. This model could correctly predict 77 % of the WDI drugs and 83 % of the ACD molecules as drugs and nondrugs, respectively. Similar results have been obtained using neural network [71, 72] and decision trees [73], using the same databases of drugs and nondrugs and the same set of descriptors. The performance of the decision tree model was comparable to that of the neural networks, correctly predicting ~83 % of a validation set not included in the initial model.

3.9 A binary QSAR model to classify natural products from synthetic molecules

Researchers within the working group of Jurgen Bajorath were able to build a model in order to distinguish between NPs from the DNP and SDs from ACD based on Shannon entropy (SE) analysis [74]. The authors computed values of 98 descriptors from 2D representations of 199,420 ACD molecules and 116,364 NPs from the DNP, respectively. SE values were then defined as in the following equation:

$$SE = - \sum p_i \log_2 p_i \qquad (3.4)$$

where p is the probability of observing a particular descriptor value, computed from the number of compounds with a descriptor value that falls within a specific histogram bin, or "count" (c), for a specific data interval i. Thus, p_i is calculated as in the following equation:

$$p_i = \frac{c_i}{\sum c_i} \qquad (3.5)$$

The SE concept, initially employed in digital communication theory, is now popularly used in molecular descriptor analysis, since it is often combined with binary QSAR methodology to correlate structural features and properties of compounds with a binary formulation of biological activity (i. e. active or inactive). The authors adapted this approach to correlate molecular features with chemical source (i. e. natural or synthetic) by applying different combinations of such descriptors and variably distributed structural keys to the training sets of natural and synthetic molecules and used it to derive predictive binary QSAR models. The derived models were then applied to predict the source of compounds >80 % prediction accuracy for the best models.

3.10 Conclusions

In the previous chapter, the fundamental physical and chemical concepts behind "drug-likeness" and "NP-likeness" were examined [75]. NPs have often been said not to abide by the RO5, as noted by Christopher Lipinski himself [76], although about 60 % of compounds from the DNP showed no violation of any of these "rules" [21]. In this work, our intention was to navigated from simple rule-based approaches for determining what could likely be orally bioavailable of "drug-like", "lead-like" and of "NP-like" for more advanced approaches like neural networks, decision trees and a combination of Shannon entropy and binary QSAR. We have shown that naturally occurring compounds represent a significant proportion of known drugs and that the chemical space occupied by natural compounds is much wider than those of synthetic compounds and known drugs, implying that a large proportion of possible "drug-like" space is yet to be investigated. While the studies add to the plethora of recently published works on the "drug-likeness" of NPs, it no doubt increases our understanding of the physicochemical properties that make NPs fall within the ranges associated with "drug-like" molecules. To gain more insight into the NP-likeness of small molecules, we encourage readers to consult NP-Scout, a recently published machine learning tool for quantifying and visualizing NP-likeness in small molecules [77].

Acknowledgements: FNK acknowledges a return fellowship and an equipment subsidy from the Alexander von Humboldt Foundation, Germany. Financial support for this work is acknowledged from a ChemJets fellowship from the Ministry of Education, Youth and Sports of the Czech Republic awarded to FNK.

List of abbreviations and definitions

ACD	Available Chemicals Directory
ADME/T	Absorption, distribution, metabolism, excretion and toxicity
ChemGPS	Chemical global positioning system
DNP	Dictionary of Natural Products
Drug-likeness	The ability to resemble drugs
FDA	Food and Drug Administration
HTS	High-throughput screening
MW	Molecular weight
NOC	Naturally occurring compound
NPs	Natural products
PCA	Principal component analysis

PDB	Protein databank
RO5	"Rule of 5"
SDs	Synthetic drugs
SE	Shannon entropy
T2DM	Type II diabetes mellitus
UNPD	Universal Natural Products Database
WDI	World Drug Index
WOMBAT	World of Molecular Bioactivity

Authors Contributions

FNK is a chemist specialized in computer-aided drug discovery from NPs. He did the conception, design and carried-out the *in silico* experimental studies. He further analyzed and interpreted the data. KDN is an organic chemist specialized in natural products/pharmaceutical chemistry and teaches DMPK studies. He contributed to the study design, drafting and correction of the manuscript. GAA is an organic chemist specialized in natural products chemistry and teaches DMPK studies. He contributed in correcting the manuscript. SS has a PhD in computer-aided drug discovery. He was involved in the conception and writing parts of the introduction.

References

[1]   Abegaz BM, Kinfe H. Secondary metabolites, their structural diversity, bioactivity, and ecological functions: an overview. Phys Sci Rev. 2018. DOI: 10.1515/psr-2018-0100.

[2]   Newman DJ, Cragg GM. Natural products as sources of new drugs from 1981 to 2014. J Nat Prod. 2016;79:629–61.

[3]   Dobson CM. Chemical space and biology. Nature. 2004;432:824–8.

[4]   Lipinski CA, Lombardo F, Dominy BW, Feeney PJ. Experimental and computational approaches to estimate solubility and permeability in drug discovery and development settings. Adv Drug Deliv Rev. 1997;23:3–25.

[5]   Oprea TI, Davis AM, Teague SJ, Leeson PD. Is there a difference between leads and drugs? a historical perspective. J Chem Inf Comput Sci. 2001;41:1308–15.

[6]   Hann MM, Oprea T. Pursuing the leadlikeness concept in pharmaceutical research. Curr Opin Chem Biol. 2004;8:255–63.

[7]   Feher M, Schmidt JM. Property distributions: differences between drugs, natural products, and molecules from combinatorial chemistry. J Chem Inf Comput Sci. 2003;43:218–27.

[8]   Saldívar-González FI, Pilón-Jiménez BA, Medina-Franco JL. Chemical space of naturally occurring compounds. Phys Sci Rev. 2018. DOI: 10.1515/psr-2018-0103.

[9]   Harvey AL, Edrada-Ebel R, Quinn RJ. The re-emergence of natural products for drug discovery in the genomics era. Nat Rev Drug Discov. 2015;14:111–29.

[10]   Rodrigues T, Reker D, Schneider P, Schneider G. Counting on natural products for drug design. Nat Chem. 2016;8:531–41.

[11]   Wani MC, Taylor HL, Wall ME, Coggon P, McPhail AT. Plant antitumor agents. VI. Isolation and structure of taxol, a novel antileukemic and antitumor agent from *Taxus brevifolia*. *J Am Chem Soc*. 1971;93:2325–7.

[12]   Walsh CT, Fischbach MA. Natural products version 2.0: connecting genes to molecules. J Am Chem Soc. 2010;132:2469–93.

[13] Stratton CF, Newman DJ, Tan DS. Cheminformatic comparison of approved drugs from natural product versus synthetic origins. Bioorg Med Chem Lett. 2015;25:4802–7.

[14] Li JWH, Vederas JC. Drug discovery and natural products: end of an era or an endless frontier? Science. 2009;325:161–5.

[15] Pan L, Chai HB, Kinghorn AD. Discovery of new anticancer agents from higher plants. Front Biosci (Schol Ed). 2013;4:142–56.

[16] Harvey AL. Natural products in drug discovery. Drug Discov Today. 2008;13:894–901.

[17] Koehn FE, Carter GT. The evolving role of natural products in drug discovery. Nat Rev Drug Discov. 2005;4:206–20.

[18] Pye CR, Bertin MJ, Lokey RS, Gerwick WH, Linington RG. Retrospective analysis of natural products provides insights for future discovery trends. Proc Natl Acad Sci USA. 2017;114:5601–6.

[19] Skinnider MA, Magarvey NA. Statistical reanalysis of natural products reveals increasing chemical diversity. Proc Natl Acad Sci USA. 2017;114:E6271–2.

[20] Grabowski K, Schneider G. Properties and architecture of drugs and natural products revisited. Curr Chem Biol. 2007;1:115–27.

[21] Quinn RJ, Carroll AR, Pham NB, Baron P, Palframan ME, Suraweera L, et al. Developing a drug-like natural product library. Nat Prod. 2008;71:464–8.

[22] Lee M-L, Schneider G. Scaffold architecture and pharmacophoric properties of natural products and trade drugs: application in the design of natural product-based combinatorial libraries. J Comb Chem. 2001;3:284–9.

[23] Hert J, Irwin JJ, Laggner C, Keiser MJ, Shoichet BK. Quantifying biogenic bias in screening libraries. Nat Chem Biol. 2009;5:479–83.

[24] Camp D, Davis RA, Campitelli M, Ebdon J, Quinn RJ. Drug-like properties: guiding principles for the design of natural product libraries. J Nat Prod. 2012;75:72–81.

[25] Ertl P, Roggo S, Schuffenhauer A. Natural product-likeness score and its application for prioritization of compound libraries. J Chem Inf Model. 2008;48:68–74.

[26] Ertl P, Schuffenhauer A. Cheminformatics analysis of natural products: lessons from nature inspiring the design of new drugs. Prog Drug Res. 2008;66:219–35.

[27] Larsson J, Gottfries J, Muresan S, Backlund A. ChemGPS-NP: tuned for navigation in biologically relevant chemical space. J Nat Prod. 2007;70:789–94.

[28] Oprea TI, Gottfries J. Chemography: the art of navigating in chemical space. J Comb Chem. 2001;3:157–66.

[29] Eriksson L, Johansson E, Kettaneh-Wold N, Wold S. PCA. In: Eriksson L, Byrne T, Johansson E, Trygg J, Vikstrom C, editor(s). *Multi- and megavariate data analysis*. Umeå: Umetrics Academy, 2001:43–69.

[30] Gu J, Gui Y, Chen L, Yuan G, Lu H-Z, Xu X. Use of natural products as chemical library for drug discovery and network pharmacology. PLoS One. 2013;8:e62839.

[31] Rosén J, Gottfries J, Muresan S, Backlund A, Oprea TI. Novel chemical space exploration via natural products. J Med Chem. 2009;52:1953–62.

[32] Gu J, Chen L, Yuan G, Xu X. A drug-target network-based approach to evaluate the efficacy of medicinal plants for type II diabetes mellitus. Evid Based Complement Alternat Med. 2013;2013:203614.

[33] Lachance H, Wetzel S, Kumar K, Waldmann H. Charting, navigating, and populating natural product chemical space for drug discovery. J Med Chem. 2012;55:5989–6001.

[34] López-Vallejo F, Giulianotti MA, Houghten RA, Medina-Franco JL. Expanding the medicinally relevant chemical space with compound libraries. Drug Discov Today. 2012;17:718–26.

[35] Pascolutti M, Quinn RJ. Natural products as lead structures: chemical transformations to create lead-like libraries. Drug Discov Today. 2014;19:215–21.

[36] Reaxys, version 1.7.8. Elsevier, 2012. RRN 969209. Accessed: 9 Jan 2014.

[37] Shen JH, Xu XY, Cheng F, Liu H, Luo XM, Shen J, et al. Virtual screening on natural products for discovering active compounds and target information. Curr Med Chem. 2003;10:2327–42.

[38] He M, Yan XJ, Zhou JJ, Xie GR. Traditional Chinese medicine database and application on the Web. J Chem Inf Comput Sci. 2001;41:273–7.

[39] Qiao XB, Hou TJ, Zhang W, Guo SL, Xu XJ. A 3D structure database of components from Chinese traditional medicinal herbs. J Chem Inf Comput Sci. 2002;42:481–9.

[40] Knox C, Law V, Jewison T, Liu P, Ly S, Frolkis A, et al. DrugBank 3.0: a comprehensive resource for "Omics" research on drugs. Nucleic Acids Res. 2011;39:D1035–41.

[41] Berman HM, Westbrook J, Feng Z, Gilliland G, Bhat TN, Weissig H, et al. The protein databank. Nucleic Acids Res. 2000;28:235–42.

[42] Kanehisa M, Goto S, Sato Y, Furumichi M, Tanabe M. Kegg for integration and interpretation of large-scale molecular data sets. Nucleic Acids Res. 2012;40:D109–14.

[43] Zuegg J, Cooper MA. Drug-likeness and increased hydrophobicity of commercially available compound libraries for drug screening. Curr Top Med Chem. 2012;12:1500–13.

[44] Shi BX, Chen FR, Sun X. Structure-based modelling, scoring, screening, and in vitro kinase assay of anesthetic pkc inhibitors against a natural medicine library. SAR QSAR Environ Res. 2017;28:151–63.

[45] Sánchez-Rodríguez A, Pérez-Castillo Y, Schürer SC, Nicolotti O, Mangiatordi GF, Borges F, et al. From flamingo dance to (desirable) drug discovery: a nature-inspired approach. Drug Discov Today. 2017;22:1489–502.

[46] Ventosa-Andrés P, La-Venia A, Ripoll CA, Hradilová L, Krchňák V. Synthesis of nature-inspired medium-sized fused heterocycles from amino acids. Chemistry. 2015;21:13112–19.

[47] Koulouridi E, Valli M, Ntie-Kang F, Bolzani VS. A primer on natural product-based virtual screening. Phys Sci Rev. 2018. DOI: 10.1515/psr-2018-0107.

[48] Chen Y, de Bruyn Kops C, Kirchmair J. Data resources for the computer-guided discovery of bioactive natural products. J Chem Inf Model. 2017;57:2099–111.

[49] Chen Y, Garcia de Lomana M, Friedrich N-O, Kirchmair J. Characterization of the chemical space of known and readily obtainable natural products. J Chem Inf Model. 2018;58:1518–32.

[50] Darvas F, Keseru G, Papp A, Dormán G, Urge L, Krajcsi P. In silico and ex silico ADME approaches for drug discovery. *Top Med Chem*. 2002;2:1287–304.

[51] Hodgson J. ADMET – turning chemicals into drugs. Nat Biotechnol. 2001;19:722–6.

[52] Navia MA, Chaturvedi PR. Design principles for orally bioavailable drugs. Drug Discov Today. 1996;1:179–89.

[53] Lombardo F, Gifford E, Shalaeva MY. In silico ADME prediction: data, models, facts and myths. Mini Rev Med Chem. 2003;3:861–75.

[54] Gleeson MP, Hersey A, Hannongbua S. In-silico ADME models: a general assessment of their utility in drug discovery applications. Curr Top Med Chem. 2011;11:358–381.

[55] DiMasi JA, Hansen RW, Grabowski HG. The price of innovation: new estimates of drug development costs. J Health Econ. 2003;22:151–85.

[56] OCHEM - A platform for the creation of in silico ADME / Tox prediction models. http://www.eadmet.com/en/ochem.php. Accessed: 16 Feb 2019.

[57] Meteor software, version 13.0.0. Lhasa Ltd, Leeds, UK, 2010.

[58] QikProp software, version 3.4. New York, NY: Schrödinger, LLC, 2011.

[59] Cruciani G, Crivori P, Carrupt PA, Testa B. Molecular fields in quantitative structure-permeation relationships: the VolSurf approach. J Mol Struc-Theochem. 2000;503:17–30.

[60] Tetko IV, Bruneau P, Mewes H-W, Rohrer DC, Poda GI. Can we estimate the accuracy of ADMET predictions? Drug Discov Today. 2006;11:700–7.

[61] Hansch C, Leo A, Mekapatia SB, Kurup A. QSAR and ADME. Bioorg Med Chem. 2004;12:3391–400.

[62] Greene N, Judson PN, Langowski JJ. Knowledge-based expert systems for toxicity and metabolism prediction: DEREK, StAR and METEOR. SAR QSAR Environ Res. 1999;10:299–314.

[63] Button WG, Judson PN, Long A, Vessey JD. Using absolute and relative reasoning in the prediction of the potential metabolism of xenobiotics. J Chem Inf Comput Sci. 2003;43:1371–7.

[64] Cronin MTD. Computer-assisted prediction of drug toxicity and metabolism. In: Hillisch A, Hilgenfeld R, editors. Modern methods of drug Discovery. Basel: Birkhäuser, 2003:259–278.

[65] Hou T, Wang J. Structure-ADME relationship: still a long way to go? Expert Opin Drug Metab Toxicol. 2008;4:759–70.

[66] Pires DEV, Blundell TL, Ascher DB. *pk*CSM: predicting small-molecule pharmacokinetic and toxicity properties using graph-based signatures. J Med Chem. 2015;58:4066–72.

[67] Jayaseelan KV, Steinbeck C, Moreno P, Truszkowski A, Ertl P. Natural product-likeness score revisited: an open-source, open-data implementation. *BMC Bioinform*. 2012;13:106

[68] Available at: http://www.myexperiment.org/packs/183.html. Accessed: 16 Feb 2019.

[69] Available at: http://sourceforge.net/projects/nplikeness/. Accessed: 16 Feb 2019.

[70] Sadowski J, Kubinyi H. A scoring scheme for discriminating between drugs and nondrugs. J Med Chem. 1998;41:3325–9.

[71] Ajay A, Walters PW, Murcko MA. Can we learn to distinguish between "drug-like" and "non drug-like" molecules? J Med Chem. 1998;41:3314–24.

[72] Frimurer TM, Bywater R, Naerum L, Lauritsen LN, Brunak S. Improving the odds in discriminating "drug-like" from "non drug-like" compounds. J Chem Inf Comput Sci. 2000;40:1315–24.

[73] Wagener M, van Geerestein VJ. Potential drugs and nondrugs: prediction and identification of important structural features. J Chem Inf Comput Sci. 2000;40:280–92.

[74] Stahura FL, Godden JW, Xue L, Bajorath J. Distinguishing between natural products and synthetic molecules by descriptor Shannon entropy analysis and binary QSAR calculations. J Chem Inf Comput Sci. 2000;40:1245–52.

[75] Athar M, Sona AN, Bkono BD, Ntie-Kang F. Fundamental physical and chemical concepts behind "drug-kiness" and "natural product-likeness. Phys Sci Rev 2019.10.1515/psr-2018-0101.

[76] Lipinski CA. Chris Lipinski discusses life and chemistry after the rule of five. Drug Discov Today. 2003;8:12–16.

[77] Chen Y, Stork C, Hirte S, Kirchmair J. NP-Scout: machine learning approach for the quantification and visualization of the natural product-likeness of small molecules. Biomolecules. 2019;9:43.

Fernanda I. Saldívar-González, B. Angélica Pilón-Jiménez
and José L. Medina-Franco

4 Chemical space of naturally occurring compounds

Abstract: The chemical space of naturally occurring compounds is vast and diverse. Other than biologics, naturally occurring small molecules include a large variety of compounds covering natural products from different sources such as plant, marine, and fungi, to name a few, and several food chemicals. The systematic exploration of the chemical space of naturally occurring compounds have significant implications in many areas of research including but not limited to drug discovery, nutrition, bio- and chemical diversity analysis. The exploration of the coverage and diversity of the chemical space of compound databases can be carried out in different ways. The approach will largely depend on the criteria to define the chemical space that is commonly selected based on the goals of the study. This chapter discusses major compound databases of natural products and cheminformatics strategies that have been used to characterize the chemical space of natural products. Recent exemplary studies of the chemical space of natural products from different sources and their relationships with other compounds are also discussed. We also present novel chemical descriptors and data mining approaches that are emerging to characterize the chemical space of naturally occurring compounds.

Keywords: biodiversity, BioFacQuim, cheminformatics, consensus diversity plots, drug discovery, foodinformatics, molecular diversity, natural products

4.1 Introduction

Chemical space is a concept that helps to address questions such as: How many compounds exist? How many components can be synthesized? Similarly, the concept of chemical space is highly attached to the relationship among collections of chemical compounds. Currently, there is no single or unique criterion to define chemical space. Dobson stated that the chemical space "encompasses all possible small organic molecules, including those present in biological systems" [1]. Lipinski and Hopkins described it "as being analogous to the cosmological universe in its vastness, with chemical compounds populating space instead of stars" [2]. The concept of chemical space has become more relevant as compounds and their information increase over time [3].

This article has previously been published in the journal *Physical Sciences Reviews*. Please cite as: Saldívar-González, F.I., Pilón-Jiménez, B.A., Medina-Franco, J.L. Chemical Space of Naturally Occurring Compounds. *Physical Sciences Reviews* [Online] **2018** DOI: 10.1515/psr-2018-0103.

https://doi.org/10.1515/9783110579352-005

For practical applications, chemical space can be used as a "tool" that helps to find associations in complex data and rapidly exploit the increasing information available for the discovery of drugs and other research areas such as food science [4]. More specific applications of the chemical space include evaluating the diversity of different data sets, exploring the relationships between compound collections and assessing the potential to cover other regions in the chemical space yet to be explored. Likewise, this tool is useful to design novel compound libraries and in the selection of compounds from existing libraries for computational and/or experimental screening [5].

To generate a visual representation of the chemical space of compound collection two main components are required: the molecular representation of the molecules to define the multidimensional descriptor space, and a visualization technique used to reduce the multidimensional space into two or three dimensions.

Descriptors based on the structure (constitution, configuration, and conformation of the molecule) or descriptors based on properties (physical, chemical and biological) can be used to represent molecules. The interpretations and predictions that can be made will depend on the type of descriptor used [6].

Regarding data visualization, there are several established methods to generate approximate representations of the chemical space [7–9]. All these methods are applicable to virtually any molecular library. However, the choice of method depends on the expected qualities of the visualization, or on the ability of the method to provide useful graphics.

In this chapter different applications of the visualization of the chemical space in the study of natural products (NPs) are discussed, as well as the cheminformatic approaches used and those that are emerging to characterize the chemical space of natural compounds. Recent exemplary studies of the chemical space of NPs products from different sources and their relationships with other compounds are also commented.

4.1.1 Importance of chemical space of natural products

NPs continue to be an important resource of drug discovery [10]. Despite the advent of efficient technologies such as combinatorial chemistry and high-throughput screening (HTS), over the past few years, NPs have attracted again the attention of academics and researchers focused on pharmaceutical chemistry. This is because NPs have proven to be a more promising source of drugs and novel structures than the compounds obtained by combinatorial chemistry [11]. Indeed, progress on technical advances and genomic and metabolomic approaches are largely contributing to further enhance the interest in natural product-based, drug discovery [12]. Figure 4.1 shows the chemical structures of approved drugs from natural origin approved in the last four years. At the time of writing, (September 2018) Migalastat (Galafold®) is the

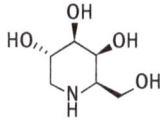

Migalastat
(2018, Amicus therapeutics)
Streptomyces lydicus
Fabry disease

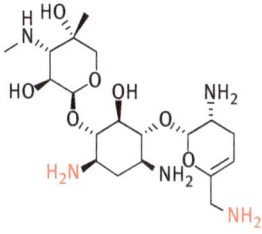

Sisomicin
Micromonospora inyoensis

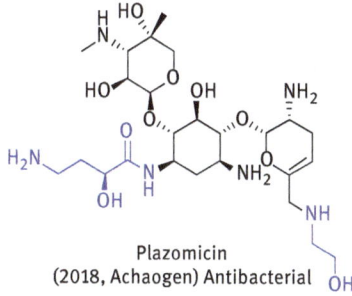

Plazomicin
(2018, Achaogen) Antibacterial

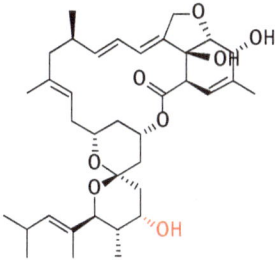

Nemadectin
Streptomyces cyaneogriseus
subsp. *noncyanogenus*

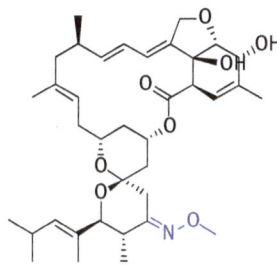

Moxidectin
(2018, Medicines Development for Global Health)
Antiparasitic

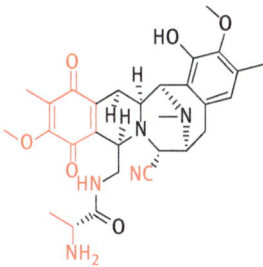

Cyanosafracin B
(2015, Pharma Mar. & Janssen)
Pseudomonas fluorescens
Chemotherapeutic agent

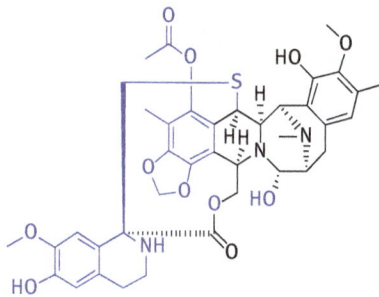

Trabectedin
(2015, Pharma Mar. & Janssen)
Ecteinascidia turbinata
Chemotherapeutic agent

Figure 4.1: Chemical structures of a natural product (top) and derivatives (right) recently approved for clinical use. The source, therapeutic indication, year of approval and pharmaceutical company are indicated. Migalastat is one of the most recent isolated natural products approved by the FDA in the past four years. In blue are the modifications that were made to compounds isolated from natural products (left) and were approved by the FDA.

most recent small molecule drug approved by the Food and Drug Administration (FDA) of the United States which was isolated from the fungus *Streptomyces lydicus* PA-5726 [13]. It was found by Amicus Therapeutics and it is used for the treatment of Fabry disease, restoring the activity of specific mutant forms of α-galactosidase [14]. The high-cost and multi-step chemical processes to synthesize migalastat led to the development of a low-cost and sustainable process of fermentation with *Streptomyces lydicus* PA-5726.

Plazomicin (Zemdri®) is a drug derived from a NP (Figure 4.1). It was developed by Achaogen as a next-generation aminoglycoside that inhibits bacterial protein synthesis. It was synthetically derived from sisomicin, which is isolated from the aerobic fermentation of *Micromonospora inyoensis*, by appending a hydroxy-aminobutyric acid substituent at position 1 and a hydroxyethyl substituent at position 6′ [15].

Moxidectin (Moxidectin®) is another case of an approved drug derived from a compound previously isolated from a NP. It is a macrocyclic lactone derived from nemadectin that was isolated from *Streptomyces cyaneogriseus* subsp. *non-cyanogenus* but Moxidectin has the addition of a methoxime moiety at C-23 [16]. It is used in the treatment of against River blindness [17] or onchocerciasis due to *Onchocerca volvulus*. Interestingly, of the first cheminformatic analysis of natural products were focused on macrocycles and macrolides [18, 19]. In those publications, authors highlighted the importance of biologically active macrocycles in drug discovery.

Trabedectin (ET-743, Yondelis®) was developed by PharmaMar [20]. It was discovered and isolated from the Caribbean Sea squirt *Ecteinascidia turbinate*. This drug is an alkylating agent that has shown significant broad-spectrum potential as a single agent second-line drug alone or in combination particularly in the treatment of liposarcomas and leiomyosarcomas [21]. PharmaMar searched for a suitable source because of the poor yields of Trabedectin [20] obtained from Mediterranean aqua farms, plus the economic impact of the extraction and purification processes. The company developed some synthesis process but, even with these improvements, it was not suitable for manufacturing ET-743 at an industrial scale. Nevertheless, PharmaMar found a semi-synthetic process starting from cyanosafracin B, an antibiotic obtained by fermentation from the bacteria *Pseudomonas fluorescens* obtaining good yields and it was economically profitable.

In this sense, the traditional approach to search for active compounds from NPs is being modified to take advantage of technological advances and explore the biologically relevant chemical space through chemometric approaches [22–24]. Also, studies of chemical space visualization have been shown to be useful when dealing with large libraries of potential bioactive molecules. For instance, it is estimated that about 250k NPs can be found in virtual libraries [25]. It is expected that this number will increase in the coming years and enrich the databases that can be currently searched.

Table 4.1: Representative studies of the chemical space of natural products.

Library	Properties calculated	Visualization method	Analysis	Ref.
NPs, drugs, and compounds from combinatorial libraries.	Topological.	PCA	Diversity.	[41]
NPs, drugs, bioactive molecules, Lipinski's rule of five compliant, compounds from DOS and molecular fragments.	Topological and physicochemical.	Radar plots and PCA	Diversity.	[42]
Merck's NP collection, the company's sample collection, and 200 top-selling drugs of 2006.	Topological and physicochemical.	PCA	Diversity.	[43]
NPs, human metabolites, drugs, clinical candidates, and known bioactive compounds.	Topological and physicochemical.	Pie charts, scatterplots	Comparative.	[44]
NPs, bioactive, and organic drug-like molecules.	Physicochemical and structural.	PCA	Diversity.	[45]
Fragment-sized and No fragment-sized NPs.	Physicochemical and structural.	PCA, SOMs	Diversity.	[46]
NPs in UNPD and, approved drugs.	Physicochemical.	PCA, Network-based	Diversity and biological activity.	[30]
Fungi metabolites, approved anticancer drugs, approved non-anticancer drugs, compounds in clinical trials, GRAS.	Physicochemical and diversity.	PCA	Diversity and complexity.	[47]
18 virtual and 9 physical NP libraries using the DNP as an encyclopedic reference.	Physicochemical.	PCA	Diversity.	[48]
Terrestrial and marine NPs.	Scaffolds.	Tree maps method	Comparative (difference of scaffolds).	[49]

So far, the characterization of the chemical space of NPs has been conducted with different approaches from which we can recognize its importance (cf. Table 4.1). As elaborated in the next sections of this manuscript, one of the most widespread uses has been to make comparisons between these compounds with other reference libraries such as synthetic compounds or drugs approved for clinical use. This type of comparisons had led to the conclusion that NPs have chemical structures with increased diversity and complexity as compared to other compound collections. In addition, the characterization of the chemical space of NPs can be used to classify bioactive compounds according to their biological properties. The rationale is that similar molecules have similar bioactivity. Based on this hypothesis, an important application in the area of NPs has been the selection of compounds for virtual screening and the emergence of biology-oriented synthesis (BIOS): BIOS is focused on the "islands of bioactivity" that have composite data sets containing central structures of compound classes that are biologically relevant [26, 27].

4.1.2 Overview of a representative cheminformatic analysis of natural products databases

4.1.2.1 Databases of natural products

Several public natural product databases are assembled, curated and maintained by academic and non-profit groups. Examples are the Traditional Chinese Medicine TCM database@Taiwan [28, 29] and the Universal Natural Product Database [30]. Other compound databases focused on different geographical regions of the globe have been developed. Hereunder are described as representative examples.

AfroDb [31], developed by Nitte-Kang et al., is a major initiative that put together a subset of compounds representative of the African medicinal plants containing around 1,000 three-dimensional structures. The same group developed the ConMedNP collection [32], an extension of the previously published database CamMedNP. The augmented library ConMedNP is a compilation of 3,177 compounds from the Central African flora. NuBBE$_{DB}$ is a database of compounds from Brazilian biodiversity [33]. NuBBE$_{DB}$ presently contains data of 2218 compounds, mainly from plants, marine organisms, and fungi [34] comprising compounds from species from all six Brazilian biomes [35].

In Mexico, an emerging in-house compound database of natural products is being assembled by an academic group putting together natural products reported over the past 10 years by the School of Chemistry of the National Autonomous University of Mexico (UNAM). At the time of writing (September 2018) the in-house collection herein referred as BioFacQuim, has 423 compounds mostly isolated from plants and fungi. In this set 316 compounds were isolated from 49 different genus of plants, 98 compounds were isolated from 19 genus of fungi and 9 compounds were isolated from Mexican propolis (sticky dark-colored hive product collected by bees from living plant sources). BioFacQuim contains the compound name, SMILES, reference, kingdom (Plantae or Fungi), genus, and species of the natural product. This collection would be part of D-TOOLS [36]. Other compound datasets of natural products have been made accessible as supporting information of peer-reviewed publications or can be requested from the authors (cf. Table 4.2).

Table 4.2: Data sets included in the visual representation of the chemical space.

Database	Size	Reference
Cyanobacteria metabolites	473	In-house
Fungi metabolites	206	[47]
Marines	6253	[64]
Semi-synthetics (NATx)	26,318	ac-discovery.com
Drug approved	1806	www.drugbank.ca

4.1.2.2 Cheminformatic analysis of databases of natural products

As discussed in Section 4.1.1., the chemical space is a multidimensional space of descriptors, which can be measured experimentally or calculated *in silico*. Multidimensional data mining tools that are available to handle this information are hierarchical clustering, decision trees, multidimensional scaling, genetic algorithms, neural networks, and support vector machines [37]. However, to navigate through the chemical space, it is required to use methods that allow projecting this multidimensional into a lower dimensional space to create graphics susceptible to visual inspection and analysis. So far, common visualization methods to represent chemical space are principal component analysis (PCA) and self-organizing maps (SOMs), also known as Kohonen networks. Other visualizations approaches are multi-fusion similarity (MFS) maps, radar plots, Sammon mapping, activity-seeded structure-based clustering, singular value decomposition, minimal spanning tree, k-means clustering, generative topographic mapping (GTM) [38], hierarchical GTM [39] and the recently developed ChemMaps [40].

Table 4.1 summarizes representative examples of studies of the chemical space covered by compound databases from NPs. Some of these examples are discussed in the next sections.

4.1.2.3 Comparison with other compounds collections

The technique most used to study the chemical space of NPs has been PCA. Several studies reported thus far are focused on diversity analysis of which very valuable conclusions or interpretations have been obtained. For example, it has been observed that NPs have a large chemical diversity and their chemical space is clearly distinguishable from the space of synthetic compounds [41, 42, 50]. Similarly, several collections of NPs such as the Dictionary of Natural Products (DNP) and the Universal Natural Product Database (UNPD) cover a much broader region of chemical space than synthesized compounds or than approved drugs [43, 48, 50]. However, when NPs are compared to collections of bioactive compounds or approved drugs, many NPs have approximately the same coverage of chemical space [30, 41, 42, 50]. These results encourage the continued used of libraries of NPs to identify bioactive compounds for later development or optimization.

4.1.2.4 Analysis of different sources of natural products

The differences in the coverage of the chemical space of NPs according to the origin of the compounds have also been evaluated. For example, Muigg et al. [51] compared the chemical space regions of NPs collected from marine and terrestrial organisms with that of synthetic compounds. They found clear differences in the regions of the chemical space covered by the compounds of these three origins. For example, NPs extracted from marine organisms tend to be large and very flexible compared to synthetic compounds. In contrast, NPs that originate from terrestrial organisms are often large and rigid. Recently, Shang et al. [49] analyzed the chemical space covered by natural marine products. They discovered that long chains and macrocyclic structures are more prominent among marine natural products than on terrestrial ones. Similarly, Saldivar-Gonzalez et al. [52] reported a comparison of NPs from marine sources, cyanobacteria and fungi metabolites. It was concluded that metabolites from cyanobacteria are remarkable for their high structural complexity and distinct profile based on molecular properties and sub-structural alerts that are different from other NPs.

4.1.3 Novel cheminformatic approaches to navigate the chemical space of natural products

In order to navigate efficiently the chemical space and the biological space associated with NPs, different research groups have developed methods toward the rapid identification of structure-activity relationships, easier navigation through the space, and facilitate the identification of new classes of compounds with a desired biological activity. To this end, ChemGPS-NP [53] was one of the first chemographic models used to describe in a global manner the physicochemical properties of NPs and has been shown to be useful for a variety of applications [51, 54, 55]. A web-based service of ChemGPS-NP model was developed to facilitate its use [56]. ChemGPS-NP is a PCA-based model of physicochemical property space, defined by training-set of carefully selected compounds acting as 'satellites'. In this model, the compound can be predicted to position and evaluated on a very large scale using PCA score prediction. A recent application of ChemGPS-NP was the assessment of datasets, characterizing their neighborhoods and then interpreting the map using the prediction of a control set of compounds with activities determined experimentally [57].

Other approaches for the visual representation of the chemical space are focused on representing the molecules beyond data points in a map to enhance the interpretation of the data. Examples of this approach are *Molecule Cloud* or

Scaffold tree. Molecule Cloud allows the visual representation of the most common structural characteristics of chemical databases in the form of a cloud diagram [58]. Scaffold tree is a method for classifying molecules based on their scaffolds. In this approach, the molecules are converted to their frameworks, then the rings are removed one by one according to a set of predefined rules, creating a hierarchy of scaffolds [59]. This approach is reminiscent of an earlier scaffold-based classification of compound data sets based on the scaffolds generated by Xu and Johnson [60].

Researchers in Dortmund, in cooperation with Novartis, investigated the scaffold content of NPs and then classified them hierarchically, by size and complexity, in a tree-like diagram. This structural classification of NPs (SCONP) [61] allows an easy and intuitive navigation in the universe of scaffolds for the identification of new regions of interest for the development of libraries inspired by NPs.

Another useful program for the analysis and visualization of scaffolds is Scaffold Hunter [62]. Unlike Scaffold tree, Scaffold Hunter allows including bioactivity data to identify new classes of scaffolds and compounds endowed with the desired activity. Using this tool and a fingerprint analysis Tao et al. [63] analyzed the distribution profiles of the natural product lead of drugs (NPLD) in chemical space, this in order to obtain useful clues to prioritize the efforts in the study of NPs. Useful information regarding the mechanisms that partially contribute to the formation of these profiles was identified. In particular, the trend of NPLD to join preferentially to privileged target sites is influenced collectively by potent binding to the target-sites and such additional factors as the optimization potential to reach the drug sweet spots in the chemical space with more adequate metabolic stability, metabolite safety, absorption, and physical forms.

4.2 Coverage of natural product chemical space from diverse sources

4.2.1 Plants, fungi, cyanobacteria, and marine

It is generally accepted that NPs are compounds with large diversity and structural complexity, however, these properties can vary according to the source of the compounds. In a previous study, Ertl et al. [50] classified the NPs of the DNP according to their origin, with the purpose of analyzing systems of rings that are typical according to the source of the compounds. However, the chemical space according to this classification was not discussed in detail.

Figure 4.2 shows an example of the visualization of the chemical space of NPs according to the origin of the compounds, which includes a database of semi-synthesis compounds and a database of approved drugs by the FDA as a reference (Table 4.2).

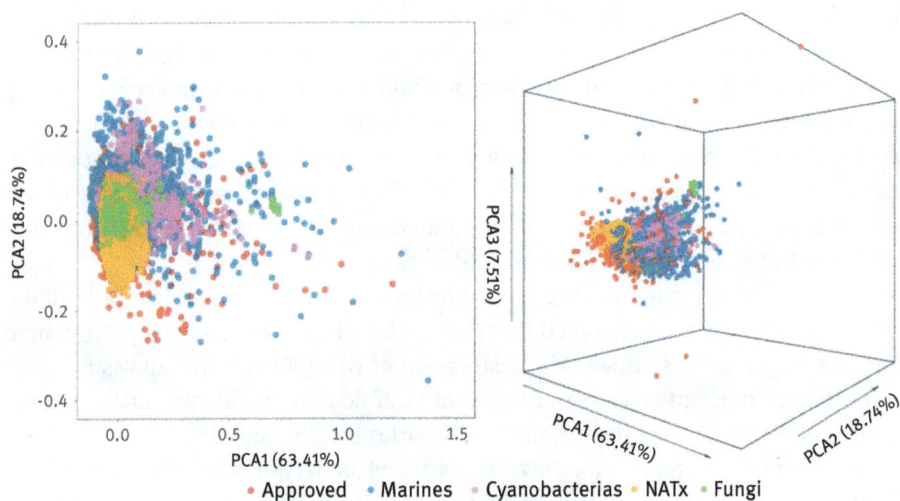

Figure 4.2: 2D and 3D visual representation of the chemical space of natural products databases of different origin (Table 4.2). The visual representation was generated with a principal component analysis of six physicochemical properties: molecular weight, hydrogen bond donors, hydrogen bond acceptors, the octanol and/or water partition coefficient, topological polar surface area and number of rotatable bonds.

As shown in Figure 4.2 the database of approved drugs covers mainly the chemical space and is also the database with the largest diversity in physicochemical properties. In contrast, the database of semi-synthetic products occupies a more restricted space, which in turn is included within the space of approved drugs. In this visualization can be observed that, in general, the NPs considered in this study occupy a space of traditional physicochemical properties, so that their study can lead to identify new compounds with possible therapeutic activity. Similarly, it is also observed that some of the compounds present in NPs collections occupy regions of the chemical space not yet explored and may be useful in virtual screening studies for therapeutic targets in which molecules with activity have not yet been found. Indeed, the approved drug that appears as outlier in the visual representation of the chemical space is trabectedin, the natural product that was recently approved as drug (Figure 4.1).

4.2.2 Comparison of natural products with themselves and other reference libraries

As in the case of maritime navigation where different tools or points of references are used such as the position of the stars to orient themselves; in the study of chemical space, it is sometimes useful to map the chemical space not in absolute coordinates (properties) but in relation to a set of reference points (molecules with well-defined properties).

The selection of reference libraries will always depend on the case study. However, there are collections that have been used frequently as a point of reference. Such is the case of the set of approved drugs, useful for determining the space of "safe" compounds with good physicochemical properties. Collections from synthesis have also been very useful to map the space feasible synthetically. Recently, libraries of NPs have been used as a reference in the design and development of libraries of NP-likeness compounds.

4.2.2.1 Commercial screening libraries of natural products

Currently, there is a large number of NP libraries, most of them can be accessed for free, for example, for virtual screening [25]. In a recent work, Chen et al. [48] analyzed the information available on the content, coverage, and relevance of individual libraries of natural products. In total 18 virtual databases (including the DNP), 9 physical libraries and the Protein Data Bank (PDB) were analyzed. Among the most important results of this study is the fact that the chemical space covered by the known NPs is substantially larger than that of readily obtainable NPs and drugs. DNP and UNPD are the known NP sets that clearly cover the larger regions of the chemical space. However, readily obtainable NPs are highly diverse (representing more than 5700 different Murcko scaffolds) and are accumulated in densely populated areas with approved drugs. With respect to individual physicochemical properties, readily obtainable NPs are substantially smaller than known NPs and drugs. In properties such as logP, the number of chiral centers and number of Csp^3 readily obtainable NPs and drugs are comparable. Distinctive features of the individual databases were also observed. That analysis provided a complete and detailed overview of the known and easily obtainable NPs, which will undoubtedly be useful in the selection of data sources for computer-guided drug discovery.

4.2.2.2 Natural products libraries versus synthetic compounds

One of the main interests in NP-based drug discovery is differentiating compounds coming from nature with compounds obtained by synthesis. In this sense, several studies confirm that chemical space of NPs is clearly distinguishable from the space of synthesis compounds [41, 42, 50]. In comparison with synthetic molecules similar to drugs, NPs stand out for their enormous structural and physicochemical diversity [61, 65, 66]. In addition, NPs have been shown a larger structural complexity, in particular with respect to the stereochemical aspects [67].

Regions that share NPs with synthetic compounds are of particular interest for drug design. The regions in chemical space where both collections overlap may be of interest for the design of NP-inspired compounds [45].

4.2.2.3 Natural product libraries versus approved drugs

Collections of drugs and other databases containing information of the bioactivity profile against one or multiple biological endpoints are useful for multiple applications, including the systematic description of structure-activity relationships (also called "activity landscape modeling") [68] and the further understanding of polypharmacology [69]. These collections are also very useful to try to delineate the boundaries of the currently explored medicinally relevant chemical space [70]. Navigation guided by the bioactivity of the chemical space allows focusing the design of libraries, which it is also known as BIOS [71].

Some prominent examples of public databases annotated with biological activity are PubChem [72], ChEMBL [73, 74], and DrugBank [75] that contain approved drugs for clinical use.

Since NPs exhibit a broad range of biological activities in different organisms, this based on their specific biological purposes in evolution, it is not surprising that drugs and NPs share parts of the chemical space. However, there is also a large part of NPs that occupy parts of the chemical space not explored yet by the available detection collections and that at the same time, are adhering to a great extent to the rule of the five [41, 76, 77].

4.2.2.4 Natural product libraries versus food chemicals

Food chemicals are a cornerstone in the food industry. Their study represents a step beyond the emerging field of "Food Informatics" [78]. In studies that directly compare the chemical structures of food chemicals with collections of natural products reveal that food chemicals have a high structural diversity, comparable to that of NPs and other reference libraries [79]. Likewise, there is a large overlap between the chemical space of food chemical products and NPs [4], this is somehow expected because several NPs are currently used as dietary sources.

4.3 Diversity in chemical space: quantification and implications

As discussed so far, the visualization of the chemical space has many applications. One of them is to determine the diversity of databases, however, it can be complemented with other cheminformatic methods for a more quantitative analysis, which provides information to prioritize the selection of libraries or sub-libraries for experimental selection. The diversity analysis helps to evaluate the structural novelty of a compound collection [80].

If the purpose of a selection project is to identify new lead compounds, then it is desirable to select collections with chemically diverse structures to increase the probability of identifying new compounds that can become leads. However, if the purpose of the campaign is to optimize one or more specific chemical scaffolds, then it is desirable to explore dense regions of the chemical space [81].

Approaches to assess the diversity can be divided into two parts. One is the analysis of structural diversity that encodes information of the structure based on fingerprints, pharmacophoric characteristics or definitions of sub-structures [82]. The second part is the analysis of chemical diversity that encodes information of macroscopic chemical properties (e.g., solubility, logP) or calculated energies, among others.

The structural diversity of NP databases using structural fingerprints and molecular scaffolds has been reported and reviewed in several papers [50, 61, 83, 84].

4.3.1 Molecular properties

Molecular descriptors capture information of the whole molecule and are usually straightforward to interpret. Physicochemical properties frequently used to describe chemical libraries include molecular weight (MW), number of rotatable bonds (RBs), hydrogen-bond acceptors (HBAs), hydrogen-bond donors (HBDs), topological polar surface area (TPSA), and the octanol/water partition coefficient (SlogP), which are properties commonly used as descriptors to represent lead-like, drug-like, or medicinally relevant chemical spaces [85, 86].

To illustrate a visual representation of the property-based chemical space Figure 4.3 depicts the comparison of 423 compounds from the current version of BioFacQuim (*vide supra*) with 2,214 compounds in NuBBE$_{DB}$ [34], 885 AfroDb [31]

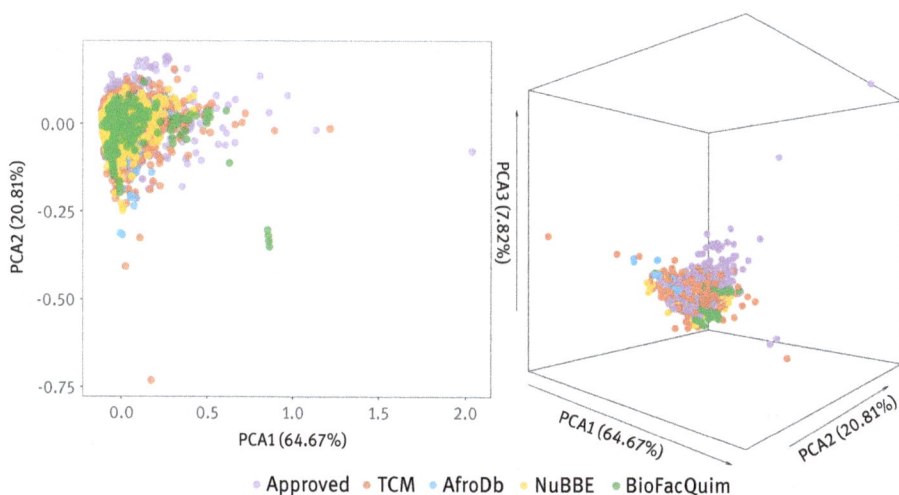

Figure 4.3: 2D and 3D visual representation of the chemical space of NP databases by geographical context. The visual representation was generated with a principal component analysis of six physicochemical properties: molecular weight, hydrogen bond donors, hydrogen bond acceptors, the octanol and/or water partition coefficient, topological polar surface area and number of rotatable bonds.

and a subset of 3000 compounds of TCM [28]. The NP databases are compared to a collection of 1806 drugs approved for clinical use obtained from DrugBank [75]. The first two principal components capture 85.48% of the variance. For the first PC, the larger loadings correspond to SlogP followed by RB, whereas for the second PC the largest loading corresponds to HBD followed by TPSA. The visual representation of the chemical space indicates that many the NPs occupy the same space as the already approved drugs while other compounds, such as some TCM's and BioFacQuim's compounds cover neglected regions of the drug–ADME space. As commented above, interestingly, one of the main outliers in the visual representation of the chemical space in Figure 4.3 is the natural product recently approved as drug trabectedin (Figure 4.1).

4.3.2 Structural fingerprints

Many structural features escape the very general information obtained with physi-cochemical and complexity descriptors. Molecular fingerprints are widely used and have been successfully applied to a number of cheminformatics and computer-aided drug design applications [87]. Fingerprints are especially useful for similarity calcu-lations, such as database searching or clustering, generally measuring similarity with the Tanimoto coefficient. A disadvantage of some fingerprints is that they are difficult to interpret. Also, it is well-known that chemical space will depend on the types of fingerprints used. Using multiple fingerprints and representations to derive consen-sus conclusions have been proposed as a solution. NP databases have been largely analyzed using structural fingerprints. Representative examples are in Table 4.1.

4.3.3 Scaffolds

A complementary approach to characterize compound databases is through mole-cular scaffolds or 'chemotypes' i.e., the core structure of a molecule [88]. Same as physicochemical properties, molecular scaffolds are easy to interpret and facilitate the communication with a scientist working in different disciplines. For instance, this representation is associated with the concepts of "scaffold hopping" [89] and "privi-leged structures" [90]. Scaffold content analysis is broadly used to compare com-pound databases, to identify novel scaffolds in a compound library, to evaluate the performance of virtual screening approaches, and to analyze the structure-activity relationships of sets of molecules with measured activity.

Measuring and comparing the scaffold diversity of compound collections depends on several factors including the specific approach to describe the scaffolds, the size of the database, and the distribution of the molecules in those scaffold classes. Often, scaffold diversity is measured based on frequency counts. While these measures are correct in the way they are defined they do not provide sufficient information concerning the specific distribution of the molecules across the different scaffolds, particularly the most populated ones. Medina-Franco et al. [91] proposed the use of an entropy-based metric to measure the distribution of the molecules across different scaffolds, particularly the most populated ones, as a complementary metric for the comprehensive scaffold diversity analysis of compound data sets.

To illustrate the scaffold content and diversity analysis, Figure 4.4 shows the ten most frequent scaffolds found in NuBBE$_{DB}$ (2,214 compounds in total, *vide supra*) and in BioFacQuim (423 compounds, *vide supra*). The recovery percentage of the cyclic systems are 31.5% for BioFacQuim and 19.70% for NUBBE$_{DB}$. Comparing the 10 most frequent scaffolds and the recovery percentage, the current version of BioFacQuim is less diverse than the NuBBE$_{DB}$, even though that 4.9% of NuBBE$_{DB}$ compounds are acyclic. This result is not that surprising due to the larger number of NPs and sources gathered thus far to build the current version of NuBBE$_{DB}$ [34].

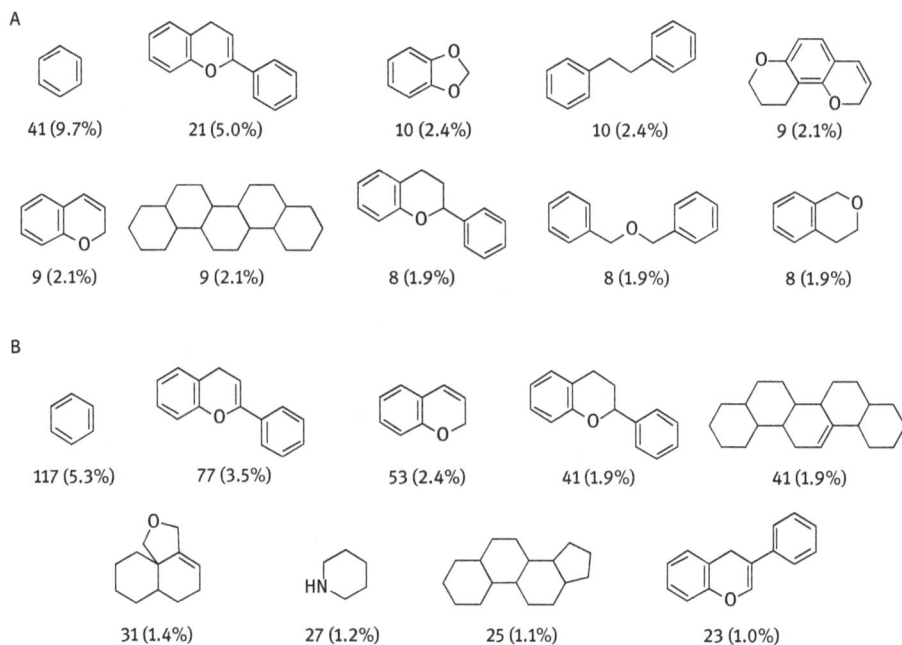

Figure 4.4: Most frequent cyclic systems scaffolds (Murcko scaffolds) found in (A) BioFacQuim (B) NuBBE$_{DB}$. The frequency and percentage are shown. Cyclic systems shown recover 31.5% and 19.70%, respectively. The second most frequent system for NuBBE$_{DB}$ was acyclic compounds with 4.5% recovered.

Another criterion to compare the scaffold diversity of the databases is the Scaled Shannon Entropy (SSE) [91]. It is used as a measure of the specific distribution of molecules in the most populated scaffolds in a compound database. SSE values closer to 1.0 indicate that the molecules are more equally distributed in the scaffolds (high diversity) and smaller SSE values indicate that most of the molecules are distributed in fewer scaffolds (low diversity). The NuBBE$_{DB}$ SSE is 0.931 being a little larger than the SSE of BioFacQuim which is 0.911. Based on these two metrics, it can be concluded that the current version of NuBBE$_{DB}$ is slightly more diverse than BioFacQuim.

4.3.4 Consensus diversity analysis: consensus diversity plots

Since, as commented above, chemical diversity and our perception of chemical space depend on the molecular representation, an intuitive two-dimensional graph so-called Consensus Diversity (CD) Plot, was developed to represent in low dimensions the diversity of chemical libraries considering simultaneously multiple molecular representations [92]. CD Plots have already been used to characterize the global diversity of fungi metabolites [93] and NPs from Panama [94]. More recently, CD Plots were employed to compare the diversity of 23,883 food chemicals with drugs approved for clinical use, Generally Regarded as Safe molecules and screening compounds from ZINC [4]. Figure 4.5 shows a CD Plot comparing the global diversity of NuBBE$_{DB}$, BioFacQuim, TCM, AfroDb and approved drugs. BioFacQuim and AfroDb are found in the bottom right quadrant (yellow area), which indicates that the scaffolds of the molecules are the main factor that contributes to the diversity, having a relatively low diversity by fingerprints. The CD Plots also indicate that AfroDb is more diverse than BioFacQuim according to the Euclidean distance between its properties. In contrast, the approved drugs have an average diversity by fingerprints and relatively low diversity by scaffolds which indicates that the chemical characteristics of the whole molecule and/or the side chains contribute significantly to the diversity, although it is not very diverse in terms of its physicochemical properties. The area where TCM is located (bottom left quadrant) indicates that the library has the relative largest fingerprint and scaffold diversity of the data sets that are being compared. In contrast, NuBBE$_{DB}$ (at the top-right quadrant) is the data set with the relative lowest fingerprint and scaffold diversity despite the fact it has a high intra-dataset diversity of its physicochemical properties.

4.4 Conclusions and future directions

NPs provide an evolutionary validated useful starting point for the design of new bioactive molecules. Several academic groups, not-for-profit, and commercial initiatives are integrating resources of NPs including compound databases. It is anticipated that more academic groups keep integrating their resources. The wealth of NPs in several countries has also motivated the interest in systematically explore the coverage of the chemical space of these collections and analyze systematically the chemical and structural diversity.

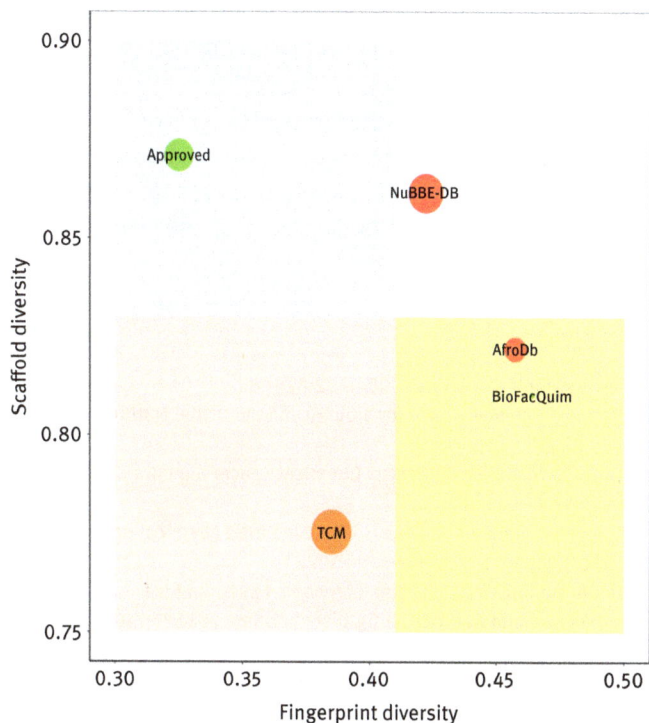

Figure 4.5: Consensus Diversity Plot comparing the diversity of BioFacQuim, AfroDb, TCM, NuBBE$_{DB}$ and approved drugs. The median Tanimoto coefficient of MACCS keys fingerprints represent the structural diversity (fingerprint diversity) of each data set and is plotted on the X-axis. The scaffold diversity of each database was defined as the area under the curve (AUC) of the respective scaffold recovery curves, and it is represented on the Y-axis. The quadrants are colored as follows: red, indicates the library is diverse considering its scaffolds and/or side chains; white, the library is not diverse; blue, the library is diverse if the chemical features of the entire molecule are considered and/or side chains contribute significantly to the diversity; yellow, the scaffolds of the molecules are the main factor contributing to the diversity and/or this set contains mostly rings with few side chains. Data points are colored by the diversity of the physicochemical properties of the data set as measured by the Euclidean distance of six properties of pharmaceutical relevance. The distance is represented with a continuous color scale from red (more diverse) to orange/brown (intermediate diversity) to green (less diverse). The relative size of the data set is represented with the size of the data point: smaller data points indicate compound data sets with fewer molecules.

Thus far, it has been quantified distinct features of NPs such as their molecular scaffolds, property diversity, and large structural complexity. The collections of NPs would keep promoting local and global initiatives to identify bioactive compounds and potentially enrich the medicinally relevant chemical space. Furthermore, if the screening data of NPs is made accessible, in preference in the public domain, the large structure-activity information for natural compounds would facilitate the development of algorithms and eventually construct predictive models using machine learning.

Acknowledgements: This work was supported by the National Council of Science and Technology (CONACyT, Mexico) grant number 282785. FIS-G is thankful to CONACyT for the granted scholarship number 629458. BAP-J is grateful for the support given by the subprogram 127 "Basic Training in Research" of the School of Chemistry, UNAM.

References

[1]   Dobson CM. Chemical space and biology. Nature. 2004;432:824–8.
[2]   Lipinski C, Hopkins A. Navigating chemical space for biology and medicine. Nature. 2004;432:855–61.
[3]   Awale M, Visini R, Probst D, Arús-Pous J, Reymond J-L. Chemical space: big data challenge for molecular diversity. Chimia. 2017;71:661–6.
[4]   Naveja JJ, Rico-Hidalgo MP, Medina-Franco JL. Analysis of a large food chemical database: chemical space, diversity, and complexity. *F1000Res*. 2018;7.
[5]   López-Vallejo F, Giulianotti MA, Houghten RA, Medina-Franco JL. Expanding the medicinally relevant chemical space with compound libraries. Drug Discov Today. 2012;17:718–26.
[6]   López-Vallejo F, Waddell J, Yongye AB, Houghten RA, Medina-Franco JL. A large scale classification of molecular fingerprints for the chemical space representation and SAR analysis. J Cheminform. 2012;4:P26.
[7]   Medina-Franco JL, Martinez-Mayorga K, Giulianotti MA, Houghten RA, Pinilla C. Visualization of the chemical space in drug discovery. Current Comput - Aided Drug Des. 2008;4:322–33.
[8]   Osolodkin DI, Radchenko EV, Orlov AA, Voronkov AE, Palyulin VA, Zefirov NS. Progress in visual representations of chemical space. Expert Opin Drug Discov. 2015;10:959–73.
[9]   Opassi G, Gesù A, Massarotti A. The hitchhiker's guide to the chemical-biological galaxy. Drug Discov Today. 2018;23:565–74.
[10]   Newman DJ, Cragg GM. Natural products as sources of new drugs from 1981 to 2014. J Nat Prod. 2016;79:629–61.
[11]   Bauer A, Brönstrup M. Industrial natural product chemistry for drug discovery and development. Nat Prod Rep. 2014;31:35–60.
[12]   Harvey AL, Edrada-Ebel R, Quinn RJ. The re-emergence of natural products for drug discovery in the genomics era. Nat Rev Drug Discov. 2015;14:111–29.
[13]   Alvarec-Ruiz E, Collis AJ, Dann AS, Forsbury AP, Reddy SJ, Vázquez Muniz MJ, Microbiological process. Patent. 2017. https://patentimages.storage.googleapis.com/96/8b/de/87242640defaa1/CN106687596A.pdf. Accessed: 30 Sep 2018.
[14]   Pereira DM, Valentão P, Andrade PB. Tuning protein folding in lysosomal storage diseases: the chemistry behind pharmacological chaperones. Chem Sci. 2018;9:1740–52.
[15]   Zhanel GG, Lawson CD, Zelenitsky S, Findlay B, Schweizer F, Adam H, et al. Comparison of the next-generation aminoglycoside plazomicin to gentamicin, tobramycin and amikacin. Expert Rev Anti Infect Ther. 2012;10:459–73.
[16]   Cobb R, Boeckh A. Moxidectin: a review of chemistry, pharmacokinetics and use in horses. Parasit Vectors. 2009;2:S5.
[17]   Ca G, Ci F, Ag P, Chen C, Tipping R, Cm C, et al. Safety, tolerability, and pharmacokinetics of escalating high doses of ivermectin in healthy adult subjects. J Clin Pharmacol. 2002;42:1122–33.

[18] Brandt W, Haupt VJ, Wessjohann LA. Chemoinformatic analysis of biologically active macro-cycles. Curr Top Med Chem. 2010;10:1361–79.

[19] Wessjohann LA, Ruijter E, Garcia-Rivera D, Brandt W. What can a chemist learn from nature's macrocycles? – A brief, conceptual view. Mol Divers. 2005;9:171–86.

[20] Cuevas C, Francesch A. Development of Yondelis (trabectedin, ET-743). A semisynthetic process solves the supply problem. Nat Prod Rep. 2009;26:322–37.

[21] Gajdos C, Elias A. Trabectedin: safety and efficacy in the treatment of advanced sarcoma. Clin Med Insights Oncol. 2011;5:35–43.

[22] Scotti L, Ferreira EI, Ms S, Mt S. Chemometric studies on natural products as potential inhibitors of the NADH oxidase from Trypanosoma cruzi using the VolSurf approach. Molecules. 2010;15:7363–77.

[23] Scotti MT, Scotti L. Editorial: chemometrics in drug discovery. Comb Chem High Throughput Screen 2015;18:702–03.

[24] Rodrigues T, Reker D, Schneider P, Schneider G. Counting on natural products for drug design. Nat Chem. 2016;8:531–41.

[25] Chen Y, de Bruyn Kops C, Kirchmair J. Data resources for the computer-guided discovery of bioactive natural products. J Chem Inf Model. 2017;57:2099–111.

[26] Maier ME. Design and synthesis of analogues of natural products. Org Biomol Chem. 2015;13:5302–43.

[27] Wilk W, Zimmermann TJ, Kaiser M, Waldmann H. Principles, implementation, and application of biology-oriented synthesis (BIOS). Biol Chem. 2010;391:491–97.

[28] Cy-C C. TCM Database@Taiwan: the world's largest traditional Chinese medicine database for drug screening in silico. PLoS One. 2011;6:e15939.

[29] Tsai T-Y, Chang K-W, Chen CY. iScreen: world's first cloud-computing web server for virtual screening and de novo drug design based on TCM database@Taiwan. J Comput Aided Mol Des. 2011;25:525–31.

[30] Gu J, Gui Y, Chen L, Yuan G, Lu H-Z XX. Use of natural products as chemical library for drug discovery and network pharmacology. PLoS One. 2013;8:e62839.

[31] Ntie-Kang F, Zofou D, Babiaka SB, Meudom R, Scharfe M, Lifongo LL, et al. AfroDb: a select highly potent and diverse natural product library from African medicinal plants. PLoS One. 2013;8:e78085.

[32] Ntie-Kang F, Onguéné PA, Scharfe M, Owono Owono LC, Megnassan E, Mbaze LM, et al. ConMedNP: a natural product library from Central African medicinal plants for drug discovery. RSC Adv. 2014;4:409–19.

[33] Valli M, Dos Santos RN, Ld F, Ch N, Castro-Gamboa I, Ad A, et al. Development of a natural products database from the biodiversity of Brazil. J Nat Prod. 2013;76:439–44.

[34] Pilon AC, Valli M, Dametto AC, Pinto MEF, Freire RT, Castro-Gamboa I, et al. NuBBE DB: an updated database to uncover chemical and biological information from Brazilian biodiversity. Sci Rep. 2017;7:7215.

[35] NuBBE - Núcleo de Bioensaios, Biossíntese e Ecofisiologia de Produtos Naturais (Nuclei of Bioassays, Ecophysiology and Biosynthesis of Natural Products Database). http://nubbe.iq. unesp.br/portal/nubbedb.html. Accessed 30 Sep 2018.

[36] Naveja JJ, Oviedo-Osornio CI, Trujillo-Minero NN, Medina-Franco JL. Chemoinformatics: a perspective from an academic setting in Latin America. Mol Divers. 2018;22:247–58.

[37] Medina-Franco JL. Chemoinformatic Characterization of the Chemical Space and Molecular Diversity of Compound Libraries. In: Trabocchi A, editor. Diversity-Oriented Synthesis. Hoboken, NJ, USA: John Wiley & Sons, Inc., 2013:325–52.

[38] Gaspar HA, Sidorov P, Horvath D, Marcou G. Generative topographic mapping approach to chemical space analysis. ACS Symp Ser. 2016. https://elibrary.ru/item.asp?id=27576908.

[39] Tino P, Nabney I. Hierarchical GTM: constructing localized nonlinear projection manifolds in a principled way. IEEE Trans Pattern Anal Mach Intell. 2002;24:639–56.

[40] Naveja JJ, Medina-Franco JL. ChemMaps: towards an approach for visualizing the chemical space based on adaptive satellite compounds. *F1000Res*. 2017;6:1134.

[41] Feher M, Schmidt JM. Property distributions: differences between drugs, natural products, and molecules from combinatorial chemistry. J Chem Inf Comput Sci. 2003;43:218–27.

[42] Shelat AA, Guy RK. The interdependence between screening methods and screening libraries. Curr Opin Chem Biol. 2007;11:244–51.

[43] Singh SB, Chris Culberson J. Chapter 2: chemical space and the difference between natural products and synthetics. In: Antony D Buss, Mark S Butler (editors). Natural product chemistry for drug discovery, Cambridge, UK: Royal Society of Chemistry. 2009:28–43.

[44] Chen H, Engkvist O, Blomberg N, Li J. A comparative analysis of the molecular topologies for drugs, clinical candidates, natural products, human metabolites and general bioactive compounds. Med Chem Commun. 2012;3:312–21.

[45] Ertl P, Schuffenhauer A. Cheminformatics analysis of natural products: lessons from nature inspiring the design of new drugs. Prog Drug Res. 2008;66:217, 219–35.

[46] Pascolutti M, Campitelli M, Nguyen B, Pham N, Gorse A-D, Quinn RJ. Capturing nature's diversity. PLoS One. 2015;10:e0120942.

[47] González-Medina M, Prieto-Martínez FD, Naveja JJ, Méndez-Lucio O, El-Elimat T, Pearce CJ, et al. Chemoinformatic expedition of the chemical space of fungal products. Future Med Chem. 2016;8:1399–412.

[48] Chen Y, Garcia de Lomana M, N-O F, Kirchmair J. Characterization of the chemical space of known and readily obtainable natural products. J Chem Inf Model. 2018;58:1518–32.

[49] Shang J, Hu B, Wang J, Zhu F, Kang Y, Li D, et al. Cheminformatic Insight into the differences between terrestrial and marine originated natural products. J Chem Inf Model. 2018;58:1182–93.

[50] Ertl P, Schuffenhauer A. Cheminformatics analysis of natural products: lessons from nature inspiring the design of new drugs. Prog Drug Res. 2008;66:217, 219–35.

[51] Muigg P, Rosén J, Bohlin L, Backlund A. In silico comparison of marine, terrestrial and synthetic compounds using ChemGPS-NP for navigating chemical space. Phytochem Rev. 2013;12:449–57.

[52] Saldívar-González FI, Valli M, Da Silva Bolzani V, Medina-Franco JL. Chemical diversity of NuBBE database: A chemoinformatic characterization 2018.

[53] Larsson J, Gottfries J, Muresan S, Backlund A. ChemGPS-NP: tuned for navigation in biologically relevant chemical space. J Nat Prod. 2007;70:789–94.

[54] Rosén J, Rickardson L, Backlund A, Gullbo J, Bohlin L, Larsson R, et al. ChemGPS-NP mapping of chemical compounds for prediction of anticancer mode of action. QSAR Comb Sci. 2009;28:436–46.

[55] Korinek M, Tsai Y-H, El-Shazly M, Lai K-H, Backlund A, Wu S-F, et al. Anti-allergic Hydroxy Fatty Acids from Typhonium blumei Explored through ChemGPS-NP. Front Pharmacol. 2017;8:356.

[56] Rosén J, Lövgren A, Kogej T, Muresan S, Gottfries J, Backlund A. ChemGPS-NP(Web): chemical space navigation online. J Comput Aided Mol Des. 2009;23:253–9.

[57] Frédérick R, Bruyère C, Vancraeynest C, Reniers J, Meinguet C, Pochet L, et al. Novel trisubstituted harmine derivatives with original in vitro anticancer activity. J Med Chem. 2012;55:6489–501.

[58] Ertl P, Rohde B. The molecule cloud - compact visualization of large collections of molecules. J Cheminform. 2012;4:12.

[59] Schuffenhauer A, Ertl P, Roggo S, Wetzel S, Koch MA, Waldmann H. The scaffold tree–visualization of the scaffold universe by hierarchical scaffold classification. J Chem Inf Model. 2007;47:47–58.

[60] Medina-Franco JL, Petit J, Maggiora GM. Hierarchical strategy for identifying active chemotype classes in compound databases. Chem Biol Drug Des. 2006;67:395–408.

[61] Koch MA, Schuffenhauer A, Scheck M, Wetzel S, Casaulta M, Odermatt A, et al. Charting biologically relevant chemical space: A structural classification of natural products (SCONP). Proc Natl Acad Sci USA. 2005;102:17272–77.

[62] Schäfer T, Kriege N, Humbeck L, Klein K, Koch O, Mutzel P. Scaffold Hunter: a comprehensive visual analytics framework for drug discovery. J Cheminform. 2017;9:28.

[63] Tao L, Zhu F, Qin C, Zhang C, Chen S, Zhang P, et al. Clustered distribution of natural product leads of drugs in the chemical space as influenced by the privileged target-sites. Sci Rep. 2015;5:9325.

[64] Pye CR, Bertin MJ, Lokey RS, Gerwick WH, Linington RG. Retrospective analysis of natural products provides insights for future discovery trends. Proc Natl Acad Sci USA. 2017;114:5601–6.

[65] Camp D, Garavelas A, Campitelli M. Analysis of physicochemical properties for drugs of natural origin. J Nat Prod. 2015;78:1370–82.

[66] Stratton CF, Newman DJ, Tan DS. Cheminformatic comparison of approved drugs from natural product versus synthetic origins. Bioorg Med Chem Lett. 2015;25:4802–7.

[67] Clemons PA, Bodycombe NE, Carrinski HA, Wilson JA, Shamji AF, Wagner BK, et al. Small molecules of different origins have distinct distributions of structural complexity that correlate with protein-binding profiles. Proc Natl Acad Sci USA. 2010;107:18787–92.

[68] Medina-Franco JL, Navarrete-Vázquez G, Méndez-Lucio O. Activity and property landscape modeling is at the interface of chemoinformatics and medicinal chemistry. Future Med Chem. 2015;7:1197–211.

[69] Reddy AS, Zhang S. Polypharmacology: drug discovery for the future. Expert Rev Clin Pharmacol. 2013;6:41–7.

[70] Medina-Franco JL, Martinez-Mayorga K, Meurice N. Balancing novelty with confined chemical space in modern drug discovery. Expert Opin Drug Discov. 2014;9:151–65.

[71] van Hattum H, Waldmann H. Biology-oriented synthesis: harnessing the power of evolution. J Am Chem Soc. 2014;136:11853–9.

[72] Kim S, Thiessen PA, Bolton EE, Chen J, Fu G, Gindulyte A, et al. PubChem substance and compound databases. Nucleic Acids Res. 2016;44:D1202–13.

[73] Gaulton A, Bellis LJ, Bento AP, Chambers J, Davies M, Hersey A, et al. ChEMBL: a large-scale bioactivity database for drug discovery. Nucleic Acids Res. 2012;40:D1100–7.

[74] Gaulton A, Hersey A, Nowotka M, Bento AP, Chambers J, Mendez D, et al. The ChEMBL database in 2017. Nucleic Acids Res. 2017;45:D945–54.

[75] Wishart DS, Feunang YD, Guo AC, Lo EJ, Marcu A, Grant JR, et al. DrugBank 5.0: a major update to the DrugBank database for 2018. Nucleic Acids Res. 2018;46:D1074–82.

[76] Boufridi A, Quinn RJ. Harnessing the properties of natural products. Annu Rev Pharmacol Toxicol. 2018;58:451–70.

[77] Rosén J, Gottfries J, Muresan S, Backlund A, Oprea TI. Novel chemical space exploration via natural products. J Med Chem. 2009;52:1953–62.

[78] Martinez-Mayorga K, Medina-Franco JL, editors. Foodinformatics: applications of chemical information to food chemistry, Switzerland: Springer. 2014. https://www.springer.com/gp/book/9783319102252.

[79]   Medina-Franco JL, Martínez-Mayorga K, Peppard TL, Del Rio A. Chemoinformatic analysis of GRAS (Generally recognized as safe) flavor chemicals and natural products. PLoS One. 2012;7: e50798.

[80]   Medina-Franco JL. Advances in computational approaches for drug discovery based on natural products. Revista Latinoamericana de Química. 2013;41:95–110.

[81]   Houghten RA, Pinilla C, Giulianotti MA, Appel JR, Dooley CT, Nefzi A, et al. Strategies for the use of mixture-based synthetic combinatorial libraries: scaffold ranking, direct testing in vivo, and enhanced deconvolution by computational methods. J Comb Chem. 2008;10:3–19.

[82]   Brown N, Jacoby E. On scaffolds and hopping in medicinal chemistry. Mini Rev Med Chem. 2006;6:1217–29.

[83]   Singh N, Guha R, Giulianotti MA, Pinilla C, Houghten RA, Medina-Franco JL. Chemoinformatic analysis of combinatorial libraries, drugs, natural products, and molecular libraries small molecule repository. J Chem Inf Model. 2009;49:1010–24.

[84]   Yongye AB, Waddell J, Medina-Franco JL. Molecular scaffold analysis of natural products databases in the public domain. Chem Biol Drug Des. 2012;80:717–24.

[85]   Lipinski CA. Lead- and drug-like compounds: the rule-of-five revolution. Drug Discov Today Technol. 2004;1:337–41.

[86]   Veber DF, Johnson SR, Cheng H-Y, Smith BR, Ward KW, Kopple KD. Molecular properties that influence the oral bioavailability of drug candidates. J Med Chem. 2002;45:2615–23.

[87]   Maldonado AG, Doucet JP, Petitjean M, Fan B-T. Molecular similarity and diversity in chemoinformatics: from theory to applications. Mol Divers. 2006;10:39–79.

[88]   Schuffenhauer A, Varin T. Rule-based classification of chemical structures by scaffold. Mol Inform. 2011;30:646–64.

[89]   Schneider G, Neidhart W, Giller T, Schmid G. "Scaffold-hopping" by topological pharmaco-phore search: a contribution to virtual screening. Angew Chem Int Ed. 1999;38:2894–96.

[90]   Evans BE, Rittle KE, Bock MG, DiPardo RM, Freidinger RM, Whitter WL, et al. Methods for drug discovery: development of potent, selective, orally effective cholecystokinin antagonists. J Med Chem. 1988;31:2235–46.

[91]   Medina-Franco JL, Martínez-Mayorga K, Bender A, Scior T. Scaffold diversity analysis of compound data sets using an entropy-based measure. QSAR Comb Sci. 2009;28:1551–60.

[92]   González-Medina M, Prieto-Martínez FD, Owen JR, Medina-Franco JL. Consensus diversity plots: a global diversity analysis of chemical libraries. J Cheminform. 2016;8:63.

[93]   González-Medina M, Owen JR, El-Elimat T, Pearce CJ, Oberlies NH, Figueroa M, et al. Scaffold diversity of fungal metabolites. Front Pharmacol. 2017;8:180.

[94]   Olmedo DA, González-Medina M, Gupta MP, Medina-Franco JL. Cheminformatic characterization of natural products from Panama. Mol Divers. 2017;21:779–89.

David J. Newman

5 From natural products to drugs

Abstract: It is frequently assumed, particularly in the last 15 plus years, that "Natural Product Structures" are no longer a source of drugs in the twenty-first century. In fact, this is not at all true. Even today, in the search for novel agents against manifold diseases, natural product structures, some quite old and some quite recent, are behind the compounds that are either recently (last 5–10 years) approved or that are now in clinical trials against manifold diseases of man. This chapter will cover agents approved since 2010 to the end of 2017 by the US FDA and its equivalent in other countries, plus selected agents that have entered clinical trials against major diseases such as cancer and infections that have "in their chemical pedigree" a natural product structure, even if the final product may be totally synthetic in nature.

Keywords: natural products, drugs, drug design

5.1 Introduction

As part of a long-standing series of reviews by Newman and Cragg, with the last published version in early 2016 [1] which covered drugs approved world-wide from 1981 through the end of 2014, we continued to show how basic natural product structures have led to approved drugs against a multitude of diseases of man. Although we have not yet published the next review, which will take the story up through the end of 2017, we do have the initial data capture and can use it to add to the figures from the 2016 review in order to "set the stage" for the chapter.

A caveat should be mentioned at this point. Over the last 3 years, approximately 230 "compounds" (including biologics and vaccines) have been approved, but when the figures are carefully parsed, 24 are actually "biosimilars" approved in the USA, the EU, Japan, Korea and Russia, and another 80 plus are biologics and vaccines of one type or another. In addition, a number (less than 10) are combinations of previously approved drugs that are now being pharmaceutically combined. When the next review (covering 1981–2017) is completed, as in this article, the percentages of compounds based on natural products will be based on "small molecule numbers", not upon "all approved drug entities". This has been done in the past, but it needs to be reemphasized.

As mentioned in the abstract, and from the caveat above, analyses and discussions will be performed on data from 2010 to 2017 for approved small molecules, and

This article has previously been published in the journal *Physical Sciences Reviews*. Please cite as: Newman, J.D. From Natural Products to Drugs. *Physical Sciences Reviews* [Online] **2018** DOI: 10.1515/psr-2018-0111.

https://doi.org/10.1515/9783110579352-006

it should also be pointed out that a drug is only counted one time, even if, as is now frequently the case, particularly with the antitumour kinase inhibitors, they are subsequently approved by other governments for different tumour types.

In the subsequent discussion the codes from Newman and Cragg will be used, but will be limited to N, and ND. An occasional S* (sometimes with the "/NM" attachment) may be commented on. The definitions of N and ND are "Natural Product" and Natural Product Derived", respectively. S* stands for "Synthetic but has an NP pharmacophore" and the "/NM" modifier means "natural product mimic" and is applied to a synthetic compound when its activity is due to a direct enzymatic and/or receptor activation or inhibition where a normal substrate is part of the assay. Thus, a compound that is a competitive inhibitor/activator of the target could fall under this category. There are many examples given in the 2016 review [1] and in earlier versions referenced therein.

5.2 Natural product and natural product-derived structures as approved drugs 2010–2017

Using the N and ND classifications and only counting small molecules (no biologics or vaccines) with an exception being made for monoclonal antibodies with a natural product- based warhead which have been classified as ND, in Table 5.1, the basic statistics for N and ND sourced drugs are compared to the overall number of approved drugs that year, world-wide. It should be emphasized that many analyses have been made over the years by various groups but most, if not all, have used only the US FDA and/or the EU MEAE lists. The figures used here cover all countries where data is available, though data from the Peoples Republic of China (PRC) is difficult to access.

Table 5.1: Sources of drugs from 2010–2017.

Year	B	N	NB	ND	S	S/NM	S*	S*/NM	Small_ Mols	%N; ND
2010	7	1		8	5	6			20	45
2011	6	2		6	5	3		11	27	30
2012	16	2	2	6	8	6	2	9	33	24
2013	8	1		3	9	6	3	9	31	13
2014	18	2	2	6	11	11	1	11	42	19
2015		1		5					40	15
2016		2		2					20	20
2017				10					39	26

Codes as in [1]

Inspection of Table 5.1 shows that even today, natural products and their derivatives are still significantly involved as sources of approved drugs, with percentages ranging from a low of 13 in 2013 to a high of 45 in 2010 (though in that year only 20 small molecules were approved world-wide). Recently in 2016 and 2017, the figures have rebounded to 20 to 26 percent of small molecules. It should be noted as mentioned earlier, that a large percentage of approved drug entities are now either biologics or vaccines.

In Table 5.2, the generic and trade names of the compounds in Table 5.1 are shown, arranged by year, and listing the disease(s) for which the compounds were initially approved. It should be noted that the country shown as the "approving country" is not necessarily where the compound was first identified and/or synthesized. Though there are 58 compounds listed in this table, only the structures of 17 have been chosen for illustrative purposes.

Table 5.2: N and ND drugs approved 2010 to 2017.

Generic name	Trade name	Disease	Year	Source	Country	Figure_#
Romidepsin	Istodax	Anticancer	2010	N	USA	1
Mifamurtide	Junovan	Anticancer	2010	ND	EU	
Vinflunine	Javlor	Anticancer	2010	ND	UK	
Cabazitaxel	Jevtana	Anticancer	2010	ND	USA	
Fingolimod HCl	Gilenya	Multiple sclerosis	2010	ND	USA	2
Zucapsacin	Civanex	Osteoarthritis	2010	ND	Canada	
Diquafosol tetrasodium	Diquas	Dry eye syndrome	2010	ND	Japan	
Laninamivir octanoate	Inavir	Antiviral	2010	ND	Japan	
Eribulin	Halaven	Anticancer	2010	ND	USA	3
Fidaxomicin	Dificid	Antibacterial	2011	N	USA	4
Spinosad	Natroba	Insecticide	2011	N	USA	
Ceftaroline fosamil acetate	Teflaro	Antibacterial	2011	ND	USA	5
Zucapsaicin	Zuacta	Antiarthritic	2011	ND	Canada	6
Abiraterone acetate	Zytiga	Anticancer	2011	ND	USA	
Brentuximab vedotin	Adcetris	Anticancer	2011	ND	USA	7
Eldecalcitol	Edirol	Osteoporosis	2011	ND	Japan	
Tesamorelin	Egrifta	Lipodystrophy syndrome	2011	ND	USA	
Ingenol mebutate	Picato	Anticancer	2012	N	USA	8
Homoharringtonine	Ceflatonin	Anticancer	2012	N	USA	9
Pasireotide	Signifor	Cushings Syndrome	2012	ND	EU	
BF-200 ALA	Ameluz	Anticancer	2012	ND	EU	
Teduglutide	Gattex	Short bowel syndrome	2012	ND	EU	

(continued)

Table 5.2 (continued)

Generic name	Trade name	Disease	Year	Source	Country	Figure_#
Linaclotide	Constella	Irritable bowel syndrome	2012	ND	USA	
Carfilzomib	Kyprolis	Anticancer	2012	ND	USA	
Hyaluronic acid crosslinked	Gel-One	Antiarthritic	2012	ND	USA	
Cholic acid	Orphacol	Errors in bile synthesis	2013	N	EU	10
Dimethyl fumarate	Tecfidera	Multiple sclerosis	2013	ND	USA	
Lixisenatide	Lyxumia	Antidiabetic diabetes type 2	2013	ND	EU	
Trastuzumab emtansine	Kadcyla	Anticancer	2013	ND	USA	11
omega-3-carboxylic acids	Epanova	Hypertriglycerid-emia	2014	N	USA	
None given	PICN	Anticancer	2014	N	India	
Vorapaxar	Zontivity	Coronary artery disease	2014	ND	USA	
Naloxegol	Movantik	Opiod induced constipation	2014	ND	USA	
Aafamelanotide	Scenesse	Phototoxicity in adults	2014	ND	EU	
Dalbavancin	Dalvance	Antibacterial	2014	ND	USA	12
Oritavancin	Orbactiv	Antibacterial	2014	ND	USA	
Cetolozane/ Tazobactam	Zerbaxa	Antibacterial	2014	ND	USA	
Ceftazidime/ Avibactam	Avycaz	Antibacterial	2015	ND	USA	
deoxycholic acid sodium salt	Kybella	Dermatological	2015	N	USA	
Padeliporfin potassium	Stakel	Anticancer	2015	ND	Mexico	
Cangrelor tetrasodium	Kengreal	Coronary artery disease	2015	ND	EU	13
Uridine triacetate	Xuriden	Orotic aciduria	2015	ND	EU	
Hyaluronate/ Triamcinolone	Cingal	Osteoarthris	2015	ND	Canada	
Prasterone	Intrarosa	Dyspareunia	2016	N	USA	
Migalastat hydrochloride	Galafold	Fabry's Disease	2016	N	EU	
Dronabinol hemisuccinate	Syndros	Anorexia	2016	ND	USA	14
Obeticholic acid	Ocaliva	Cirrohsis, primary	2016	ND	USA	
Meropenem/ Vaborbactam	Vabomere	Antibacterial	2017	ND	USA	

(continued)

Table 5.2 (continued)

Generic name	Trade name	Disease	Year	Source	Country	Figure_#
Naldemedine tosylate	Symproic	Opiod induced constipation	2017	ND	USA	
Semaglutide	Ozempic	Antidiabetic diabetes type 2	2017	ND	USA	
Lantoprostene bunod	Vyzulta	Glaucoma	2017	ND	USA	
Dinalbuphine sebacate	Naldebain	Analgesia	2017	ND	Taiwan	
Plecanatide	Trulance	Irritable bowel syndrome	2017	ND	USA	
Inotuzumab ozogamicin	Besponsa	Anticancer	2017	ND	USA	
Abaloparatide	Tymlos	Osteoporosis	2017	ND	USA	15
Forodesine HCL	Mundesine	Anticancer	2017	ND	Japan	16
Midostaurin	Rydapt	Anticancer	2017	ND	USA	17

(Data from Newman & Cragg 2018 in preparation)

5.3 Discussion on the selected compounds in Table 5.2

In all cases where the data is available, the generic name followed by the trade name will be given in order to better identify the compound(s) being discussed. The structures are given under their Figure number below.

5.3.1 Romidepsin (Istodax®; Figure 5.1)

This compound is unusual in one particular respect in that it was placed into preclinical and clinical trials before its mechanism of action was known (or even suspected). It finally was shown to be an HDAC inhibitor [2]. Interestingly, some 8 years later, it was identified from a plant sample that was a part of the NCI's plant

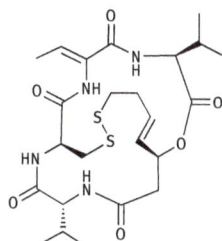

Figure 5.1: Romidepsin.

collection programme, and though attempts were made to isolate the probable producing bacterium from the plant extract, it was not successful.

5.3.2 Fingolimod (Gilenya®; Figure 5.2) and Derivatives

This compound has a very interesting chemical lineage as it was developed over many years from the fungal metabolite myriocin, orignally isolated from *Mycelia sterilia*. The stories behind the identification and then subsequent chemical manipulations to produce an orally effective drug against multiple sclerosis were reported in two papers in 2010 [3] and 2011 [4], with a much more recent discussion of the future of such compounds directed against the sphingosine-1-phosphate pathway published as a perspective article in the Journal of Medicinal Chemistry by Dyckman in 2017 [5]. Table 5.1 in the Dyckman paper should be consulted to see the multiplicity of molecules that have been "spun-off" as a result of the success of fingolimod, with a recent publication demonstrating the scale-up methodology of BMS-960 (Figure 5.2a) [6], a back-up compound and Novartis has a "chemical relative" of fingolamid known as siponimod (Figure 5.2b) that is currently in Phase III trials [7]. under the EXPAND study (NCT01665144).

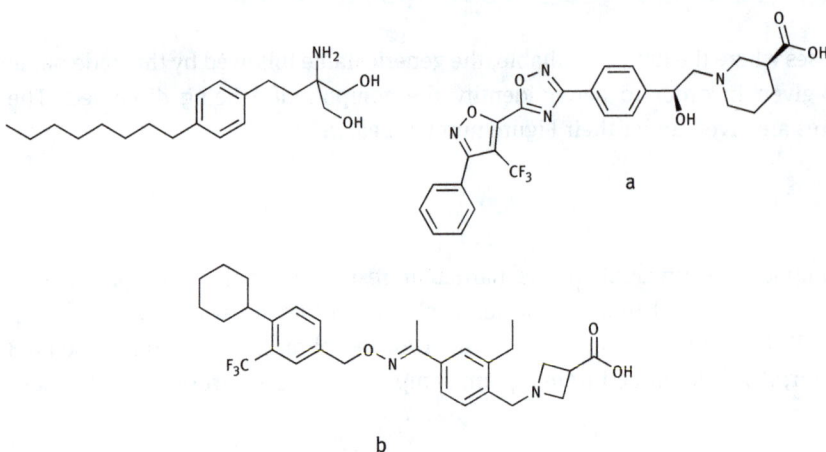

Figure 5.2: Fingolimod; Figure 5.2a: BMS-960; Figure 5.2b: Siponimod.

5.3.3 Eribulin (Halaven®; Figure 5.3)

This compound is the most complex natural product derivative ever synthesized, and originated from halichondrin B, a sponge metabolite isolated in Japan, and then many years later and with over 200 derivatives synthesized, approved in the USA in

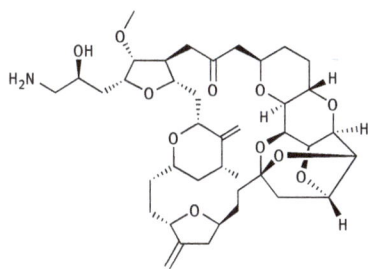

Figure 5.3: Eribulin.

2010. The complex story behind the evolution of this derivative has been presented in a number of publications by the original group [8] and then by other researchers [9, 10], demonstrating the capabilities of modern synthetic chemistry when shown an active structure, and then an example of the amount of work that has to go into clinical trials to further identify different tumours that may respond [11].

5.3.4 Fidaxomicin (Dificid®; Figure 5.4)

One of the areas that is low in numbers in approved drugs for the last 10 plus years, is that covering antimicrobials and in particular, antibacterial agents. In the time frame of this chapter, there have been seven compounds approved but what is significant is that only one (fidaxomicin), is actually a natural product, the others are derivatives of older base structures and two more will be covered later on. A recent paper by McAlpine [12] gives a short history of the problems involved with this compound, as the first reports were from Lepetit in a US patent in 1976. The group of similar compounds had a series of names and "parents" as it kept moving around as companies were absorbed by others. Finally, it was approved for use against a very specific pathogen *Clostridium difficile* and is one of the very few cases where an

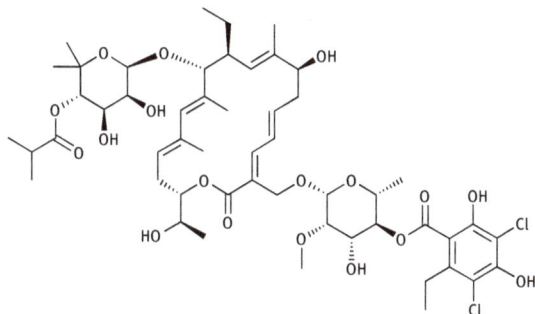

Figure 5.4: Fidaxomicin.

antibiotic is approved for just one indication, *Clostridium difficile*-induced colitis, the third most common nosocomial infection in the US, with the Centers for Disease Control (CDC) estimating 29,000 deaths per year in the US as of 2017.

5.3.5 Ceftaroline fosamil acetate (Teflaro®; Figure 5.5) and Cefozopran

By making a structural modification of the fourth-generation cephalosporin cefozopran (Figure 5.5a) which was launched in 1995, Takeda Pharmaceutical developed the orally active prodrug ceftaroline (Figure 5.5) which has broad spectrum activity against Gram-positive pathogens, in particular methicillin resistant *Staphylococcus aureus* (MRSA) and multidrug resistant *Streptococcus pneumoniae* (MDRSP), plus common Gram-negative microbes. This compound can be considered a fifth-generation cephalosporin. The methodology for large scale production is given in the 2013 paper by Ding et al., which should be consulted for further information [13]. The compound was developed and launched by Cerexa, Inc. and Forest Laboratories, with FDA approval for skin infections and bacterial pneumonia in 2011. Though a long way from the initial cephalosporin C structure, the basic β-lactam and six-membered ring is still present, as in that first cephalosporin.

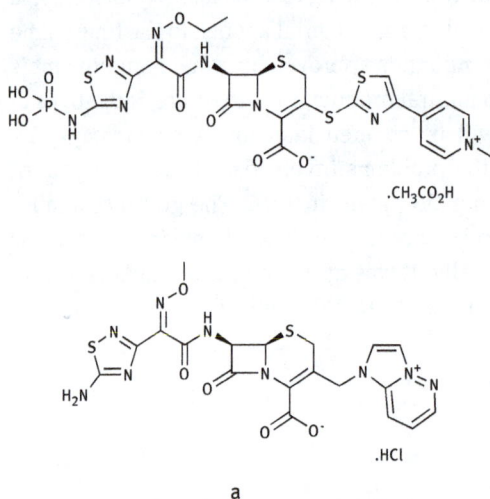

Figure 5.5: Ceftaroline fosamil acetate; Figure 5.5a: Cefozopran.

5.3.6 Zucapsaicin (Zuacta®; Figure 5.6)

It was tempting to assign this molecule to the "N" category as it is the *cis*-isomer of the natural product capsaicin and its activity like that of the parent compound, is

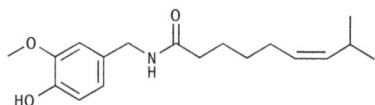

Figure 5.6: Zucapsacin.

mediated through the transient receptor potential vanilloid type 1 (TRPV1) channel, a ligand-gated ion channel expressed in the spinal cord, brain, and localized on neurons in sensory projections to the skin, muscles, joints, and gut. Synthesis is quite straightforward using a four step process starting with 6-bromohexanoic acid and ending up with recovery via crystallization giving better than 50 % yield [13].

5.3.7 Brentuximab vedotin (Adcetris®; Figure 5.7)

This antibody-drug-conjugate (ADC) was the second agent approved by the US FDA where a natural product-based warhead was attached to a specific monoclonal antibody. The first was Mylotarg® which used a modified version of the microbial product calicheamicin but was later withdrawn in the USA, and then was recently (2017) reintroduced at almost the same time as Besponsa® which used the same warhead but a different monoclonal. The ND definition was used for these agents in addition to brentuximab vedotin as the "operative" end of the complex was based on a natural product derivative. The story leading to this latter agent, and further use of structures based on dolastatin 10 as warheads, were covered up through early 2016 in a review by Newman and Cragg that should be consulted for further information [14]. Suffice to say that there are a significant number of agents using variants of the base dolastatin 10 structure in Phase I to Phase III clinical trials at the time of writing and a further discussion on ADCs is given later in the chapter.

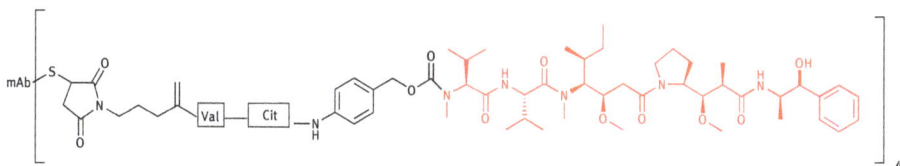

Figure 5.7: Brentuximab vedotin.

5.3.8 Ingenol mebutate (Picato®; Figure 5.8)

This is one of the very few drugs nowadays that have come directly from "folk medicine" as it was first described in the 1970s in Australia and then later was developed by the Australian biotech Peplin. The initial report was by Weedon and

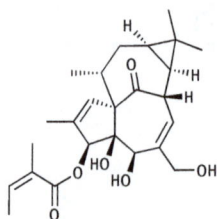

Figure 5.8: Ingenol mebutate.

Chick [15] on the use by native Australians and others who had learned from them. A major problem was obtaining enough of the material from plant sources, but it was initially solved by use of another closely related plant that produced the diterpene ingenol as part of a generalized cultivation to provide lamp oil [16]. Work by chemists at Leo Pharmaceuticals (who took over Peplin) led to a three step process to provide the correct ester [17]. Subsequent work by the Baran group led to a successful synthesis from (+)-Carene, a constituent of the "lamp-oil" referred to earlier [18] and the biosynthetic pathway to this terpene was recently described [19]. Further work by the Baran group has led to methods that demonstrate ways to both synthesize the desired product but also how to extend these methods to investigate other related compounds for their potential [20, 21]. A more comprehensive report was given by Newman in 2016 which should be consulted for the history and subsequent development of this agent [22].

5.3.9 Omacetaxine mepesuccinate (homoharingtonine) (Ceflatonin®; Figure 5.9)

Another "outlier" would be homoharringtonine (Figure 5.9), which though first isolated in the USA was used in clinical trials and possibly as a formal treatment in the PRC, before its approval in the USA in 2012. The compound was first isolated by researchers at the Northern Regional Research Laboratory (US Department of Agriculture) and reported in 1970 [23], with a review of the alkaloids from the producing genus being published in 1976 by the discovery chemistry group [24]. In

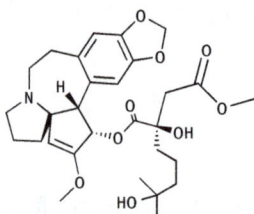

Figure 5.9: Omacetaxine mepsuccinate.

the US the molecule lingered with little work being done with it for a significant number of years. However, in the People's Republic of China (PRC), the compound was used quite extensively in the 1970s to the early 1980s in clinical settings [25]. In addition, in the middle 1980s, NCI scientists also provided information in the Journal of Clinical Oncology [26]. Following many trials and use in the PRC, the FDA approved the compound as Ceflatonin® in 2012. An excellent discussion of the history and problems with this compound was published by Kantarjian et al. in 2013, which should be read to see the manifold problems that had to be overcome before approval in the USA [27]. A final "twist" to this very long story is that in 2012, Chinese investigators published a report demonstrating that the endophytic fungus *Alternaria tenuissima* isolated from the plant source *Cephalotaxus mannii* Hook. F., could produce this compound under a variety of fermentation conditions [28].

5.3.10 Cholic acid (Orphacol®; Figure 5.10)

Cholic acid is the product of the natural degradation of cholesterol via the P450 enzyme cholesterol 7α-hydroxylase (CYP7A1) which produces chenodeoxycholic acid. Then another P450 enzyme in the pathway, sterol 12α-hydroxylase (CYP8B1) converts the intermediate product to cholic acid. This compound is a natural ligand for the nuclear receptor "farnesoid X" or FXR an essential part of the subsequent transcription cascade via the retinoic X receptor (RXR) [29]. The compound was approved in the EU in 2013 and launched in France in 2014 to treat pediatric patients with inborn errors of bile acid synthesis. In addition, a recent paper in PLoS One demonstrated evidence that this agent might also protect against biofilm formation by Gram-negative pathogens in the host [30]. This drug is defined as N as it is a product of regular metabolism in eukaryotes.

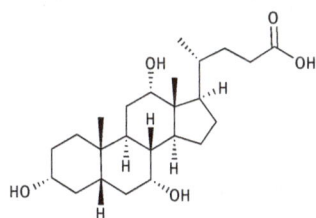

Figure 5.10: Cholic acid.

5.3.11 Trastuzumab emtansine (Kadcyla®; Figure 5.11) and Linkers

This ADC is the third one to be approved by the FDA since the 2000 approval of Mylotarg®. It has as its "warhead" the antitubulin compound maytansine, attached via

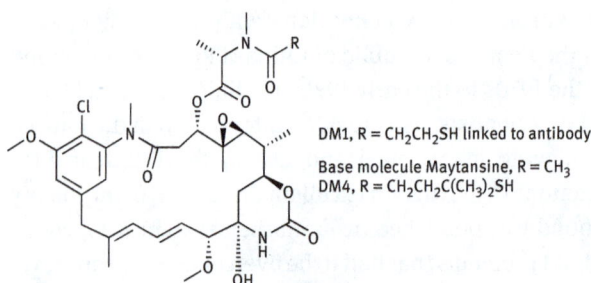

DM1, R = CH$_2$CH$_2$SH linked to antibody
Base molecule Maytansine, R = CH$_3$
DM4, R = CH$_2$CH$_2$C(CH$_3$)$_2$SH

Figure 5.11: Trastuzumab emtansine.

a relatively simple thio-based linker. This particular ADC has two components of distinct interest. The first is that the antibody is a drug in its own right (Herceptin®) also known as trastuzumab, which is specifically directed towards the *her2neu* locus, and the warhead was a derivative of the "nominal plant-sourced" compound maytansine. One later advantage of the use of the particular antibody was that even in *her2neu* refractory patients, the antibody would still bind and allow the entrance of the warhead into the tumour. The second is that it has now been proven that maytansine is not a plant product, but is a product of endophytic microbes in the rhizosphere of the plant (the area around the roots) and that maytansine is then transported into the plant for use as a deterrent factor, probably against attacking fungi [31].

Although it might be thought that delivery of such payloads would avoid toxicity due to the targeting aspect of the ACD, a recent meta-analysis of available toxicity data show that off-target toxicity may occur, probably due to hydrolysis of the linker en route to the desired targeted organelle [32]. Hopefully advances in linkers and warheads will reduce such factors in the future, as currently there are over 130 ACDs in various stages of clinical studies (preclinical to phase III trials [data from the Integrity® database in mid-March, [32] including "warheads" from natural products obtained from terrestrial and marine eukaryote and prokaryote sources.

5.3.12 Dalbavancin (Dalvance®; Figure 5.12)

In 1995, Malabarba et al. reported the isolation and structures of a series of glycopeptide antibiotics originally isolated as part of the A40926 series of compounds that closely resembled the teicoplanin complex [33]. The length of time from the initial work on this series of compounds that finally led to the approval of dalvabancin in 2014, is an example of the problems due to the takeover climate of the late 1900s and early 2000s amongst pharmaceutical houses. It was originally developed at the Italian Lepitit Research Centre then owned by Marion Merrell Dow (which was a US-based conglomerate from the 1980s). The Lepitit Centre was purchased by Hoechst in 1995 and spun off as Research Italia SPA 2 years later. In 2003 Versicor acquired Research

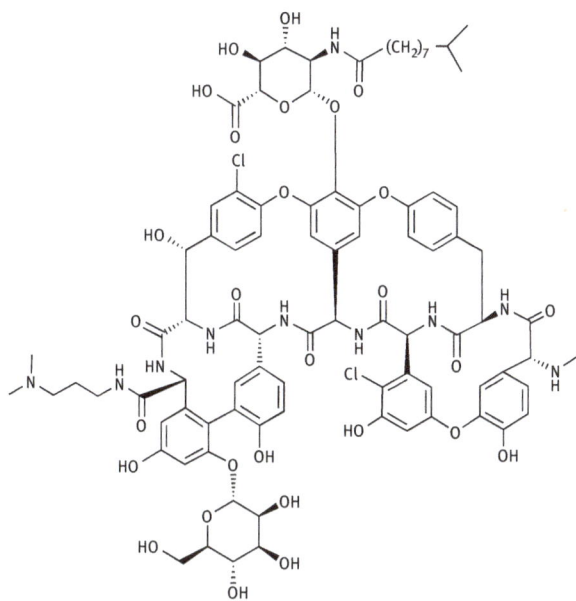

Figure 5.12: Dalbavancin.

Italia and renamed the combined company as Vicuron Pharmaceuticals. In turn, Vicuron was purchased by Pfizer in 2005. Durata Pharmaceuticals then acquired the compound after three Phase III trials, as the FDA required yet more testing at the Phase III level. This was to establish a non-inferiority endpoint by comparing two doses of IV dalbavancin given 7 days apart, against two daily doses of vancomycin for 14 days. Although only the *vanA* phenotype showed resistance, the data were adequate enough to permit approval in 2014 as treatment of methicillin resistant *Staphylococcus aureus* and methicillin-resistant *Staphylococcus epidermidis* infections. Later, approval was given for a single dose treatment against "acute bacterial skin and skin structure infections" (ABSSSI) in adults [34].

5.3.13 Cangrelor tetrasodium (Kengreal®; Figure 5.13)

This agent is based on the structure of ATP and was designed to bind to a platelet specific ADP-receptor known as the P2Y12 subtype. That particular receptor will bind ATP as a competitive antagonist of the target, but ATP is rapidly degraded to ADP, thus removing any antagonistic activity. Cangrelor was deliberately designed to mimic ATP by replacing the anhydride groups by methylenes, with a halogen added for longer half-life and with addition of sulfide linked chains, one of which had a trifluoromethyl substituent which increased the potency by six-fold, and not to

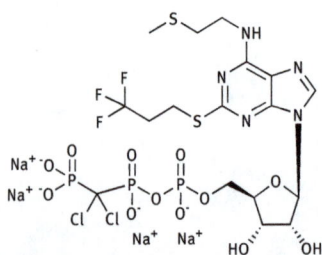

Figure 5.13: Cangrelor tetrasodium.

be degradable. The compound was approved for coronary artery disease and launched in the EU in 2015. Due to its chemical provenance, there is no doubt in listing this agent as an ND.

5.3.14 Dronabinol hemisuccinate (Syndros®; Figure 5.14)

This compound was launched in 2017, though approved in the middle of 2016 by the US FDA for the treatment of anorexia associated with weight loss in patients with AIDS. Also for the prevention of nausea and vomiting associated with cancer chemotherapy, in patients who failed to respond adequately to conventional antiemetic treatments. This is a prodrug of dronabinol (synthetic Δ-tetrahydro-cannabinol or THC) which was approved for use in the USA in 1986 and is an ND.

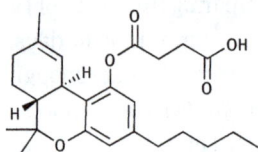

Figure 5.14: Dronabinol hemisuccinate.

5.3.15 Abaloparatide (Tymlos®; Figure 5.15)

Abaloparatide is a 34 amino acid analog of native human parathyroid hormone and is a parathyroid hormone receptor 1 (PTHR1) agonist. It was launched in 2017 in the U.S. for the subcutaneous treatment of postmenopausal women with osteoporosis at high risk for fracture. Nominally it is coded as [Glu(22,25),Leu(23,28,31),Aib(29)Lys(26,30)] hPTHrP(1–34)NH2 where the "Aib" at position 29 is "2-aminoisobutyryl" and the C-terminal alanine is amidated. Since it is directly derived from the structure of the human hormone, it is classified as an ND.

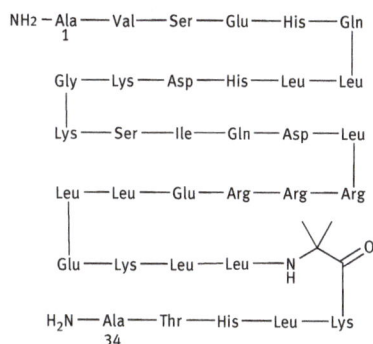

Figure 5.15: Abaloparatide.

5.3.16 Forodesine hydrocholoride (Mundesine®; Figure 5.16)

This compound is a mimic of the transition-state of the target enzyme purine nucleo-side phosphorylase (PNP) and as can be seen from its structure, it is based on a purine structure. It was known during its discovery and early development at the Albert Einstein College of Medicine in New York City as Immucillin-H. The origin of the molecule resulted from the recognition that an inherited deficiency in the enzyme PNP markedly depleted T-cells and suppressed T-cell immunity. Therefore, there was speculation that specific inhibitors of PNP might have potential as a treatment for T-cell malignancies. Forodesine treatment increases plasma 2-deoxyguanosine (dGuo) and intracellular dGTP levels through its action as a PNP inhibitor, which then causes T-cell apoptosis, which means that it is an effective treatment for peripheral T-Cell lymphoma. Inspection of the structure and its target enzyme defines this agent as an "ND".

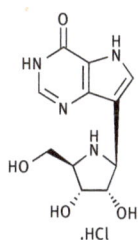

Figure 5.16: Forodesine hydrochloride.

5.3.17 Midostaurin (Rydapt®; Figure 5.17)

Although there have been many analogues of the base staurosporin structure synthesized since its first identification in 1977 by the Omura group [35], it was not until 2017

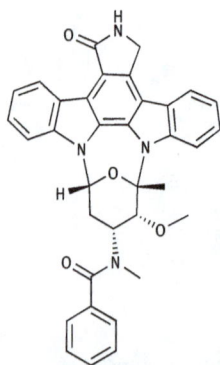

Figure 5.17: Midostaurin.

that any of the molecules derived from this agent were approved as human use drugs. Thus midostaurin, was reported by Amon et al. in 1993 to have *in vitro* selectivity for protein kinase C [36], under its original Ciba Geigy code number of CGP 41,251, and the formal synthesis of this compound and others related to it, were published the following year [37]. It was not for 24 more years before midostaurin was approved as a treatment for acute myeloid leukaemia by the FDA in 2017. Recent data covering the clinical trials and their results for this agent have been published by Stansfield and Pollyea and this review gives an excellent overview of the trials and tribulations leading to the recent approval [38].

Currently, as of March 2018, only one staurosporin derivative, a pegylated version of a compound whose biological activity was reported in 1988, K-252a [39], is currently in Phase II clinical trials under the name pegcantratinib, though over 400 different derivatives are listed in the Integrity® database as having been in some form of biological assay from preclinical to clinical. This compound is only one synthetic step removed from staurosporin so is classified as an "ND".

5.4 Some selected agents in clinical use and/or trials that are np-based

5.4.1 Antibody-drug conjugates

Currently there are over 120 agents with a natural product-based warhead that are in some form of preclinical or clinical trial, with almost all directed against some form of cancer. The "warheads" are from a variety of sources with the early agents being nominally from plant or marine sources. I use the term "nominally" since it is now well established that the dolastatins, which are the base structure(s) for the Seattle Genetics MMAE (Figure 5.18) and MMAF (Figure 5.19), are based on dolastatin 10 (Figure 5.20),

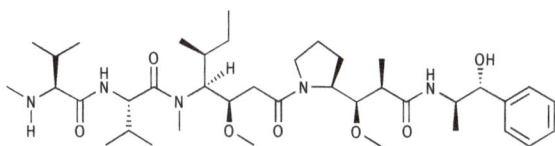

Figure 5.18: Monomethylauristatin E.

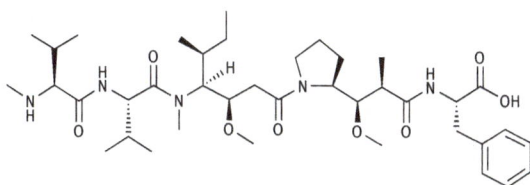

Figure 5.19: Monomethylauristatin F.

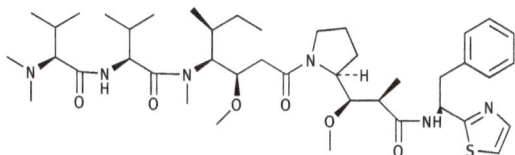

Figure 5.20: Dolastatin 10.

which though originally isolated from the marine nudibranch *Dolabella auricularia*, has been proven to be produced by a cyanophyte of the genus *Symploca* that the sea hare consumes as part of its diet. References and discussion of most of the current marine-based warheads were given in the review by Newman and Cragg in 2017 [40], with a coverage of these and other sources discussed in the 2017 review by Beck et al. in Nature Reviews of Drug Discovery [41]. Similarly, as referred to in the discussion of Kadcyla® (Figure 5.11) above, the actual source of maytansine is microbial.

Of the six that are currently in Phase III trials, against cancer, one (Sacituzumab govitecan) uses a derivative of the "plant product" camptothecin which is another metabolite that might have an endophytic microbe in its background. Two (Polatuzumab vedotin and Depatuxizumab mafodotin) use MMAE and MMAF, respectively. One (Mirvetuximab soravtansine) uses maytansine with a different thio-linker to the one used for Kadcyla®. Of the remaining two, one (Trazuzumab duocarmazine) uses a duocarmycin-based warhead linked to the *her2neu* targeted Mab used in Kadcyla®, and the remaining Phase III ACD (Rovalpituzumab tesirine) uses a variation on the microbial product anthramycin, known as pyrrolobenzodiazepine (PBD). An excellent paper in 2017 by Mantaj et al. describes these agents and their derivation from anthramycin which should be consulted for fuller details [42].

Moving to the 18 that are in Phase II trials, only two do not use the agents noted above or variations thereon, particularly those based originally on dolastatin 10. The two outliers have "taken lessons" from some of the early attempts to discover warhead molecules with one, Radretumab contains ^{131}I as the "warhead" and is directed against a fibronectin target, the other outlier, is known as DT-2219 and contains variations designed around diptheria toxins.

Of those in Phase I and Phase I/II trials, most have warhead entities that are based upon the natural products referred to above, but there are also some interesting variations. These include doxorubicin, a phthalocyanine-related photo-dyestuff (causally related to the base porphyrin structure), and the use of the totally synthetic molecule eribulin (Figure 5.3) that is listed as an "ND" due to its base structure coming from the marine sponge metabolite halichondrin B. Another is based on tubulysins, (with the structure of tubulysin A shown; Figure 5.21) an extremely potent microbial product, with another using the chloro-derivative (PM050489; Figure 5.22) of a PharmaMar compound that is in Phase II trials as a standard antitumour agent. Both of the PharmaMar compounds were originally isolated from a marine sponge but are now synthesized due to lack of the natural source [43–46]. Unusually there is an antibiotic related molecule based on rifamycin (listed as DSTA-4637S) that is designed to be an antibiotic drug delivery system, with another based on a steroid-linked Mab for treatment of rheumatoid arthritis and other diseases in Phase I under AbbVie (ABBV-3373).

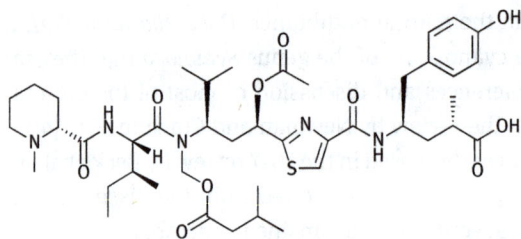

Figure 5.21: Tubulysin A.

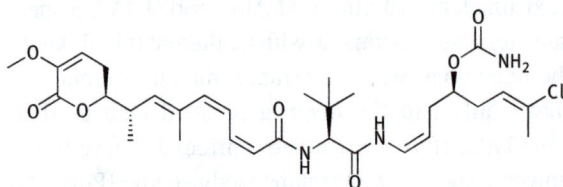

Figure 5.22: PM-050489.

5.4.2 Nucleoside-based agents

5.4.2.1 Anti-hepatitis c agents

One agent that is of significant importance is the masked nucleotide known as Sofosbuvir (Sovaldi®; Figure 5.23) which was approved by the US FDA in 2013. This agent could effectively "cure" hepatitis C infections in one dose series over a few weeks. Though there was significant "push-back" due to the cost of the treatment, since it was an effective cure, the compound has had a significant impact on this disease.

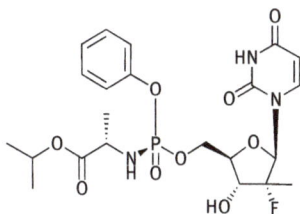

Figure 5.23: Sofosbuvir.

Two other nucleosidic compounds that can also be thought of as "masked nucleotides" are currently in Phase II clinical trials against HCV infection. These are Uprifosbuvir (Figure 5.24) from Merck and Adafosbuvir (Figure 5.25) from Alios BioPharma. Both are inhibitors of RNA-Directed RNA Polymerase (NS5B) (HCV) and are based upon an uridine scaffold.

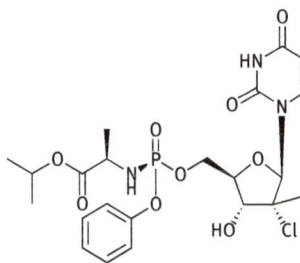

Figure 5.24: Uprifosbuvir.

5.4.2.2 Adenosine receptor agonist

Piclidenoson (Figure 5.26), a potent and selective compound based upon a substituted adenosine nucleus, is in Phase III trials under Can-Fite Biopharma for the oral treatment of rheumatoid arthritis and it is also in Phase II trials as a potential treatment for glaucoma. Interestingly, this compound was directly licenced from the NIH by Can-Fite.

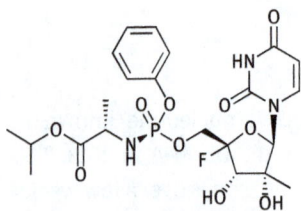

Figure 5.25: Adafosbuvir.

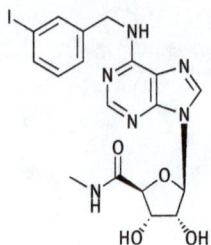

Figure 5.26: Piclidenoson.

5.4.2.3 Antitumour active compounds

Although there are a number of agents based upon nucleosides at varying phases of clinical trials, only a relative few will be discussed.

Acelarin (Figure 5.27) is another of the masked nucleotide type of molecule, but this time rather than being develop as an antiviral, it is now reported to be in Phase II trials for ovarian cancer and in Phase III trials for pancreatic cancer under the auspices of NuCana BioMed. It is a pyrimidine-based molecule with a benzoylated alanine linked to a monophosphate group.

Sapacitabine (Figure 5.28) is a relatively simple substituted cytosine arabinoside that is currently in Phase III clinical trials under the auspices of Cyclacel for the oral

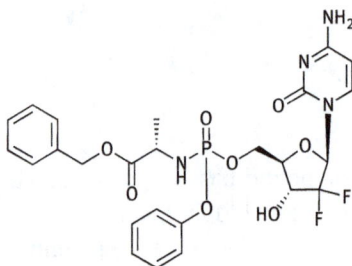

Figure 5.27: Acelarin.

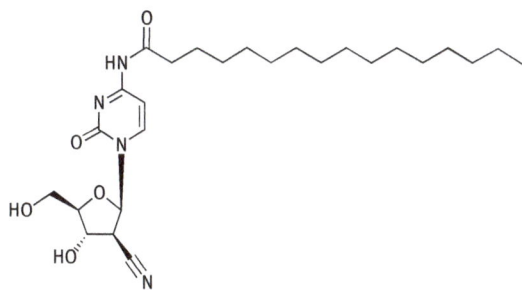

Figure 5.28: Sapacitabine.

treatment of newly diagnosed acute myeloid leukaemia. In addition, there are Phase II trials against a number of other leukaemias also underway.

Guadecitabine sodium (Figure 5.29) is an unusual combination of an aza-substituted cytidine and guanosine, that is scheduled to have an NDA submitted by Astex in 2018, potentially as a treatment for acute myeloid leukaemia. Its mechanism of action is as a DNA methyltransferase inhibitor (DNMT).

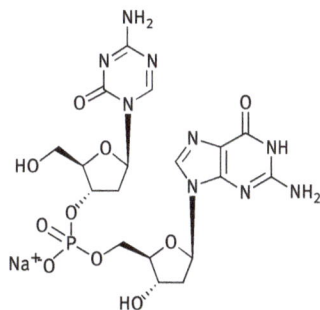

Figure 5.29: Guadecitabine.

T-dCyd (Figure 5.30) is an unusual nucleoside derivative with a thio-substituted sugar that is also a DNMT inhibitor that is currently in Phase I trials against a variety of solid tumours under the auspices of the NCI. This compound was first synthesized by the Southern Research Institute in the USA.

5.4.2.4 Shingles and post-herpetic pain

A compound that could be considered as having evolved from acyclovir, famiciclovir and valcyclovir, is the orally active bicyclic compound FV-100 (Valnivudine®; Figure 5.31). This compound is currently in Phase III studies under ContrVir Pharmaceuticals for the treatment of *Varicella zoster* infections and related pain management.

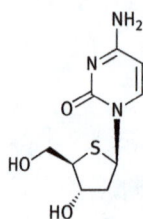

Figure 5.30: T-dCyd.

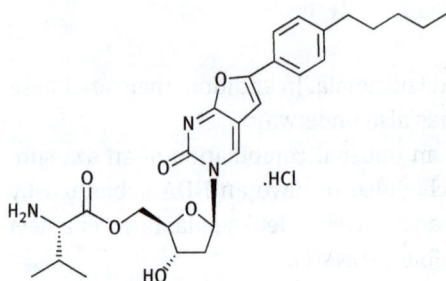

Figure 5.31: Valnivudine.

5.4.2.5 Respiratory syncytial virus

Lumicitabine (Figure 5.32) is a nucleoside analogue that is currently in Phase II clinical trials against this virus and is designed to inhibit RSV replication by inhibiting the RNA-directed RNA polymerase of the virus itself.

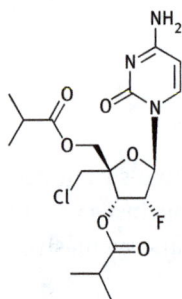

Figure 5.32: Lumicitabine.

5.4.2.6 Cytomegalovirus

Finally, in what might be considered "a return to the past", meaning before the realization that nucleosides with arabinose instead of ribose had potent biological activities,as at that time chemists used almost any nitrogen base as long as it had a ribose-based molecule as the sugar, to synthesize what could be called "pseudo-nucleosides".

Currently a dichlorobenzimidazole with a ribose substituent Maribavir (Camvia®; Figure 5.33), a serine/threonine protein kinase UL97 inhibitor, is in Phase III clinical trials for treatment of transplant patients with CMV infections under the auspices of Shire, now the third company to undertake development of this compound.

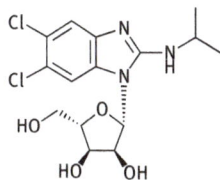

Figure 5.33: Maribavir.

5.4.2.7 Comment on nucleoside-based compounds recently approved and/or under clinical trials

Even though nucleoside-based drugs have been around from the 1950s as antitumour and antiviral drugs, even today, close to 65 years after the introduction of such compounds, the basic scaffolds are still in contention and have demonstrated that these basic building blocks of life still have a lot to teach scientists involved in drug discovery and development.

5.4.3 Beta-lactam antibiotics

Though penicillin went into use during World War II and cephalosporin C was identified in the early 1950s, these basic antibiotics are still in use with novel variations still in clinical trials in addition to ceftaroline (Figure 5.5) which was discussed earlier in the chapter.

Cefilavancin trihydrochloride (Figure 5.34) is a complex compound consisting of a vancomycin-based compound linked to a cephalosporin that is now in Phase III trials in Russia for complex skin and skin structure infections. It was originally developed by Theravance but was then licenced to R-Pharm for future development.

BAL-30,072 (Figure 5.35) is an interesting monobactam that has entered Phase I clinical trials under the Novartis spin-off company Basilea, against community acquired respiratory tract infections (CARTI) and Gram-negative multidrug resistant infections.

Cefiderocol (Figure 5.36) is in Phase III clinical trials at Shionogi for the intravenous treatment of serious infections in adult patients caused by carbapenem-resistant

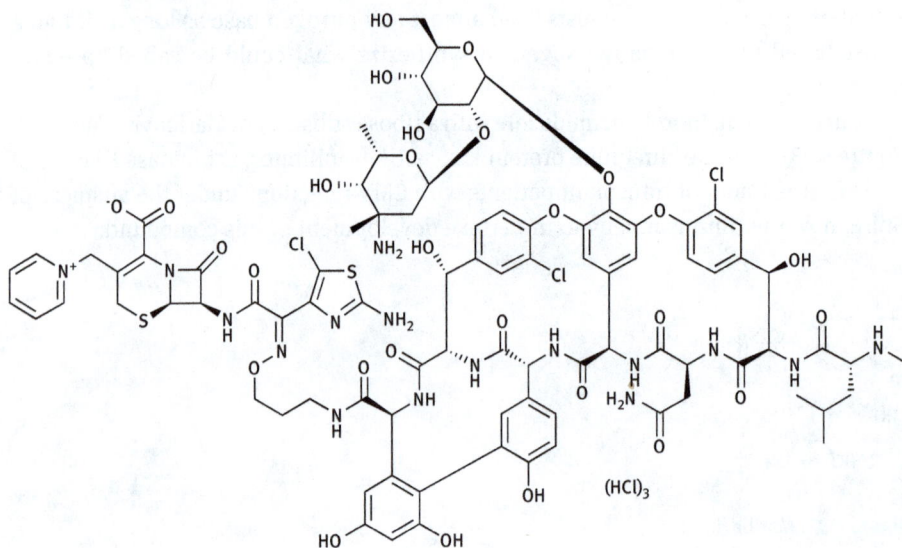

Figure 5.34: Cefilavancin trihydrochloride.

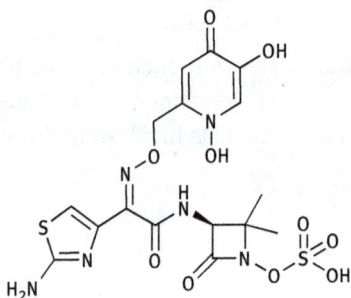

Figure 5.35: BAL-30072.

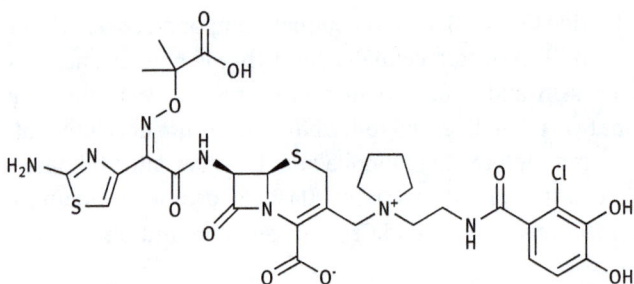

Figure 5.36: Cefiderocol.

Gram-negative pathogens. Originally developed by Glaxo SmithKline it was licenced to Shionogi in 2010 as part of a joint development project.

LYS-228 (Figure 5.37) like BAL-30,072 above, is a monobactam antibiotic, current in early clinical development (Phase I clinical trials) at Novartis for the treatment of Enterobacteriaceae infections.

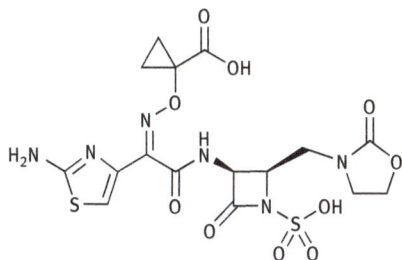

Figure 5.37: LYS-228.

Cefepime (Axepime®; Figure 5.38)/AAI-101 (Figure 5.39), a combination of a beta-lactam antibiotic with a novel extended spectrum beta-lactamase inhibitor, was placed into Phase II clinical trials by Allecra Therapeutics, and is targeted for the treatment of Gram-negative multi drug-resistant infections. Cefepime was approved as a single agent in 1995, and the novel inhibitor may well extend the range of the cephalosporin as has occurred with combinations such as amoxicillin and clavulanate.

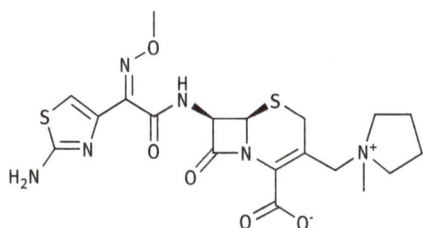

Figure 5.38: Cefepime.

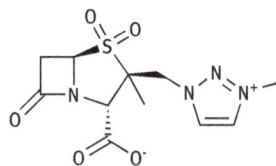

Figure 5.39: AAI-101.

5.4.4 One S*/NM. pitavastatin magnesium (zipitamag®; Figure 5.40)

This compound is somewhat unusual as it is simply the magnesium salt of a compound approved as the calcium salt by Japan in 2003 (Livalo®) and by the US FDA in August of 2009. The compound effectively has the "warhead" present in all of the "statins," that was originally identified in lovastatin (Mevacor®) which was first introduced in 1987, coupled to a different series of cyclic structures. As with all "statins" from Lipitor® onwards, these have been classified as an S*/NM using the Newman and Cragg conventions.

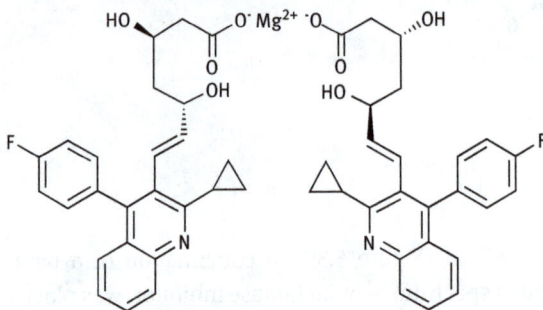

Figure 5.40: Pitavastatin Magnesium.

The US FDA approved the magnesium salt in July of 2017 and listed it as a "new active ingredient". There is little data on Zipitamag®, simply that it was first developed by the Indian company Dr Reddy's, though there is substantial data on the original pitavastatin synthesis and biological data under the code number NK-104, as shown in the paper published without a listed author in Drugs of the Future in 1998 [47].

Whether this recent compound should have been considered as the equivalent to a "biosimilar" by the US FDA, is not mentioned, but this appears to be the first time that this has occurred in the years from 1981, thus it will be listed as an approved drug in 2017.

5.5 In conclusion

Hopefully the reader can see from the examples given above, that even towards the end of the second decade of the twenty-first Century, natural products and their basic structures, are still demonstrating their worth.

When it comes to deciding which compounds should be utilized as "warheads" for site-directed monoclonal antibodies, except for a few that are designed as imaging agents with radionuclides as the payload, a substantial number of those that are in clinical trials are utilizing some version of marine-derived compounds. The majority of the "base compounds such as the dolastatins", are now known to be produced by free-living microbes, irrespective of what organism they were first isolated from (nudibranch, sponges etc.). Other warheads come from terrestrial microbes or from nominal plant products, and as mentioned earlier maytansine is definitively from endophytic microbes, and other plant derived compounds such as camptothecin have been reported to have "a microbe in their background."

Although in the cancer area, there are significant numbers of protein tyrosine kinase inhibitors that are totally synthetic, they are in fact in a significant number of cases, structural isosteres of ATP. The papers by Fabbro, who can be considered the "father of Gleevec®" the first of the PTK inhibitors to be approved, should be consulted by the interested reader as in these, he demonstrates how closely most of the approved agents up through 2015 fit this description [48–50].

It should also be recognized that *de novo* combinatorial chemistry as a source of drug leads has not lived up to the hype and massive amounts of money expended on it in the 1990s – early 2000s, but as a developmental tool, once you have the active structure, combinatorial systems are very successful in the development of a lead. Only three compounds were ever found by *de novo* combinatorial chemistry up through the end of 2014, and one of these was from a fragment system using NMR [1].

Thus, natural products be they the drug compound itself, a lead to what could be considered a semi-synthetic such as the pitavastatin molecule referred to above, or even wholly synthetic compounds (e. g. eribulin), are still a viable series of structural leads from which to gain information for future generations of drug candidates, particularly in the areas of infectious diseases (bacterial, fungal and viral) and cancer in all of its manifestations in man.

References

[1]   Newman DJ, Cragg GM. Natural products as sources of new drugs 1981 to 2014. J Nat Prod. 2016;79:629–61.
[2]   Nakajima H, Kim YB, Terano H, Yoshida M, Horinouchi S. FR901228, a potent antitumor antibiotic, is a novel histone deacetylase inhibitor. Exp Cell Res. 1998;241:126–33.
[3]   Brinkmann V, Billich A, Baumruker T, Heining P, Schmouder R, Francis G, et al. Fingolimod (FTY720): discovery and development of an oral drug to treat multiple sclerosis. Nat Revs Drug Discov. 2010;9:883–97.
[4]   Strader CR, Pearce CJ, Oberlies NH. Fingolimod (FTY720): A recently approved multiple sclerosis drug based on a fungal secondary metabolite. J Nat Prod. 2011;74:900–07.
[5]   Dyckman AJ. Modulators of sphingosine-1-phosphate pathway biology: recent advances of sphingosine-1-phosphate receptor 1 (S1P1) agonists and future perspectives. J Med Chem. 2017;60:5267–89.
[6]   Hou X, Zhang H, Chen B-C, Guo Z, Singh A, Goswami A, et al. Regioselective epoxide ring opening for the stereospecific scale-up synthesis of BMS-960, a potent and selective isoxazole-containing S1P1 receptor agonist. Org Process Res Dev. 2017;21:200–207.

[7]   Pan S, Gray NS, Gao W, Mi Y, Fan Y, Wang X, et al. Discovery of BAF312 (siponimod), a potent and selective S1P receptor modulator. ACS Med Chem Lett. 2013;4:333–337.

[8]   Yu MJ, Kishi Y, Littlefield BA. Discovery of E7389, a fully synthetic macrocyclic ketone analog of Halichondrin B. In: Cragg GM, Kingston DGI, Newman DJ, editor(s). Anticancer agents from natural products, 1st ed. Boca Raton: Taylor and Francis, 2005:267–280.

[9]   Wang Y, Serradell N, Bolos J, Rosa E. Eribulin mesilate. Drugs Fut. 2007;32:681–98.

[10] Jackson KL, Henderson JA, Phillips AJ. The halichondrins and E7389. Chem Rev. 2009;109:3044–79.

[11] Twelves C, Cortes J, Vahdat LT, Wanders J, Akerele C, Kaufman PA. Phase III trials of Eribulin Mesylate (E7389) in extensively pretreated patients with locally recurrent or metastatic breast cancer. Clin Breast Can. 2010;10:160–63.

[12] Mcalpine JB. The ups and downs of drug discovery: the early history of Fidaxomicin. J Antibiot (Tokyo). 2017;70:492–94.

[13] Ding HX, Liu KK-C, Sakya SM, Flick AC, O'donnell CJ. Synthetic approaches to the 2011 new drugs. Bioorg Med Chem. 2013;21:2795–825.

[14] Newman DJ, Cragg GM. Drug candidates from marine sources: an assessment of the current "state of play. Planta Medica. 2016;82:775–89.

[15] Weedon D, Chick J. Home treatment of basal cell carcinoma. Med J Aust. 1976;1:928.

[16] Appendino G, Tron GC, Cravotto G, Palmisano G, Jakupovic J. An expeditious procedure for the isolation of ingenol from the seeds of *Euphorbia lathyris*. J Nat Prod. 1999;62:76–79.

[17] Liang X, Grue-Sorensen G, Petersen AK, Hogberg T. Semisynthesis of ingenol 3-angelate (PEP005): efficient stereoconservative angeloylation of alcohols. Synlett. 2012;23: 2647–52.

[18] Jørgensen L, Mckerrall SJ, Kuttruff CA, Ungeheuer F, Felding J, Baran PS. 14-Step synthesis of (+)-Ingenol from (+)-3-Carene. Science. 2013;341:878–82.

[19] Luo D, Callari R, Hamberger B, Wubshet SG, Nielsen MT, Andersen-Ranberg J, et al. Oxidation and cyclization of casbene in the biosynthesis of Euphorbia factors from mature seeds of *Euphorbia lathyris* L. Proc Natl Acad Sci USA. 2016;113:E5082–E5089.

[20] Michaudel Q, Ishihara Y, Baran PS. Academia–Industry symbiosis in organic chemistry. Acc Chem Res. 2015;48:712–21.

[21] Mckerrall SJ, Jørgensen L, Kuttruff CA, Ungeheuer F, Baran PS. Development of a concise synthesis of (+)-Ingenol. J Am Chem Soc. 2014;136:5799–5810.

[22] Newman DJ. Developing natural product drugs: supply problems and how they have been overcome. Pharmacol Therap. 2016;162:1–9.

[23] Powell RD, Weisleder D, Smith Jr CR, Rohwedder WK. Structures of harringtonine, isoharringtonine, and homoharringtonine. Tet Letts. 1970;11:815–18.

[24] Smith Jr CR, Powell RG, Mikolajczak KL. The genus *Cephalotaxus*: source of homoharringtonine and related anticancer alkaloids. Cancer Treat Rep. 1976;60:1157–70.

[25] Zhang ZY. Clinical analysis of the therapeutic effect of semisynthetic harringtonine in treating 55 cases of nonlymphocytic leukemia. Zhonghua Nei Ke Za Zhi [Chinese Journal of Internal Medicine]. 1981;20:667–69.

[26] O'Dwyer PJ, King SA, Hoth DF, Suffness M, Leyland-Jones B. Homoharringtonine–perspectives on an active new natural product. J Clin Oncol. 1986;4:1563–68.

[27] Kantarjian HM, O'Brien S, Cortes J. Homoharringtonine/Omacetaxine Mepesuccinate: the long and winding road to food and drug administration approval. Clin Lymph Myel Leuk. 2013;13:530–33.

[28] Liu Y, Liu S-X, Li Y-C, Li C-F. Optimization of homoharringtonine fermentation conditions for *Alternaria tenuissima* CH1307, an endophytical fungus of *Cephalotaxus mannii* Hook F. J Trop Med. 2012;3:236–42.

[29] Zhang Y. Farnesoid X receptor: acting through bile acids to treat metabolic disorders. Drugs Fut. 2010;35:635–41.

[30] Sanchez LM, Cheng AT, Warner CJA, Townsley L, Peach KC, Navarro G, et al. Biofilm formation and detachment in Gram-negative pathogens Is modulated by select bile acids. PLoS ONE. 2016;11:e0149603.

[31] Kusari S, Lamshoft M, Kusari P, Gottfried S, Zuhlke S, Louven K, et al. Endophytes are hidden producers of maytansine in *Putterlickia* roots. J Nat Prod. 2014;77:2577–84.

[32] Masters JC, Nickens DJ, Xuan D, Shazer RJ, Amantea M. Clinical toxicity of antibody drug conjugates: a meta-analysis of payloads. Invest New Drugs. 2018;36:121–35.

[33] Malabarba A, Ciabatti R, Scotti R, Goldstein BP, Ferrari P, Kurz M, et al. New semisynthetic glycopeptides MDL 63,246 and MDL 63,042, and other amide derivatives of antibiotic A-40,926 active against highly glycopeptide-resistant VanA enterococci. J Antibiot (Tokyo). 1995;48:869–83.

[34] Garnock-Jones KP. Single-dose dalbavancin: A review in acute bacterial skin and skin structure infections. Drugs. 2017;77:75–83.

[35] Omura S, Iwai Y, Hiroano A, Nakagawa A, Awaya J, Tsuchiya H, et al. A new alkaloid AM-2282 of *Streptomyces* origin, taxonomy, fermetnation, isolation and preliminary characteristics. J Antibiot (Tokyo). 1977;30:275–82.

[36] Amon U, Von Stebut E, Subramanian N, Wolff HH. CGP 41251, a novel protein kinase inhibitor with *in vitro* selectivity for protein kinase C, strongly inhibits immunological activation of human skin mast cells and human basophils. Pharmacol. 1993;47:200–08.

[37] Caravatti G, Meyer T, Fredenhagen A, Trinks U, H. M, Fabbro D. Inhibitory activity and selectivity of staurosporine derivatives towards protein kinase C. Bioorg Med Chem Letts. 1994;4:399–404.

[38] Stansfield LC, Pollyea DA. Midotaurin: A new oral agent targeting FMS-like tyrosine kinase 3-mutant acute myeloid leukemia. Pharmacother. 2017;31:1586–99.

[39] Hashimoto S. K-252a, a potent protein kinase inhibitor, blocks nerve growth factor-induced neurite outgrowth and changes in the phosphorylation of proteins in PC12h cells. J Cell Biol. 1988;107:1531–39.

[40] Newman DJ, Cragg GM. Current status of marine-derived compounds as warheads in anti-tumor drug candidates. Mar Drugs. 2017;15:99.

[41] Beck A, Goetsch L, Dumontet C, Corvaïa N. Strategies and challenges for the next generation of antibody–drug conjugates. Nat Revs Drug Discov. 2017;16:315–37.

[42] Mantaj J, Jackson PJM, Rahman KM, Thurston DE. From anthramycin to pyrrolobenzodiazepine (PBD) containing antibody–drug conjugates (ADCs). Angew Chem Int Ed. 2017;56:462–88.

[43] Prota AE, Bargsten K, Fernando Diaz J, Marsh M, Cuevas C, Liniger M, et al. A new tubulin-binding site and pharmacophore for microtubule-destabilizing anticancer drugs. Proc Natl Acad Sci USA. 2014;111:13817–21.

[44] Martín MJ, Coello L, Fernández R, Reyes F, Rodríguez A, Murcia C, et al. Isolation and first total synthesis of PM050489 and PM060184, two new marine anticancer compounds. J Am Chem Soc. 2013;135:10164–71.

[45] Pera B, Barasoain I, Pantazopoulou A, Canales A, Matesanz R, Rodriguez-Salarichs J, et al. New interfacial microtubule inhibitors of marine origin, PM050489/PM060184, with potent anti-tumor activity and a distinct mechanism. ACS Chem Biol. 2013;8:2084–94.

[46] Aviles PM, Guillen MJ, Dominguez JM, Muñoz-Alonso MJ, Garcia-Fernandez LF, Garranzo M, et al. MI130004, an antibody-drug conjugate including a novel payload of marine origin: evidences of *in vivo* activity. Europ J Cancer. 2014;50:502.

[47] ANON. NK-104. Drugs Fut. 1998;23:847–59.

[48] Fabbro D, Cowan-Jacob SW, Moebitz H. Ten things you should know about protein kinases: IUPHAR Review 14. Brit J Pharmacol. 2015;172:2675–700.
[49] Rask-Andersen M, Zhang J, Fabbro D, Schioth HB. Advances in kinase targeting: current clinical use and clinical trials. Trends Pharmacol Sci. 2014;35:604–20.
[50] Fabbro D, Ruetz S, Buchdunger E, Cowan-Jacob SW, Fendrich G, Liebetanz J, et al. Protein kinases as targets for anticancer agents: from inhibitors to useful drugs. Pharmacol Therap. 2002;93:79–98.

Part II: **Chemoinformatics Tools and Methods for Natural Product Structure Elucidation**

Aurélien F. A. Moumbock, Fidele Ntie-Kang, Sergi H. Akone,
Jianyu Li, Mingjie Gao, Kiran K. Telukunta and Stefan Günther

6 An overview of tools, software, and methods for natural product fragment and mass spectral analysis

Abstract: One major challenge in natural product (NP) discovery is the determination of the chemical structure of unknown metabolites using automated software tools from either GC–mass spectrometry (MS) or liquid chromatography–MS/MS data only. This chapter reviews the existing spectral libraries and predictive computational tools used in MS-based untargeted metabolomics, which is currently a hot topic in NP structure elucidation. We begin by focusing on spectral databases and the general workflow of MS annotation. We then describe software and tools used in MS, particularly those used to predict fragmentation patterns, mass spectral classifiers, and tools for fragmentation trees analysis. We then round up the chapter by looking at more advanced approaches implemented in tools for competitive fragmentation modeling and quantum chemical approaches.

Keywords: cheminformatics, fragment analysis, drug discovery, natural products, software

6.1 Introduction

Metabolomics deals with the comprehensive profiling of metabolites within an animal, plant, and microbial metabolome. It can lead to the discovery of novel specialized metabolites or natural products (NPs), which are of particular interest in small molecule drug discovery [1, 2]. Metabolomics strategies can be classified broadly as targeted [1] and untargeted [3]. The former aims to quantify specific metabolites within a metabolome, with the help of internal standards. The latter entails the extensive chemical identification and characterization of metabolites within a metabolome [1–3].

Several hyphenated analytic platforms have been developed in metabolomics [4], involving the coupling of a separation technique like high-performance liquid chromatography (HPLC), gas chromatography (GS), supercritical fluid chromatography, and capillary electrophoresis and a detection technique like mass spectrometry

This article has previously been published in the journal *Physical Sciences Reviews*. Please cite as:
Moumbock, A.F.A., Ntie-Kang, F., Akone, H.S., Li, J., Gao, M., Telukunta, K.K., Günther, S. An overview of tools, software and methods for natural product fragment and mass spectral analysis. *Physical Sciences Reviews* [Online] **2019** DOI: 10.1515/psr-2018-0126.

https://doi.org/10.1515/9783110579352-007

(MS) and nuclear magnetic resonance (NMR) [5–7]. MS is advantageous over NMR spectroscopy, due to its relatively faster data acquisition and higher sensitivity [7]. The two most employed hyphenated techniques in MS-based metabolomics are GC–MS, for volatile metabolites analysis, and HPLC–MS (or simply LC–MS) involving chiefly tandem MS (MS/MS) or multiple-stage MS {MS(n)}, for non-volatile metabolites analysis [8].

Generally, the main information obtained from mass spectra includes elemental composition, molecular mass, and molecular formula [9]. The most abundant chemical elements in NPs are C, H, O, N, S, and P. However, in contrast to proteins, NPs do not have fixed building blocks and as a result, complete structure elucidation of unknown NPs with the single use of MS data is seemingly impossible [10]. Furthermore, a very large amount of data is generated from untargeted metabolomics analyses, which require expert knowledge to gain structural insights of metabolites, in a time-consuming process. The straightforward approach to identify unknown NPs is *via* an MS library search, where the MS spectra of unknown NPs are compared to those in annotated MS spectral libraries, recorded with similar instruments [10–12].

However, the number of NPs whose spectra are available in spectral libraries is relatively minute as compared to the number of NPs that populate the chemical space. As a result, computer-assisted structure elucidation (CASE) has emerged as a prominent strategy to handle this issue [13]. *De novo* CASE approach involves the automatic generation of candidate structures followed by the prediction of fragmentation spectra and the subsequent ranking of the structures based on the concurrence of computed spectra (fragmentation trees) and experimental spectra (spectral trees) [11, 14]. The first endeavor in this field was the DENDRAL project, which began in the late 1960s. It aimed to establish the automated generation of candidate structures as well as the rule-based fragmentation spectrum prediction for the various metabolite classes. Unfortunately, the program was discontinued [15–18].

Nevertheless, the DENDRAL project paved the way to the development of the currently large number of open source and commercial software tools for MS fragment analysis, which are either GUI-driven/command-line programs or web servers. In order to evaluate the currently available tools for *de novo* CASE and equally foster the development of new tools, the Critical Assessment of Small Molecule Identification contest was initiated in 2012 [19–21]. It is an open contest for the determination of the molecular formula and structure of unknown metabolites, with automated software tools from either GC–MS or LC–MS/MS data only. This chapter reviews the existing spectral libraries and predictive computational tools used in MS-based untargeted metabolomics, which is currently a hot topic in NP structure elucidation. In a series of publications, Böcker and coworkers have extensively reviewed the existing methods used in NP computational MS [9, 12, 22, 23]. Blaženović et al. also reviewed the literature and identified four main approaches used by *in silico* fragmentation tools [24], beyond the straightforward spectral matching, briefly summarized by Figure 6.1.

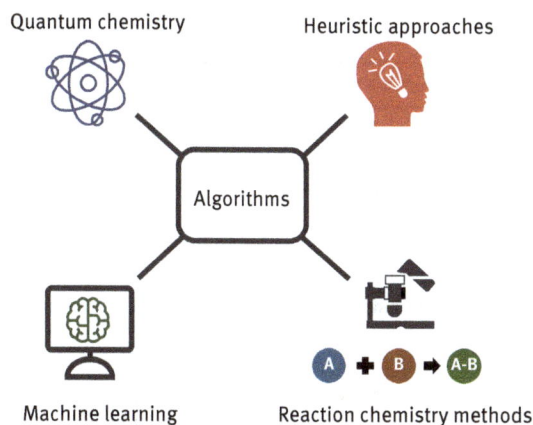

Figure 6.1: Summary of the various methods implemented in the tools described in this work [24]. Figure previously published under a Creative Commons License.

The first part will focus on spectral databases and the general workflow of MS annotation. This is followed by a description of some of the established methods and software used in MS, particularly to follow and/or predict fragmentation patterns. The last paragraphs deal with more advanced approaches, e. g. a description of tools for competitive fragmentation modeling (CFM) and quantum chemical (QC) approaches.

6.2 Mass spectral databases

6.2.1 Available databases

A summary of mass spectral databases for the identification of fragments has been provided in Table 6.1.

Table 6.1: Selected mass spectral databases for NPs dereplication.

Database	Web accessibility
European MassBank (NORMAN MassBank)	https://massbank.eu/MassBank/
MoNA	http://mona.fiehnlab.ucdavis.edu/
METLIN Metabolomics Database	http://metlin.scripps.edu/
Wiley Registry™ of Mass Spectral Data, 11th Edition	https://www.sisweb.com/software/wiley-registry.htm/
NIST Mass Spectral Library (NIST 17)	http://nistmassspeclibrary.com/
HMDB	http://www.hmdb.ca/

(continued)

Table 6.1 (continued)

Database	Web accessibility
MMCD	http://mmcd.nmrfam.wisc.edu/
GMD	http://gmd.mpimp-golm.mpg.de/
PRiMe	http://prime.psc.riken.jp/
FiehnLib Library	https://fiehnlab.ucdavis.edu/projects/fiehnlib/
MetaboLights Database	https://www.ebi.ac.uk/metabolights/
mzCloud	https://www.mzcloud.org/
AIST: Spectral Database for Organic Compounds (SDBS)	https://sdbs.db.aist.go.jp/sdbs/cgi-bin/cre_index.cgi/
LMSD	http://www.lipidmaps.org/data/structure/
LipidBlast	https://fiehnlab.ucdavis.edu/projects/LipidBlast/
GNPS Spectral libraries	https://gnps.ucsd.edu/ProteoSAFe/libraries.jsp/
ReSpect	http://spectra.psc.riken.jp/
DIMEdb	https://dimedb.ibers.aber.ac.uk/

MoNA: Massbank of North America; HMDB: Human Metabolome Database; MMCD: Madison Metabolomics Consortium Database; GMD: Golm Metabolome Database; PRiMe: Platform for RINKEN Metabolomics; LMSD: LIPID MAPS Structure Database.

It is well known that the identification of the chemical structure of a substance a compound can be done by MS based on the separation of its ions according to their mass and charge (m/z). This identification may imply the use of databases for the detection of the known compounds in substances that had never been analyzed before. In fact, classification and annotation are the main objectives for searching databases. In recent years, several freely accessible and commercially available databases have been developed for the fast and reliable identification of metabolites (Table 6.1).

Those databases found applications ranging from plant-based and microbial secondary metabolites to pharmaco-metabolomics [25]. The database searching for natural compounds identification uses mass spectrum without interpretation to query a database of a theoretical spectrum [26].

Nowadays, a common variant consists of using libraries of previously identified spectra instead of a database of theoretically derive spectra. In a broad sense, database searching is a comparative spectral analysis which entails searching for entries in the database that are most similar to a query spectrum, especially in terms of some similarities criterion such as retention time, correlation coefficients, fragmentation mass spectrum, etc. [26, 27].

Databases are usually used to compare the spectra of the compounds with theoretical spectra. Meanwhile, libraries are collections of experimental spectra already assigned to a specific compound. In NP discovery workflows, dereplication (which is the process of identifying known molecules), plays an important role and uses databases and spectral libraries to save time and resources by quickly and efficiently triggering known compounds [28, 29].

6.2.2 The general workflow of MS annotation

Usually, the mass spectrometer is coupled with liquid chromatography with a diode array detector, and then the database compares the corresponding mass spectrum for the identification of the compound. Many algorithms have been developed and have commonalities such as pre-grouping and indexing of the data [30], a screening method for the identification of a smaller number of potential positives [31], scoring procedures based on probabilistic quantities [32], and ranking methodologies [33]. The most commonly used databases in NPs discovery include MarinLit [34], AntiBase [35], Dictionary of Natural Products [36], and SciFinder, to name a few. These databases contain organic compounds and biological molecules that are searchable by compound name or exact mass.

Those databases are not commonly used in metabolomics due to the lack of mass spectral information and most of the molecules have limited biological relevance. Having reference compound data and reference spectral data is critical in metabolomics to annotate metabolites and the autonomous metabolomic workflow for compound annotation and identification includes data acquisition (MS and MS/MS), followed by quantitative comparative analysis and metabolite identification. In this regards, the most widely used mass spectral databases in metabolomics contain GC–MS or LC–MS/MS spectra giving additional information on known metabolites including physiological concentration, metabolic reaction, and biological role [37, 38]. It is worth mentioning that database search is usually straightforward for electron ionization (EI) fragmentation spectra than for collision-induced dissociation (CID) tandem MS spectra due to its high reproducibility across instruments and data have been collected for over many years.

The Human Metabolome Database (HMBD) is a web-accessible database with the most current and comprehensive human metabolites, covering 114,100 metabolite entries in its latest version HMBD 4.0 [39]. These metabolites include peptides, amino acids, carbohydrates, lipids, cosmetics, organic acids, biogenic amines, vitamins, minerals, drugs, contaminants, and pollutants [40]. Another widely used database is MELTIN which incorporates MS/MS mass spectra of over 240,000 metabolites from diverse origins including plant, bacteria, and human [41].

Open community mass spectra repositories have also surfaced in the past years, including the European Massbank (NORMAN MassBank) and MassBank of North America (MoNA) based on consortium members from different research groups in the European Union, Japan, Brazil, the United States, China, and Switzerland [38, 42]. The distinctive feature of Massbank is the possibility to identify metabolites independently of the instrument configuration and MS manufacturer [42]. This database is the longest standing community database. mzCloud is another open community database for free identification and annotation of LC–MS/MS and LC–MS(n) data sets of more than 3,000 compounds (Figure 6.2 [25]) [38]. The distinctive feature of mzCloud is the possible annotation and identification of the compounds even in the case that they are not present in the library through substructure search [25]. Online access to the database is free of charge and no registration is required.

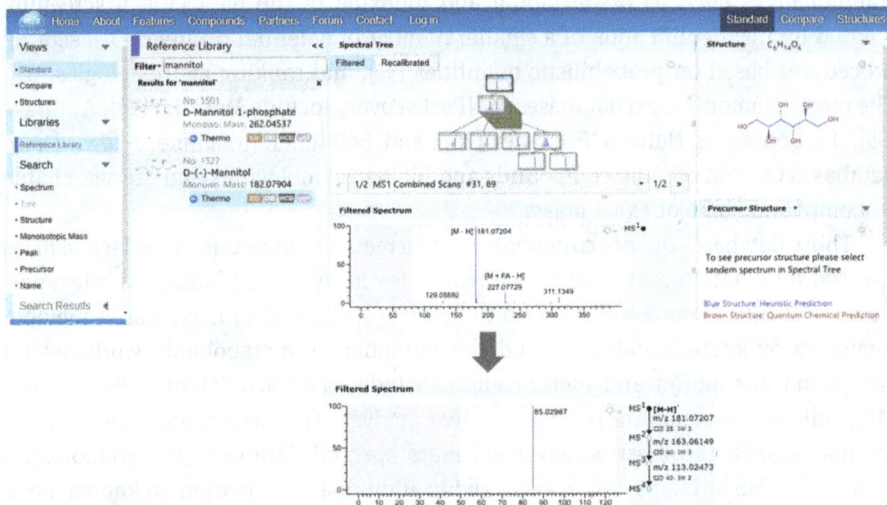

Figure 6.2: A screenshot of the mzCloud tool for compound annotation [25]. Figure reproduced with permission.

The National Institute of Standards and Technology (NIST) Mass Spectral Library is another widely used well-curated and comprehensive mass spectral database containing spectral data relevant for GC–MS (EI fragmentation) and also MS/MS spectra relevant for LC–MS (CID fragmentation) in its last version (NIST 17) to facilitate metabolite annotation. NIST 17 EI library is a repository of 306,622 EI spectra of 267,376 unique compounds, while the NIST 17 Tandem Library contains 574,826 spectra of 13,808 unique compounds, consisting of plant and human metabolites, drugs, as well as industrial and environmental compounds [38].

Fihn library is a GC–EI mass spectral database containing over 2,200 EI mass spectra of over 1,000 authentic chemical standards of primary metabolites below 550 Da molecular weight [43]. Another GC–EI mass spectral database is the Golm Metabolome Database which is hosted by the Max Plank Institute for Molecular Plant Physiology and is a repository of plant metabolites [44].

6.3 Software used in computational MS

6.3.1 Rule-based fragmentation prediction

The structural dereplication of NPs through MS library search is rapid and straight-forward, whereas chemical libraries, conversely, do not exhaustively cover the chemical space and have a significant preponderance of known NPs without anno-tated MS data from standards. In addition, experimental mass spectra recorded on different instruments may be dissimilar, especially in the case of tandem MS [45, 46]. These factors limit the use of MS library search, not only for the structural dereplica-tion of "known unknown" NPs but also for the *de novo* CASE of "unknown unknown" NPs. To solve this bottleneck, the strategy of the pioneering DENDRAL project was borrowed, whereby computationally predicted spectra are used in place of experi-mental spectra in spectral matching. In general, the molecular mass and/or molecu-lar formula of an *unknown* NP is used as input to retrieve candidate structures from a large chemical database such as PubChem or ChemSpider or from a molecular structure generator such as OMG [47] or MOLGEN 5.0 [48]. Subsequently, software tools can be used to generate the *in silico* mass spectra for the candidate structures, on the basis of fragmentation rules for specific metabolites classes, obtained from data mining of spectral literature. The structure of an *unknown* NP can, thus, be elucidated from the spectral matching of an experimental spectrum with those of an *in silico* MS library of candidate structures. CONGEN [49] and MASSIMO [50] are the earliest representatives of software operating on the above principle. They are both in-house software and their efficiency in large-scale studies could not be evaluated.

Presently, the three predominant commercial software (Table 6.2) for rule-based fragmentation are Mass Frontier (ThermoFinnigan, San Jose, CA, USA), ACD/MS

Table 6.2: Summary of tools for rule-based fragmentation spectrum prediction.

Tool	Web accessibility	Reference
CONGEN	Not available	[49]
MASSIMO	Not available	[50]
Mass Frontier	https://www.thermofisher.com/de/de/home/industrial/ mass-spectrometry/liquid-chromatography-mass-spectro metry-lc-ms/lc-ms-software/multi-omics-data-analysis/ mass-frontier-spectral-interpretation-software.html	–
ACD/MS Fragmenter	https://www.acdlabs.com/products/adh/ms/ms_frag/	–
MOLGEN-MS; MOLGEN-MSF	http://www.mathe2.uni-bayreuth.de/markus/ei-ms/; https://molgen.de/products.html/	[51]
MASSIS (*CNRS in-house software*)	Not available	[52]

(continued)

Table 6.2 (continued)

Tool	Web accessibility	Reference
HAMMER	https://more.bham.ac.uk/viant/hammer/	[53]
MS-Finder	http://prime.psc.riken.jp/Metabolomics_Software/MS-FINDER/index.html/	[54]

Fragmenter (Advanced Chemical Development, Toronto, ON, Canada), and MOLGEN-MS(F) [51, 55]. Chen et al. [11, 52–54, 56–64] developed the MASSIS algorithm which uses a combination of special cleavage rules (such as the McLafferty rearrangement, the retro-Diels–Alder reaction, neutral losses, and oxygen migration) and statistical fragment-peak intensity relationships for mass spectral simulation.

Although these software were originally developed to predict EI fragmentation spectra, great advances have been made to extend their use to the prediction of CID-based fragmentation as well. Zhou et al. [56] reported HAMMER, an open source software which controls the commercial Mass Frontier software, for the automated and high throughput construction of *in silico* MS(n) libraries for spectral matching. Its use was validated in two successful case studies. MS-Finder is rule-based and combinatorial fragmenter. The approach uses 9 rules of hydrogen rearrangement learned from CID-based fragmentations, to predict MS/MS spectra of candidate structures filtered from its 15 incorporated molecular databases [57, 58].

Ruled-based fragmentation prediction still presents some limitations. Experimental EI fragmentation spectra are highly reproducible, albeit with some-what complex rearrangements which are very challenging to predict. Also, NPs have high structural diversity and the rules learned from recorded spectra are not exhaustive, while "unknown" NPs with probable hitherto uncharacterized fragmentation patterns are frequently discovered.

6.3.2 Combinatorial fragmentation prediction

Combinatorial fragmentation tools try to explain the peaks in the measured fragmentation spectrum of the query compound, as opposed to rule-based fragmentation tools which aim at simulating a fragmentation spectrum of a query compound. These tools simulate the fragmentation of a candidate structure by breaking molecular bonds and using the substructures to explain the peaks in a query fragmentation spectrum [11, 59].

The first step is to assign costs for bond dissociation based on the principle that some bonds break more easily than others. Here, a suitable cost function is necessary to approximate the cost of bond cleavages, such as bond dissociation energies, standard bond energies, or the type of chemical bonds (single, multiple, or aromatic).

In the next step, each peak in the measured spectrum will be explained by matching with the substructure peak of minimal cost.

Early methods such as EPIC [58] and FiD [61] enumerate all possible fragments of candidate structures by systematic bond disconnection and rank the resulting substructures according to the cost for generating the corresponding fragments (Table 6.3). The exhaustive enumeration is computationally challenging even for medium-sized compounds and hence cannot be used for a large set of candidate molecular structures.

Table 6.3: Summary of tools for combinatorial fragmentation.

Tool	Web accessibility	References
EPIC (*Merk in-house software*)	Not available	[60]
FiD	https://www.cs.helsinki.fi/group/sysfys/software/fragid/	[61]
MetFrag; MetFrag CL; MetFragR	https://msbi.ipb-halle.de/MetFrag; http://ipb-halle.github.io/MetFrag/projects/metfragcl/; https://github.com/ipb-halle/MetFragR	[62, 63]
MIDAS; MIDAS-G	http://facultyweb.mga.edu/yingfeng.wang/Assets/midas/midas.html/; http://facultyweb.mga.edu/yingfeng.wang/Assets/midasg/midas-g.htm	[64, 65]
MAGMa; MAGMa+	http://www.emetabolomics.org/magma/; https://github.com/savantas/MAGMa89-plus/	[66, 67]

In order to avoid combinatorial explosion, most current approaches use heuristics to restrict the number of generated molecular substructures and the number of predicted unlikely fragments. The most common combinatorial fragmenter MetFrag [62, 63] uses three heuristics: by setting a maximum search tree depth, the number of bond cleavage allowed is limited; fragments with a mass lower than the lowest mass of any fragment peaks are not considered; and finally, redundants are removed. Therefore, it is fast enough to screen large chemical databases, such as KEGG, PubChem, or ChemSpider, with a measured fragmentation spectrum of an unknown compound as input. MetFrag has been actively developed and improved, allowing web-based use (MetFragBeta), local use with a command line interface (MetFrag CL), or R-Package (MetFragR) [19, 20].

The software tool MIDAS [64] also uses the systematic bond cleavage but differs from MetFrag in the way that it scores the predicted fragments based on the actual mass of its precursor in the measured spectrum and the number of bond dissociations. It has been reported that MIDAS outperforms MetFrag with regards to the number of correctly identified metabolites [64]. Ridder et al. [66] developed a method for the substructure-based annotation of high-resolution MS(n) spectral trees (MAGMa), which uses the output hierarchical data to account for the fragment peaks observed at consecutive levels of the MS(n) spectral tree.

All the above-mentioned software suffer from low predictive accuracy of fragments, yielding numerous bogus fragments. In a further development, the hybrid approach MAGMa+ [67] was developed, which incorporates both MIDAS and MAGMa, meanwhile using metabolite-dependent optimized parameters obtained with machine learning (ML) techniques, resulting in an increment in fragment predictive accuracy as compared to each individual algorithm. Most recently, Wang et al. introduced MIDAS-G as a C++ implementation of MIDAS, which uses edge-based graph grammars formalism for the prioritization of bond types of interest and their adjacent substructures [65].

The main drawback of combinatorial fragmentation is that fragments predicted from structural rearrangements such as hydrogen rearrangement can be covered only in a limited way. An additional point of concern is the development of a good cost function to score the predicted substructures.

6.3.3 Mass spectral classifiers

Mass spectral classifiers are algorithms used for the automatic recognition of substructures and structural properties in the molecular structure of the investigated compound [68, 69]. In its simplest form, it is to confirm, when given the spectrum of an unknown compound, if a particular substructure (or more general chemical property) is present or not. Thus, a straightforward classifier response is a binary *yes/no* answer [12, 70]. In general, a fixed set of numerical features that are closely related to molecular structures are generated for a specific spectrum which are essential for good performance of the classifier, followed by the application of methods from multivariate statistics [12]. In another sense, prediction of fingerprints such as substructures of an unknown molecule is done by the initially use of MS data. One important tool for substructures prediction is ML. In the past years, many software tools for mass spectral classifiers have been developed (Table 6.4).

Table 6.4: Summary of tools for mass spectral classifiers.

Tool	Web accessibility	Reference
CSI:FingerID	https://www.csi-fingerid.uni-jena.de/	[71]
NIST MS Search	https://chemdata.nist.gov/mass-spc/ms-search/	–
AMDIS	http://www.amdis.net/	[72]
MS2Analyzer	https://fiehnlab.ucdavis.edu/projects/MS2Analyzer/	[73]
FingerID	https://github.com/icdishb/fingerid/	[74]
ChemDistiller	https://bitbucket.org/iAnalytica/chemdistillerpython	[75]
MetExpert	https://sourceforge.net/projects/metexpert/	[76]
LipidBlast	https://fiehnlab.ucdavis.edu/projects/LipidBlast/	[77]
SIRIUS	https://bio.informatik.uni-jena.de/sirius/	[78]

One of such is the Automatic Mass Spectral Deconvolution and Identification System (AMDIS), which has been developed for automatically finding distinct chemical components in the GC/MS data file followed by their comparison to a library of spectra of chemicals [79]. One important application of AMDIS is the chemical analysis during the Chemical Weapons Convention inspection allowing the detection of well-known chemical weapons along with their decomposition products by the use of database such as the Organization for the Prohibition of Chemical Weapons (OPCW) Central Analytical Database [79]. This database is a repository of over 4,900 mass spectra of over 3,500 distinct chemicals. Another well-known software classifier is CSI: FingerID which has been developed to predict substructure from electrospray ionization mass spectra followed by the comparison with candidate molecules originated from database queries [80].

One current state-of-art method for the correct identification is MetExpert made available to assist researchers especially in the field of NPs with GC/MS data interpretation and metabolites identification without the use of spectral libraries [76]. This software has the advantage of allowing *in silico* derivatization increasing the number of identified molecules compared to existing molecular databases.

6.3.4 Fragmentation trees analysis

Fragmentation tree analysis is an essential aspect of MS. Among the tools for fragmentation trees analysis (Table 6.5), clustering of MS^2 spectra for metabolite identification (CluMSID) is an R package useful for tandem mass spectra and neutral loss pattern similarities as a part of the metabolite annotation workflow [81]. This package has the benefit of offering functions for all analysis steps from the import of raw data to data mining by unsupervised multivariate methods, together with respective (interactive) visualizations. Several methods have been implemented in fragmentation analysis, including fragmentation trees for the identification pipeline for unknown metabolites, as implemented in the FT-BLAST tool [82], and calculation of the similarity between high-resolution mass spectral fragmentation trees, as implemented in MetiTree tool [83]. FT-BLAST is used to query databases using fragmentation tree alignment, resulting in hit lists containing compounds with

Table 6.5: Summary of tools for fragmentation trees analysis.

Tool	Web accessibility	Reference
CluMSID	https://hithub.com/tdepke/CluMSID/	[81]
FT-BLAST	https://bio.informatik.uni-jena.de/research/	[82]
MetiTree	http://www.metitree.nl/ https://github.com/ NetherlandsMetabolomicsCentre/metitree/wiki	[83]
SIRIUS	https://bio.informatik.uni-jena.de/software/sirius/	[84]

large structural similarity to the unknown metabolite. The characteristic substructure of the molecules in the hit list may be a key structural element of the unknown compound and might be used as a starting point for structure elucidation. Meanwhile, the approach in MetiTree is used to query multiple-stage mass spectra in MS spectral libraries.

SIRIUS [84] implements a computationally efficient approach for orthogonal time-of-flight MS, used to determine sum formulas of molecules, a crucial step in the identification of an unknown metabolite. The latter is able to correctly identify sum formulas for molecules, ranging in mass up to 1,000 Da.

6.3.5 Fingerprint fragmentation (advanced mass spectral classifiers)

Tools for the identification of fingerprints from generated fragments are provided in Table 6.6. Particularly, intelligent Metabolomic (iMet) is a freely available computational tool that facilitates the structural annotation of metabolites not described in databases [85]. iMet makes use of MS/MS spectra and the exact mass of an unknown metabolite to identify metabolites in a reference database that are structurally similar to the unknown metabolite. An added advantage is that its algorithm suggests the chemical transformation that converts the known metabolites into the unknown one.

Table 6.6: Summary of tools for fingerprint fragmentation.

Tool	Web accessibility	Reference
iMet	http://imet.seeslab.net	[85]
PIF (commercial software)	Not available	[86]

The commercial tool, Precursor Ion Fingerprinting (Figure 6.3), distributed by Thermo Scientific (https://www.mzcloud.org/ToDownload/PIFApplicationnote. pdf), makes use of a technique which readily lends itself to routine automation and that offers the advantage of metabolite identification with no *a priori* knowledge of the active pharmaceutical ingredient (or any biotransformation products [86]).

6.3.6 Competitive fragmentation modeling

The *in silico* identification software (ISIS) [87] and the CFM-ID are web servers [88], based on the competitive fragmentation modeling (CFM) algorithm, which are very useful in modeling competitive fragmentation processes in MS analysis (Table 6.7). The freely available software ISIS is useful in generating *in silico* spectra of lipids by implementing a ML algorithm that helps to find accurate bond cleavage rates in a

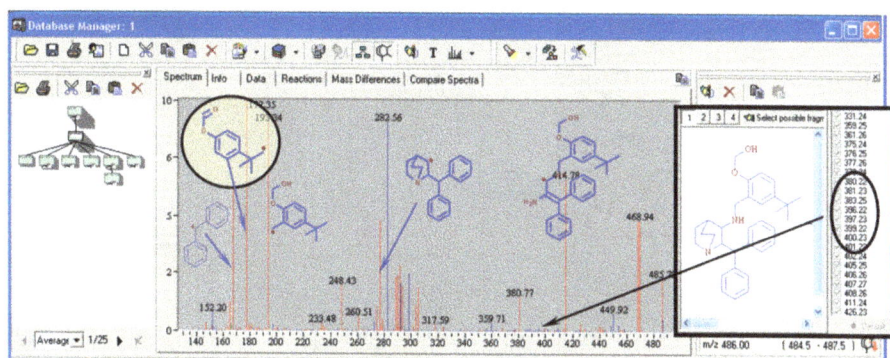

Figure 6.3: Snapshot of the PF tool showing annotated MS2 spectrum for a metabolite M7 with key fragment at *m/z* 177 highlighted [86].

Table 6.7: Summary of tools for competitive fragmentation modeling.

Tool	Web accessibility	Reference
ISIS	http://omics.pnl.gov/	[87]
CFM-ID	http://cfmid.wishartlab.com/	[88]

mass spectrometer by employing CID tandem MS instead of using chemical reaction rate equations or rule-based fragmentation libraries [87]. Meanwhile, the web server CFM-ID is useful for automated metabolite identification. The algorithm helps in the annotation of the peaks in a spectrum for a known chemical structure, prediction of spectra for a given chemical structure and putative metabolite identification and attempts a predicted ranking of possible candidate structures for a target spectrum [88]. This last approach was shown to outperform existing methods such as MetFrag and FingerID when tested on multiple datasets. Additionally, the web server provides a simple interface for using these algorithms and a graphical display of the resulting annotations, spectra, and structures (Figure 6.4).

6.3.7 Quantum chemistry fragmentation

Quantum chemistry (QC) fragmentation approaches are the most advanced and most computationally expensive methods for MS spectral analysis. These have been summarized in Table 6.8.

QC packages for spectral analysis include the computer program QCEIMS [89–91], which performs all necessary computations except the QC part for which the efficient programs ORCA [97, 98], DFTB+ [99], and MNDO [100] are used. The Quantum Chemical Fragment Precursor Tests [93] approaches have been implemented in

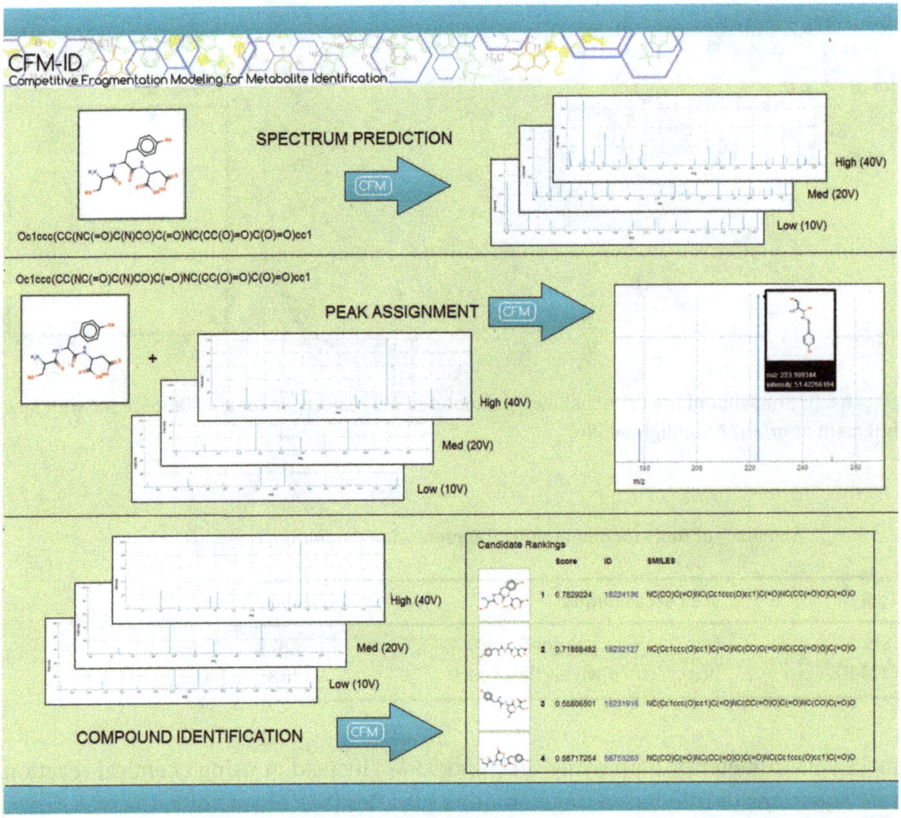

Figure 6.4: Summary of the three tasks provided by the CFM-ID web server: Spectrum prediction, peak assignment, and compound identification. Examples of possible inputs and the corresponding graphical output are shown for each [88]. The original figure published under a Creative Commons License.

Table 6.8: Summary of tools and methods for DFT quantum chemical fragmentation.

Tool	Web accessibility	References
QCEIMS	https://www.chemie.uni-bonn.de/pctc/mulli ken-center/software/qmdff/qmdff	[89–91]
QC-FPT	Not available	[92]
DFT applications	Not available	[93–95]
ChemFrag	Not available	[96]

several commercial software packages, including Mass Frontier (Thermo Scientific, www.thermoscientific.com) and MS Fragmenter (ACD Labs, www.acdlabs.com).

Density functional theory has been implemented in the determination of tandem mass spectrometric fragmentation pathways [93], in the prediction of EI [91], and in the verification and extension of the mobile proton model [95].

Most recently, the ChemFrag tool [96], which implements QC for the detection of fragmentation pathways and the annotation of fragment ions with chemically reasonable structures, was published by Schüler et al. The advantage of this tool is that it combines QC approach with a rule-based approach. When applied on several doping substances as test cases, the approach was able to correctly annotate fragment ions, predicting fragments that are chemically more realistic than those from purely combinatorial approaches, or approaches based on ML. Additionally, the output from ChemFrag often coincides with manually annotated spectra by experts.

6.4 Conclusions

Other chapters in this book have focused on computational tools/methods for the structure elucidation of secondary metabolites by focusing on NMR [101–103]. It has been our aim to assist NP chemists by providing an overview of the methods and (freely available and commercial) tools useful to assist in structure elucidation analysis of MS data. In this chapter, we have briefly summarized the methods and described the available resources, including databases, web servers, and software, to assist in MS analysis. Several web links have also been embedded in the text to enable the reader to access the servers, databases and useful tools for MS analysis.

Acknowledgements: AFAM was supported by a doctoral research grant from the German Academic Exchange Service (DAAD). FNK acknowledges a return fellowship and an equipment subsidy from the Alexander von Humboldt Foundation, Germany. Financial support for this work is acknowledged from a ChemJets fellowship from the Ministry of Education, Youth and Sports of the Czech Republic awarded to FNK. JL was supported by the German National Research Foundation [DFG, Research Training Group 1976] and by the Baden-Württemberg Foundation [BWST_WSF-043].

References

[1]   Roberts LD, Souza AL, Gerszten RE, Clish CB. Targeted metabolomics. Curr Protoc Mol Biol. 2012;30:1–24.
[2]   Schrimpe-Rutledge AC, Codreanu SG, Sherrod SD, McLean JA. Untargeted metabolomics strategies-challenges and emerging directions. J Am Soc Mass Spectrom. 2016;27:1897–905.
[3]   Vinayavekhin N, Saghatelian A. Untargeted metabolomics. Curr Protoc Mol Biol. 2010;30:1–24.
[4]   Kind T, Fiehn O. Advances in structure elucidation of small molecules using mass spectrometry. Bioanal Rev. 2010;2:23–60.
[5]   Ren JL, Zhang AH, Kong L, Wang XJ. Advances in mass spectrometry-based metabolomics for investigation of metabolites. RSC Adv. 2018;8:22335–50.

[6]   Koyama N, Tomoda H. MS network-based screening for new antibiotics discovery. J Antibiot.
       2019;72:54–6.
[7]   Gowda GA, Djukovic D. Overview of mass spectrometry-based metabolomics: opportunities
       and challenges. Methods Mol Biol. 2014;1198:3–12.
[8]   Dettmer K, Aronov PA, Hammock BD. Mass spectrometry-based metabolomics. Mass Spectrom
       Rev. 2007;26:51–78.
[9]   Scheubert K, Hufsky F, Böcker S. Computational mass spectrometry for small molecules.
       J Cheminform. 2013;5:12.
[10]   Vinaix M, Schymanski EL, Neumann S, Navarro M, Salek RM, Yanes S. Mass spectral databases
       for LC/MS- and GC/MS-based metabolomics: state of the field and future prospects. Trends
       Analyt Chem. 2016;78:23–35.
[11]   Vaniya A, Fiehn O. Using fragmentation trees and mass spectral trees for identifying unknown
       compounds in metabolomics. Trends Analyt Chem. 2015;69:52–61.
[12]   Hufsky F, Scheubert K, Böcker S. Computational mass spectrometry for small-molecule
       fragmentation. Trends Analyt Chem. 2014;53:41–8.
[13]   Kerber A, Meringer M, Rücker C. CASE via MS: ranking structure candidates by mass spectra.
       Croatica Chemica Acta. 2006;79:449–64.
[14]   Su BH, Shen MY, Harn YC, Wang SY, Schurz A, Lin C, et al. An efficient computer-aided structural
       elucidation strategy for mixtures using an iterative dynamic programming algorithm. J
       Cheminform. 2017;9:57.
[15]   Smith DH, Gray NA, Nourse JG, Crandell CW. Analytica chimica acta the dendral project: recent
       advances in computer- assisted structure elucidation. Anal Chim Acta. 1981;133:471–97.
[16]   Gray NA. Chemometrics and intelligent laboratory systems dendral and meta-dendral - the
       myth and the reality. Chemom Intell Lab Syst. 1988;5:11–32.
[17]   Djerassi C, Smith DH, Crandell CW, Gray NA, Nourse JG, Lindley MR The DENDRAL project:
       computational aids to natural products structure elucidation. Pure Appl Chem. 1982;54:2425–42.
[18]   Lindsay RK, Buchanan BG, Feigenbaum EA, Lederberg J. Artificial intelligence DENDRAL:
       a case study of the first expert system for scientific hypothesis formation. Artif Intell.
       1993;61:209–61.
[19]   Nishioka T, Kasama T, Kinumi T, Makabe H, Matsuda F, Miura D, et al. Winners of CASMI2013:
       automated tools and challenge data. Mass Spectrom. 2014;3:S0039.
[20]   Nikolić D. CASMI 2016: a manual approach for dereplication of natural products using tandem
       mass spectrometry. Phytochem Lett. 2017;21:292–6.
[21]   Schymanski EL, Ruttkies C, Krauss M, Brouard C, Kind T, Dührkop K, et al. Critical assessment of
       small molecule identification 2016: automated methods. J Cheminform. 2017;9:22.
[22]   Hufsky F, Böcker S. Mining molecular structure databases: identification of small molecules
       based on fragmentation mass spectrometry data. Mas Spectrom Rev. 2017;36:624–33.
[23]   Hufsky F, Scheubert K, Böcker S. New kids on the block: novel informatics methods for natural
       product discovery. Nat Prod Rep. 2014;31:807–17.
[24]   Blaženović I, Kind T, Ji J, Fiehn O. Software tools and approaches for compound identification of
       LC-MS/MS data in metabolomics. Metabolites. 2018;8:318.
[25]   Misra BB, van der Hooft JJ. Updates in metabolomics tools and resources: 2014–2015.
       Electrophoresis. 2016;37:86–110.
[26]   Gu WY, Li N, Leung EL, Zhou H, Luo GA, Liu L, et al. Metabolites software-assisted flavonoid
       hunting in plants using ultra-high performance liquid chromatography-quadrupole-time of
       flight mass spectrometry. Molecules. 2015;20:3955–71.
[27]   Dunn WB, Erban A, Weber RJ, Creek DJ, Brown M, Breitling R, et al. Mass appeal: metabolite
       identification in mass spectrometry-focused untargeted metabolomics. Metabolomics.
       2013;9:44–66.

[28] Lang G, Mayhudin NA, Mitova MI, Sun L, van der Sar S, Blunt JW, et al. Evolving trends in the dereplication of natural product extracts: new methodology for rapid, small-scale investigation of natural product extracts. J Nat Prod. 2008;71:1595–9.

[29] Bugni TS, Harper MK, McCullochand MW, Whitson EL. Natural product chemistry for drug discovery. London, UK: Royal Society of Chemistry; 2010. p. 272–98.

[30] Li Y, Chi H, Wang LH, Wang HP, Fu Y, Yuan ZF, et al. Speeding up tandem mass spectrometry based database searching by peptide and spectrum indexing. Rapid Commun Mass Spectrom. 2010;24:807–14.

[31] Craig R, Beavis RC. A method for reducing the time required to match protein sequences with tandem mass spectra. Rapid Commun Mass Spectrom. 2003;17:2310–6.

[32] Fenyö D, Beavis RC. A method for assessing the statistical significance of mass spectro- metry-based protein identifications using general scoring schemes. Anal Chem. 2003;75:768–74.

[33] Sonego P, Kocsor A, Pongor S. ROC analysis: applications to the classification of biological sequences and 3D structures. Brief Bioinform. 2008;9:198–209.

[34] In MarinLit; Department of Chemistry, University of Canterbury. http://www.chem.canterbury. ac.nz/marinlit/marinlit.shtml, 2013.

[35] Laatsch H. Antibase, a data base for rapid dereplication and strcuture determination of microbial natural products. Weinheim, Germany: Wiley-VCH; 2013.

[36] DN P. Chapman and Hall Chemical Database 2013. Dictionary of Natural Products.

[37] Bouslimani A, Sanchez LM, Garg N, Dorrestein PC. Mass spectrometry of natural products: current, emerging and future technologies. Nat Prod Rep. 2014;31:718–29.

[38] Vinaixa M, Schymanski EL, Neumann S, Navarro M, Salek RM, Yanes O. Mass spectral data- bases for LC/MS- and GC/MS-based metabolomics: state of the field and future prospects. Trends Anal Chem. 2016;78:23–35.

[39] Wishart DS, Feunang YD, Marcu A, Guo AC, Liang K, Vázquez-Fresno R, et al. HMDB 4.0: the human metabolome database for 2018. Nucleic Acids Res. 2017;46:608–17.

[40] Wishart DS. Metabolomics: applications to food science and nutrition research. Trends Food Sci Technol. 2008;19:482–93.

[41] Smith CA, O'Maille G, Want EJ, Qin C, Trauger SA, Brandon TR, et al. METLIN: a metabolite mass spectral database. Ther Drug Monit. 2005;27:747–51.

[42] Horai H, Arita M, Kanaya S, Nihei Y, Ikeda T, Suwa K, et al. MassBank: a public repository for sharing mass spectral data for life sciences. J Mass Spectrum. 2010;45:703–14.

[43] Kind T, Wohlgemuth G, Lee DY, Lu Y, Palazoglu M, Shahbaz S, et al. FiehnLib: mass spectral and retention index libraries for metabolomics based on quadrupole and time-of-flight gas chro- matography/mass spectrometry. Anal Chem. 2009;81:10038–48.

[44] Kopka J, Schauer N, Krueger S, Birkemeyer C, Usadel B, Bergmüller E, et al. GMD@ CSB. DB: the Golm metabolome database. Bioinformatics. 2004;21:1635–8.

[45] Chaleckis R, Meister I, Zhang P, Wheelock CE. Challenges, progress and promises of metabolite annotation for LC–MS-based metabolomics. Curr Opin Biotechnol. 2019;55: 44–50.

[46] Wolfender J-L, Nuzillard J-M, van der Hooft JJ, Renault J-H, Bertrand S. Accelerating metabolite identification in natural product research: toward an ideal combination of liquid chromato- graphy–high-resolution tandem mass spectrometry and NMR profiling, *in silico* databases, and chemometrics. Anal Chem. 2019;91:704–42.

[47] Peirocely JE, Rojas-Chertó M, Fichera D, Reijmers T, Coulier L, Faulon J-L, et al. OMG: open molecule generator. J Cheminform. 2012;4:21.

[48] Gugisch R, Kerber A, Kohnert A, Laue R, Meringer M, Rücker C, et al. MOLGEN 5.0, a molecular structure generator. Adv Math Chem Appl. 2015;26:113–38.

[49] Gray NA, Carhart RE, Lavanchy A, Smith DH, Varkony T, Buchanan BG, et al. Computerized mass spectrum prediction and ranking. Anal Chem. 1980;52:1095–102.

[50] Gasteiger J, Hanebeck W, Schulz KP. Prediction of mass spectra from structural information. J Chem Inf Model. 1992;32:264–71.

[51] Kerber A, Laue R, Meringer M, Varmuza K. MOLGEN-MS: evaluation of low resolution electron impact mass spectra with MS classification and exhaustive structure generation. Adv Mass Spectrom. 2001;15:939–40.

[52] Chen H, Fan B, Xia H, Petitjean M, Yuan S, Panaye A, et al. MASSIS: a mass spectrum simulation system. 1. Principle and method. Eur J Mass Spectrom. 2003;9:175–86.

[53] Chen H, Fan B, Petitjean M, Panaye A, Doucet J-P, Li F, et al. MASSIS: a mass spectrum simulation system. 2: procedures and performance. Eur J Mass Spectrom. 2003;9: 445–57.

[54] Fan B, Chen H, Petitjean M, Panaye A, Doucet J, Xia H, et al. New strategy of mass spectrum simulation based on reduced and concentrated knowledge databases. Spectrosc Lett. 2005;38:145–70.

[55] Schymanski EL, Meringer M, Brack W. Matching structures to mass spectra using fragmentation patterns: are the results as good as they look? Anal Chem. 2009;81:3608–17.

[56] Zhou J, Weber RJ, Allwood JW, Mistrik R, Zhu Z, Ji Z, et al. HAMMER: automated operation of mass frontier to construct in silico mass spectral fragmentation libraries. Bioinformatics. 2014;30:581–3.

[57] Tsugawa H, Kind T, Nakabayashi R, Yukihira D, Tanaka W, Cajka T, et al. Hydrogen rearrangement rules: computational MS/MS fragmentation and structure elucidation using MS-FINDER software. Anal Chem. 2016;88:7946–58.

[58] Vaniya A, Samra SN, Palazoglu M, Tsugawa H, Fiehn O. Using MS-FINDER for identifying 19 natural products in the CASMI 2016 contest. Phytochem Lett. 2017;21:306–12.

[59] Domingo-Almenara X, Montenegro-Burke JR, Benton HP, Siuzdak G. Annotation: a computational solution for streamlining metabolomics analysis. Anal Chem. 2018;90:480–9.

[60] Hill AW, Mortishire-Smith RJ. Automated assignment of high-resolution collisionally activated dissociation mass spectra using a systematic bond disconnection approach. Rapid Commun Mass Spectrom. 2005;19:3111–8.

[61] Heinonen M, Rantanen A, Mielikäinen T, Kokkonen J, Kiuru J, Ketola RA, et al. FiD: a software for *ab initio* structural identification of product ions from tandem mass spectrometric data. Rapid Commun Mass Spectrom. 2008;22:3043–52.

[62] Wolf S, Schmidt S, Müller-Hannemann M, Neumann S. *In silico* fragmentation for computer assisted identification of metabolite mass spectra. BMC Bioinform. 2010;11:148.

[63] Ruttkies C, Schymanski EL, Wolf S, Hollender J, Neumann S. MetFrag relaunched: incorporating strategies beyond in silico fragmentation. J Cheminform. 2016;8:3.

[64] Wang Y, Kora G, Bowen BP, Pan C. MIDAS: a database-searching algorithm for metabolite identification in metabolomics. Anal Chem. 2014;86:9496–503.

[65] Wang Y, Wang X, Zeng X. MIDAS-G: a computational platform for investigating fragmentation rules of tandem mass spectrometry in metabolomics. Metabolomics. 2017;13:116.

[66] Ridder L, van der Hooft JJ, Verhoeven S. Automatic compound annotation from mass spectrometry data using MAGMa. Mass Spectrom. 2014;3:S0033.

[67] Verdegem D, Lambrechts D, Carmeliet P, Ghesquière B. Improved metabolite identification with MIDAS and MAGMa through MS/MS spectral dataset-driven parameter optimization. Metabolomics. 2016;12:98.

[68] Hewlins MJE. Computers in mass spectrometry Academic Press, London, 1978.265 pp. $9.80. ISBN 0-12-168750-3. Biomed Biol Mass Spec. 1979 3;6:III–IV.

[69] Varmuza K. Pattern recognition in analytical chemistry. Anal Chim Acta. 1980;122:227–40.

[70] Varmuza K, Werther W. Mass spectral classifiers for supporting systematic structure elucidation. J Chem Inf Comput Sci. 1996;36:323–33.

[71] Dührkop K, Shen H, Meusel M, Rousu J, Böcker S. Searching molecular structure databases with tandem mass spectra using CSI:fingerID. Proc Natl Acad Sci USA. 2015;112:12580–5.

[72] Stein SE. An integrated method for spectrum extraction and compound identification from gas chromatography/mass spectrometry data. J Am Soc Mass Spectrom. 1999;10:770–81.

[73] Ma Y, Kind T, Yang D, Leon C, Fiehn O. MS2Analyzer: a software for small molecule substructure annotations from accurate tandem mass spectra. Anal Chem. 2014;86:10724–31.

[74] Heinonen M, Shen H, Zamboni N, Rousu J. Metabolite identification and molecular fingerprint prediction through machine learning. Bioinformatics. 2012;28:2333–41.

[75] Laponogov I, Sadawi N, Galea D, Mirnezami R, Veselkov KA. ChemDistiller: an engine for metabolite annotation in mass spectrometry. Bioinformatics. 2018;34:2096–102.

[76] Qiu F, Lei Z, Sumner LW. MetExpert: an expert system to enhance gas chromatography–mass spectrometry-based metabolite identifications. Anal Chim Acta. 2018;1037:316–26.

[77] Kind T, Liu KH, Lee DY, DeFelice B, Meissen JK, Fiehn O. LipidBlast *in silico* tandem mass spectrometry database for lipid identification. Nat Methods. 2013;10:755–8.

[78] Dührkop K, Fleischauer M, Ludwig M, Aksenov AA, Melnik AV, Meusel M, et al. SIRIUS 4: a rapid tool for turning tandem mass spectra into metabolite structure information. Nat Methods. 2019. doi:10.1038/s41592-019-0344-8.

[79] Mallard WG. AMDIS in the chemical weapons convention. Anal Bioanal Chem. 2014;406:5075–86.

[80] Dührkop K, Shen H, Meusel M, Rousu J, Böcker S. Searching molecular structure databases with tandem mass spectra using CSI: fingerID. Proc Natl Acad Sci USA. 2015;112:12580–5.

[81] Depke T, Franke R, Broenstrup M. ClusMSID: an R package for similarity-based clusterering of tandem mass spectra to aid feature annotation in metabolomics. Bioinformatics. 2019. doi:10.1093/bioinformatics/btz005.

[82] Ludwig M, Hufsky F, Elshamy S, Böcker S. Finding characteristic substructures for metabolite classes. German Conf Bioinform. 2012. doi:10.4230/OASIcs.GCB.2012.23.

[83] Rojas-Cherto M, Peironcely JE, Kasper PT, van der Hooft JJ, de Vos RC, Vreeken R, et al. Metabolite identification using automated comparison of high-resolution multistage mass spectral trees. Anal Chem. 2012;84:5524–34.

[84] Böcker S, Letzel MC, Lipták Z, Pervukhin A. SIRIUS: decomposing isotope patterns for metabolite identification. Bioinformatics. 2009;25:218–24.

[85] Aguilar-Mogas A, Sales-Pardo M, Navarro M, Guimerà R, Yanes O. iMet: a network-based computational tool to assist in the annotation of metabolites from tandem mass spectra. Anal Chem. 2017;89:3474–82.

[86] Horner JA, Thakur RA, Mistrik R. PIF: precursor ion fingerprinting – searching for a structurally diagnostic fragment using combined targeted and data dependent MSn. San Jose, CA, USA: Thermo Fisher Scientific, 2008.

[87] Kangas LJ, Metz TO, Isaac G, Schrom BT, Ginovska-Pangovska B, Wang L, et al. *In silico* identification software (ISIS): a machine learning approach to tandem mass spectral identification of lipids. Bioinformatics. 2012;28:1705–13.

[88] Allen F, Pon A, Wilson M, Greiner R, Wishart D. CFM-ID: a web server for annotation, spectrum prediction and metabolite identification from tandem mass spectra. Nucleic Acids Res. 2014;42:W94–9.

[89] Grimme S. QCEIMS—a general program to compute EI-MS with quantum chemistry, version 2.17. Bonn: University of Bonn, 2013.

[90] Grimme S. Towards first principles calculation of electron impact mass spectra of molecules. Angew Chem Int Ed Engl. 2013;52:6306–12.

[91]   Ásgeirsson V, Bauer CA, Grimme S. Quantum chemical calculation of electron ionization mass spectra for general organic and inorganic molecules. Chem Sci. 2017;8:4879–95.

[92]   Janesko BG, Li L, Mensing B. Quantum chemical fragment precursor tests: accelerating de novo annotation of tandem mass spectra. Anal Chim Acta. 2017;995:52–64.

[93]   Alex A, Harvey S, Parsons T, Pullen FS, Wright P, Riley J-A. Can density functional theory (DFT) be used as an aid to a deeper understanding of tandem mass spectrometric fragmentation pathways? Rapid Commun Mass Spectrom. 2009;23:2619–27.

[94]   Cautereels J, Claeys M, Geldof D, Blockhuys F. Quantum chemical mass spectrometry: *ab initio* prediction of electron ionization mass spectra and identification of new fragmentation pathways. J Mass Spectrom. 2016;51:602–14.

[95]   Cautereels J, Blockhuys F. Quantum chemical mass spectrometry: verification and extension of the mobile proton model for histidine. J Am Soc Mass Spectrom. 2017;28:1227–35.

[96]   Schüler J-A, Neumann S, Müller-Hannemann M, Brandt W. ChemFrag: chemically meaningful annotation of fragment ion mass spectra. J Mass Spectrom. 2018;53:1104–15.

[97]   Neese F. ORCA—an ab initio, density functional and semiempirical program package, Verion 2.9 (Rev. 0). Germany: Max Planck Institute for Bioinorganic Chemistry, 2011.

[98]   Neese F. The ORCA program system. WIREs Comput Mol Sci. 2012;2:73–8.

[99]   Frauenheim T DFTB + (Density Functional based Tight Binding), 2008. http://www.dftb.org/.

[100] Thiel W. *MNDO2005 version 7.0,*. Mülheim, Germany: MPI für Kohlenforschung, 2005

[101] Valli M, Russo HM, Pilon AC, Pinto MEF, Dias NB, Freire RT, et al. Computational methods for NMR and MS for structure elucidation I: software for basic NMR. Phys Sci Rev. 2018.DOI: 10.1515/psr-2018-0108.

[102] Valli M, Russo HM, Pilon AC, Pinto MEF, Dias NB, Freire RT, et al. Computational methods for NMR and MS for structure elucidation II: database resources and advanced methods. Phy Sci Rev. 2019.DOI: 10.1515/psr-2018-0167.

[103] Bitchagno GTM, Tanemossu SAF, et al. Computational methods for NMR and MS for structure elucidation III: more advanced approaches. Phys Sci Rev. 2019.DOI: 10.1515/psr-2018-0109.

Marilia Valli, Helena Mannochio Russo, Alan Cesar Pilon,
Meri Emili Ferreira Pinto, Nathalia B. Dias, Rafael Teixeira Freire,
Ian Castro-Gamboa and Vanderlan da Silva Bolzani

7 Computational methods for NMR and MS for structure elucidation I: software for basic NMR

Abstract: Structure elucidation is an important and sometimes time-consuming step for natural products research. This step has evolved in the past few years to a faster and more automated process due to the development of several computational programs and analytical techniques. In this paper, the topics of NMR prediction and CASE programs are addressed. Furthermore, the elucidation of natural peptides is discussed.

Keywords: CASE, mass spectrometry, nuclear magnetic resonance

7.1 Introduction

Structure elucidation has been important to the state-of-the-art research on analytical chemistry, and so far of several scientific discoveries involving organic chemistry. It is also an essential convergence in all studies of the identification of natural products as well as drug discovery. Organic compounds constitute a particularly complex universe, with an estimate that the number of compounds with molecular weights below 500 can reach 10^{60} structures, which is incomprehensibly large [1]. This uncalculated number of possibilities is the result of complex hierarchical layers that organic compounds are part of. All organic compounds have in their structure carbon (C) and hydrogen (H) and may also contain other atoms such as S, N, P, halogens, metals, etc. They might have different and multiple organic functions, bonding types; single (sp^3), double (sp^2) and triple (sp), and stereochemistry (spatial arrangement and chirality) to form complex skeletons that allow structural diversification making structure elucidation a real challenge.

It is unquestionable that living organisms provide a large variety of secondary metabolites that might be of interest for researchers to develop new drugs, cosmetics or functional foods. These compounds can be classified into natural products, synthetic compounds based on secondary metabolites and semisynthetic natural derivatives [2]. Natural products contain a large number of pharmacophore groups and stereogenic centers, providing a robust source of structural characteristics useful for

This article has previously been published in the journal *Physical Sciences Reviews*. Please cite as:
Valli, M., Russo, H.M., Pilon, A.C., Pinto, M.E.F., Dias, N.B., Freire, F.T., Bolzani, V.S. Computational Methods for NMR and MS for Structure Elucidation I: Software for basic NMR. *Physical Sciences Reviews* [Online] **2019** DOI: 10.1515/psr-2018-0108.

https://doi.org/10.1515/9783110579352-008

pharmacologic studies and drug discovery. Numerous substances identified in natural product collections are important targets to identify hits and leads, even the most challenging ones, such as protein-protein interactions [3].

For many decades, the search for novel chemical structures with new biological activities has been the focus of many research groups around the globe. Approximately 300,000 secondary metabolites from biodiversity have been identified to date. However, although many of these compounds have been elucidated, some of them were evaluated only for a specific target or not even biologically evaluated at all. Additionally, new biological activities for already known compounds are also the focus of recent studies [4].

Over the past few years, many computer programs have been developed to aid in the structure elucidation process. Nuclear magnetic resonance (NMR) has experienced great development in recent years, which was possible due to advances in software and computers. In the past, processing NMR data were only possible at the equipment workstation. Recent software has enabled researchers to process, analyze, predict and elucidate NMR data in a fast and robust manner. A variety of NMR processing programs have been developed for this purpose, such as Agilent (formerly Varian) VNMRJ [5], Bruker Topspin [6], JEOL Delta [7], SpinWorks [8], Mestrelab [9] and ACD/Labs [10].

Artificial intelligence (AI) has become one of the most widespread topics in our society and is related to machine capacity for problem-solving and learning. AI development is based on human behavior and mimics the process done by humans (experts) [11].

7.2 NMR predictions

NMR chemical shift predictions became popular during the last two decades with quantum mechanical and density functional theory (DFT) methods capable of predicting NMR spectra with good accuracy. Many programs for NMR prediction are available, mainly for ^{1}H and ^{13}C spectra but also for 2D spectra, both commercial and freely available online. The methods used for calculating or predicting ^{13}C NMR spectra can be empirical [12–14], algorithms based on experimental data [10, 15–17] and neural networks [18, 19]. The outdated additive model was suggested [20] for calculating ^{13}C chemical shifts accounting for increments of atoms near the prediction center. This approach was very limited since only sp^3 carbons could be predicted. Fragment-based methods rely on databases with assigned chemical shifts for the carbons of the chemical structure. The carbon of each fragment is later regarded for shift prediction, either the exact fragment or by interpolation. The accuracy of the results depends on the robustness of the database used for creating fragments.

A combination of these methods is also available [21]. The HOSE (Hierarchically Ordered Spherical Description of Environment) code was developed to assign single atoms and ring systems. The topology of chemical structures was the basis for automatically generating this code and was very useful to match chemical structures to spectral data [22]. The ACD/Labs prediction software is based on the fragment method with a database of more than 2 million experimental data. Many parameters can influence NMR prediction, such as solvent, temperature, sample concentration, and shimming. Some programs enable the user to account for solvent influence, which provides a more accurate prediction [23].

The prediction of ^{1}H NMR signals is considered more complex than that of ^{13}C NMR signals because ^{1}H NMR consists of multiplets, and comparisons of spectra are also more difficult. The software is based on incremental methods of linear models or a structural database approach. Partial least squares and neural network methods are also used for calculating ^{1}H NMR shifts [24–26]. The ACD/Labs HNMR predictor is based on a database of more than 1.5 million ^{1}H NMR shifts (experimental). The predicted data are provided with 95 % confidence and take into account spectrometer frequency. 2D NMR spectra can also be predicted and may aid in the structure elucidation process. Programs for this task are based on ^{1}H and ^{13}C prediction programs that, after decades of study on DFT methods and their improvement, have become even more accurate to the point that a standard deviation of proton chemical shift is approximately 0.1 ppm and is less than 2.0 ppm for carbon chemical shift [27]. A database of 2D NMR predicted spectra was created [28] using the ACD program, which is available for structure searching [23].

Structure verification is based on comparing experimental and predicted spectra by visual inspection or by calculating deviations. The comparison of chemical shifts is the core of structure elucidation. "SpecSolv" is a computer-assisted structure elucidation (CASE) approach using ^{13}C NMR that is part of the SpecInfo database [29]. SpecInfo is a database for calculating chemical shifts using spectral information for thousands of compounds. SpecSolv can elucidate the structures of most organic molecules up to a molecular weight of approximately 1000 Da. In SpecSolv, it is possible to retrieve matching structures with given ^{13}C NMR shifts. The program is encoded to find substructures by searching parts of the spectrum, and then it groups the substructures, calculates ^{13}C NMR shifts and returns the probable candidate. There are two limitations to this program. The first refers to different molecular conformation, which chemical shift calculated values are affected. The second is that unusual natural products, having complex structural features, do not have substructure matches offered by this program [29, 30].

In the automation of the structural elucidation, several steps are required and structure generation is one of these steps. Thus, ASSEMBLE algorithm was one of the first structure generators developed 40 years ago, and generate structures by expansion using connectivity matrices [31]. CONGEN and GENOA were also developed as molecular builders [32]. The COCOA program is equivalently a structure generator, however, works with a structure reduction approach, while the previously cited programs are structure assembly based. Elucidation systems were then developed, such as CHEMICS [33], SESAMI [34], ACCESS [35], CSEARCH [36], EPIOS [37], and SpecSolv [29], and all these link the structure generation process with spectrum interpretation [34].

The first step in computational structure elucidation by NMR is peak assignment. The second step is molecular fragments generation, such as -CH_3 and -$NHCH_2$, and the final step is the generation of the possible structures from these components. The structures may also be checked for consistency with 2D NMR data when available [30]. In general, the main information extracted from the obtained NMR spectra is chemical shifts. ^1H–^1H, ^1H–^{13}C, and ^{13}C–^{13}C J-couplings and additional NMR information, such as spatial (Nuclear Overhauser effect spectroscopy - NOESY), long-range couplings [LR-HSQMBC for $^{4,5}J_{CH}$ [38]] or long-range carbon-carbon interactions [INADEQUATE [39]], can contribute to providing the correct molecular structure [40].

There are some commercial and freely available software for NMR prediction that are described further. The chemistry software ChemDraw (PerkinElmer) has a user-friendly interface for drawing molecules and has an NMR prediction tool. NMRDDB is a freely available online software for NMR prediction with the tool of the FCT-Universidade NOVA de Lisboa developed by Yuri Binev and João Aires-de-Sousa [41]. The well-known Mestrenova and ACD/Labs also commercialize software for NMR prediction.

7.3 Computational methods for structure elucidation

CASE is an intelligent tool created to provide suggestions for the molecular structures of compounds based on spectroscopic data, database information and computational programs. Such programs are designed to aid chemists, spectroscopists, and researchers with the task of elucidating a chemical structure. Although research in this area was initiated in the 1970s, the software has only very recently been offered for scientists and research labs [42].

One of the main uses of CASE is for dereplication, i.e., to quickly detect known compounds to avoid the possibly redundant step of elucidating an already known molecule. NMR and MS data are mainly used for searching spectral similarity in databases [42]. Details of this topic are described further in this chapter.

The objective of CASE programs is to provide all possible chemical structures consistent with a set of spectroscopic data. This aim is better achieved and becomes more standardized and reliable when human intervention is minimal [30]. Nevertheless, an interaction between the researcher and a CASE program after the proposed structures are generated is still important to judge the correct structure, given that there may be more than one option. CASE programs significantly reduce the effort and time taken to determine complex natural product structures [30]. CASE systems complement the skills of the spectroscopist and may even replace them in the future. It may seem unlikely that this important step could be automated by a computer, but this is becoming a reality in many fields. With these techniques, the bottleneck of structure elucidation is being removed from the natural product chemistry.

This paper is dedicated to natural product researchers, and therefore, the approach discussed here is the use of CASE programs to elucidate complex natural product structures. The spectroscopic methods used for successful structure elucidation comprise mainly one- and two-dimensional NMR methods and mass spectrometry. The final objective of a structure generator is to use the generated components to produce a list of all possible structures without missing any possible structures.

Database projects (SpecInfo [15, 43]), CSearch [36] and researchers [44] made spectral and structural data available, which enabled the creation of programs for structure elucidation of compounds. CASE systems are only possible because of such databases containing chemical structures and spectra [42]. Details of such databases are discussed further in this chapter.

7.3.1 CASE programs – examples

To demonstrate CASE programs and their potential to contribute to structure elucidation, some software and applications are mentioned to highlight the importance of these programs.

7.3.1.1 Mestrelab Mnova software

The company Mestrelab provides several tools to aid structure elucidation by NMR and MS The formerly MestRe-C [45] was replaced by Mnova, currently in version 12, a multiplatform software that can be used for visualization, processing, analysis and elucidation of 1D and 2D NMR data. With Mnova, it is possible to assign NMR spectra to their corresponding chemical structure (which can be done manually or automatically by using a robust algorithm) [46], such as to draw the target molecule, predict NMR spectra, and finally compare with experimental data [45, 47]. Peak-picking efficiently is one of the most critical steps when analyzing data automatically. Many methods are based on finding maxima and minima values but may present issues regarding inaccurate peaks, coupling constants, and overlapping multiplets. Mnova uses a robust peak-picking algorithm (Global Spectral Deconvolution) for auto-assignments and multiplet analysis that also provides information on the origin of the peak, i.e. if it is from solvents, compounds, and artifacts.

Mnova enables the processing of many spectra in a consistent and automatic way, generation of reports and creation of procedures for customary processing and analysis. It is also possible to process and analyze LC/GC/MS data in one single document within the same Mnova interface. Mnova is available for Windows, Linux and Mac OSX [48].

Mnova Structure Elucidation is a tool in Mnova 12 that provides a robust and simple way to elucidate structures by NMR. The company has implemented a CASE system, with a workflow simple to use and learn, that determines structures from NMR data. This program combines several modern methods of peak-picking strategies (GSD deconvolution, filtering, smoothing), constraint generation, and structure generation [49].

7.3.1.2 ACD/structure elucidator suite software

One of the most practical and robust CASE systems was developed by the Canadian company ACD/Labs. With 20 years of history and more than 10 software versions, this system has been helping many studies around the globe that make use of structure elucidation at any point.

StrucEluc enables scientists to quickly find and verify a previously described chemical structure since it is connected to a great number of spectral libraries (including the ChemSpider database). For this, it automatically compares the predicted spectra with the experimental NMR data and generates a match factor. This program may also be applied to assist the elucidation process of a new compound due to the possibility of entering advanced algorithms to quantitatively evaluate the match factor between the proposed structure and its experimental spectra. For the most challenging cases, it allows *de novo* structure elucidation, in which it is possible to generate a set of structures that could fit the observed correlations in experimental data and to include other analytical information (such as MS, UV/Vis, FT-IR, and chromatographic results) to reduce the structural possibilities. Furthermore, it is also possible to enter NOESY and/or ROESY (rotating-frame Overhauser spectroscopy) data to determine the relative stereochemistry [50].

The first article regarding the use of this software was written in 1999; in this article, the authors mention that this software is an innovative tool for structure elucidation [51]. Over two years, the improvement in the software and the addition of new tools allowed it to determine a chemical structure using 2D NMR, making possible the elucidation of even heavier molecules [52]. In 2002, 60 natural products (ranging from 15 to 65 skeletal atoms) were analyzed by this software, of which 58 were unambiguously determined [53], demonstrating its power and robustness. Since then, many applications have been successfully employed, such as (1) solving structures up to 100 atoms [54], (2) assisting in the literature revisions of already published structures and (3) avoiding incorrect hypotheses and helping in the evaluation of proposed ones [55]. Some examples of its use are discussed later in this chapter.

An interesting application of this program was performed by Li and coworkers [56], in which a new triterpenoid aldehyde was isolated from *Adansonia digitate* (Malvaceae), popularly known as Baobab. This new compound presented an unusual 5/6-ring skeleton that was relatively deficient in hydrogen atoms and a singlet signal at 7.94 ppm with a total of 8 HMBC ^{1}H–^{13}C interactions, making it difficult to determine its structure. Therefore, the authors resorted to the ACD/Structure Elucidator program and entered the molecular formula, ^{1}H, ^{13}C, HSQC, and HMBC data. Initially, the program generated 152,400 possible structures in only 7 min and 32 s that could be filtered to 61 structures by entering a "fast increment-deviation" statistic, d_{FI} (^{13}C) $\leq$ 4 ppm/carbon. These structures were ranked using *neural net statistics* [$d_N(^{13}$C)] by comparing predicted ^{13}C chemical shifts with the obtained ones. From these data, it was possible to select the 8 most likely structures based on the best match of the second interaction of the obtained ^{13}C chemical shift data with the predicted ones. These proposed (**1–8**) candidates are shown in Figure 7.1. It was possible to observe that the top candidate (**1**) was, indeed, the most likely structure; by adding COSY (correlation spectroscopy) data to the elucidation process, the other candidates were eliminated, and **1** was confirmed as the new natural product 5-[1-(3,4-dihydroxyphenyl)ethyl]cyclopenta[c]pyran-7-carbaldehyde (**1**) obtained from *A. digitata* [56].

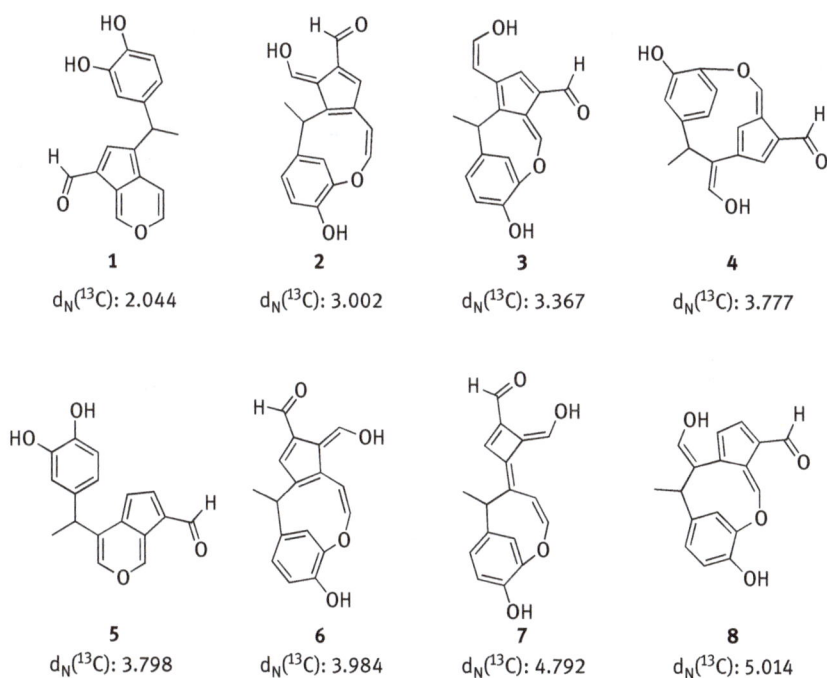

1

$d_N(^{13}$C): 2.044

2

$d_N(^{13}$C): 3.002

3

$d_N(^{13}$C): 3.367

4

$d_N(^{13}$C): 3.777

5

$d_N(^{13}$C): 3.798

6

$d_N(^{13}$C): 3.984

7

$d_N(^{13}$C): 4.792

8

$d_N(^{13}$C): 5.014

Figure 7.1: Top eight candidates of the possible structure generated by ACD/structure elucidator, in which **1** was confirmed to be the new natural product 5-[1-(3,4-dihydroxyphenyl)ethyl]cyclopenta[c] pyran-7-carbaldehyde. The ranking is based on the differences between predicted and obtained ^{13}C chemical shifts: the better the ranking, the lower the $d_N(^{13}$C).

Another enlightening example of the use of this program was described by Buevich and Elyashberg [40], in which a combination of the CASE approach and DFT methods of chemical predictions was applied for the determination of the correct chemical structure even in challenging cases. Although quantum mechanics (QM)-based chemical shift prediction (e. g. DFT methods) takes much more time than empirical methods (CASE), in the cases whereby only CASE is applied and is insufficient to converge the possibilities to a single structure, the application of QM on the limited number of top-ranked structures can assist in the elucidation process.

Buevich and Elyashberg [40] have given an example of Aquatolide, a natural product whose structure was initially erroneously described and revised years later. Aquatolide is a sesquiterpenoid lactone isolated from *Asteriscus aquaticus* (Asteraceae), whose original (erroneous) structure (**9**) was proposed based on 1D and 2D NMR data [57]. Twenty-three years later, studies comparing experimental and DFT-predicted chemical shifts revealed significant discrepancies between those values, leading to the revised structure **10**. Structure **10** was confirmed by X-ray crystallography [58] and total synthesis [59]. To evaluate whether a CASE approach would be successfully employed in this case, ^1H, ^{13}C, COSY, HSQC, and HMBC data were used by the ACD/Structure Elucidator program, which provided only three possible structures ranked considering ^{13}C average deviations d_A, d_N, and d_I. The most likely structure was identical to the Aquatolide structure (**10**), and interestingly, the original structure **9** was not even considered due to the high deviations. QM calculations of the ^{13}C chemical shifts were performed for these three structures, and the DFT-predicted ^{13}C chemical shifts showed that structure **10** was the most likely, proving that the combination of CASE and DFT can provide the correct structure of an unusual structure [40]. An illustrative workflow comparing the original and the CASE + DFT approaches is shown in Figure 7.2.

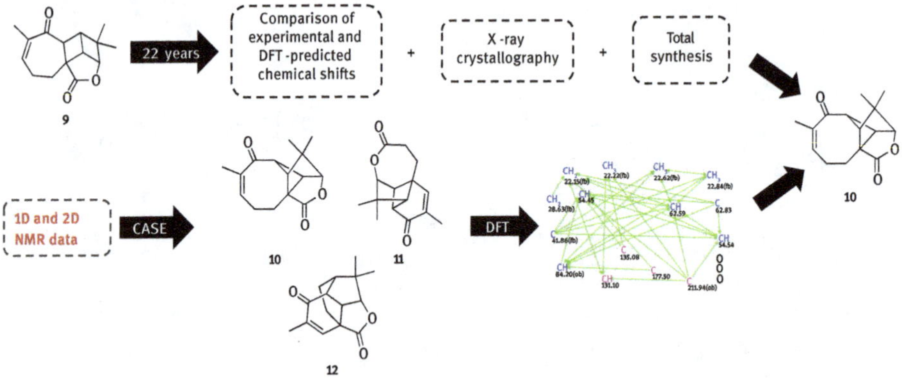

Figure 7.2: Comparison of the two approaches used in the structure elucidation of Aquatolide: using NMR data, DFT, X-ray crystallography and total synthesis or using the CASE approach combined with the DFT method. Adapted, reprinted with permission from [40]. Copyright 2016 American Chemical Society.

Considering the abovementioned examples and many others that can be found in the literature, it can be said that *StrucEluc* is, indeed, one robust and powerful tool for structure elucidation that can assist researchers in this process, making it possible to save time and optimize this important step in the natural products workflow.

7.4 Mass spectrometry and NMR for peptide sequencing

7.4.1 Fundamental aspects in MS of peptides and proteins

Mass spectrometry has been used for the structural investigation of peptides and proteins since 1980 [60, 61]. Through the use of ionization techniques, such as electrospray ionization (ESI) [62, 63] and matrix-assisted laser desorption/ionization (MALDI) [64], in combination with the development of sequential spectrometry (MS^n), the investigation of large molecules is now accessible in several laboratories [65].

The advantages of MS^n techniques with respect to speed, sensitivity, and applicability in complex peptide mixtures have gradually led to the substitution of techniques such as Edman by LCMS/MS [66]. Hybrid mass spectrometers, combining different types of mass analyzers, such as "Quadrupole-Time-of-Flight" (Q-TOF), have been increasingly used in a large number of applications due to the speed of data acquisition, high mass accuracy, resolution power, and high sensitivity [67]. The incorporation of ion-trap (IT) into TOF (IT-TOF) analyzers led not only to an increase in resolution power but also to high performance sequential spectrometric experiments since the IT system is capable of MS^n analysis.

Spectra that come from an ESI type usually form multiple charge ions (+2, +3, +4, etc.), especially when the experiment is performed on ion-trap instruments. Due to its functional construction architecture, this type of analyzer provides low-energy collisions between the ions, favoring the transfer of charge between them and consequently the ions containing more than one charge [68].

To interpret the MS and MS^n spectra, it is necessary to deconvolve and reconstruct the original spectrum to the form of monoprotonated ions $[M+H]^{+1}$. This work can be performed using mass table tools present in mass spectrometer software. The mass table provides information for m/z, intensity and charge values of each ion present in the MS/MS spectrum, in addition to the information of which ions are monoisotopic. Eq. (7.1) is used for the conversion of all the multi-charge ions into monoprotonated ions:

$$\text{Mass (Da)} = (m/q \times q) - q \tag{7.1}$$

m = mass; q = charge; and m/q = m/z (m/z ratio obtained in the mass spectrum)

7.4.2 Peptide fragmentation and sequencing

The fragmentation of peptides through mass spectrometry for the subsequent amino acid sequence elucidation is commonly performed by collision-induced dissociation (CID) [69]. Although other methodologies for the fragmentation of peptides, such as electron capture dissociation (ECD) and electron transfer dissociation (ETD) [70], have been developed, CID is undoubtedly the most widely used method and is routinely obtained in triple-quadrupole, ion-trap, and Q-ToF instruments using electrospray ionization [71]. The newest trend in *de novo* sequencing is the combination of different MS/MS activation techniques.

During the fragmentation process under CID conditions, the fragment ions receive a specific nomenclature depending on the region of the molecule that retains the residual charge (proton). When the residual charge remains on the N-terminal side, *a-,b-* and *c-* ions are generated (depending on which chemical bond was broken); on the other hand, when the residual charge remains on the C-terminal side, the *x-*, *y-* and *z-* ions are generated (depending on which chemical bond was fragmented). The pairs of ions *a/x*, *b/y* and *c/z* correspond to opposite fragments and are complementary to each other. Through fragmentation simulations, the complementarity and the presence of several series of ions prove if the obtained sequence is correct. The most accepted nomenclature for ions belonging to the peptide sequences is represented in Figure 7.3, as originally proposed by Roepstorff and Fohlman [72] and subsequently modified by Biemann [73].

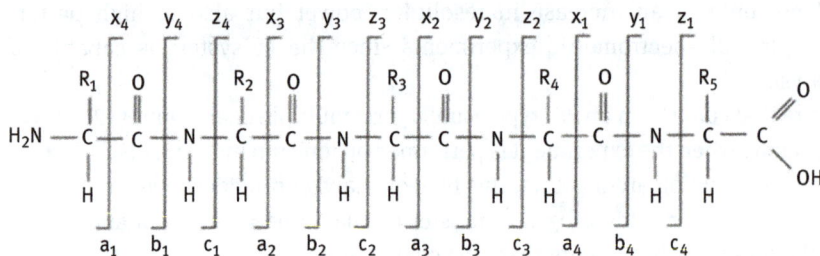

Figure 7.3: Peptide fragmentation scheme. Nomenclature of the major series of ions formed by peptide fragmentation: *a-, b-, c-, x-, y-,* and *z-ions*. Hydrogen rearrangements are omitted in this simplified annotation. R1, R2, and R3 represent the side chains of the amino acid residues. Source: Adapted from Biemann [72].

Preferentially, CID allows peptide fragmentation at the amide bonds along the chain, generating a sequence of *b*-ion series (if the charge is in the N-terminal portion of the fragment) or *y*-ion series (if the charge is in the C-terminal portion) [74]. Ideal experimental conditions (e. g. energy and collision gas) must be determined to generate fragmentation of the molecules in their peptide bonds. As a result of this fragmentation (series of *b*- and *y*-ions complementary to each other), the difference of *m/z* values between two consecutive ions in the same series reveals the mass of an amino acid residue and consequently, the identity of the same [65, 69].

The MS/MS spectra of tryptic peptides (with R or K in the C-terminal position) are often dominated by *y*-ions, which indicate higher stability of *y*-ions compared to *b*-ions due to the high basicity of these amino acid residues [69]. In the fragmentation of multiply charged ions, *b*- and *y*-ions are often formed simultaneously, but experimental data show that the balance between *b*- and *y*-ions is strongly dependent on the peptide sequence, particularly on the location of the basic residues [75].

Although the determination of the amino acids in a peptide is possible by simply calculating the mass difference between neighboring peaks in a specific series of ions, such work is quite difficult due to a series of factors:

–   The set of expected fragment ions can be incomplete, or in other words, there may be an absence of some ions in the *b*- and *y*-series;
–   Some fragments may undergo internal rearrangements and subsequent fragmentation;
–   The ions may be present with different states of charge, hindering the correct assignment of the ions (such difficulty applies in the interpretation of spectra that is not deconvoluted);
–   Some fragments may undergo neutral rearrangement of hydrogens during fragmentation.

Thus, the sum of these factors may induce the erroneous assignment of the ion series, making interpretation of the spectrum quite challenging. Most of the time, data interpretation must be performed by a combination of automated sequencing and manual evaluation [76–78].

7.4.3 Important rules for peptide spectrum interpretation

At the time of determining a sequence in a mass spectrum, there is a series of relevant information that benefits the correct determination of the amino acid sequence. As an example, there are some common losses or ions that give clues to the presence of certain amino acids. The following are some of these clues:

–   When there is a cleavage before or after R, the peak of –17 (loss of ammonia) may be greater than the corresponding *y*- or *b*-ion.
–   The series of ions may be incomplete when there is aspartic acid in the sequence.
–   Peptide fragments containing the amino acid residues R, K, Q, and N can lose ammonia (–17).
–   Peptide fragments containing the amino acid residues S, T, and E may lose water (–18). For glutamic acid, E must be at the N-terminus of the fragment for this observation to be made.

– The *b*-ion intensity will drop when the next residue is P, G, H, K, or R.
– Internal cleavages can occur at P and H residues. These are the result of a double cleavage event.

7.4.3.1 Immonium ions

When fragmentation occurs simultaneously at the amino- and carboxy-terminal positions of the same amino acid residue, immonium ions are produced. The immonium ions can be found in the low *m/z* range of the spectrum, and they can serve as diagnostic ions, indicating the presence or absence of certain amino acids in the sequence.

The intensity of immonium ions is strongly dependent on the position of the related amino acid within the peptide. Since an N-terminal position is particularly favorable, direct cleavage from the N-terminus is considered an important mechanism for the formation of immonium ions originating from amino acids located within the peptide chain that require medium to high offset values for their optimal formation since the fission of two covalent bonds is required [79, 80].

Immonium ions are often observed for aromatic amino acids (F, Y, pY, H, W), in addition to most aliphatic residues (L, I, V, P, camC, Met), and for D, E, R, and K. G, A, S, T, and C, immonium ions are normally not observed. Amino acids positioned at the N-terminus can generate abundant immonium ions, except for G, A, S, T, and C, as mentioned above. Consequently, a sequence-dependent effect or an insufficient collision offset may be the cause of the absence of immonium ions [75].

7.4.3.2 Isobaric mass: I/L ambiguities

The amino acids I and L are isobaric and cannot be differentiated using CID as a dissociation mechanism. When this mass difference is verified in the spectrum, it is recommended to use the symbol X or Lxx (L/I is another commonly used notation), according to the Hunt nomenclature [60].

The distinction between I and L residues can be accomplished by observing the *d*- and/or *w*- ions with the use of high collision intensity during the acquisition of the CID spectra of the peptides. Thus, it becomes possible to analyze ions resulting from partial fragmentation of the side chains of the amino acid residues, determined by the *d*- and/or *w*- ions [81].

The I/L ambiguities in the C- and N-terminal residues can be solved by synthesizing the sequences for each peptide (with I or L positioned at each terminus), which must be submitted to chromatographic analysis (in the same conditions as the initial experiment). The retention times of synthetic peptides can be compared to those of natural peptides [82].

7.4.3.3 Acetylation of lysine residues

K and Q are practically isobaric amino acids with a mass of 128.095 and 128.058, respectively. If a spectrometer is capable of generating results with high mass accuracy and resolution (such as Q-TOF, Orbitrap, and FT-ICR), the mass difference of 0.03638 u can be used to differentiate K and Q.

The acetylation of K residues can be used to distinguish between the isobaric amino acid residues of K and Q. For this experiment, peptides are derived with acetic anhydride and submitted to mass spectrometric analysis under CID conditions. The ε-amino group from the side chain of K residues and the α-amino group of the N-terminal residue of each peptide become acetylated, contributing in increments of 42 mass units per acetyl group that was incorporated in the peptide chain. The side chain of the Q residue does not react with acetic anhydride [82].

7.4.3.4 Posttranslational modifications (PTMs)

The identification and localization of certain PTMs can be performed through fragmentation rules for modified amino acid side chains (producing neutral loss or marker fragment ions). The PTMs occurring in the side chains of certain amino acids, such as S and T phosphorylation, glycosylation and/or oxidation of M make such side groups labile so that the neutral loss of these ions can be observed. As an example, it is possible to verify whether the constituent S of a given peptide has a phosphorylation. For this, it is necessary to verify whether there is an ion with a mass of 98 u inferior to the corresponding ion (b- or y-).

ETD is appropriate for fragmentation of large peptides, consequently improving sequence coverage over the use of CID alone [83]. ETD is advantageous for the study of PTMs, such as phosphorylation [84]. However, CID-based database search algorithms often underperform when analyzing ETD spectra, thus necessitating ETD-specific algorithms [85, 86]. For example, while CID fragmentation results in predictable b- and y-ion series, ETD fragmentation produces the more complex z-, c-, a-, and y-ion series.

7.4.4 Software and database tools

Since huge amounts of information are easily generated, data analysis is a key step in "-omics" research. Experiments involving peptidomic and/or proteomic analysis can be optimized with the use of different software for data acquisition, processing, analysis, and representation. Tandem mass spectrometry (MS/MS) is used as a major tool for peptide identification in current proteomics.

In a typical MS/MS experiment, protein mixtures are first digested into suitably sized peptides, and then the peptides are ionized via an ionization process. After that, selected peptides are further broken into fragment ions, and their tandem mass spectra (MS/MS spectra) are collected [87].

The large number of fragmentation spectra generated by the mass spectrometers requires automated search apparatuses capable of identifying and quantifying the analyzed peptides. In Table 7.1, there is a list of websites of databases, database

Table 7.1: Software and database tools for peptide and protein analysis (adapted from Szabo and Janaky [88]).

Software/database	Website
Tools for mass spectrometry	
Expasy tools	www.expasy.org
EBI tools, databases	www.ebi.ac.uk
UniProt, SwissProt, neXtProt	www.uniprot.org, www.nextprot.org
NCBI database, tools	www.ncbi.nlm.nih.gov
HUPO, c-HPP	www.hupo.org, www.c-hpp.org
Tools for peptide sequencing	
Lutefisk	www.hairyfatguy.com/Lutefisk
Novor	https://www.rapidnovor.com/download/
PepNovo	proteomics.ucsd.edu/Software/PepNovo/
PEAKS	www.bioinformaticssolutions.com
pNovo	pfind.ict.ac.cn/software/pNovo/index.html
Tools for database search	
SEQUEST	thermo.com
SpectrumMill	www.chem.agilent.com
ProteinLynxGlobalServer	www.waters.com
ProteinPilot	www.absciex.com
ProteinProspector	prospector.ucsf.edu
MASCOT	matrixscience.com
ProbID	tools.proteomecenter.org/wiki/index.php?title= Software:ProbID
X! Tandem (+the GPMdb database)	www.thegpm.org
MS-GF+	proteomics.ucsd.edu/software-tools/ms-gf/
Morpheus	morpheus-ms.sourceforge.net/
MS Amanda	ms.imp.ac.at/?goto=msamanda
Sequence tag and combined approaches	
InsPecT	proteomics.ucsd.edu/Software/Inspect.html
Popitam	code.google.com/p/popitam
TagRecon, DirectTag	fenchurch.mc.vanderbilt.edu/software.php
ByOnic	www.proteinmetrics.com/products/byonic/
Spectral library search	
SpectraST	www.peptideatlas.org/spectrast/
X! P3	p3.thegpm.org
BiblioSpec	skyline.gs.washington.edu
Postidentification treatment	
PeptideProphet/ProteinProphet	www.proteomecenter.org/software.php
Percolator	noble.gs.washington.edu/proj/percolator/
Scaffold	www.proteomesoftware.com/
MassSieve	www.ncbi.nlm.nih.gov/staff/slottad/MassSieve/
PeptideClassifier	http://www.mop.uzh.ch/software.html

(continued)

Table 7.1 (continued)

Software/database	Website
Spectral libraries/data management	
PeptideAtlas	www.peptideatlas.org
Proteios	www.proteios.org
SBEAMS	sbeams.org
CPAS	www.labkey.org/
PRIDE	www.ebi.ac.uk/pride/
MASPECTRAS 2	genome.tugraz.at/maspectras
ProteomXchange	www.proteomexchange.org
Multifunctional frameworks and pipelines	
MaxQuant (Andromeda search)	www.biochem.mpg.de/en/rd/maxquant/
EasyProt (Phenyx search)	easyprot.unige.ch/
VEMS 5.0	portugene.com/vems.html
Trans-proteomic pipeline	www.proteomecenter.org/software.php

searching, *de novo* sequencing, spectral library-searching algorithms, tools for post identification processing, MS data management, and other software for peptide structure elucidation.

The first post-MS step is to produce peak lists from MS raw data consisting of MS scans with three-dimensional axes (time, m/z, and ion counts) and MSMS scans with parent ion masses, acquired time, and fragment ions with ion counts.

7.4.5 Automated *de novo* sequencing

Proteomics investigation frequently requires *de novo* sequencing of peptides from tandem mass spectrometry (MS/MS). The MS/MS data size has grown; subsequently, *de novo* sequencing analyses are carried out more regularly with computer software than by a human expert.

De novo sequencing algorithms have some advantages over database search algorithms; for example, they can be used to identify peptides not contained in the database. This advantage is especially important for the identification of protein variants not represented in the database (Hoopmann and Moritz 2013). *De novo* peptide sequencing is a challenging and computationally intensive problem that includes both pattern recognition and global optimization on noisy and incomplete data.

Typically available *de novo* sequencing algorithms include PepNovo [89], pNovo [90], Novor [91], NovoHMM [92], Lutefisk [93] and the PEAKS software [94]. *De novo* sequencing is used to predict full or partial sequences. However, the prediction of peptide sequences from MS/MS spectra is dependent on the quality of the data, and this results in well-predicted sequences only for very high-quality data, while the results for mid- to low-quality data can sometimes be very bad.

Even when a protein database is available, *de novo* sequencing has been employed to assist the database search analysis. It was used to increase database search sensitivity and accuracy by confirming database search results [95] and to speed up database search by using *de novo* sequence tags as a filter [95–98].

7.4.6 Protein identification in data banks

Experimentally derived tandem mass spectra (MS/MS) are generally searched against a protein database of theoretically derived peptides, which provides identification of peptide sequences if the corresponding proteins are present in the databases.

Search algorithms aim to explain a recorded fragmentation spectrum by a peptide sequence from a predefined database, returning a list of peptide sequences that fit the experimental data. Some studies have examined scoring criteria for deciding whether to accept or reject a peptide identification from the different database search algorithms [99–104]. While the issue is still debated, two methods have been adopted generally. One method calculates a probability score for peptide or protein identification, and the second method estimates the false discovery rate (FDR) after the database search. For both methods, a value of 0.05 is typically a good starting place for initially accepting peptide identification.

The mathematical and statistical methods used for comparing and scoring the theoretical values and experimental data constitute the primary differences among the different database search programs [104]. Several strategies have been described to reduce the FDR of such matching approaches both at the peptide identification and protein assembly levels [105, 106].

The most commonly used programs and representative database search software packages include Mascot [107], SEQUEST [108], X!Tandem [109], OMSSA [110], Protein Prospector [111], MaxQuant [112], and MS-GFDB [113].

In Sequest, a signal processing technique called autocorrelation is used to mathematically determine the overlap between the theoretical spectrum, derived from each sequence obtained in the database in question, and the experimental spectrum. The result of such overlap is expressed quantitatively in terms of a score for each peptide (Xcorr). Xcorr is a parameter that depends on several factors, such as the state of charge of the peptide as well as the size of the data that is being used for the search [104, 108].

Mascot also involves the calculation of fragments theoretically predicted for all peptides of a database, according to the mass of the precursor ion previously determined. The m/z values of the predicted fragments are compared to the experimental fragments, and the comparison starts based on the more intense *b*- and *y*-ions. The probability that the m/z value of a theoretically obtained fragment randomly coincides with the m/z value of an experimentally obtained fragment is calculated and expressed as the negative of the log of that number (score). Thus, the higher the value obtained, the lower the probability that this result is a "coincidence". The manual interpretation of spectra is recommended and indispensable in some situations [114].

Other programs, such as STRING, provide a network view of functional protein associations based on direct (physical) and indirect (functional) protein-protein interactions [115]. This tool uses both known and predicted interactions to which a confidence score is attributed by comparing to a set of trusted associations (KEGG database). In other words, the protein-protein interaction score is the probability of the existence of such an interaction in the KEGG database [116]. This tool can be very useful when the objective is to gain insight into the biological processes that might be involved in the pathogenesis of diseases because it provides the gene ontology (GO) annotation of the inputted proteins.

7.4.7 Peptide identification using NMR

NMR experiments with proteins and peptides have been a long-standing goal in structural biology since chemical shifts are measurable under very general conditions and with precision [117, 118]. In the peptide identification procedure, the assignment was performed through a process in which each resonance must be associated with a specific nucleus in the investigated molecule. Before the 1980s, the assignment was established through NMR based on the assumption that the structure of the protein in solution was the same as in the X-ray structure. In the early 1980s, after the introduction of 2D NMR techniques such as COSY and NOESY, a sequential assignment procedure was developed for the assignment that relied only on information about the amino acid sequence [119]. However, only in the late 1980s was the first structure of a small globular protein published [120], and the credibility of NMR as a structural tool for peptides and proteins was strengthened over the years as its performance increased: 3D NMR was introduced first on unlabeled proteins followed quickly by a new set of triple resonance experiments [121] using ^{15}N- and ^{13}C-labeled samples.

In 2002, the Nobel Prize in Chemistry was awarded to Kurt Wüthrich for his development of NMR spectroscopy for determining the three-dimensional structure of biological macromolecules in solution. It has been a concern for spectroscopists to improve the understanding of minute details about the complicated conformational dependencies of the chemical shifts and the spin system, i.e., a set of nuclei connected by chemical bonds [117].

It is a consensus in NMR analysis of peptides and proteins that the order of difficulty increases as size increases. Peptides are small linear chains of amino acids. As a general rule, peptides are defined as molecules that consist of between 2 and 50 amino acids, whereas proteins are made up of 50 or more amino acids [122]. In this chapter, we address peptide NMR.

Currently, NMR experiments focusing on peptides start with a process called assignment, where the resonance should be associated with a specific nucleus in the investigated molecule. The data are assigned by one-dimensional (1D) ^{1}H, ^{13}C, and ^{15}N and two-dimensional (2D) techniques such as COSY, TOCSY (total correlation spectroscopy), NOESY and ROESY NMR. Several overlaps may occur in ^{1}H and ^{13}C NMR spectra due to the similarity of the chemical shifts from different amino acid residues, which is the reason 2D NMR techniques became very suitable for structure determination [123–125]. In summary, the assignment process can be performed using homonuclear spectra in three steps:

(a) Groups of scalar-coupled protons are first identified from TOCSY and COSY spectra and assigned to distinct amino acid types, with a characteristic pattern of cross signals. For example, Gly is the only residue in which two protons interact with the amide proton. Val, Leu, and Ile residues can be recognized by their two methyl groups, which give a characteristic row of double signals. In contrast, His, Trp, Tyr, and Phe residues are difficult to distinguish because Hα interacts with two methylene protons that do not show any J coupling to any of the ring protons. Therefore, these residues are also difficult to distinguish from Asp, Asn, Cys, and Ser residues [118, 126, 127].

(b) The next step in the assignment procedure concerns the identification of the sequence-specific position of each amino acid. This process can be performed preliminarily using NOESY. The connection of an amino acid in sequence (i) to its following sequence (i+1) can be observed because the distance of the amide proton of (i+1) to the Hα, Hβ or Hγ protons of (i) is frequently smaller than 5 Å. An exception is proline residues, which have no amide proton and for which no HN(i)-Hα(i-1) cross signal can be observed. However, if proline (i) is in its *trans* conformation, sequential HN(i-1)-Hδ(i) and Hα(i-1)-Hδ(i) cross signals can be observed [128]. It is important to highlight that the complete assignment of ^{13}C and ^{1}H resonances of peptides must be achieved via scalar coupling constants exclusively, using heteronuclear $^1J_{CH}$, $^2J_{CH}$, and $^3J_{CH}$ correlations from heteronuclear single-quantum coherence (HSQC) and heteronuclear multiple-bond correlation (HMBC) spectra (sequential assignment) to avoid bias by mixing assignment and determination of the stereostructure [127].

(c) The stereospecific assignments of diastereotopic methylene protons and methyl groups, which are mostly based on homo- and heteronuclear ^3J coupling constants, are established [127].

In practice, in peptide NMR, a spin system is usually defined as an amino acid residue (AA-fragment) - N, HN, Cα, Hα, C', and all nuclei of the side chain. Some amino acids have very characteristic Cα and Cβ chemical shifts (e. g. Ala, Gly, Ser, and Thr), while others can be easily mixed up (Phe/Tyr/Asp/Asn/Cys or Gln/Glu/Met) [129].

Recently, NMR spectrometers have benefited from several technological advances, such as higher magnetic field (≥ 950 MHz), cryoprobes and spectrometer electronics that lead to superb experimental long-term stability, and alternate processing methods are possible with the increased power of computers [130].

The demand for rapid, automated structure methods has brought forward refined new software for analyzing large molecules by spectroscopic techniques. NMR processing programs have been developed over the past several years. Here, we show some software used to process data, assignment and the molecular view of peptides (Table 7.2).

Table 7.2: Software used for peptide NMR analysis.

Software	Website	Reference
Processing data		
NMRPipe	https://www.ibbr.umd.edu/nmrpipe/	[131]
TopSpin	https://www.bruker.com/products/mr/nmr/nmr-software/software/topspin/overview.html	
Felix	https://www.FelixNMR.com	
CcpNmr	https://www.ccpn.ac.uk	[132]
Assignment		
CcpNmr	https://www.ccpn.ac.uk	[132]
Sparky	www.cgl.ucsf.edu/home/sparky/	[133]
AUTOASSIGN	http://nmr.cabm.rutgers.edu/autoassign/cgi-bin/aaenmr.py	[134]
CARA	http://cara.nmr-software.org/portal/	
Molecular viewers		
Pymol	https://pymol.org/2/	[135]
Rasmol	http://www.umass.edu/microbio/rasmol/	[136]
VMD	https://www.ks.uiuc.edu/Research/vmd/	[137]
Swiss-PdbViewer	https://spdbv.vital-it.ch	[138]

Users must decide between numerous software suites available to do processing data, such as NMRPipe, TopSpin, Felix and CcpNMR. These software programs increase the resolution of the spectrum, convert the time domain function to the frequency domain function, remove truncation artifacts, correct the phases of the spectrum and increase the signal-to-noise ratio. In Protein Data Bank (PDB), it is noted that approximately 40% of all NMR structures accepted are processed by NMRPipe. This software is a UNIX-based collection of programs and scripts for manipulating multidimensional NMR data. NMRPipe started with a spectral processing engine, and over the years, it has been augmented with a variety of facilities for spectral analysis and quantification, extraction of structural information from NMR data, and manipulation of molecular structures [139].

CcpNMR is a suite of programs, congregating from Format Converter (for data exchange with common textual NMR formats) to SpecView (for spectrum visualization, resonance assignment, and analysis). The CcpNmr Analysis program has been developed to replace NMR assignment applications such as ANSIG and Sparky. It's a complete program, with all new computer code; moreover, it provides novel ways to approach the analysis of NMR data, with particular emphasis on improving user productivity. Assignment in this software allows the user to represent anonymous but connected assignment states and allows atomic assignment to be made to several peaks at once [140].

There are free and open-source molecular graphics systems for visualization, animation, editing, and publication-quality imagery, as shown in Table 7.2. PyMOL is highlighted by its wide use and is still free for academic users. The academic version has limitations that may lag somewhat behind the most recent version that Schrödinger maintains, and no official support is offered. However, it is possible to find answers to questions on the web.

7.5 Conclusions and future directions

The development of several computational programs, databases, and analytical techniques made structure elucidation a faster process. The automation of this important step is in the process and could be a remarkable improvement for natural product research. NMR prediction of organic compounds is currently a well-established method available online for free or commercially by some companies. CASE programs are beginning to provide good results in structure elucidation, but in many cases, they are still not entirely automated. This field is rather new, and therefore, many new algorithms and technologies are to be developed in the near future. In fact, the elucidation procedure of peptides differs from other organic small molecules. NMR spectroscopy and mass spectrometry have been applied successfully to structural proteomics studies over the past few years. Certainly, understanding the assignment, fragmentation rules from peptides and how to interpret and annotate these data are important steps for all involved in proteomic research. Furthermore, advances in hardware design, data acquisition methods and automation of data analysis have been developed and successfully applied to high-throughput structure determination techniques. The set of rules and information compiled in this text, as well as the use of computer resources for this research, helps in understanding the ways to work on the elucidation of biological structures.

Acknowledgements: The authors acknowledge Fundação de Amparo à Pesquisa do Estado de São Paulo (FAPESP) grants #2013/07600-3 (CIBFar-CEPID), #2014/50926-0 (INCT BioNat CNPq/FAPESP), Coordenação de Aperfeiçoamento de Pessoal de Nível Superior (CAPES), Conselho Nacional de Desenvolvimento Científico e Tecnológico (CNPq) and Termo de Execução Descentralizado Arbocontrol #74/2016 for grant support and research fellowships. Authors acknowledge scholarships: MV (CNPQ #167874/2014-4 and #152243/2016-0; Finatec #120/2017), HMR (CNPQ #142014/2018-4), ACP (Fapesp #2016/13292-8), MEFP (Fapesp #2017/17098-4).

References

[1] Dobson CM. Chemical space and biology. Nature. 2004;432:824–8.

[2] Cragg GM, Newman DJ, Snader KM. Natural products in drug discovery and development. J Nat Prod. 1997;60:52–60.

[3] Harvey AL, Edrada-Ebel R, Quinn RJ. The re-emergence of natural products for drug discovery in the genomics era. Nat Rev Drug Discov. 2015;14:111–29.

[4] Hubert J, Nuzillard JM, Renault JH. Dereplication strategies in natural product research: how many tools and methodologies behind the same concept? Phytochem Rev. 2017;16:55–95.

[5] Agilent VnmrJ. Software for nuclear magnetic resonance spectroscopy. https://www.agilent.com/cs/library/flyers/public/VnmrJ4_Core_Features.pdf.

[6] Bruker. NMR software & Downloads. https://www.bruker.com/service/support-upgrades/software-downloads/nmr.html.

[7] JEOL. Delta™ NMR data processing software. https://www.jeolusa.com/PRODUCTS/Nuclear-Magnetic-Resonance/Delta-NMR-Software.

[8] Marat K. SpinWorks program. Nuclear magnetic resonance lab. https://home.cc.umanitoba.ca/~wolowiec/spinworks/.

[9] Mestrelab Research. Mnova12. http://mestrelab.com/.

[10] ACD/Labs. Platforms and products. https://www.acdlabs.com/products/.

[11] Langley P. The computational support of scientific discovery. Int J Human-Comput Stud. 2000;53:393–410.

[12] Clerc JT, Sommerauer HA. A minicomputer program based on additivity rules for the estimation of 13c-nmr chemical shifts. Anal Chim Acta. 1977;95:33–40.

[13] Fürst A, Retsch E. A computer program for the prediction of ^{13}C-NMR chemical shifts of organic compounds. Anal Chim Acta. 1990;229:17–25.

[14] Jurs PC, Ball JL, Anker LS, Friedman TL. Carbon-13 nuclear magnetic resonance spectrum simulation. J Chem Inf Comput Sci. 1992;32:272–8.

[15] Neudert R, Penk M. Enhanced structure elucidation. J Chem Inf Comput Sci. 1996;36:244–8.

[16] Bremser W. Expectation ranges of 13C NMR chemical shifts. Magn Reson Chem. 1985;23:271–5.

[17] Chen L, Robien W. The CSEARCH-NMR data base approach to solve frequent questions concerning substituent effects on ^{13}C NMR chemical shifts. Chemom Intell Lab Syst. 1993;19:217–23.

[18] Gasteiger J, Zupan J. Neural networks in chemistry. AngewChem. 1993;32:503–27.

[19] Clouser DL, Jurs PC. Simulation of ^{13}C nuclear magnetic resonance spectra of tetrahydropyrans using regression analysis and neural networks. Anal Chim Acta. 1994;295:221–31.

[20] Grant DM, Paul EG. Carbon-13 magnetic resonance. II. Chemical shift data for the alkanes. J Am Chem Soc. 1964;86:2984–90.

[21] Schweitzer RC, Small GW. Automated spectrum simulation methods for carbon-13 nuclear magnetic resonance spectroscopy based on database retrieval and model-building strategies. J Chem Inf Comput Sci. 1997;37:249–57.

[22] Bremser W. Hose - a novel substructure code. Anal Chim Acta. 1978;103:355–65.

[23] Elyashberg ME, Williams A, Blinov K. Contemporary computer-assisted approaches to molecular structure elucidation. Cambridge, U.K: Royal Society of Chemistry, 2012.

[24] Aires-de-Souza J, Hemmer MC, Gasteiger J. Prediction of 1H NMR chemical shifts using neural networks. Anal Chem. 2002;74:80–90.

[25] Binev Y, Aires-de-Souza J. structure-based predictions of 1H NMR chemical shifts using feed-forward neural networks. J Chem Inf Model. 2004;44:940–5.

[26] Binev Y, Corvo M Aires-de-Souza J. The impact of available experimental data on the prediction of 1H NMR chemical shifts by neural networks. J Chem Inf Model. 2004;44:946–9.

[27] Lodewyk MW, Siebert MR, Tantillo DJ. Computational prediction of 1H and 13C chemical shifts: a useful tool for natural product, mechanistic, and synthetic organic chemistry. Chem Rev. 2012;112:1839–62.

[28] Simpson AJ, Lefebvre B, Moser A, Williams AJ, Larin N, Kvasha M, et al. Identifying residues in natural organic matter through spectral prediction and pattern matching of 2D NMR datasets. Magn Reson Chem. 2004;42:14–22.

[29] Will M, Fachinger W, Richert JR Fully automated structure elucidation - A spectroscopist's dream comes true. J Chem Inf Comput Sci. 1996;36:221–7.

[30] Jaspars M. Computer assisted structure elucidation of natural products using two-dimensional NMR spectroscopy. Nat Prod Rep. 1999;16:241–8.

[31] Shelley CA, Hays TR, Munk ME, Roman RV. An approach to automated partial structure expansion. Anal Chim Acta. 1978;103:121–32.

[32] Carhart RE, Smith DH, Gray NA, Nourse JG, Djerassi C. Applications of artificial intelligence for chemical inference. 37. GENOA: a computer program for structure elucidation utilizing overlapping and alternative substructures. J Org Chem. 1981;46:1708–18.

[33] Kudo Y, Sasaki S. Principle for exhaustive enumeration of unique structures consistent with structural information. J Chem Inf Comput Sci. 1976;16:43–56.

[34] Munk ME, Velu VK, Madison MS, Robb EW, Baderstscher M, Christie BD, et al. Chemical Information Processing in Structure Elucidation. Recent advances in chemical information II. Cambridge, UK: Royal Society of Chemistry. In: Collier, H., editor 1993:247–63.

[35] Bremser W, Fachinger W. Multidimensional spectroscopy. Magn Reson Chem. 1985;23:1056–71.

[36] Schutz V, Purtuc V, Felsinger S, Robien W. CSEARCH-STEREO: A new generation of NMR database systems allowing three-dimensional spectrum prediction. Fresenius' J Anal Chem. 1997;359:33–41.

[37] Carabedian M, Dagane I, Dubois J-E. Elucidation by progressive intersection of ordered substructures from carbon-13 nuclear magnetic resonance. Anal Chem. 1988;60:2186–92.

[38] Williamson RT, Buevich AV, Martin GE, Parella T. LR-HSQMBC: A sensitive NMR technique to probe very long-range heteronuclear coupling pathways. J Org Chem. 2014;79:3887–94.

[39] Uhrín D. Recent developments in liquid-state INADEQUATE studies. In Annual Reports on NMR Spectroscopy. [s.l.]. Cambridge, MA, US: Academic Press, 2010:1–34.

[40] Buevich AV, Elyashberg ME. Synergistic combination of CASE algorithms and DFT chemical shift predictions: a powerful approach for structure elucidation, verification, and revision. J Nat Prod. 2016;79:3105–16.

[41] Binev Y, Marques MM, Aires-de-Sousa J. Prediction of 1H NMR coupling constants with associative neural networks trained for chemical shifts. J Chem Inf Model. 2007;47:2089–97.

[42] Steinbeck C. Recent developments in automated structure elucidation of natural products. Nat Prod Rep. 2004;21:512–18.

[43] Canzler D, Hellenbrandt M. SPECINFO - the spectroscopic information system on STN international. Fresen J Anal Chem. 1992;344:167–72.

[44] Steinbeck C, Kuhn S, Krause S. NMRShiftDB constructing a free chemical information system with open-source components. J Chem Inf Comput Sci. 2003;43:1733–9.

[45] Cobas JC, Sardina FJ. Nuclear magnetic resonance data processing. MestRe-C: Software Package Desktop Comput ConceptsMagnReson Part A. 2003;19A:80–96.

[46] Cobas C, Seoane F, Vaz E, Bernstein MA, Dominguez S, Péreza M, et al. Automatic assignment of 1H-NMR spectra of small molecules. Magn Reson Chem. 2013;51:649–54.

[47] Mestrelab Research NMR Predict. 2018. http://mestrelab.com/software/mnova/nmr-predict/.

[48] Mestrelab Research Resources. 2018. http://resources.mestrelab.com/mestrecvsmnova.

[49] Mestrelab Research Mnova Structure Elucidation. 2018. http://mestrelab.com/software/mnova/structure-elucidation/.

[50] ACD/Structure Elucidator Suite. 2018. https://www.acdlabs.com/products/com_iden/elucida tion/struc_eluc/.

[51] Elyashberg M, Blinov K, Martirosian E. A new approach to computer-aided molecular structure elucidation: the expert system structure elucidator. Lab Autom Inform Manag. 1999;34:15–30.

[52] Blinov KA, Elyashberg ME, Molodtsov SG, Williams AJ, Martirosian ER. An expert system for automated structure elucidation utilizing ^{1}H-^{1}H, ^{13}C-^{1}H and ^{15}N-^{1}H 2D NMR correlations. Fresen J Anal Chem. 2001;369:709–14.

[53] Elyashberg ME, Blinov KA, Williams AJ, Martirosian ER, Molodtsov SG. Application of a new expert system for the structure elucidation of natural products from their 1D and 2D NMR data. J Nat Prod. 2002;65:693–703.

[54] Elyashberg ME, Blinov KA, Williams AJ, Molodtsov SG, Martin GE. Are deterministic expert systems for computer-assisted structure elucidation obsolete? J Chem Inf Model. 2006;46:1643–56.

[55] Elyashberg M, Williams AJ, Blinov K. Structural revisions of natural products by computer-assisted structure elucidation (CASE) systems. Nat Prod Rep. 2010;27:1296.

[56] Li X-N, Ridge CD, Mazzola EP, Sun J, Gutierrez O, Moser A, et al. Application of a computer-assisted structure elucidation program for the structural determination of a new terpenoid aldehyde with an unusual skeleton. Magn Reson Chem. 2017;55:210–13.

[57] San Feliciano A, Medarde M, Del Corral JM, Aramburu A, Gordaliza M, Barrero AF. Aquatolide. A new type of humulane-related sesquiterpene lactone. Tetrahedron Lett. 1989;30:2851–4.

[58] Lodewyk MW, Soldi C, Jones PB, Olmstead MM, Rita J, Shaw JT, et al. The correct structure of Aquatolide—experimental validation of a theoretically-predicted structural revision. J Am Chem Soc. 2012;134:18550–3.

[59] Saya JM, Vos K, Kleinnijenhuis RA, van Maarseveen JH, Ingemann S, Hiemstra H. Total synthesis of Aquatolide. Org Lett. 2015;17:3892–4.

[60] Hunt DF, Yates JR, Shabanowitz J, Winston S, Hauer CR. Protein sequencing by tandem mass spectrometry. Proc Nat Acad Sci USA. 1986;17:6233–7.

[61] Biemann K, Scoble HA. Characterization by tandem mass spectrometry of structural modifications in proteins. Science. 1987;237:992–8.

[62] Dole M, Mack LL, Hines RL, Mobley RC, Ferguson LD, Alice MB. Molecular beams of macroions. J Chem Phys. 1968;49:2240–9.

[63] Fenn JB, Mann M, Meng CK, Wong SF, Whitehouse CM. Electrospray ionization for mass spectrometry of large biomolecules. Science. 1989;246:64–71.

[64] Hillenkamp F, Karas M, Beavis RC, Chait BT. Matrix-assisted laser desorption/ionization mass spectrometry of biopolymers. Anal Chem. 1991;15:1193–203.

[65] Dongre AR, Jones JL, Somogyi A, Wysocki VH. Influence of peptide composition, gas-phase basicity, and chemical modification on fragmentation efficiency: evidence for the mobile proton model. J Am Chem Soc. 1996;118:8365–74.

[66] Seidler J, Zinn N, Boehm ME, Lehmann WD. De novo sequencing of peptides by MS/MS Proteomics. 2010;10:634–49.

[67] Steen H, Kuster B, Mann M. Quadrupole time-of-flight versus triple-quadrupole mass spectrometry for the determination of phosphopeptides by precursor ion scanning. J Mass Spectrom. 2001;36:782–90.

[68] Prentice BM, Xu W, Ouyang Z, Mcluckey SA. Dc potentials applied to an end-cap electrode of a 3-D ion trap for enhanced MS functionality. Int J Mass Spectrom Amsterdam. 2011;306: 114–22.

[69] Tabb DL, Smith LL, Breci LA, Vysocki VH, Lin D, Yates JR Statistical characterization of ion trap tandem mass spectra from doubly charged tryptic peptides. Anal Chem. 2003;75:1155–63.

[70] Syka JE, Coon JJ, Schroeder MJ, Shabanowitz J, Hunt DF. Peptide and protein sequence analysis by electron transfer dissociation mass spectrometry. Proc Nat Acad Sci USA. 2004;101:9528–33.

[71] Bauer MD, Sun Y, Keough T, Lacey MP. Sequencing of sulfonic acid derivatized peptides by electrospray mass spectrometry. Rapid Commun Mass Spectrom. 2000;14:924–9.

[72] Roepstorff P, Fohlman J. Proposal for a common nomenclature for sequence ions in mass spectra of peptides. Biomed Mass Spectrom. 1984;11:601.

[73] Biemann K. Appendix 5. Nomenclature for peptide fragment ions (positive ions). Methods Enzymol. 1990;193:886–7.

[74] Adamczy KM, Gebler JC, Wu J. Charge derivatization of peptides to simplify their sequencing with an ion trap mass spectrometer. Rapid Commun Mass Spectrom. 1999;13:1413–22.

[75] Lehmann WD. Protein phosphorylation analysis by electrospray mass spectrometry: a guide to concepts and practice/Wolf D. Lehmann, vol. xiv. Cambridge: Royal Society of Chemistry, 2010:379.

[76] Jia C, Hui L, Cao W, Lietz CB, Jiang X, Chen R, et al. High-definition de novo sequencing of crustacean hyperglycemic hormone (CHH)-family neuropeptides. Mol Cell Proteomics. 2012;11:1951–64.

[77] Medzihradszky KF, Bohlen CJ. Partial de novo sequencing and unusual CID fragmentation of a 7 kDa, disulfide-bridged toxin. J Am Soc Mass Spectrom. 2012;23:923–34.

[78] Samgina TY, Artemenko KA, Gorshkov VA, Ogourtsov SV, Zubarev RA, Lebedev AT. De novo sequencing of peptides secreted by the skin glands of the caucasian green frog rana ridibunda. Rapid Commun Mass Spectrom. 2008;22:3517–25.

[79] Falick AM, Hines WM, Medzihradszky KF, Baldwin MA, Gibson BW. Low-mass ions produced from peptides by high-energy collision-induced dissociation in tandem mass spectrometry. J Am Soc Mass Spectrom. 1993;4:882–93.

[80] Hohmann LJ, Eng JK, Gemmill A, Klimek J, Vitek O, Reid GE, et al. Quantification of the compositional information provided by immonium ions on a quadrupole-time-of-flight mass spectrometer. Anal Chem. 2008;80:5596–606.

[81] Mendes MA, De Souza BM, Santos LD, Palma MS. Structural characterization of novel chemotactic and mastoparan peptides from the venom of the social wasp Agelaia pallipes pallipes by high-performance liquid chromatography/electrospray ionization tandem mass spectrometry. Rapid Commun Mass Spectrom. 2004;18:636–42.

[82] Dias NB, De Souza BM, Gomes PC, Brigatte P, Palma MS. Peptidome profiling of venom from the social wasp Polybia paulista. Toxicon. 2015;107:290–303.

[83] Molina H, Matthiesen R, Kandasamy K, Pandey A. Comprehensive comparison of collision induced dissociation and electron transfer dissociation. Anal Chem. 2008;80:4825–35.

[84] Mikesh LM, Ueberheide B, Chi A, Coon JJ, Syka JE, Shabanowitz J, et al. The utility of ETD mass spectrometry in proteomic analysis. Biochim Biophys Acta. 2006;1764:1811–22.

[85] Liu X, Shan B, Xin L, Ma B. Better score function for peptide identification with ETD MS/MS spectra. BMC Bioinformatics. 2010;11:S1–S4.

[86] Sadygov RG, Good DM, Swaney DL, Coon JJ. A new probabilistic database search algorithm for ETD spectra. J Proteome Res. 2009;8:3198–205.

[87] Wysocki VH, Resing KA, Zhang Q, Cheng G. Mass spectrometry of peptides and proteins. Methods. 2005;35:211–22.

[88] Szabo Z, Janaky T. Challenges and developments in protein identification using mass spectrometry. TrAC Trends Anal Chem. 2015;69:76–87.

[89] Frank A, Pevzner P. PepNovo: de novo peptide sequencing via probabilistic network modeling. Anal Chem. 2005;77:964–73.

[90] Chi H, Chen H, He K. pNovoþ: de novo peptide sequencing using complementary HCD and ETD tandem mass spectra. J Proteome Res. 2013;12:615–25.

[91] Ma B. Novor: real-time peptide de novo sequencing software. J Am Soc Mass Spectrom. 2015;26:1885–94.

[92] Fischer B, Roth V, Roos F, Grossmann J, Baginsky S, Widmayer P, et al. NovoHMM: A hidden Markov model for de novo peptide sequencing. Anal Chem. 2005;77:7265–73.

[93] Taylor JA, Johnson RS. Sequence database searches via de novo peptide sequencing by tandem mass spectrometry. Rapid Commun Mass Spectrom. 1997;11:1067–75.

[94] Ma B, Zhang K, Hendrie C, Liang C, Li M, Doherty-Kirby A, et al. PEAKS: powerful software for peptide de novo sequencing by tandem mass spectrometry. Rapid Commun Mass Spectrom. 2003;17:2337–42.

[95] Zhang J, Xin L, Shan B, Chen W, Xie M, Yuen D, et al. PEAKS DB: de novo sequencing assisted database search for sensitive and accurate peptide identification. Mol Cell Proteom. 2012;11:1–8.

[96] Tanner S, Shu H, Frank A, Wang LC, Zandi E, Mumby M, et al. InsPecT: identification of post translationally modified peptides from tandem mass spectra. Anal Chem. 2005;77:4626–39.

[97] Liu C, Yan B, Song Y, Xu Y, Cai L. Peptide sequence tag-based blind identification of post-translational modifications with point process model. Bioinformatics. 2006;22:e307–13.

[98] Han X, He L, Xin L, Shan B, Ma B. Peaks PTM: mass spectrometry-based identification of peptides with unspecified modifications. J Proteome Res. 2011;10:2930–6.

[99] Washburn MP, Wolters D, Yates JR Large-scale analysis of the yeast proteome by multidimensional protein identification technology. Nat Biotechnol. 2001;19:242–7.

[100] Keller A, Nesvizhskii AI, Kolker E, Aebersold R. Empirical statistical model to estimate the accuracy of peptide identifications made by MS/MS and database search. Anal Chem. 2002;74:5383–92.

[101] MacCoss MJ, Wu CC, Yates JR Probability-based validation of protein identifications using a modified SEQUEST algorithm. Anal Chem. 2002;74:5593–9.

[102] Nesvizhskii AI, Keller A, Kolker E, Aebersold R. A statistical model for identifying proteins by tandem mass spectrometry. Anal Chem. 2003;75:4646–58.

[103] Peng J, Elias JE, Thoreen CC, Licklider LJ, Gygi SP. Evaluation of multidimensional chromatography coupled with tandem mass spectrometry (LC/LC-MS/MS) for large-scale protein analysis: the yeast proteome. J Proteome Res. 2003;2:43–50.

[104] Sadygov RG, Liu H, Yates JR Statistical models for protein validation using tandem mass spectral data and protein amino acid sequence databases. Anal Chem. 2004;76:1664–71.

[105] Nesvizhskii AI, Aebersold R. Interpretation of shotgun proteomic data: the protein inference problem. Molecular & Cellular Proteomics. 2005;4:1419–40.

[106] Nesvizhskii AI. A survey of computational methods and error rate estimation procedures for peptide and protein identification in shotgun proteomics. J Proteomics. 2010;73:2092–123.

[107] Perkins DN, Pappin DJ, Creasy DM, Cottrell JS. Probability-based protein identification by searching sequence databases using mass spectrometry data. Electrophoresis. 1999;20:3551–67.

[108] Eng J, McCormack AL, Yates JR An approach to correlate tandem mass spectral data of peptides with amino acid sequences in a protein database. J Am Soc Mass Spectrom. 1994;5:976–89.

[109] Craig R, Beavis RC. TANDEM: matching proteins with tandem mass spectra. Bioinformatics. 2004;20:1466–7.

[110] Geer LY, Markey SP, Kowalak JA, Wagner L, Xu M, Maynard DM, et al. Open mass spectrometry search algorithm. J Proteome Res. 2004;3:958–64.

[111] Chalkley RJ, Baker PR, Huang L, Hansen KC, Allen NP, Rexach M, et al. Comprehensive analysis of a multidimensional liquid chromatography mass spectrometry dataset acquired on a quadrupole selecting quadrupole collision cell, time-of-flight mass spectrometer: II. New developments in protein prospector allow for reliable and comprehensive automatic analysis of large datasets. Mol Cell Proteomics. 2005;4:1194–204.

[112] Cox J, Mann M. MaxQuant enables high peptide identification rates, individualized p.p.b.-range mass accuracies and proteome-wide protein quantification. Nat Biotechnol. 2008;26:1367–72.

[113] Kim S, Mischerikow N, Bandeira N, Navarro JD, Wich L, Mohammed S, et al. The generating function of CID, ETD and CID/ETD pairs of tandem mass spectra: applications to database search. Mol Cell Proteomics. 2010;9:2840–52.

[114] Cantú MD, Carrilho E, Wulff NA, Palma MS Sequenciamento de peptídeos usando espectrometria de massas: um guia prático. Quim Nova. 2008;31:669–75.

[115] Szklarczyk D, Franceschini A, Wyder S, Forslund K, Heller D, Huerta-Cepas J, et al. STRING v10: protein-protein interaction networks, integrated over the tree of life. Nucleic Acids Res. 2015;43:D447–52.

[116] von Mering C, Jensen LJ, Snel B, Hooper SD, Krupp M, Foglierini M, et al. STRING: known and predicted protein–protein associations, integrated and transferred across organisms. Nucleic Acids Res. 2005;33:D433–7.

[117] Kohlhoff KJ, Robustelli P, Cavalli A, Salvatella X, Vendruscolo M. Fast and accurate predictions of protein NMR chemical shifts from interatomic distances. J Am Chem Soc. 2009;7:13894–5.

[118] Wuthrich K. NMR spectra of proteins and nucleic acids in solution. In: Wuthrich K, editor. NMR of proteins and nucleic acids. New York: Wiley, 1986:1–320.

[119] Redfield C. The Application of 1H Nuclear Magnetic Resonance Spectroscopy to the Study of Enzymes. In: Cooper A, Houben J, Chien L, editor(s). The Enzyme Catalysis Process: Energetics, Mechanism and Dynamic. Boston, MA: Springer, 1989:141–58.

[120] Williamson MP, Havel TF, Wüthrich K. Solution confor-mation of proteinase inhibitor IIA from bull seminal plasma by 1H nuclear magnetic resonance and distance geometry. J Mol Biol. 1985;182:295–315.

[121] Ikura M, Kay LE, Bax A. A novel approach for sequential assignment of proton, carbon-13, and nitrogen-15 spectra of larger proteins: heteronuclear triple-resonance three-dimensional NMR spectroscopy. Application to calmodulin. Biochemistry. 1990;29:4659–67.

[122] Clayden J, Greeves N, Warren S, Wothers P. Organic chemistry. New York: Oxford University Press Inc., 2001:1133.

[123] Pomilio AB, Battista ME, Vitale AA. Naturally-occurring cyclopeptides: structures and bioactivity. Curr Org Chem. 2006;10:2075–121.

[124] Kaas Q, Craik DJ. NMR of plant proteins. Prog Nucl Magn Reson Spectrosc. 2013;71:1–34.

[125] Machado A, Liria CW, Proti PB, Remuzgo C, Miranda MT Chemical and enzymatic peptide synthesis: basic aspects and applications. Quim Nova. 2004;27:781–9.

[126] Bax A. Two-dimensional NMR and protein structures. Annu Rev Biochem. 1989;58:223–56.

[127] Beck JG, Frank AO, Kessler H. NMR of peptides. In: Bertini I, McGreevy KS, Parigi G, editors. NMR of biomolecules. Weinheim: Wiley-VCH Verlag GmbH & Co. KGaA, 2012:328–44.

[128] Bax A. Multidimensional nuclear magnetic resonance methods for protein studies. Curr Opin Struc Biol. 1994;4:738–44.

[129] Cavanagh J, Fairbrother WJ, Palmer AG, Skelton NJ. Protein NMR spectroscopy. Principles and practice, 2nd ed. USA: Academic Press, 2006: 912 pp.

[130] Marion D. An introduction to biological NMR spectroscopy. Mol Cell Proteomics. 2013; 12:3006–25.

[131] Delaglio F, Grzesiek S, Vuister GW, Zhu G, Pfeifer J. NMRPipe: a multidimensional spectral processing system based on UNIX pipes. J Biomol NMR. 1995;6:277.

[132] Vranken WF, Boucher W, Stevens TJ, Fogh RH, Pajon A, Llinas M, et al. The CCPN Data Model for NMR Spectroscopy: development of a Software Pipeline. Proteins. 2005;59:687–96.

[133] Lee W, Tonelli M, Markley JL. NMRFAM-SPARKY: enhanced software for biomolecular NMR spectroscopy. Bioinformatics. 2015;31:1325–7.

[134] Zimmerman DE, Kulikowski CA, Huang Y, Feng W, Tashiro M, Shimotakahara S, et al. Automated analysis of protein NMR assignments using methods from artificial intelligence. J Mol Biol. 1997;269:592–610.

[135] DeLano WL. ThePyMOL molecular graphics system. San Carlos, CA: DeLano Scientific, 2002.

[136] Sayle R, Milner-White EJ. RasMol: biomolecular graphics for all. Trends Biochem Sci (TIBS). 1995;20:374.

[137] Humphrey W, Dalke A, Schulten K. VMD - visual molecular dynamics. J Molec Graphics. 1996;14:33–8.

[138] Guex N, Peitsch MC. SWISS-MODEL and the Swiss-PdbViewer: an environment for comparative protein modeling. Electrophoresis. 1997;18:2714–23.

[139] NMRipe. NMRPipe: a multidimensional spectral processing system based on UNIX pipes. 2018. https://www.ibbr.umd.edu/nmrpipe/index.html.

[140] CcpNMR. Collaborative computing project for NMR. 2018. https://www.ccpn.ac.uk/about.

Marilia Valli, Helena Mannochio Russo, Alan Cesar Pilon,
Meri Emili Ferreira Pinto, Nathalia B. Dias, Rafael Teixeira Freire,
Ian Castro-Gamboa and Vanderlan da Silva Bolzani

8 Computational methods for NMR and MS for structure elucidation II: database resources and advanced methods

Abstract: Technological advances have contributed to the evolution of the natural product chemistry and drug discovery programs. Recently, computational methods for nuclear magnetic resonance (NMR) and mass spectrometry (MS) have speeded up and facilitated the process of structural elucidation even in high complex biological samples. In this chapter, the current computational tools related to NMR and MS databases and spectral similarity networks, as well as their applications on dereplication and determination of biological biomarkers, are addressed.

Keywords: NMR and MS databases, dereplication, spectral similarity networks

8.1 Introduction

In the beginning of the era of structural identification, the analysis of structures was performed by chemical reactions to verify the presence of carbons, types of bonds, and finally, when scientists supposedly had an idea about the structure of a molecule they synthesized. One example is the famous isolation of morphine in 1804 by Friedrich Sertürner, which was only determined half a century later by Robert Robinson (1952) using organic synthesis and complementary approaches, such as physicochemical analysis, melting point, infrared analysis and optical rotation [1].

Since then, technological advances have contributed to the evolution of natural product chemistry and drug discovery programs. Currently, state-of-the-art chromatography (gas chromatography – GC, liquid chromatography – LC and Capillary Electrophoresis – CE) allows the separation of hundreds of compounds in a single analysis. Infrared and ultraviolet techniques bestow information on organic functions and unsaturated systems. ToF (Time of Flight), Orbitrap and FT-ICR mass spectrometers (Fourier-transform ion cyclotron resonance mass spectrometry) provide up to 1 million mass resolution power and scanning speeds that allow the development of hyphenated approaches, such as LC coupled to mass spectrometry

This article has previously been published in the journal *Physical Sciences Reviews*. Please cite as:
Valli, M., Russo, H.M., Pilon, A.C., Pinto, M.E.F., Dias, N.B., Freire, F.T., Bolzani, V.S. Computational Methods for NMR and MS for Structure Elucidation II: Database Resouces and Advanced Methods. *Physical Sciences Reviews* [Online] **2019** DOI: 10.1515/psr-2018-0167.

https://doi.org/10.1515/9783110579352-009

(LC-MS) and GC-MS, as well as fragmentation techniques that aid in the structural determination of compounds in complex mixtures [2]. Modern analytical techniques, such as nuclear magnetic resonance (NMR), also contribute to the determination of organic structures, establishing spatial arrangements between atoms and their neighbors through magnetic fields (1D and 2D NMR experiments), as well as vibrational chiroptic spectroscopic methods – electronic circular dichroism (ECD), vibrational circular dichroism (VCD) and Raman optical activity (ROA) – that have aided in conformational and configurational tasks [2, 3].

Over the past few years, many programs have been developed to aid in the structure elucidation process. Databases and molecular networking (MN) have been increasingly used for structure elucidation and are in constant development.

8.2 Databases for assisting structure elucidation

Analytical and computational advances (computational empowerment, expansion of the internet, secondary memory storage capacity and information system) brought forward the development of several fields of chemical biology, including the –omics approaches, biochemistry, natural products and specifically, structure elucidation processes [4, 5]. However, the unprecedented increasing amount of data generated in a postgenomic era has made the management of data storage a real challenge for any research laboratory.

To address these issues with data management, scientists initially decided to share and store raw data and spectral information in online data centers, with the primary goal of ensuring long-term data availability and providing shortcuts to other researchers [6]. One of the pioneering examples of a web-based identification tool was the Protein Database (PDB), a repository composed of Cartesian coordinates of proteins and enzymes designed and developed for biologists and chemists. Similarly, natural products were later compiled by the Dictionary of Natural Products (DNPs), one of the first to organize organic compounds in books and manuals and currently one of the main catalogs of natural products with more than 290,000 substances in its web platform [6].

As previously mentioned, these databases were initially developed to dam the flow of "omics" science data, ensuring free access to experimental results and long-term data availability [7]. Over time, bioinformatics, computer science, and statistics have been slowly incorporated into easy-to-use online interfaces, offering new forms of database design, content access (uploads and downloads), and pattern recognition tools, in addition to expanding the storage capacity and management of large quantities of heterogeneous data. Thus, the former repositories uniquely associated with data compilation lost their value and were gradually transformed into a new database category, known as "knowledge discovery databases (KDDs)" or data mining databases [4].

KDDs are characterized by multiple layers of interactive, nontrivial, and iterative processes to identify comprehensible, reliable, and potentially useful patterns of large amounts of data [8]. These KDDs, combined with state-of-the-art artificial intelligence tools, have played a key role in the current lifestyle of our contemporary society. Maps, stores and services are easily available and are specifically targeted to the user based on his or her pattern of choices and preferences. Likewise, identifying as many patterns as possible in genetic and transcribed sequences, proteins and metabolites of organisms will bring a new way of understanding nature and its possible effects. The integration of "omics" aspects into environmental/genetic/ temporal phenomena will allow the understanding of the physiological dynamics between genes, proteins and metabolic processes in a broader context [9].

For small molecules, two fields notably have used large amounts of data for research and development: (i) drug discovery, encompassing a variety of subareas such as natural products, combinatorial chemistry and molecular modeling, and (ii) metabolomics, consisting of a multidisciplinary field to understand organisms, comparing different temporal/environmental/genetic conditions using metabolic profiles.

For drug discovery, there are several KDDs with the aim of storing, distribut- ing and discovering information from different types of data. Some of them are very general, while others report specific information about molecular pathways, structural features, side effects, drug targets or even a specific target (an enzyme, for example). The number of drug design databases also increased as a result of high-throughput screening approaches developed by pharmaceutical enterprises. These advances allowed us to estimate the structure-activity relationships, mechanisms of action and absorption, distribution, metabolism, excretion and toxicity (ADMET) studies based on molecular characteristics. ChemChem, ChemBank, ChemProt, PDBind and Zinc bioassays are among the most prominent drug discovery databases [4, 6].

One of the main concerns involving public domain databases (KDDs) is data curation. Artificial intelligence curation, limitations on chemical structure software, and differences in individual database applications result in a lack of chemical equivalence between public domain repositories. For the purposes of chemical structure elucidation, different representations of chemical structures were devel- oped to minimize the lack of cohesion related to the chemical representation of enantiomers, tautomers, charged species and stereogenic centers and other conflict- ing problems. CTAB (connection-table, a table of atoms and its corresponding bond- ing table) was the first molecular representation symbolizing the chemistry valence model (nodes and edges as atoms and bonds). With the expansion of computational science, this model presented some problems: (1) inefficient: CTAB requires one or two dozen bytes per atom (modern notation uses 4–5 bytes per atom); (2) lack of specificity: absence of valence assumptions and discarding of hydrogen may bring molecular ambiguity; (3) problems with the notation of connectivity: the notation of the conformers requires the complete repetition of all the parameters of the molecular structure, although they are identical. However, many public domain databases are still using similar extensions, such as v2000 molfiles [10].

Line notations – single-line string – came to simplify and improve the representation of chemical structures. SMILES and InChI are well-known examples of line notation and allow canonical forms to be processed by computers (canonization is the process in which several forms of a name/structure are treated in a unique way). For example, ethanol in SMILES is OCC and in InChI is 1S/C2H6O/c1-2–3/h3H,2H2,1H3. InChI has an advantage over SMILES since it solves many chemical ambiguities, particularly in relation to the stereo centers, tautomers and other problems of the valence model [10].

ChemSpider (65 million compounds), PubChem (96 million compounds) and ChEMBL (1.8 million compounds) are among the most commonly used compound-centric databases based on SMILES/InChI and other forms of chemical structures. The DNPs in its online platform offers more than 150 thousand molecular structures containing several descriptors, including ontological and taxonomic parameters [6].

In the area of natural products, many of the databases were designed to minimize the time spent in structure identification, to converge the scattered information about the biological activity of the compounds and to describe and classify chemotaxonomic data. Since the 1990s, dereplication strategies have been developed and applied using web services through chemical representations, biological data, biogeography, taxonomic information (gender and family), etc.

NuBBE$_{DB}$, NAPRALERT, Marinlit (marine natural products), Antibase (microorganisms and fungi) and TCM-database@ are some other examples of compound-species databases with more specific data. NuBBE$_{DB}$, for example, was created in 2013 and was initially intended to provide molecular descriptors and chemical structures of natural products for molecular modeling and medicinal chemistry studies (*ca.* 600 compounds). Since 2015, ongoing efforts have been made to expand its content and include a greater diversity of natural sources to establish it as a comprehensive compendium of available biogeochemical information on Brazilian biodiversity [11]. Currently, NuBBE$_{DB}$ has more than 2000 compounds and provides validated multidisciplinary information, chemical descriptors, species sources, geographic locations, spectroscopic data (NMR) and pharmacological properties [4].

Owing to their high-throughput nature, most metabolomics studies involve the use of cutting-edge analytical instruments to collect and analyze large amounts of data on complex biological mixtures. Therefore, for a reliable interpretation of the results, it is necessary to have not only the aid of databases centered on compounds (metabolites) but also an entire organizational system that takes into account the collection, sample preparation, extraction, metabolic pathways and so on. In this sense, several types of metabolomics databases have emerged in recent years that can be classified into four categories:

1. Compound-centric databases, which are repositories of metabolites with respective biological activities or associations of drug discovery. Here, we can cite PubChem, the Human Metabolome DataBase (HMDB), the DNPs and DrugBank.
2. Databases of metabolic pathways, which are associated with regulatory pathways, including proteins and genetic interactions. This group includes KEGG, WikiPathways, BioCyc, HumanCyc and Reactome.
3. LIMSs (Laboratory Information Management Systems) which are repositories relative to the experimental projects, raw data, sample details and laboratory information. We can cite SetupX, LIMS Sesame, Metabo LIMS, MeMo and Metabolomics Workbench.
4. Spectral reference databases, which are commonly used to identify and classify metabolites. NIST, GMD, MassBank, GNPS and MetLin are some examples of this group [6].

Although all public domain tools and databases are relevant, the identification of metabolites remains the most time-consuming stage for both metabolomics and natural product analysis [7]. It is especially difficult to identify a compound based only on the manual inspection of LC-MS, GC-MS or NMR characteristics without considering possible biases and failures.

Due to difficulty in metabolic identification and classification (standardization of identification levels), the Metabolomics Standard Initiative (MSI), along with other scientific communities involved in the identification of a large number of molecules, has issued a guide to standardize the level of metabolite identity based on four criteria:

– Level 1 is when a metabolite determined by manual inspection, software or web service is compared with an authentic standard under the same experimental and laboratory conditions. As many of the metabolites found in metabolomics and natural products are not commercially available as authentic standards, the majority of structural identifications follow lower-level criteria.
– Level 2, so-called putative annotation, occurs in the absence of an authentic standard (orthogonal comparison).
– Lower levels 3 and 4 are associated with metabolite class and completely unknown, respectively. In addition, a series of computational tools and web services have been developed to facilitate metabolomics studies, mainly for the structure elucidation process [7, 12].

Table 8.1 summarizes the most commonly used packages, software and web services for structure elucidation in drug discovery, natural products and metabolomics.

Table 8.1: Database and software used for structure elucidation of organic molecules.

Database/Software	Content	Type	Webpage
Mass Spectrometry			
7-Golden Rules	Molecular formula prediction from MW and isotopic patterns	Package–Open Access	http://fiehnlab.ucdavis.edu/projects/seven-golden-rules
ACD Mass Fragmenter	Predicts MS/MS fragmentation patterns from literature rules	Software–Commercial	https://www.acdlabs.com
BinBase	A database system for automated metabolite annotation	Open Access, queryable	http://eros.fiehnlab.ucdavis.edu:8080/binbase-compound/
FiehnLib	GC-MS spectra with RI data for 1000+ metabolites	Commercial	http://fiehnlab.ucdavis.edu/projects/fiehnlib
Global Natural Products Network (GNPS)	LC-MS and MS/MS from natural products and peptides	Open Access, queryable	https://gnps.ucsd.edu/ProteoSAFe/static
Golm Metabolome (GMD)	GC-MS spectra for plant and animal metabolites	Open Access, queryable	http://gmd.mpimp-golm.mpg.de/
LipidBank	LC-MS using Orbitrap Technology from lipid species	Open Access, queryable	http://lipidbank.jp/
MaConda	LC-MS OD contaminants	Open Access, queryable	http://www.maconda.bham.ac.uk/
MMD database	GC-MS and MS/MS data on small metabolites	Open Access, queryable, Downloadable	http://dbkgroup.org/MMD/
Mass Bank	MS/MS and EI-MS spectra from organic compounds	Open Access, queryable, Downloadable	http://www.massbank.jp
MassBase	LC-MS, GC-MS, and CE-MS for biological samples	Open Access, queryable	http://webs2.kazusa.or.jp/massbase/
METFRAG	Software for the annotation of high precision tandem mass spectra of metabolites	Open Access, queryable	https://msbi.ipb-halle.de/MetFragBeta/

Name	Description	Access	URL
METLIN	LC-MS and MS/MS (FT-MS spectra) for biological samples	Open Access, queryable	https://metlin.scripps.edu/landing_page.php?pgcontent=mainPage
MoNA	CE-MS, LC-MS, GC-MS including MS/MS spectra from biological samples	Open Access, queryable	http://mona.fiehnlab.ucdavis.edu/
MzCloud	ESI-MS, APCI-MS and MS/MS spectra for organic and inorganic compounds	Open Access, queryable	https://www.mzcloud.org/
Mzmine		Software – Open Access	http://mzmine.github.io/
NIST MS Library and GC retention Index Database	EI-MS spectra and RI values for compounds	Commercial	http://nistmassspeclibrary.com
Thermo Scientific &HighChem Mass Frontier	Predicts MS/MS and MSn fragmentation patterns from literature rules	Commercial	https://www.thermofisher.com
Thermo Scientific Fragment Library	Literature-derived MS fragment trees	Commercial	https://www.thermofisher.com
XCMS	Online platform to upload and process LC-MS data	Open Access, queryable	https://xcmsonline.scripps.edu/landing_page.php?pgcontent=mainPage

NMR

Name	Description	Access	URL
ACD/Labs Aldrich NMR library	^{13}C and ^{1}H NMR spectra for compounds	Commercial	https://www.acdlabs.com/
ACD/Labs HNMR DB and CNMR DB	^{13}C and ^{1}H NMR spectra for > 200,000 compounds	Commercial	https://www.acdlabs.com/products/dbs/nmr_db)
BioMagResBank - BMRB - metabolomics	^{1}H and ^{13}C NMR spectra (1D and 2D) of plant and animal metabolites	Open Access, queryable, Downloadable	http://www.bmrb.wisc.edu/
Bruker AMIX	1D and 2D NMR spectra of metabolites at multiple pH values	Software – Commercial	https://www.bruker.com
Chenomx Inc.	1D NMR spectra of metabolites at multiple pH values	Software – Commercial	https://www.chenomx.com/
MestreNova	Predicts 1D ^{1}H and ^{13}C NMR spectra based on structure	Software – Commercial	http://mestrelab.com/
MetaboMiner	1D and 2D NMR spectra of metabolites	Open Access, queryable, Downloadable	http://wishart.biology.ualberta.ca/metabominer/

(continued)

Table 8.1 (continued)

Database/Software	Content	Type	Webpage
NAPROC-13	^{13}C NMR spectra from > 6000 natural products	Open Access, queryable, Downloadable	http://c13.usal.es/c13/usuario/views/inicio.jsp?lang=es&country=ES
NMRShiftDB	NMR spectra of natural products and organic compounds	Open Access, queryable, Downloadable	https://nmrshiftdb.nmr.uni-koeln.de/
PerCH	Predicts 1D ^{1}H and ^{13}C NMR spectra based on structure	Commercial	http://www.perchsolutions.com/
TopSpin Bruker	Software package for NMR data analysis and the acquisition and processing of NMR spectra	Commercial	https://www.bruker.com/
General			
ChemBank	Database for small molecules and biomedically relevant assays	Open Access, queryable, Downloadable	https://pubchem.ncbi.nlm.nih.gov/source/ChemBank
Chemical Entities of Biological Interest (ChEBI)	Dictionary of molecular entities focused on chemical compounds	Open Access, queryable, Downloadable	https://pubchem.ncbi.nlm.nih.gov/source/ChemBank
ChemSpider	Database of more than 34 million structures, properties and associated information	Open Access, queryable, Downloadable	http://www.chemspider.com/
Human Metabolome Database	1D and 2D NMR, MS/MS, GC-MS spectra of metabolites	Open Access, queryable, Downloadable	http://www.hmdb.ca/
NuBBEDB	A natural product database from Brazilian biodiversity	Open Access, queryable, Downloadable	http://nubbe.iq.unesp.br/portal/nubbedb.html
PubChem	Database of chemical molecules and their activities against biological assays	Open Access, queryable, Downloadable	https://pubchem.ncbi.nlm.nih.gov/

8.3 Peak pattern recognition (PPR): a method of analysis and pattern recognition applied to NMR

Natural products are an important source of diverse molecules in which many of them have high pharmacological potential as well as potential for the design of new drugs. In the reductionist view, in which classical phytochemistry studies are included, one of the most exhaustive and difficult steps is the structure elucidation and identification of compounds present on these organisms. There are many techniques to help with the rapid characterization of known compounds (dereplication), but the lack of organized data are one of the major drawbacks in natural products and medicinal chemistry research. Based on this bottleneck, a pattern recognition and dereplication software were created that was initially applied to NMR data.

Current technologies for carrying out the identification of chemicals based on NMR signals identify the substance through a system of pattern recognition. If a group of peaks inserted into the software do not belong to a compound present in the database, no information or conflicting information is generated for the user, making it impossible to identify the class or even the chemical compound [13–18].

To overcome this drawback, a function of pattern recognition was developed using the basic principle of the Euclidean distance between peaks of two samples along with the creation of a similarity matrix to be treated by multivariate and network analysis. This procedure allowed grouping of the samples in a multivariate space, clusters and networks according to their chemical structures.

This function considers the spatial distance between the peaks of the analysis of two samples. Then, calculations identify whether peaks are similar between each other and, at the end, a similarity score is attributed for the correlation between two samples. Each peak consists of a point in a Cartesian space with coordinates. For one-dimensional NMR, the x-axis is the chemical shift (δ) of a nucleus, and the y-axis is the intensity. In two-dimensional NMR, both the x and y axes are chemical shifts of the selected nuclei in the experiment, and the z-axis is the intensity. Figure 8.1 depicts gHSQC and gHMBC overlapping spectra from roridin A and verrucarin A. Both compounds are trichothecene with similar (but not identical) structures.

Considering one sample as A and the other sample as B, the correlation peak n from each sample can be described as A_peak_n = (A δx, A δy) ppm; B_peak_n = (B δx, B δy) ppm.

The Euclidean distance between An and Bn is given by Equation 8.1:

Euclidian distance between two peaks

$$= \sqrt{\left(\mathbf{B}nx_{\delta\,\mathrm{ppm}} - \mathbf{A}nx_{\delta\,\mathrm{ppm}}\right)^2 + \left(\mathbf{B}ny_{\delta\,\mathrm{ppm}} - \mathbf{A}ny_{\delta\,\mathrm{ppm}}\right)^2} \tag{8.1}$$

Equation 8.1. Euclidean distance between two peaks from two different samples.

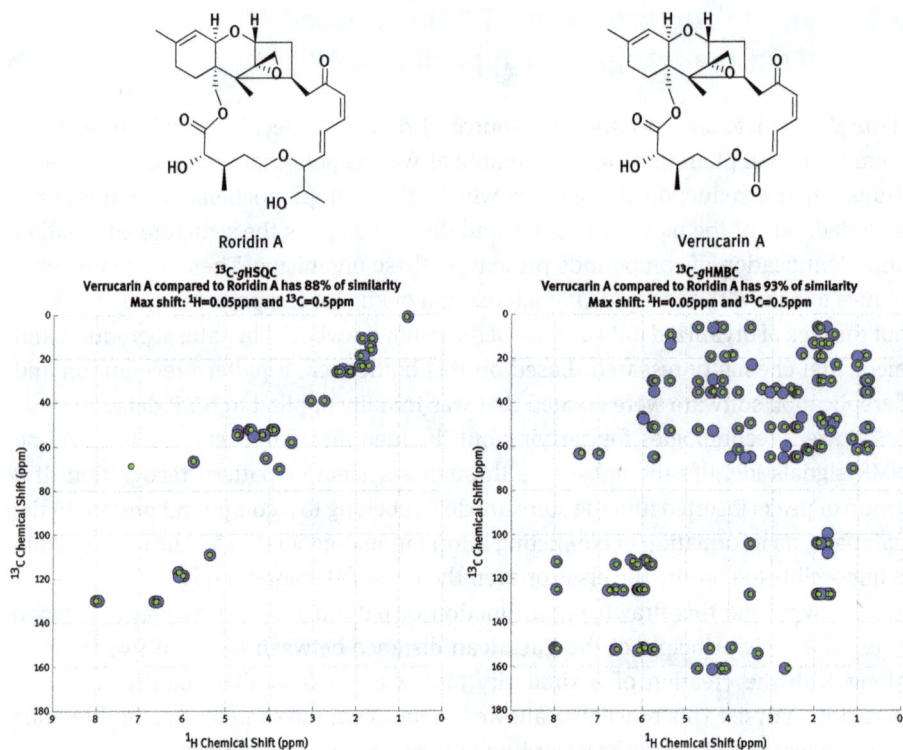

Figure 8.1: Overlapping spectra of ^{13}C-gHSQC and ^{13}C-gHMBC, respectively, with 88 % and 93 % similarity, from roridin A (green) with verrucarin A (blue).

By using Equation 1 and setting a maximum value for the Euclidian distance, it is possible to assign two peaks in different samples as similar to each other. This can be indicative of peaks from the same chemical environment that may represent the same part of a molecule. Considering the total number of peaks in an NMR analysis and taking into consideration the total peaks of A that are found in B, it is possible to generate a direct similarity coefficient represented by Equation 8.2. By performing the opposite correlation based on the same equation, that is, correlating sample B with sample A, the algorithm generates an inverse similarity coefficient. If both coefficients are 1, we can assume that these samples have the same peaks and are the same.

$$Similarity = \frac{\sum n\_sim\,B \times 100}{n\_total\,A} \tag{8.2}$$

$\sum$n_simB = Number of peaks found similar in sample B; n_totalA = Number of peaks present in sample A

Equation 8.2. Equation for the final calculation of similarity between two samples. This function uses only the peak's coordinates to perform all the calculations, which is an advantage over conventional methods that use the entire spectrum. For instance, 2D NMR experiments typically have at least one million data points, and 1D NMR spectra have approximately 64 k data points. Using the peak coordinates dramatically reduces the number of points, which affects the time and storage needed to perform the computation. This procedure allows the software to run in real time on a simple personal computer. In addition, this algorithm can be applied to any kind of analysis that generates peaks, providing a very broad methodology to be used in many fields.

When this algorithm is performed on one sample, it will generate a direct and inverse similarity score. If the algorithm is run with a database, it calculates the similarity score of one sample with all others in the database, indexing the results in a vector and creating a similarity profile for that sample. An example of the similarity profiles can be seen in Figure 8.2. This profile quantitatively represents a set of characteristics determined by the PPR algorithm and can be considered a new technique to fingerprint samples once each sample has its own specific similarity profile. Stacking all similarity profiles calculated for each sample in a database in each row creates a square matrix where pair indices (i.e., (1,1), (2,2) and (3,3)) have values of 1 due to sample correlation with itself. The rows of this matrix represent the direct similarity profiles and show how similar the selected sample is

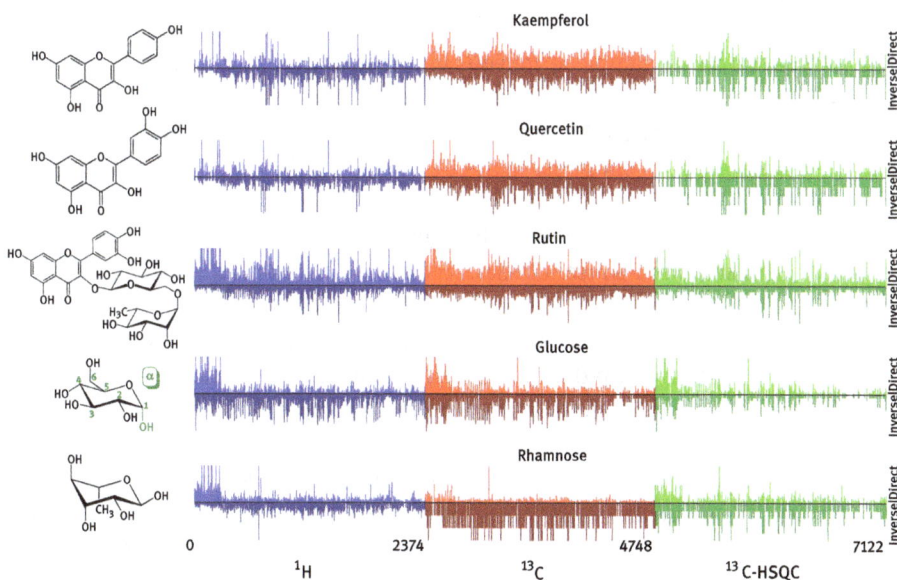

Figure 8.2: Direct and inverse similarity profiles for kaempferol, quercetin, rutin, glucose and rhamnose seen in light color and dark color, respectively. The color represents the NMR experiments: ^{1}H-NMR in blue, ^{13}C-NMR in red and ^{13}C-HSQC-NMR in green.

to the database. The columns represent the inverse similarity profiles and show how similar the database is to the sample. These two concepts are very important because they allow the creation of a pattern recognition model to observe groups of alike samples and identify sample interactions in two different manners. The major advantage of this matrix is the possibility of performing computational and multivariate analysis, grouping the samples in multivariate spaces, clusters and networks as the first methodology of its kind. An example of the function output can be seen in Figure 8.3, where it is possible to observe the network, HCA and PCA methodologies applied for structural identification. In addition, Figure 8.4 shows how the algorithm organizes classes of terpenes.

Pattern recognition can be a useful tool for structure elucidation studies. This tool is rather new and can be explored for further applications. This interdisciplinary study with theories and practice in natural products chemistry, computations, chemometrics and spectroscopy can lead to better understanding and deployment of pattern recognition software.

With this innovative software applied to chemical NMR data, the process of structure elucidation and identification of molecules became easier and faster, requiring only a few steps. The pattern recognition function shows great potential when applied to natural product studies. After isolating a compound and performing NMR analysis, the NMR peaks can be inserted in the software, compared with a database and used to identify by peak matches which substances these peaks came from.

With the use of this function, it is possible to identify the compounds that have high similarity among all those present in a database. When a database does not have the same compound as the sample, erroneous information may be obtained by the user, which can be confusing in the process of molecular elucidation.

The development of a similarity profile, which is a fingerprint technique, together with the development of similarity matrices may be a solution to this problem. The application of HCA, PCA, and networks in the similarity matrices allowed us to group the NMR peaks from an unknown substance in clusters that are represented by neighbors with similar chemical classes. In this new technique, the substances are grouped by their structural similarities.

If the databases do not have the compound that is sought, with this new algorithm, it is possible to group this substance in a cluster with compounds that have similar chemical structures. As we introduce more compounds in the database, the more accurate the algorithm becomes.

The technique of pattern recognition can be applied to any kind of data from any kind of detector with an analysis result obtained in a form of peaks that will generate a similarity matrix. These matrices, for each technique, can be concatenated in a final matrix to obtain a final substance match based on all analyses of your choice. This concatenation makes the substance match more accurate and more reliable. With the use of coupled systems, such as LC-NMR-MS-DAD, the identification of a chemical class or even the molecule itself can be done in real time.

A)

B)

C)

1265-)NDB361₂d)Flavonoid glycoside)Quercetin 3-O-a-L-rhamnopyranosyl|
1263-)NDB360₂d)Flavonoid glycoside)Kaempferol 3-O-a-L-rhamnopyranosyl|
1264-)NDB360₃d)Flavonoid glycoside)Kaempferol 3-O-a-L-rhamnopyranosyl|
1119-)NDB286₃d)Flovonol (flavonoid gylcoside rhamnoside)Kaempferol-3-
1118-)NDB286₂d)Flovonol (flavonoid gylcoside rhamnoside)Kaempferol-3-
1482-)NDB46₂d)Flovonol (flavonoid gylcoside rhamnoside)Rutin
1110-)NDB282₂d)Flovonol (flavonoid gylcoside rhamnoside)Quercetin-3-O
1108-)NDB281₂d)Flovonol (flavonoid gylcoside rhamnoside)Kaempferol-3-
1106-)NDB280₂d)Flovonol (flavonoid gylcoside rhamnoside)Kaempferol-3-
1100-)NDB278₂d)Biflavonol (Bilfavonoid gylcoside rhamnoside)Chimarrho

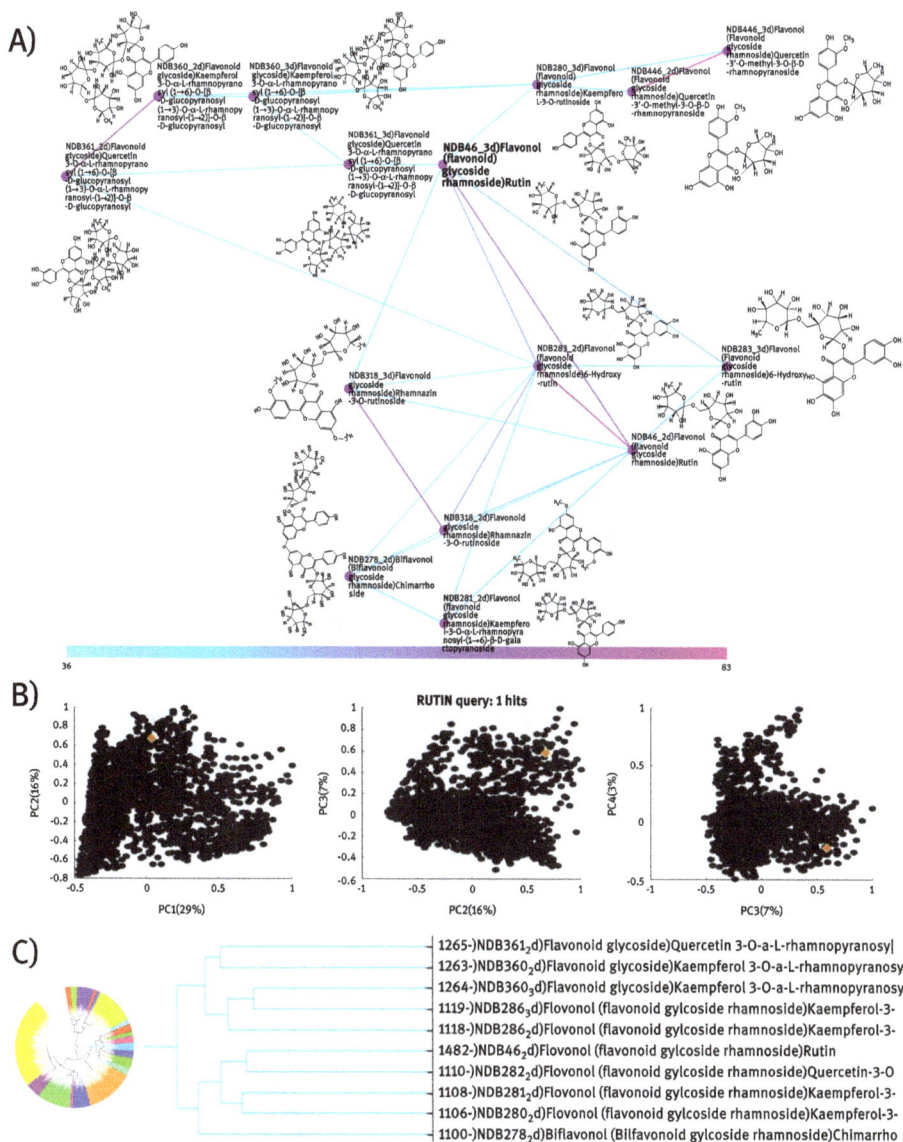

Figure 8.3: (A) general molecular network, (B) Molecular PCA, and (C) molecular HCA for the predicted NMR spectrum of rutin. These results are based on the concatenated similarity matrix containing proton, ^{13}C and HSQC NMR predicted and experimental analyses.

This technique can be considered a screening method because the size of the database used is considered a limiting factor. A good example of the potential of this pattern recognition is in the search for a specific class of compounds that have pharmacological interests. Finally, the main advantage of this methodology is the

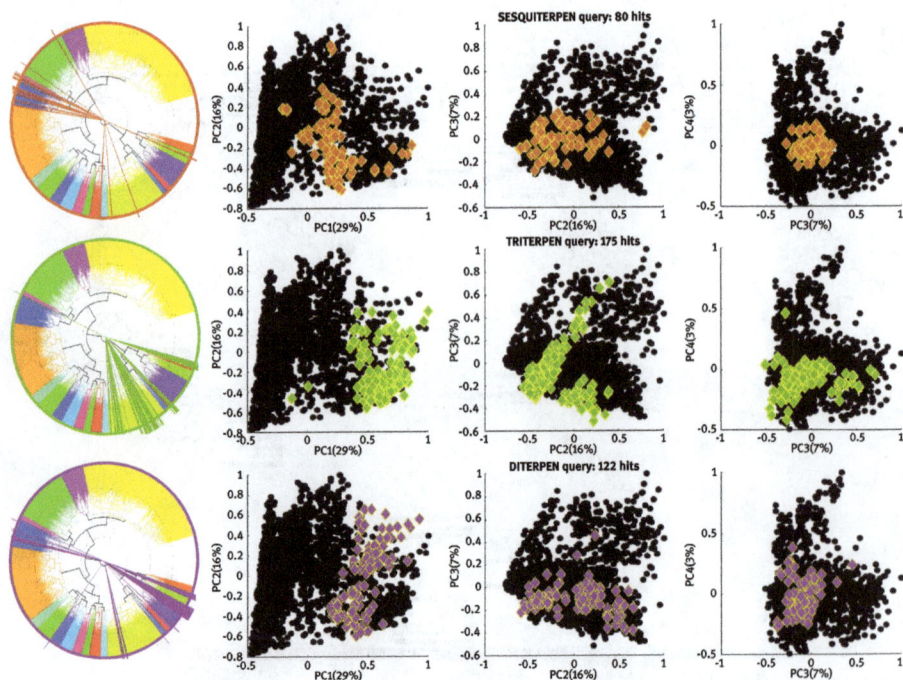

Figure 8.4: Database query for sesquiterpenes, diterpenes and triterpenes using the concatenated similarity matrix containing Proton, ^{13}C and HSQC predicted NMR.

possibility of grouping clusters and networks for molecules or mixtures not present in the database according to their chemical structures [19].

8.4 Dereplication methods and MN

Taking into account the thousands of known natural compounds, new strategies have been employed to save the time and cost spent on the research process. One of the most used strategies is called "dereplication", which consists of the preliminary chemical analysis of a complex sample (e.g. an extract) to detect the pre-existing known compounds. In this way, it is possible to focus the research on new natural compounds or even to isolate an interesting known compound for the identification of new biological activities [20, 21].

In 1978, the concept of dereplication was first described [22], followed by a 12-year hiatus, when the second manuscript containing this topic was published in 1990 [23]. Beginning 3 years later, the dereplication strategy has been increasingly employed until the present day, as shown in the data presented in Figure 8.5.

After 2012 (Figure 8.5), a large increase in the number of publications containing the dereplication topic was observed, mainly due to recent advances in

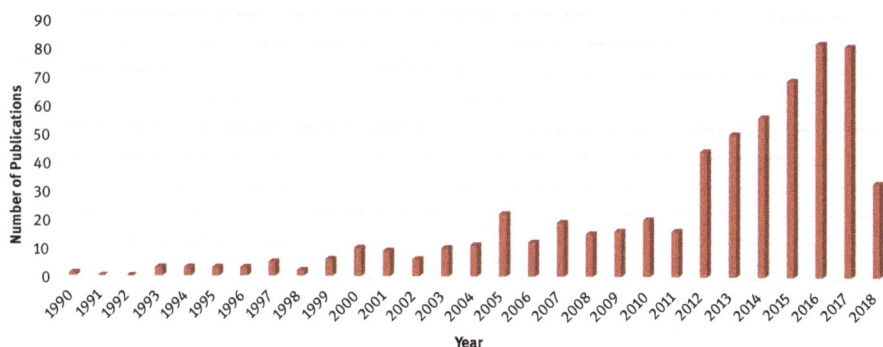

Figure 8.5: Report of the number of publications per year containing the topic "dereplication" from Thomson Reuters Web of Science (from 1990 to June 2018).

hyphenated techniques, such as HPLC–MS, HPLC–NMR, HPLC–NMR–MS and HPLC–SPE–NMR [24]. These analytical techniques have become increasingly sensitive through the years, and smaller amounts of sample can be used to achieve the expected result, which is excellent for scarce samples. It should also be mentioned that many authors might use a dereplication strategy but do not mention this exact term in the published articles. Therefore, the number of publications might be even higher than that reported in Figure 8.5.

The analysis of complex mixtures (from plants, fungi and bacteria) has provided valuable information on the chemical characteristics, minimizing time, cost and effort spent in the natural product research workflow. In general, the dereplication strategy is based on the physical similarities of molecules, such as UV–Vis profiles, chromatographic retention times, mass spectra, NMR chemical shifts and biological activities [24].

Dereplication strategies have been successfully employed in many studies covering different (but related) goals [21], such as (1) identifying the known compounds in a single extract [25], (2) accelerating bio-guided fractionation [26], (3) chemical profiling of crude extract collections (an interesting approach to build databases) [27, 28], or even (4) chemotaxonomic studies of species [29, 30].

Although dereplication methods are extremely important and effective for many purposes, the identification of known compounds in a complex matrix could still be time-consuming if made manually. It is estimated that, when considering MS/MS data, approximately only 1.8% of the spectra of an untargeted metabolomics experiment can be determined by comparing with existing data [31]. Therefore, approximately 98% of the spectra obtained in these analyses can be considered "dark matter", as described by da Silva and co-workers [32].

Considering the huge amount of information obtained from the analysis of the samples, computational methods for treating these data have been developed during the last few years to assist and accelerate the data processing.

One of the most sensitive techniques used in natural product chemistry research is mass spectrometry (MS). However, the identification of a determined substance might be inaccurate if the molecular ion mass is the only information used to search in databases, such as the NuBBE$_{DB}$ [11], SciFinder, ChemSpider, Beilstein DNPs, Antibase, MarinLit and SuperNatural [33, 34], due to the high amount of results returned. However, the MS/MS profile is a definitive characteristic of a molecule due to the uniqueness of its structure, chemical stability and functional groups. These data dictate the reactivity of the gas phase dissociation by induced collision of a determined compound; therefore, similarities in MS/MS fragmentation patterns can be used as predictive data of structural similarity [24].

A novel approach that has been widely employed is called MN, a computational strategy that assists the visualization and interpretation of obtained MS/MS data. Once similar natural products share similar MS/MS fragmentation, it is possible to build a molecular network that aids the visualization of these similarities by measuring the correlations among them using Cytoscape® software [31, 35]. Therefore, similar molecules tend to stay close at the same net in the same cluster. A cluster is formed by nodes, and each node represents an MS/MS spectrum, while the thicknesses of the edges that connect these nodes indicate the similarity between them: the thicker they are, the more similar they are [24, 31].

The next step is to correlate the obtained MS/MS molecular network with the previously described compounds in the literature. Although several natural product databases are available and assist in the dereplication strategy (such as the above-mentioned ones), many of them are not open-source or even do not process MS data. On the other hand, MS databases, such as MassBank [36] and ReSpect [37], can host a limited number of MS/MS spectra.

Within this scenario, an open-source platform called Global Natural Product Social Molecular Networking (GNPS, http://gnps.ucsd.edu) was built to provide the possibility to store, analyze and share MS/MS data, as well as compare them with all publicly available information, making it faster and easier to identify molecules in complex mixtures [38]. A brief representation of this workflow is described in Figure 8.6.

At first contact, one might be confused with the involved steps of the MN approach. Therefore, a simplified tutorial is presented at the GNPS website platform (http://gnps.ucsd.edu).

With this approach, it is possible to visually organize classes of compounds closely to determine known compounds early in the workflow and to focus on the desired types of compounds. If a large number of samples is analyzed using the data obtained using the MN strategy and the GNPS platform, it is possible to find correlations among classes of molecules with the presence or absence of genes that codify them, which might be useful for biosynthetic studies [24].

Although MN is not a computational method for structure elucidation itself, it should be considered that it can be a valuable tool in the elucidation process of unknown molecules due to the possibility of comparing the spectroscopic and

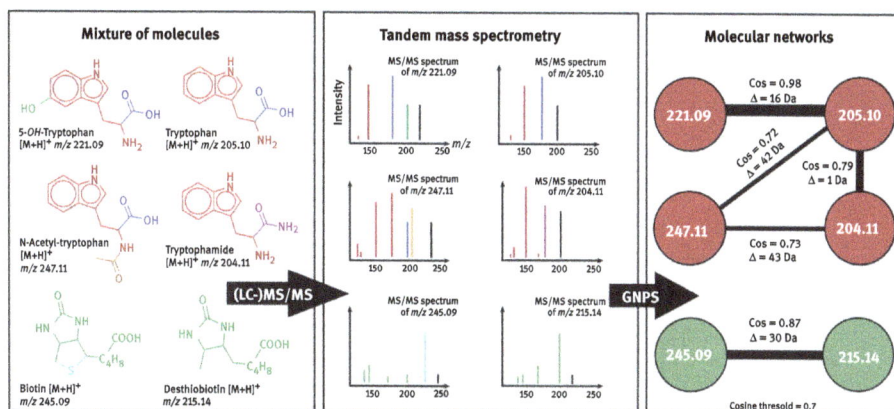

Figure 8.6: Full molecular network obtained using MS/MS data of *Geissospermum laeve* (Apocynaceae) bark's alkaloid extract. Green circles indicate data obtained from the extract, yellow circles indicate the in-house database compounds, and red squares indicate the overlapping of green and yellow circles. Blue color indicates new natural products isolated. Adapted, reprinted with permission from [40].
Copyright 2017 American Chemical Society.

spectrometric data (UV, NMR, MS) of a known compound with data generated by a new compound.

Considering that MN is a relatively recent approach, three examples of the valuable information obtained from it are presented to highlight its importance.

8.4.1 MN for reinvestigating a determined species

The Apocynaceae plant family has been known for decades to present indole alkaloids in its chemical composition [39]; however, many of these compounds have not been deeply evaluated biologically. Within this context, Fox Ramos et al. (2017) have restudied a species named *Geissospermum laeve*, an Amazonian tree in which water bark decoction is consumed by the local population [40]. *G. laeve* is the most studied species of the genus, with 22 indole alkaloids isolated comprising several biological activities, such as anti-inflammatory, anticholinesterase [41] and antiplasmodium [42].

Therefore, Fox Ramos et al. (2017) described the dereplication procedure of the known alkaloids from the alkaloid extract from the bark of *G. laeve* using the MN approach, followed by the MS/MS guided isolation and structure elucidation of three new monoterpene indole alkaloids and three known alkaloids, as illustrated in Figure 8.7. The MN (processed in the GNPS platform) result shows green circles (corresponding to the molecules present in the alkaloid extract), yellow circles (corresponding to the in-house alkaloid database initially built) and red squares

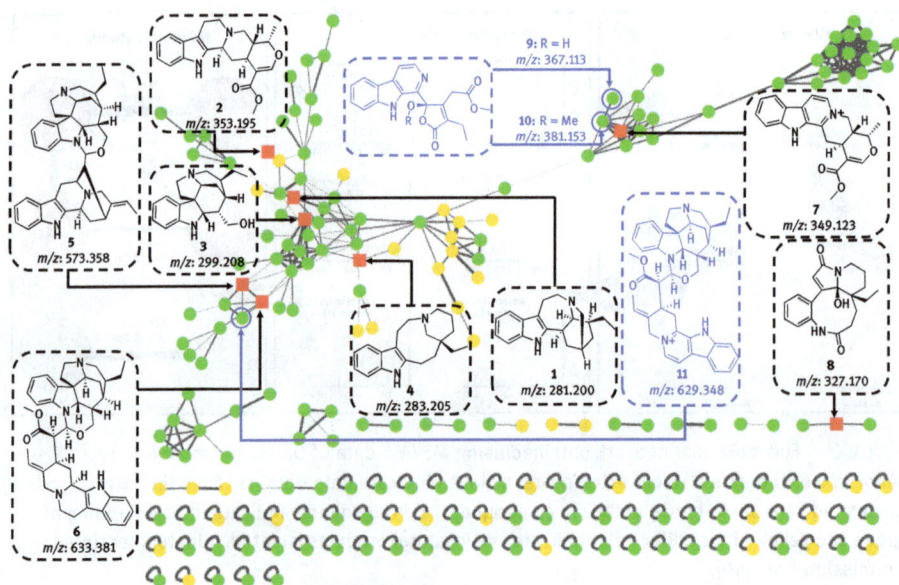

Figure 8.7: Mixture of molecules analyzed by LC-MS/MS provides MS/MS spectra that can be organized in a molecular network. Adapted, reprinted from [31]. Copyright 2017, with permission from Elsevier.

(overlap of the alkaloid extract and the in-house alkaloid database). The red squares could be determined as ibogamine (**1**), raubasine (**2**), geissoschizoline (**3**), quebrachamine (**4**), geissolosimine (**5**), geissospermine (**6**), serpentine (**7**) and leuconolam (**8**). Therefore, considering the molecule skeleton of geissolosimine (**5**) and geissospermine (**6**), one close node was chosen for the isolation, such as two nodes close to the Serpentine (**7**) red square. Additionally, the known compounds leuconolam (**8**), geissolosimine (**5**) and geissospermine (**6**) were isolated.

The isolated compounds were determined by NMR techniques as geissolaevine (**9**), *O*-methylgeissolaevine (**10**) and 3',4',5',6'-tetradehydrogeissospermine (**11**), which were new natural products elucidated by NMR techniques. Finally, compounds **5, 8** and **9–11** were evaluated *in vitro* for their antiparasitic activity against *Plasmodium falciparum* strain (FcB1), for the axenic and intramacrophage amastigote forms of *Leishmania donovani* strain (LV9) and for cytotoxicity against human fetal lung fibroblasts (MRC-5). The known compound **5** presented the highest activity for both antiplasmodial and antileishmanial activity, though, low selectivity. The new compounds **9** and **10** presented low activity and low cytotoxicity, while compound **11** presented moderate antiplasmodial activity and cytotoxicity [40]. Considering the results obtained by Fox Ramos et al. (2017), the MN approach proved to be very useful for the identification of new natural products from a well-known species and to evaluate their biological activity.

8.4.2 MN for the study of extract libraries

The MN tool has been increasingly (and successfully) applied for extract libraries, from natural products of marine microorganisms [43] to bacterial strains [44], and has contributed to the identification of numerous molecular families.

A recent study performed by Olivon et al. [45] investigated a library of 292 EtOAc extracts obtained from 107 species of the New Caledonian Euphorbiaceae family. Initially, these extracts were analyzed by UHPLC-HRMS^2 and processed using MZmine 2 open-source software [46] to determine the peak area information and for molecular formula annotation. The MN of this large number of samples resulted in 17,387 nodes, 28,307 edges and 1,231 clusters containing 3 (or more) nodes.

The next step consisted of merging taxonomical details with the network visualization to build a multi-informative molecular map. For this, it was assumed that spectral differences within taxonomically close species (e.g., all from Euphorbiaceae family) could be a synonym of chemical singularity. Therefore, a taxonomical mapping was built assigning each genus with a specific color, in which these structurally different secondary metabolites could be distinguished from the omnipresent ones. Considering this assumption, a cluster formed by unicolored nodes could indicate a specific chemical family of a specific genus, as shown in Figure 8.8.

From these data, it was possible for the authors to focus the study on a desired cluster belonging to the bark extract of *Codiaeum peltatum*, indicated as MN1, and perform an MN-guided isolation of five alkaloids, four of them chlorinated [quinolin-4-one isochloroaustralasine A (**12**), chloroaustralasine A (**13**), chloroaustralasine B (**14**), chloroaustralasine C (**15**) and *trans*-erioaustralasine hydrate (**16**)]. This unusual class of compounds was only described three times in species from the Rutaceae family, corroborating the fact that the MN strategy has great potential to orientate the (re)discovery of original scaffolds. The total synthesis of these compounds was performed to confirm their chemical structures [45].

8.4.3 MN for the study of bioactive compounds

Traditionally, when a natural extract shows an interesting bioactivity, a bioassay-guided purification is performed to obtain a pure bioactive compound. After the extraction procedure, a fractionation by liquid-liquid extraction or by chromatographic methods can provide fractions that will be tested separately for the desired target. Once the bioactive fractions are selected, the compounds present in it may be purified, identified and evaluated again for their bioactivity. Although this approach has been successfully and widely employed for decades and has led to the discovery of many important bioactive compounds, it is quite common to observe a loss of activity later in the workflow. This can be observed due to the possible degradation of the bioactive compounds during the purification process, the low concentration of

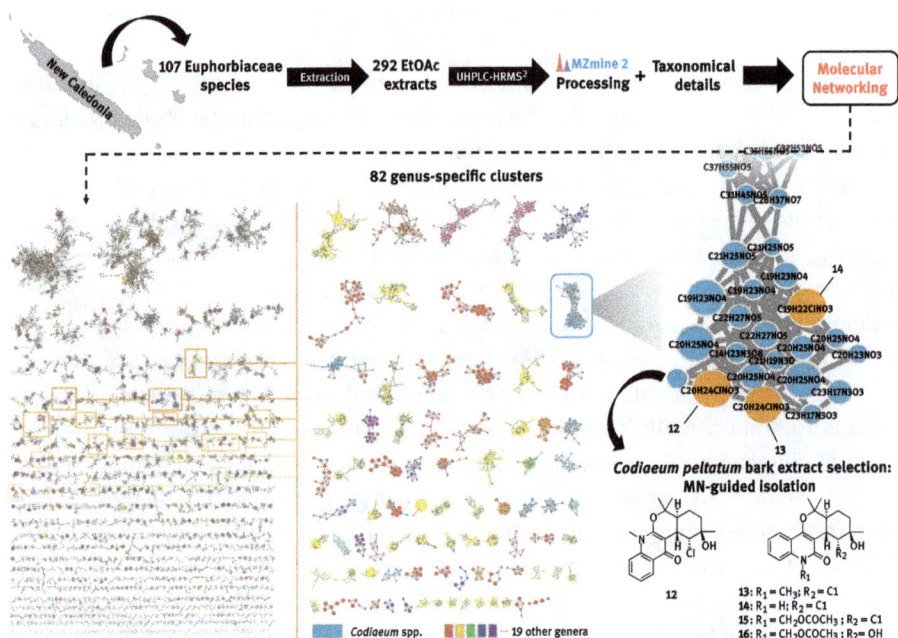

Figure 8.8: (A) MN2 cluster obtained from the bioactive MN of *Euphorbia dendroides* (Euphorbiaceae) latex extract, in which the largest nodes represent molecules with statistically significant bioactivity scores. (B) Chemical structures of the four isolated deoxyphorbol ester derivatives. Adapted, reprinted with permission from [47]. Copyright 2018 American Chemical Society.

the bioactive compound in the extract (hindering its isolation), or even the correlation of bioactivity with synergistic effects [47].

One example of the loss of bioactivity is described by Esposito et al. [48], in which the focus of the study was the latex extract of *Euphorbia dendroides* (Euphorbiaceae) that presented a potent and selective bioactivity against the chikungunya virus (CHIKV) [49]. A fractionation of this selected extract was performed by Esposito et al. [48], leading to 18 fractions, of which three showed a selective inhibition of CHIKV. Finally, 19 diterpenoid jatrophane esters were isolated; however, none of them showed the expected selective antiviral activity, leading to the assumption that the compounds responsible for this activity had not yet been determined.

In this context, a new approach for MN was recently described as "Bioactivity-Based Molecular Networking", in which biochemometric analysis can be merged with MN to observe clusters with the most possible bioactivity [47]. Since the bioassays of fractions and pure compounds have been performed, a bioactivity score (the probability of a compound being bioactive) for each molecule can be calculated based on its relative abundance between samples (Kurita et al. 2015), and a multivariate statistical model can be used to estimate this score from LC-MS data [50].

Briefly, to build the Bioactivity-Based MN, three steps were necessary: (1) processing LC-MS data (using MZmine 2 or Optimus software) to detect and relatively quantify ions across the fractions; (2) calculating the bioactivity score for each spectral feature with a Jupyter notebook using the R language; and (3) analyzing the MS/MS on the GNPS platform and visualizing the networks using Cytoscape. A tutorial containing the involved steps is freely available (https://github.com/DorresteinLaboratory/Bioactive_Molecular_Networks).

Using this approach, two clusters were observed. The major cluster (MN1) showed the previously isolated and tested diterpenoid jatrophane esters [48], while in the smaller cluster (MN2 – shown in Figure 8.9), there were no isolated or identified compounds in spectral libraries on the GNPS platform. Therefore, the authors focused on the MN2 cluster, and it was possible to isolate four deoxyphorbol esters (**17, 18, 19** and **20**). These compounds were re-evaluated for antiviral activity against CHIKV, and compounds **18** and **19** revealed potent and selective activity with EC_{50} values of 0.40±0.02 µM and 0.60±0.06 µM and SI = 34 and 41, respectively. Therefore, MN was successfully employed in the discovery of bioactive natural products on the selected extract, which was not possible to determine using traditional procedures.

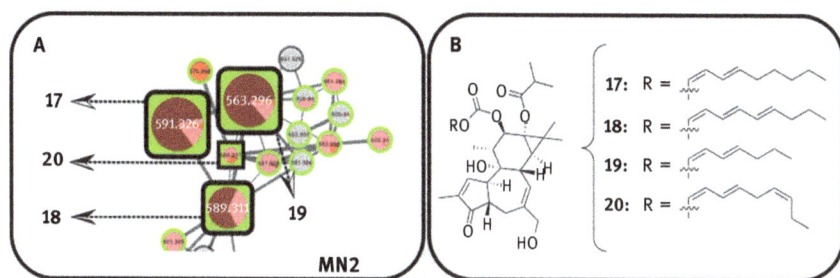

Figure 8.9: Illustrative workflow strategy employed for the MN of 292 extracts of 107 species of the Euphorbiaceae. UHPLC-MS^2 data were preprocessed using MZmine 2, merged with taxonomical information, and the MN was generated, revealing chlorinated compounds at the selected cluster. The node size is proportional to the MS^1 peak area at the crude extracts. Adapted, republished with permission of Royal Society of Chemistry, from [45]; permission conveyed through Copyright Clearance Center, Inc.

8.5 Conclusions and future directions

Databases of NMR and MS were essential for the development of software for structure elucidation and are a continuing need for sustaining this field. Dereplication methods have been largely employed on natural product discovery in recent decades to avoid re-isolation of known compounds. Within this scenario, computational methods were developed to analyze the enormous amount of data generated by increasingly sensitive analysis. Recently, MN is one of the most employed computational methods, and many examples of its use can be found in recent literature. MN can assist in the structure elucidation of new natural compounds, allowing to observe families of compounds in different clusters. Public MS/MS libraries have been increasing, and the identification of compounds using these data can become even faster and more efficient. Additionally, these tools proved to be efficient for predicting the bioactivity of compounds by merging the bioactivity of known molecules or fractions. As MN is a new kind of approach, it should be emphasized that several other applications are yet to be discovered.

Acknowledgements: The authors acknowledge Fundação de Amparo à Pesquisa do Estado de São Paulo(FAPESP) grants#2013/07600-3 (CIBFar-CEPID), #2014/50926-0 (INCT BioNat CNPq/FAPESP), Coordenação de Aperfeiçoamento de Pessoal de Nível Superior (CAPES), Conselho Nacional de Desenvolvimento Científico e Tecnológico (CNPq) and Termo de Execução Descentralizado Arbocontrol #74/2016 for grant support and research fellowships. Authors acknowledge scholarships: MV (CNPQ #167874/2014-4 and #152243/2016-0; Finatec #120/2017), HMR (CNPQ #142014/ 2018-4), ACP (Fapesp #2016/13292-8), MEFP (Fapesp #2017/17098-4).

References

[1] Duarte DF. Opium and opioids: a brief history. Rev Bras Anestesiol. 2005;55:135–46.
[2] Dunn WB, Broadhurst DI, Atherton HJ, Goodacre R, Griffin JL. System level studies of mammalian metabolomes: the roles of mass spectrometry and nuclear resonance spectroscopy. Chem Soc Rev. 2011;40:387–426.
[3] Batista JM, Jr, Blanch EW, Bolzani VS. Recent advances in the use of vibrational chiroptical spectroscopy methods for stereochemical characterization of natural products. Nat Prod Rep. 2015;32:1280–302.
[4] Pilon AC, Valli M, Dametto AC, Pinto MEF, Freire RT, Castro-Gamboa I, et al. NuBBEDB: an updated database to uncover chemical and biological information from Brazilian biodiversity. Sci Rep. 2017;7:7215.
[5] Griffiths J. A brief history of mass spectrometry. Anal Chem. 2008;80:5678–83.
[6] Pilon AC, Paez-Garcia A, Pavarini DP, Scotti MT. Chemical biology databases in mass spectrometry. Iin: chemical biology: evolving applications. Cambridge, UK: Royal Society of Chemistry, 2018:221–63.
[7] Whisart DS. Computational strategies for metabolite identification in metabolomics. Bioanalysis. 2009;1:1579–96.
[8] Fayyad U, Stolorz P. Data mining and KDD: promise and challenges. Fut Gen Comp Syst. 1997;13:99–115.
[9] Capra F. The web of life: a new scientific understanding of living systems. New York-USA: Anchor Books, 1996:1–347.
[10] O'Boyle NM, Banck M, James C, Morley C, Vandermeersch T, Hutchison GR. Open Babel: a chemical toolbox. J Cheminform. 2011;3:33.
[11] Valli M, Dos Santos RN, Figueira LD, Nakajima CH, Castro-Gamboa I, Andricopulo AD, et al. Development of a natural products database from the biodiversity of Brazil. J Nat Prod. 2013;76:439–44

[12]   Weber RJM, Lawson TN, Salek RM, Ebbels TMD, Glen RC, Goodacre R, et al. Computational tools and workflows in metabolomics: an international survey highlights the opportunity for harmonisation through Galaxy. Metabolomics. 2017;13:12.

[13]   Lewis IA, Schommer SC, Markley JL. rNMR: open source software for identifying and quantifying metabolites in NMR spectra. Mag Res Chem. 2009;47:123–6.

[14]   Plainchont B, Nuzillard JM, Rodrigues GV, Ferreira MJ, Scotti MT, Emerenciano VP. New improvements in automatic structure elucidation using the LSD (Logic for Structure Determination) and the SISTEMAT expert systems. Nat Prod Comm. 2010;5:763–70.

[15]   Steinbeck C. Recent developments in automated structure elucidation of natural products. Nat Prod Rep. 2004;21:512–18.

[16]   Robinette SL, Zhang F, Brüschweiler-Li L, Brüschweiler R. Web server based complex mixture analysis by NMR. Anal Chem. 2008;80:3606–11.

[17]   Xia J, Bjorndahl TC, Tang P, Wishart DS. MetaboMiner–semi-automated identification of metabolites from 2D NMR spectra of complex biofluids. BMC Bioinformatics. 2008;28:507.

[18]   Tulpan D, Léger S, Belliveau L, Culf A, Cuperlović-Culf M. MetaboHunter: an automatic approach for identification of metabolites from ^{1}H-NMR spectra of complex mixtures. BMC Bioinformatics. 2011;14:400.

[19]   Freire RT, Castro-Gamboa I São Paulo State University, assignee. "Method of analysis' pattern recognition and computer program". Patent WIPO WO2016183647A1 PCT/BR2015/000075. 2015.

[20]   Gaudêncio SP, Pereira F. Dereplication: racing to speed up the natural products discovery process. Nat Prod Rep. 2015;32:755–878.

[21]   Hubert J, Nuzillard JM, Renault JH. Dereplication strategies in natural product research: how many tools and methodologies behind the same concept? Phytochem Rev. 2017;16:55–95.

[22]   Hanka LJ, Kuentzel SL, Martin DG, Wiley PF, Neil GL. Detection and assay of antitumor antibiotics. In: Carter SK, Umezawa H, Douros J, Sakurai Y, editor(s). Antitumor antibiotics. Berlin, Heidelberg: Springer Berlin Heidelberg, 1978:69–76.

[23]   Beutler JA, Alvarado AB, Schaufelberger DE, Andrews P, McCloud TG. Dereplication of phorbol bioactives: lyngbya majuscula and croton cuneatus. J Nat Prod. 1990;53:867–74.

[24]   Yang JY, Sanchez LM, Rath CM, Liu X, Boudreau PD, Bruns N, et al. Molecular networking as a dereplication strategy. J Nat Prod. 2013;76:1686–99.

[25]   Queiroz MMF, Queiroz EF, Zeraik ML, Ebrahimi SN, Marcourt L, Cuendet M, et al. Chemical composition of the bark of Tetrapterys mucronata and identification of acetylcholinesterase inhibitory constituents. J Nat Prod. 2014;77:650–6.

[26]   Zhang J-G, Huang X-Y, Ma Y-B, Zhang X-M, Chen -J-J, Geng C-A. Dereplication-guided isolation of a new indole alkaloid triglycoside from the hooks of Uncaria rhynchophylla by LC with ion trap time-of-flight MS. J Sep Sci. 2018;41:1532–8.

[27]   Nielsen KF, Smedsgaard J. Fungal metabolite screening: database of 474 mycotoxins and fungal metabolites for dereplication by standardised liquid chromatography-UV-mass spectrometry methodology. J Chrom A. 2003;1002:111–36.

[28]   Fraige K, Dametto AC, Zeraik ML, de Freitas L, Saraiva AC, Medeiros AI, et al. Dereplication by HPLC-DAD-ESI-MS/MS and screening for biological activities of Byrsonima Species (Malpighiaceae). Phytochem Anal. 2017;29:196–204.

[29]   Brkljača R, Göker ES, Urban S. Dereplication and chemotaxonomical studies of marine algae of the ochrophyta and rhodophyta phyla. Mar Drugs. 2015;13:2714–31.

[30]   Dos Santos VS, Macedo FA, Vale JS, Silva DB, Carollo CA. Metabolomics as a tool for understanding the evolution of Tabebuia sensu lato. Metabolomics. 2017;13:72.

[31]   Quinn RA, Nothias LF, Vining O, Meehan M, Esquenazi E, Dorrestein PC. Molecular networking as a drug discovery, drug metabolism, and precision medicine strategy. Trend Pharm Sci. 2017;38:143–54.

[32] Da Silva RR, Dorrestein PC, Quinn RA. Illuminating the dark matter in metabolomics. Proc Nat Acad Sci USA. 2015;112:12549–50.

[33] Dunkel M, Fullbeck M, Neumann S, Preissner R. SuperNatural: a searchable database of available natural compounds. Nucleic Acids Res. 2006;34:D678–83.

[34] Banerjee P, Erehman J, Gohlke BO, Wilhelm T, Preissner R, Dunkel M. Super Natural II-a database of natural products. Nucleic Acids Res. 2015;43:D935–39.

[35] Shannon P, Markiel A, Ozier O, Baliga NS, Wang JT, Ramage D, et al. Cytoscape: asoftware environment for integrated models of biomolecular interaction networks. Gen Res. 2003;13:2498–504.

[36] Horai H, Arita M, Kanaya S, Nihei Y, Ikeda T, Suwa K, et al. MassBank: a public repository for sharing mass spectral data for life sciences. J Mass Spec. 2010;45:703–14.

[37] Sawada Y, Nakabayashi R, Yamada Y, Suzuki M, Sato M, Sakata A, et al. RIKEN tandem mass spectral database (ReSpect) for phytochemicals: A plant-specific MS/MS-based data resource and database. Phytochemistry. 2012;82:38–45.

[38] Wang M, Carver JJ, Phelan VV, Sanchez LM, Garg N, Peng Y, et al. Sharing and community curation of mass spectrometry data with global natural products social molecular networking. Nature Biotech. 2016;34:828–37.

[39] Szabó LF. Rigorous biogenetic network for a group of indole alkaloids derived from strictosidine. Molecules. 2008;13:1875–96.

[40] Fox Ramos AE, Alcover C, Evanno L, Maciuk A, Litaudon M, Duplais C, et al. Revisiting previously investigated plants: a molecular networking-based study of geissospermum leave. J Nat Prod. 2017;80:1007–14

[41] Lima JA, Costa TWR, Silva LL, Miranda ALP, Pinto AC. Antinociceptive and anti-inflammatory effects of a Geissospermum vellosii stem bark fraction. Anal Acad Bras Cien. 2016;88:237–48.

[42] Mbeunkui F, Grace MH, Lategan C, Smith PJ, Raskin I, Lila MA. In vitro antiplasmodial activity of indole alkaloids from the stem bark of Geissospermum vellosii. J Ethnopharm. 2012;139:471–7.

[43] Floros DJ, Jensen PR, Dorrestein PC, Koyama N. A metabolomics guided exploration of marine natural product chemical space. Metabolomics. 2016;12:145.

[44] Crüsemann M, O'Neill EC, Larson CB, Melnik AV, Floros DJ, Da Silva RR, et al. Prioritizing natural product diversity in a collection of 146 bacterial strains based on growth and extraction protocols. J Nat Prod. 2017;80:588–97.

[45] Olivon F, Apel C, Retailleau P, Allard PM, Wolfender JL, Touboul D, et al. Searching for original natural products by molecular networking: detection, isolation and total synthesis of chloroaustralasines. Org Chem Front. 2018;5:2171–8.

[46] Pluskal T, Castillo S, Villar-Briones A, Orešič M. MZmine 2: modular framework for processing, visualizing, and analyzing mass spectrometry-based molecular profile data. BMC Bioinformatics. 2010;11:395.

[47] Nothias L-F-F, Nothias-Esposito M, Da Silva R, Wang M, Protsyuk I, Zhang Z, et al. Bioactivity-based molecular networking for the discovery of drug leads in natural product bioassay-guided fractionation. J Nat Prod. 2018;81:758–67.

[48] Esposito M, Nothias L-F, Nedev H, Gallard J-F, Leyssen P, Retailleau P, et al. Euphorbia dendroides latex as a source of jatrophane esters: isolation, structural analysis, conformational study, and Anti-CHIKV activity. J Nat Prod. 2016;79:2873–82.

[49] L-F N-S, Dumontet V, Neyts J, Roussi F, Costa J, Leyssen P, et al. LC-MS2-Based dereplication of Euphorbia extracts with anti-Chikungunya virus activity. Fitoterapia. 2015;105:202–9.

[50] Kellogg JJ, Todd DA, Egan JM, Raja HA, Oberlies NH, Kvalheim OM, et al. Biochemometrics for natural products research: comparison of data analysis approaches and application to identification of bioactive compounds. J Nat Prod. 2016;79:376–86.

Gabin T. M. Bitchagno and Serge A. F. Tanemossu

9 Computational methods for NMR and MS for structure elucidation III: More advanced approaches

Abstract: The structural assignment of natural products, even with the very sophisticated one-dimensional and two-dimensional (1D and 2D) spectroscopic methods available today, is still a tedious and time-consuming task. Mass spectrometry (MS) is generally used for molecular mass determination, molecular formula generation and MS/MS^n fragmentation patterns of molecules. In the meantime, nuclear magnetic resonance (NMR) spectroscopy provides spectra (e. g. ^{1}H, ^{13}C and correlation spectra) whose interpretation allows the structure determination of known or unknown compounds. With the advance of high throuput studies, like metabolomics, the fast and automated identification or annotation of natural products became highly demanded. Some growing tools to meet this demand apply computational methods for structure elucidation. These methods act on characteristic parameters in the structural determination of small molecules. We have numbered and herein present existing and reputed computational methods for peak picking analysis, resonance assignment, nuclear Overhauser effect (NOE) assignment, combinatorial fragmentation and structure calculation and prediction. Fully automated programs in structure determination are also mentioned, together with their integrated algorithms used to elucidate the structure of a metabolite. The use of these automated tools has helped to significantly reduce errors introduced by manual processing and, hence, accelerated the structure identification or annotation of compounds.

Keywords: automated structure elucidation, MS, NMR, CASE, natural products

9.1 Introduction

9.1.1 Nuclear magnetic resonance (NMR)

NMR, one of the most used and universal experiments for the structural description of unknown compounds, is based on the Boltzmann equilibrium of nuclei endowed with a spin in a radiofrequency field. Certain isotopes (with odd mass or charge number), when placed in a strong magnetic field, rearrange their spin orientation in more than one shell depending on their spin number. When nuclei

This article has previously been published in the journal *Physical Sciences Reviews*. Please cite as: Bitchagno, G.T.M., Tanemossou, S.A.F. Computational Methods for NMR and MS for Structure Elucidation III. *Physical Sciences Reviews* [Online] **2019** DOI: 10.1515/psr-2018-0109.

https://doi.org/10.1515/9783110579352-010

are irradiated at a certain radio frequency, the most stable spins absorb then release the energy corresponding to the previously observed transition. The frequency of a resonant nucleus is proportional to the strength of the magnetic field. The relative frequencies of nuclei have been defined in comparison to a reference frequency mostly tetramethylsilane (TMS) and designated as chemical shifts. The chemical shift of a nucleus is a function of the screening constant (shielding) of the observed nucleus in the molecule and can provide information about the molecular structure or functional groups. Thus, chemical shifts vary with the chemical environment, the geometry of the molecule, the s-character of a carbon atom or the anisotropy of the magnetic field around a π-system. Resonant nuclei can interact either through covalent bonds (scalar coupling) or through space (dipolar coupling). The nuclear Overhauser effect (NOE) experiment, which is based on spatial interactions between nuclei, is important to define the relative configuration at some stereogenic centers [1, 2].

9.1.2 Mass spectrometry (MS)

Besides NMR spectroscopy, MS is a powerful analytical technique used for the structure elucidation of natural products. Before assigning the structure of an unknown compound, a chemist usually starts with the determination of its molecular formula to search for analogues in the literature or to match MS information (e. g. the number of carbons in the molecule) with NMR ones (e. g. the number of observed carbon signals in the ^{13}C NMR spectrum). Only MS experiments are able to provide accurate molar mass or molecular formulas. The classical step-by-step procedures for the measurement in MS include injection, vaporization, ionization, and detection of a sample which could be liquid, soluble or gas. The ions formed are accelerated by a high electric field and then separated in a magnetic field according to their mass-to-charge (*m/z*) ratio and spectra are recorded as relative intensities *versus m/z* ratios. The relative intensity of a given ion is proportional to its abundance, which also depends on the stability of the fragment before being collected as well as the type of ionization technique used to form that ion [2].

9.1.3 Rationale of this work

Most labs process NMR or MS data either manually or semi-automatically with the help of visualization tools. These steps can take experienced spectroscopists weeks or even months. Recently, attention has been paid to developing computational methods that can significantly accelerate data processing and reduce errors introduced by manual processing [2–13]. The present work review reputed and mostly used empirical methods for structure determination. Furthermore, fully automated pipelines involved in structure determination of metabolites are also presented herein.

9.2 Sequence in NMR structure elucidation

9.2.1 Steps in NMR structure determination

The structure elucidation of an unknown compound, and natural products, in particular, is a quite tedious task which also fascinates chemists. Depending on the structural complexity of the compound in question, the structure determination can take hours, days, months or even years. After the free induction decays (FIDs) are recorded and processed, the NMR spectra are collected. NMR structure determination usually involves several steps that start with picking all the peaks corresponding to signals in the 1D and 2D NMR spectra followed by defining all the homonuclear and heteronuclear interactions. Finally, the chemical shifts are assigned and the functional groups or the chemical structure are also defined as well as how these groups are connected (Figure 9.1). Each step highly relies on the previous one. However, this time-consuming work can be automated or semi-automated using computational methodologies.

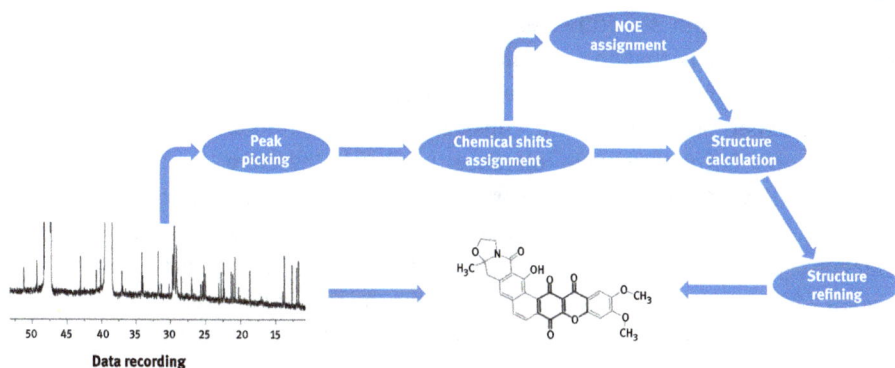

Figure 9.1: Step-by-step sequence in chemical structure elucidation by empirical considerations.

9.2.2 Peak picking

Peak picking is the first step in structure determination after recording and processing NMR spectra. It is a prerequisite for the other steps listed above (Figure 9.1), and thus requires much attention from spectroscopists. The goal of peak picking is to identify cross-signals, which contain the chemical shift information of the spin systems, from the noisy NMR spectra [14–16]. Peak picking is very sensitive and it has, so far, been difficult to design automatic methods that can deal with this sensitivity. There are two reasons for this difficulty.

- Firstly, the outputs of peak picking serve as the inputs for both the assignment and structure calculation steps. Any practical peak-picking method must, therefore, be very accurate.
- Secondly, there are various sources of errors in NMR spectra, including random noise, sample impurities, artifacts and water/solvents bands, which make peak picking a very challenging problem [14–20].

Among the existing automatic methods, AUTOPSY [16], PICKY [15] and WaVPeak [14] are the most accurate. The first and second methods attempt to estimate the noise level of a given spectrum while the third is a wavelet-based smoothing and volume-based filtering. The later overcomes the two bottlenecks; all the data points with low intensities and false positive peaks are eliminated when compared with the two other methods [14].

9.2.3 Resonance assignment

Once all peaks are identified and well picked, each of them has to be assigned a chemical shift. A nearly complete sequential assignment of resonance peaks is crucial for successful structure determination. The underlying principle for resonance assignment is that the chemical shift values for any specific nucleus at a certain position in the molecule are the same across multiple spectra. However, taking the experimental errors into consideration, these values are typically not the same but only fall within some tolerance threshold. Therefore, for every observed chemical shift value, identifying its corresponding nucleus in the target compound is not trivial. The goal of resonance assignment is to map all of these resonance peaks to their corresponding nuclei. By this mapping, the nuclei in the target molecule will be labeled with chemical shift values which enable the local structural restraint extraction for the 3D structure calculation [18]. Several software programs including AutoLink, RANDOM, PACES, MARS, CISA, IPASS, GARANT, ARIA, CANDID or ATNOS are devoted to this task.

With the development of 2D NMR and other state-of-the-art spectroscopic techniques since the 1970s, the structural calculation became easier to operate. In association with high-resolution MS (HRMS), automated methods for structure determination were quite sufficient in solving structure elucidation problems for small molecules. A signal in a one-dimension proton NMR spectrum (^{1}H NMR), at a particular frequency, can arise from a number of different proton types, in a number of different chemical environments. A cross signal in a 2D NMR spectrum clearly links two nuclei depending on the kind of experiment that is performed. Unlikely to the "fuzzy" or "analog" information in many 1D NMR spectra, signals in 2D NMR spectra can be termed as "digital" because a cross signal tells the spectroscopist that the proton resonating at a frequency f1 is either separated by two, three, or sometimes four bonds from a carbon atom resonating at a frequency f2. Since some ^{1}H NMR spectra are very poor in terms of signals (mainly due to the small amount of naturally occurring substances), the 2D NMR experiments can be unexploitable and the expert system should take this limitation into account in order to increase its efficacy [11].

9.3 Computer-assisted structure elucidation (CASE)

Computational methods used for the structure elucidation of natural products are termed the CASE expert system. These methods attempt to digitalize the way human determine structures of unknown compounds from spectroscopic data information. As reported by Elyashberg and collaborators, important outputs are expected by using CASE expert systems [19]. CASE can be used when the manual interpretation of data is impractical and outcomes are unreliable using certain techniques, such as artificial intelligence, pattern recognition, library search and spectral simulation (Figure 9.2). Moreover, databases of natural products should be part of the developed empirical method or the implementation procedure to easy structure elucidation.

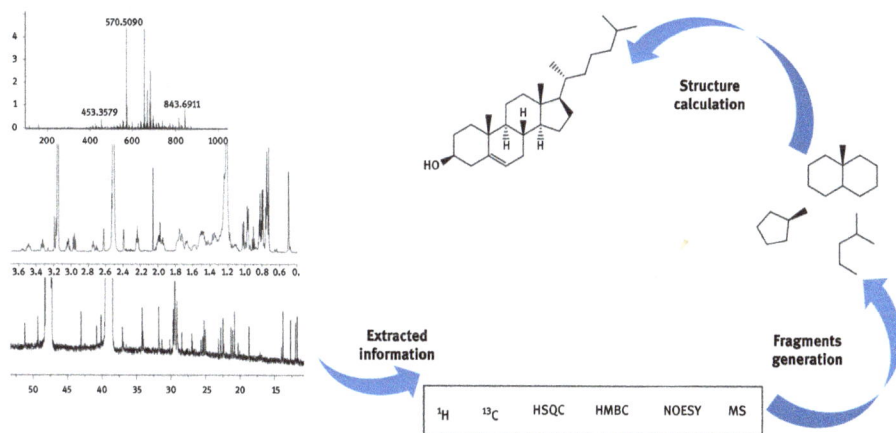

Figure 9.2: General processing on CASE system methods.

Several works are listed in the literature related to the design and use of the CASE system for the structure elucidation of unknown molecules. Since the advent of the GENOA system in 1981 by Carhart and co-workers [3], progress has been made in this area. CASE has been assessed by Shelley and Munk [4] and was followed by DARC, a system developed by Dubois et al. [5]. CHEMICS [6, 7] and Specsolv [8] later made their appearance. After that, reports on new developments have continued with LSD

[9], SESAMI [10], LUCY then SENECA [11], CAST\CNMR Structure Elucidator [12], Bruker CMC-se and more recently ACD/Structure Elucidator [13]. They are either based on 2D-NMR spectra or on ^{13}C spectra.

9.3.1 Logic for structure determination (LSD)

LSD is a program published in 1991 by Nuzillard and Massiot [9]. It is mainly related to 2D NMR data and is limited by the size of the compound (it can be applied to a maximum of 30 heavy atoms). The program is able to handle groups of superimposed carbon signals if the status of each member of the group is known. That consists of its order number (arbitrary and given by the user), its hybridization state (currently possible for sp^3 or sp^2 hybridization, but not sp^1), its valency and multiplicity (number of bounded hydrogen atoms). Special properties of atoms, deduced from elementary spectral analysis, may be indicated, mainly consisting of specifications about the status of the neighboring atoms. Correlations between groups are treated as special properties of the atoms belonging to these groups. If fully assigned substructural units have been recognized, their bonds should be added to the database as starting points for the resolution process. The molecular formula has to be known as well as the status description for each non-hydrogen atom.

Almanza et al. have used LSD approach for the structure elucidation of two clerodanes (**1** and **2**) from *Salvia haenkei* [21]. The data set used for the LSD computer analysis of these compounds did not include any information from the correlation spectroscopy (COSY) spectrum but contained the multiplicity and hybridization state of all the carbons and oxygens, heteronuclear multiple quantum correlation (HMQC) and heteronuclear multiple bond correlation (HMBC) data and bonds deduced from the presence of a ketonic carbonyl and three ester/lactone groups. The prediction gave nine structure candidates. Four of these contained a cyclobutadiene ring and no furan and could be discarded immediately. The correct structure for **1** was given twice but with different ^{13}C signal assignments. The comparison of their data with those published at the time of investigation led the authors to confirm these structures to be salviandulines A (**1**) and B (**2**) (Figure 9.3).

The LSD software code has been improved so that it recognizes a wider set of atom types to build molecules. More flexibility has been given in the interpretation of 2D NMR data, including the automatic detection of very long-range correlations. A program named pyLSD was written to deal with problems in which atom types are ambiguously defined. It has been modified to accept C, N, O, S, F, Cl, Br, I, P, Si and B atoms, sp, sp^2 and sp^3 hybridization states and electrically charged atoms. PyLSD was able to predict the correct structure of hexacyclinol (**3**), 2-*O,N*-dimethylliriodendronine (**4**) and azadirachtin (**5**) (Figure 9.3) [22].

Nevertheless, LSD lacks knowledge of chemical shift values and natural product chemistry. It can be used in combination with other expert systems for better efficacy.

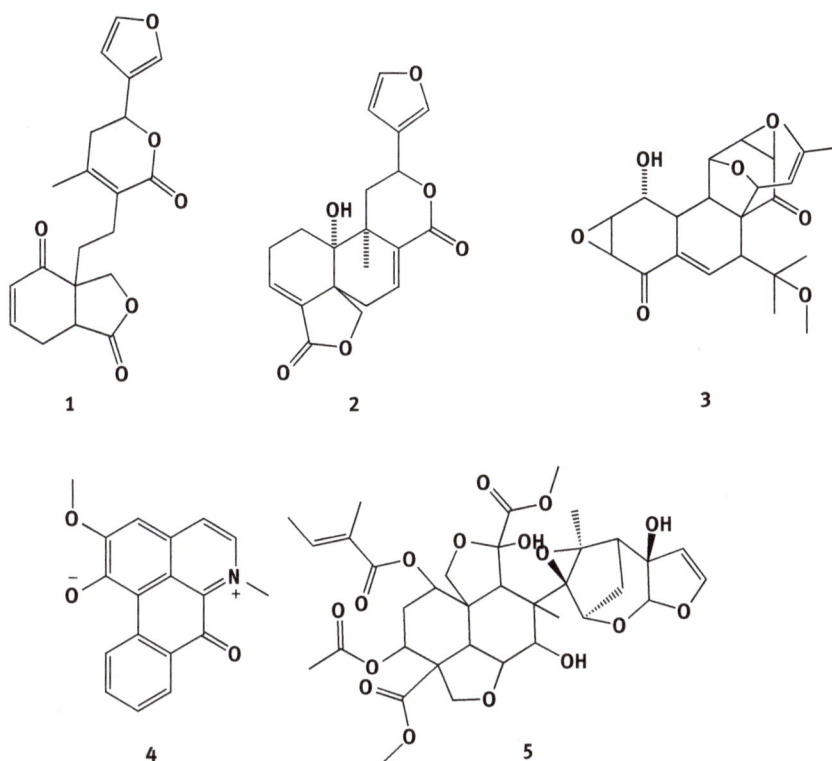

Figure 9.3: Chemical structures of compounds **1–5** either found or confirmed after computation using LSD system.

In this view, Nuzillard and Emerenciano reported the use of LSD and SISTEMAT for the structure elucidation of natural products [23]. SISTEMAT is a database of about 20,000 natural products and contains information about their ^{13}C NMR data and botanical origins. Given the mostly relying correlation of secondary metabolites and their botanical sources, the introduction of natural products databases to structure predictors has a bright future. Nuzillard and Emerenciano also reported the use of such systems in the structure elucidation of several terpenes [23].

9.3.2 SENECA

Steinbeck published the use of LUCY as a new tool for structure elucidation [24]. Some bottlenecks have, however, been associated with this program. For instance, LUCY took into account only ^2J and ^3J HMBC correlations, putting back nonstandard interactions like ^4J or ^5J. Another limitation is the number of heavy atoms that LUCY can consider, which at the moment is 30, for instance. Five years later, SENECA was developed by the same author to overcome these limitations. This program was applied to determine the structures of the sesquiterpene eurabidiol (**6**), isolated from the plant *Euryops arabicus* [25] and the fungal metabolite monochaetin (**7**) [26], as well as polycarpol (**8**), a triterpene isolated from *Onychopetalum amazonicum* [27] (Figure 9.4).

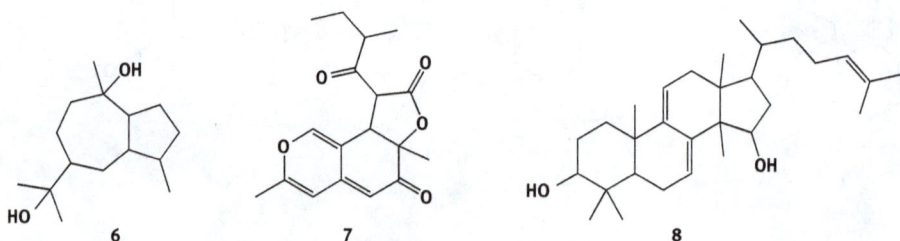

Figure 9.4: Chemical structures of compounds **6–8** confirmed after computation using the SENECA system.

9.3.3 ACD/structure elucidator

Also known as StrucEluc, ACD/Structure Elucidator is recognized as the most effective CASE program developed so far. In its first version, the program was using just ^{13}C NMR data, molecular mass and molecular formula as initial data inputs. When necessary, IR and ^{1}H NMR data were also introduced as additional data to filter the structure determination process. But gradually, the program was adapted to 2D NMR spectral data. After the user has checked and entered the input data, the system queries a series of databases to sort all possible fragments in relation to the uploaded data. Finally, the program proposes predicted structures based on these fragments. That is the so-called common mode and the system switch automatically to the classic mode if it is not able to propose one or more molecules. At this level, the system attempts to generate structures by checking the functional groups of the molecule and looking the frequencies at which those groups appear. Two lists of functional groups can be created: the "good list", containing functional groups that the system detected within the input data and the "bad list" which contains functional groups not found within these data. By generating fragments from this information the final output file contains only structures that have persisted through a filtering process in which spectral–structural correlation libraries have been established using NMR and IR spectra [28]. The selection of the most probable structure is performed on the basis of the ^{13}C chemical-shift prediction using three algorithms implemented into the system, HOSE code based [29] and neural networks and additivity rules [30]. The program is capable of elucidating the structure of an unknown compound in the presence of an unknown number of nonstandard correlations of unknown length. The software is commercially available and was used for solving many complex analytical problems. As a case study, the StrucEluc was used to determine the structure of the complex alkaloid quindolinocryptotackieine [19, 31] in an interactive mode, allowing for step-by-step resolution of many ambiguous correlations, which otherwise would have taken scientists weeks of work. It was the first time that the program was employed to solve a structure that had been unsolvable by experienced spectroscopists. The software was recently applied for the structure elucidation of armeniaspirols A−C (**9–11**), Figure 9.5 [32], new chlorinated spiro[4.4]non-8-enes exhibiting *in vitro* and *in vivo* antibacterial properties from *Streptomyces armeniacus*, a microbial species, whose occurrence had been described by Dafour and colleagues [31].

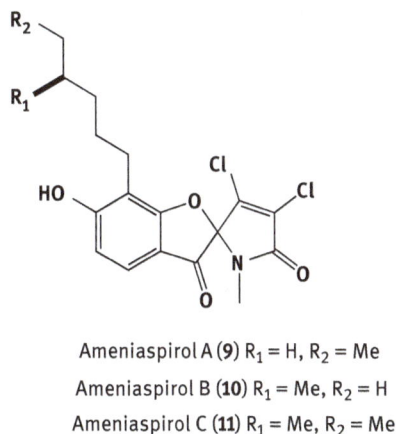

Ameniaspirol A (**9**) R_1 = H, R_2 = Me

Ameniaspirol B (**10**) R_1 = Me, R_2 = H

Ameniaspirol C (**11**) R_1 = Me, R_2 = Me

Figure 9.5: Chemical structures of ameniaspirol **A–C** confirmed by computation using StrucEluc system.

During the course of structure elucidation, the authors faced challenges in the structural assignment by NMR experiments due to the lack of protons in the spiro-ring system and consequently the very few correlations in the HMBC spectrum. StrucEluc program was, therefore, utilized by using ^{1}H and ^{13}C chemical shifts, multiplicities, carbon hybridization states, as well as COSY, HSQC and HMBC connectivities as inputs to build a molecular connectivity diagram (MCD) (Figure 9.6), which listed ^{1}J, ^{2}J, and ^{3}J bond correlations.

As inputs, a total of 19,834 molecules were generated by using the MCD, from which 9 molecules passed all filtering criteria (including a check for chemical feasibility, ring rules, removal of duplicates, etc.). In further steps, ^{13}C-spectra were predicted by increment methods, neural network, hierarchical organization of spherical environments (HOSE). The structure with the smallest deviations between

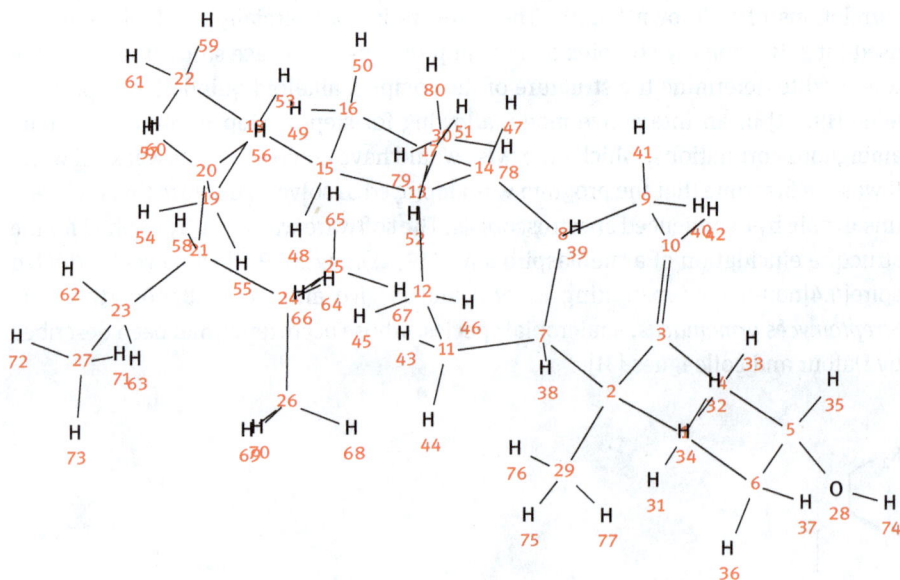

Figure 9.6: Overview of Molecular Connectivity Diagram (MCD). MCD is the method employed to interpret 2D NMR spectra. That means, from one proton, the software check ^1J from HSQC spectrum then ^2J, ^3J or ^4J from HMBC spectrum to gradually build the chemical structure of an unknown compound. By exploiting interactions from each proton signal in the molecule, fragments are proposed. These fragments serve to completely elucidate the structure.

predicted and experimental chemical shifts corresponds to **10** (Figure 9.5). Similarly, Kummerlöwe et al. investigated one of the products obtained by reacting an azide-containing 1,5-enyne in the presence of electrophilic iodine sources [33]. After having recorded and interpreted HRMS, IR, 1D NMR, COSY, HSQC, HMBC, and even 1,2-ADEQUATE spectra, the authors were not even able to determine the structure of the emerging product of this reaction. Then, they recorded the residual dipolar coupling (RDC) and obtained the structure **12** (Figure 9.7) which matched all other data. But the

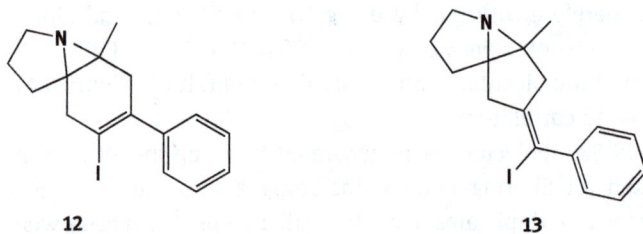

12

13

Figure 9.7: Chemical structures of compounds **12** and **13**.

proposed structure was not in agreement with the ^4J and ^5J interactions in the HMBC spectrum of the studied compound. The highly complex nature of the 2D NMR data led the authors to conclude that the problem could not be solved by a classical approach. In making this decision, they only considered the NMR data in isolation from algorithmic-assisted approaches such as those available in CASE software, like the structure elucidator. All HMBC correlations, without any exclusion and including the set of nine NSCs were used. 1,1-ADEQUATE correlations were also added to the 2D NMR data. Fuzzy Structure Generation was run with the following result: only one correct structure, **13** (Figure 9.7), was generated in 0.7 s.

StrucEluc and LSD are the most used programs to date. SESAMI system is able to search for substructures in a database containing assigned ^{13}C NMR spectra (^{13}C NMR interpretive library-search system (INFERCNMR)) [34]. The search result in a set of substructures predicted to be present in the unknown compound, each of which is assigned and estimated with very good accuracy. This method is quite different compared to LSD and StrucEluc and reduces significantly the number of proposed fragments generated by the system. To the best of our knowledge, the programs Bruker CMC-se and CAST/CNMR Structure Elucidator have never been reported for the structure elucidation of natural products and can therefore not be rated [34].

9.4 Computation in MS

MS is a key analytical technique for detecting and identifying small molecules like secondary metabolites. It is universally recognized as more sensitive than NMR. However, unlike NMR, it highly depends on the instrument used and required more steps (e. g. calibration and sample preparation and derivatization). In addition to CASE methods used in structural identification of small molecules, it is common in the natural product chemistry research area to hear about dereplication (identification of known compounds) by MS. One of the most recent platforms is the Global Natural Product Social Molecular Networking (GNPS), which is a data-driven platform for the storage, analysis and knowledge dissemination of MS/MS spectra that provide the scientific community with the ability to analyze a data set and compare it to all publicly available data [35]. The GNPS enables online dereplication through automated molecular networking analysis [35]. One of the advantages of MS lies on the fact that MS spectra are relatively easy to acquire and there are automated techniques used for the identification of substances based on their MS data by comparison to those available in databases [36]. In fact, both CASE and MS dereplication platforms are hardly related to databases. In this section, it will be difficult to distinguish between both processes knowing that the aim is the same: identification of small molecules from MS data.

The idea behind computed MS methods is to annotate mass spectra using the most probable elemental compositions found in public databases, most likely from ion fragments, and to add additional orthogonal filters to decrease the number of structure hits. Two main computational principles are known and implemented in much existing software to produce those fragments [37, 38]:

- The *fragmenter,* with a rule set model, generates fragments based on cleavage known rules (McLafferty rearrangements, retro-Diels-Alder reactions, neutral losses, oxygen migration, etc.)
- The *combinatorial Fragmenter* model, which tries to predict the fragmentation tree based on the metabolite molecular structures and tandem mass spectra.

Below are more computational aspects for identifying small molecules by matching their spectra with those of references from spectral libraries. The easiest approach for the structure identification of a metabolite is to check a match to its spectrum in a spectral library. However, some challenges exist. MS spectra vary with the ionization techniques and analyzers. Libraries should, therefore, be constructed for the uniqueness of each of the mentioned parameters. For MS/MS, libraries should address the issue of collision energy, because changes in this energy cause changes in the acquired spectra. Problems related to contaminations, noises or solvent can lead to false identification by searching for matches in a database.

9.4.1 Databases in MS: Importance and uses

Many databases have been developed to assist the identification of compounds by MS. The size of each is highly influenced by the ionization techniques used. Libraries of electronic ionization (EI) spectra are the most spread. For example, the PubChem database allows free access to more than 35 million molecular structures (https://pubchemdocs.ncbi.nlm.nih.gov/about). The National Institute of Standards and Technology (NIST) mass spectral library (https://chemdata.nist.gov/dokuwiki/doku.php?id=chemdata:start) is a huge collection of libraries built up of EI and tandem MS spectra (small molecules and peptides) and GC retention index. They are also freely available data analysis tools including AMDIS (Automated Mass Spectral Deconvolution and Identification System for GC/MS), Mass Spectrum Interpreter (connects chemical structures with mass spectra) and Mass Spectral Digitizer Program. The version 11 of NIST contained EI spectra of more than 200,000 compounds. The Wiley Registry (11th edition) contains over 775,500 EI mass spectra, 741,000 searchable chemical structures, and 599,700 unique compounds. In 2005, the Golm Metabolome Database (GMD) was made up of EI fragmentation mass spectra of about 1600 compounds [39] and the Fiehn Lab library contains about thousand metabolites. The size of tandem MS libraries is small compared to EI libraries. The NIST 11 contains collision cell spectra for only 4000 compounds (www.chemdata.nist.gov). The Wiley Registry of Tandem Mass Spectral Data comprises positive and negative mode spectra of more than 1200 compounds (www.wiley.com)

[38]. Some databases address specific research interests. The Human Metabolome Database (HMDB) comprises reference MS/MS spectra for more than 2500 metabolites found in the human body [38]. The Platform for RIKEN Metabolomics (PriMe) collects MS^n spectra for research on plant metabolomics [38]. METLIN contains high-resolution tandem mass spectra for more than 10,000 metabolites for diagnostics and pharmaceutical biomarker discovery and allows to build a personalized metabolite database from its content [38]. MassBank is a public repository with more than 30,000 spectra of about 4,000 compounds collected from different consortium members. The MMCD is a hub for NMR and MS spectral data containing about 2000 mass spectra from the literature and collected under defined conditions [38].

Nevertheless, the sizes of existing spectral libraries are smaller than those of molecular structure databases: Just to illustrate, the CAS Registry of the American Chemical Society currently contains about 25 million compounds, while PubChem alone has surpassed 50 million entries [40].

9.4.2 Empirical methods using rule-based fragmentation model

Fragmentations rules have been proposed for almost each of the existing molecular functional groups. However, the implementation of each rule to a functional group is not so easy. Several systems have been proposed for mass spectrum simulation based on described rules [41]. Among them, some (MassFrontier, ACD/MS, and MOLGEN) are more popular than others (CONGEN and MASSIS). CONGEN and MASSIS (Mass Spectrum Simulation System) are only applied for EI-MS. CONGEN is an automated tool to predict mass spectra of a given molecular structure using general models of fragmentation as well as class-specific fragmentation rules. Intensities for EI spectra were modeled with equations found by multiple linear regression analysis of experimental spectra and molecular descriptors. MASSIS combines cleavage knowledge (McLafferty rearrangements, retro-Diels-Alder reactions, neutral losses, oxygen migration) of functional groups, small fragments (end-point and pseudo-end-point fragments) and fragment-intensity relationships for simulating EI-MS spectra.

9.4.2.1 Mass Frontier
Mass Frontier (HighChem, Ltd Bratislava, Slovakia; versions after 5.0 available from Thermo Scientific, Waltham, USA), a commercially available software, contains fragmentation reactions collected from MS data in the literature [42]. Besides predicting a spectrum from a molecular structure, it can also explain an experimental fragmentation spectrum. It is limited on fragmentation pathways already published and which can be consulted from the literature, meaning that new rules cannot be considered for better comprehension. The software is based on observed experimental gas-phase fragmentation reactions, discarding data from large molecular weight and most polar substances. It contains basic fragmentation rules as well as an exhaustive library of

over 100,000 known fragmentation rules collected from published data which also allows the fragmentation prediction and annotation of unknown compounds [43].

9.4.2.2 ACD/MS

The *ACD/MS Fragmenter* (Advanced Chemistry Labs, Toronto, Canada) can only interpret a given fragmentation spectrum using a known molecular structure. Initially, these programs were designed for the prediction and interpretation of fragmentation by EI, but recently, there has been a tendency to interpret tandem MS data with ACD/MS fragmenter [39, 44]. Both programs are commercial, and no algorithmic details have been published. They are very user-friendly. The process starts by drawing a structure and selecting an ionization technique and polarity, then the software will take few seconds to propose predictions from established MS fragmentation rules based on the available spectra from the literature [39].

9.4.2.3 MOLGEN

In 2012, a concept was initiated to compare the efficiency of available computational techniques in identifying small molecules named CASMI (Critical Assessment of Small Molecule Identification). The concept was built around a program known as MOLGEN. To date, there is more than five versions of the program sorted in three CASMI categories [45]. The first category is composed by MOLGEN-MS/MS, a program which aims at proposing molecular formula using HRLC-MS/MS data as input. The second category takes as input the same data in the previous category. The versions currently available for this category are MOLGEN 3.5 and MOLGEN 5.0. Both generate structures that match the molecular formula(s) and optional structural restrictions provided by the user but are implemented differently. The third category is the most different among the three categories as it produces structures of unknown compounds by the means of low-resolution electron impact. MOLGEN-MS is used as a computed system. In addition, MOLGEN program is implemented together with already described fragmenters like Metfrag or FiD (Fragment iDentificator) and MS databases like NIST since MOLGEN as a stand-alone system failed in structure elucidation of a number of candidates [45].

9.4.3 Empirical methods using combinatorial fragmentation model

Fragmentation and ion trees aim to identify the molecular formulas of compounds, elemental compositions of fragment ions and neutral losses to perform automatic annotations on MS or MS^n spectra. A typical process visible on Figure 9.8 shows how structure can be elucidated by a combination fragmentation model. Briefly, structure fragments are primarily built from MS data, then fragmentation trees are developed for MS and MS^n spectra annotation by representing each peak by a node with a molecular formula and the output is a list of some possible structure containing that of the unknown compound being studied [46].

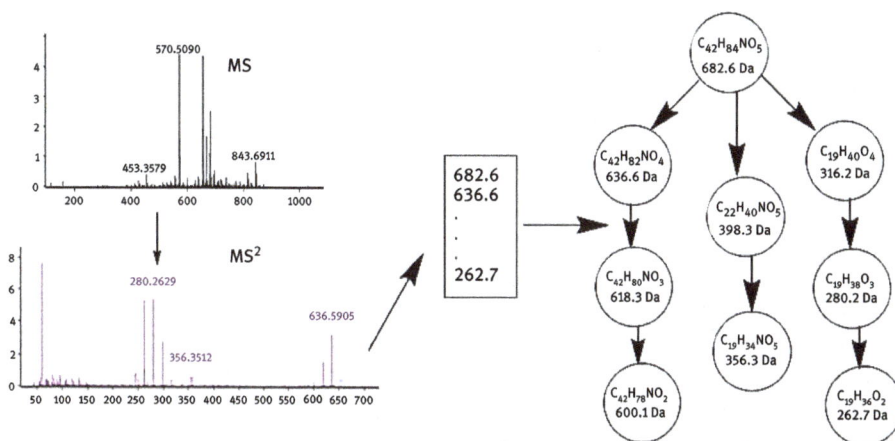

Figure 9.8: Typical combinatorial fragmentation model. After recording the MS followed by MS^n spectra, important peaks are sorted and listed mainly based on their relative intensity to the base peak and following the graduation in fragmentations (from MS to MS^n). Then, the software proceeds to calculate the molecular formula of each peak to come out with the final formula of the unknown compound.

9.4.3.1 Fragment iDentificator (FiD)

FiD is a software tool for the structural identification of product ions produced with tandem mass spectrometric measurement of low molecular weight organic compounds. FiD conducts a combinatorial search over all possible fragmentation paths and outputs to provide a ranked list of possible structures. This gives the user an advantage in situations where the MS/MS data of compounds with less well-known fragmentation mechanisms are processed. FiD software implements two fragmentation models, the single-step model that ignores intermediate fragmentation states and the multi-step model, which allows for complex fragmentation pathways [47]. The FiD software is free for academic use and is available for download from www.cs. helsinki.fi/group/sysfys/software/fragid.

9.4.3.2 MetFrag

MetFrag is a software created to obtain a candidate list from compound libraries based on the precursor mass, subsequently ranked by the agreement between measured and *in silico* fragments. During its evaluation MetFrag was able to rank most of the correct compounds within the top three candidates returned by an exact mass query in KEGG database [36, 48]. Compared to a previously published study, MetFrag obtained better results especially for large compound libraries. The candidates with a good score show a high structural similarity or just different stereochemistry, a subsequent clustering based on chemical distances reduces this redundancy by the help of InChIKey filtering. MetFrag performs a search in KEGG, ChemSpider or

PubChem on an average within 30–300 s on an average desktop PC [36, 48]. MetFrag has also been extended to analyze EI fragmentation. It can also be used in combination with CFM-ID (competitive fragmentation modeling), a suite of software tools that can perform spectra prediction and compound identification. It is based on a machine-learning approach including chemical rules and is available for ESI MS/MS data as well as EI mass spectra [43].

9.4.3.3 CSI:FingerID

CSI:FingerID (Compound Structure Identification) [40] is a freely available web-service searching molecular structure databases using tandem MS data of small molecules as input. The method computes a fragmentation tree that best explains the fragmentation spectrum of an unknown molecule. The fragmentation tree is then used to predict the molecular structure fingerprint of the unknown compound using machine learning. This fingerprint is then used to search for a molecular structure. These are three main steps computed in this software which can be run with PubChem as a database. SIRIUS GUI software is nowadays integrated into CSI: FingerID to calculate fragmentation trees and molecular formulas. It is actually coupled to the CSI:FingerID online server that matches fingerprints against a database and retrieves ranked structure candidates [43].

9.4.3.4 ChemDistiller

ChemDistiller is a recent software and hopefully a reliable one [49]. It uses structural fingerprints and fragmentation patterns together with a machine learning algorithm to annotate unknown compounds. ChemDistiller is related to many databases which provide access to more than 150 million known compounds with pre-calculated "fingerprints" and fragmentation patterns [43]. The software is also using SIRIUS for molecular mass calculation, XCMS and mzMine for peak-picking to come out with a list of the best candidate compounds from the query tandem MS spectra [49].

9.5 Conclusion

The structure elucidation of pure natural products or metabolites from complex mixtures (e. g. plant and microorganism extracts) is a tedious task that constitutes a bottleneck in bioactive compound discovery or metabolomics. However, the past years have seen an emergence of new computational methods and automated software for structure elucidation based on MS or NMR analyses. These methods have been successfully used for the annotation or structure determination of metabolites and are nowadays of great interest in natural product research and metabolomics.

Acknowledgements: The authors are grateful to the German Academic Exchange Service (DAAD) and the Alexander von Humboldt Foundation for financial support.

References

[1]   Macomber RS. A complete introduction to modern NMR spectroscopy. New York: John Wiley & Sons, 1998:357

[2]   Pavia DL, Lampman GM, Kriz GS. Introduction to spectroscopy: A guide for students of Organic Chemistry. Thomson Learning, 3rd ed. Fort Worth: Harcourt College Publishers, 2001:680.

[3]   Cahart RE, Smith DH, Gray NAB, Nourse JG, Djerassi C. GENOA: a computer program for structure elucidation utilizing overlapping and alternative substructures. J Org Chem. 1981;46:1708–18.

[4]   Shelley CA, Munk ME. CASE, a computer model for structure elucidation process. Anal Chem Acta. 1981;133:507–16.

[5]   Dubois JE, Carabedian M, Dagane I. Computer-aided elucidation of structures by carbon-13 nuclear magnetic resonance. Anal Chem Acta. 1984;158:217–33.

[6]   Funatsu K, Miyabayashi N, Sasaki S. Further development of structure generation in the automated structure elucidation system CHEMICS. J Chem Inf Comput Sci. 1988;28:18–28.

[7]   Funatsu K, Susuta Y, Sasaki S. Introduction of two-dimensional NMR spectral information to an automated structure elucidation system CHEMICS. J Chem Inf Comput Sci. 1989;29:6–11.

[8]   Bremser W, Grzonka M. SpecInfo-A multidimentional spectroscopic interpretation system. Microchim Acta. 1991;2:483–91.

[9]   Nuzillard J-M, Massiot G. Logic for structure determination. Tetrahedron. 1991;47:3655–64.

[10]   Munk ME. Computer-based structure determination: then and now. J Chem Inf Comput Sci. 1998;38:997–1009.

[11]   Steinbeck C. SENECA: A platform-independent, distributed, and parallel system for computer-assisted structure elucidation in organic chemistry. J Chem Inf Comput Sci. 2001;41:1500–7.

[12]   Koichi S, Arisaka M, Koshino H, Aoki A, Iwata S, Uno T. Chemical structure elucidation from ^{13}C NMR chemical shifts: efficient data processing using bipartite matching and maximal clique algorithms. J Chem Inf Model. 2014;54:1027–35.

[13]   Elyashberg ME, Williams AJ. Computer-based structure elucidation from spectral data. The art of solving problems. Heidelberg: Springer, 2015.

[14]   Liu Z, Abbas A, Jing B-Y, Gao X. WaVPeak: picking NMR peaks through wavelet-based smoothing and volume-based filtering. Bioinformatics. 2012;28:914–20.

[15]   Alipanahi B, Gao X, Karakoc E, Donaldson L, Li M. PICKY: a novel SVD-based NMR spectra peak picking method. Bioinformatics. 2009;25:268–75.

[16]   Koradi R, Billeter M, Engeli M, Güntert P, Wüthrich K. Automated peak picking and peak integration in macromolecular NMR spectra using AUTOPSY. J Magn Reson. 1998;135:288–97.

[17]   Korzhnev D, Ibraghimov IV, Billeter M, Orekhov VY. MUNIN: application of three-way decomposition to the analysis of heteronuclear NMR relaxation data. J Biomol NMR. 2001;21:263–8.

[18]   Wan X, Lin G. CISA: combined NMR resonance connectivity information determination and sequential assignment. IEEE/ACM Trans Comput Biol Bioinform. 2007;4:336–48.

[19]   Elyashberg ME, Blinov KA, Molodtsov SG, Williams AJ. Elucidating "undecipherable" chemical structures using computer assisted structure elucidation approaches. Magn Reson Chem. 2012;50:22–7.

[20]   Groscurth S. Tools for structure elucidation. Brucker, Accessed: 26 Jan 2019.

[21]   Almanza G, Balderrama L. Clerodane diterpenoids and an ursane triterpenoid from *Salvia haenkei* computer-assisted structural elucidation. Tetrahedron. 1997;53:14719–28.

[22]   Plainchonta B, Emerenciano VDP, Nuzillard J-M. Recent advances in the structure elucidation of small organic molecules by the LSD software. Magn Reson Chem. 2013;51:447–53.

[23]   Nuzillard J-M, Emerenciano VDP. Automatic structure elucidation through data base search and 2D NMR spectral analysis. Nat Prod Comm. 2016;1:57–64.

[24] El-Sayed AM, Al-Yahaya MA, Shah AH, Hartmann R, Breitmaier E. Eurabidiol and other sesqui-terpenes from *Euryops arabicus*. Chem Ztg. 1990;114:159–60.

[25] Christie BD. The role of two-dimensional nuclear magnetic resonance spectroscopy in compu-ter-enhanced structure elucidation. J Am Chem Soc. 1991;113:3750–7.

[26] Hellwig V. Einige Alkaloide aus *Schizantus litoralis* und ein Triterpen aus *Onychopetalum amazonicum*. Diploma Thesis, Bonn, 1994.

[27] Blinov KA, Carlson D, Elyashberg ME, Martin GE, Martirosian ER, Molodtsov S, et al. Computer-assisted structure elucidation of natural products with limited 2D NMR data: application of the StrucEluc system. Magn Reson Chem. 2003;41:359–72.

[28] Bremser W. HOSE – a novel substructure code. Anal Chim Acta. 1978;103:355–65.

[29] Elyashberg ME, Williams AJ, Martin GE. Computer-assisted structure verification and elucida-tion tools in NMR-based structure elucidation. Prog NMR Spectrosc. 2008;5:31–104.

[30] Blinov KA, Elyashberg ME, Martirosian ER, Molodtsov SG, Williams AJ, Sharaf MMH, et al. Quindolinocryptotackieine: the elucidation of a novel indoloquinoline alkaloid structure through the use of computer-assisted structure elucidation and 2D NMR. Magn Reson Chem. 2003;41:577–84.

[31] Dufour C, Wink J, Kurz M, Kogler H, Olivan H, Sabl S, et al. Isolation and structural elucidation of armeniaspirols A-C: potent antibiotics against gram-positive pathogens. Chem Eur J. 2012;18:16123–8.

[32] Kummerlöwe G, Crone B, Kretschmer M, Kirsch SF, Luy B. Residual dipolar couplings as a powerful tool for constitutional analysis: the unexpected formation of tricyclic compounds. Angew Chem Int Ed Engl. 2011;50:2643–5.

[33] Penchev PN, Schulz P-K, Munk ME. INFERCNMR: a ^{13}C NMR interpretive library search system. J Chem Inf Model. 2012;52:1513–28.

[34] Elyashberg M. Identification and structure elucidation by NMR spectroscopy. Trends Anal Chem. 2015;69:88–97.

[35] Wang M, Carver JJ, Phelan VV, Sanchez LM, Garg N, Peng Y, et al. Sharing and community curation of mass spectrometry data with Global Natural Products Social Molecular Networking. Nat Biotechnol. 2016;34:828–37.

[36] Wolf S, Schmidt S, Müller-Hannemann M, Neumann S. In silico fragmentation for computer assisted identification of metabolite mass spectra. BMC Bioinform. 2010;11:148.

[37] Nguyen DH, Nguyen CH, Mamitsuka H. Recent advances and prospects of computational methods for metabolite identification: a review with emphasis on machine learning approaches. Brief Bioinform. 2018;0:1–16.

[38] Scheubert K, Hufsky F, Bocker S. Computational mass spectrometry for small molecules. J Cheminform. 2013;5:12.

[39] Kopka J, Schauer N, Krueger S, Birkemeyer C, Usadel B, Bergmuller E, et al. Steinhauser D: GMD@CSB.DB: the Golm Metabolome Database. Bioinformatics. 2005;21:1635–8.

[40] Dührkop K, Shenb H, Meusela M, Rousub J, Böcker S. Searching molecular structure databases with tandem mass spectra using CSI:fingerID. Proc Natl Acad Sci USA. 2015;112:12580–5.

[41] Lindsay R, Buchanan B, Feigenbaum E, Lederberg J. Applications of artificial intelligence for organic chemistry: the DENDRAL project. New York: McGraw-Hill; 1980.

[42] Böcker S. Searching molecular structure databases using tandem MS data: are we there yet? Curr Opin Chem Biol. 2017;36:1–6.

[43] Blaženovic I, Kind T, Ji J, Fiehn O. Software tools and approaches for compound identification of LC-MS/MS data in metabolomics. Metabolites. 2018;8:31.

[44] Rojas-Cherto M, Kasper PT, Willighagen EL, Vreeken RJ, Hankemeier T. Reijmers TH: elemental composition determination based on MS^n. Bioinformatics. 2011;27:2376–83.

[45] Meringer M, Schymanski EL. Small molecule identification with MOLGEN and mass spectro-
      metry. Metabolites. 2013;3:440–62.
[46] Vaniya A, Fiehn O. Using fragmentation trees and mass spectral trees for identifying unknown
      compounds in metabolomics. Trends Anal Chem. 2015;69:52–61.
[47] Heinonen M, Rantanen A, Mielikäinen T, Kokkonen J, Kiuru J, Ketola RA, et al. FiD: a software for
      ab initio structural identification of product ions from tandem mass spectrometric data. Rapid
      Commun Mass Spectrom. 2008;22:3043–52.
[48] Ruttkies C, Schymanski EL, Wolf S, Hollender J, Neumann S. MetFrag relaunched: incorporating
      strategies beyond in silico fragmentation. J Cheminform. 2016;8:3.
[49] Laponogov I, Sadawi N, Galea D, Mirnezami R, Veselkov KA. ChemDistiller: an engine for
      metabolite annotation in mass spectrometry. Bioinformatics. 2018;34:2096–102.

Part III: **Chemoinformatics Tools and Methods for
Lead Compound Discovery and Development**

Eleni Koulouridi, Marilia Valli, Fidele Ntie-Kang
and Vanderlan da Silva Bolzani

10 A primer on natural product-based virtual screening

Abstract: Databases play an important role in various computational techniques, including virtual screening (VS) and molecular modeling in general. These collections of molecules can contain a large amount of information, making them suitable for several drug discovery applications. For example, vendor, bioactivity data or target type can be found when searching a database. The introduction of these data resources and their characteristics is used for the design of an experiment. The description of the construction of a database can also be a good advisor for the creation of a new one. There are free available databases and commercial virtual libraries of molecules. Furthermore, a computational chemist can find databases for a general purpose or a specific subset such as natural products (NPs). In this chapter, NP database resources are presented, along with some guidelines when preparing an NP database for drug discovery purposes.

Keywords: databases, library design, virtual screening, natural products, natural products databases

10.1 Introduction to databases (definition/construction)

A chemical database is a collection of data concerning chemical compounds. A virtual chemical database, for use in cheminformatics and in other disciplines of computational chemistry, is a collection of small molecules presented by the use of computer and their related information. Nowadays, there is a plethora of information concerning chemical compounds and their properties such as biological activities and resources. These kind of data can be very useful if these are interpreted accurately. Nevertheless, this large amount of information should be organized in a way that will help scientists to get full benefit of it using various methods. Hence, the need for storage and management of information has emerged from the "Big Data" challenge, which means to take advantage of the existing data in order to make the right decisions for the next step of a research project. The growth of scientific data has indicated the organization of data supported by computers [1].

Databases of various sizes exist. Databases of various content can also be found. There are databases of zero-dimensional (0D) or one-dimensional (1D) structures

This article has previously been published in the journal *Physical Sciences Reviews*. Please cite as: Koulouridi, E., Valli M., Ntie-Kang, F., Bolzani, V.S. A primer on natural product-based virtual screening. *Physical Sciences Reviews* [Online] **2019** DOI: 10.1515/psr-2018-0105.

https://doi.org/10.1515/9783110579352-011

such as the NCI Open Database of compounds which was released in 1999. Moreover, there are databases for two-dimensional (2D) or three-dimensional (3D) structures of small molecules. Repositories for structural data concerning biological macromolecules have been constructed and are available to the scientific community such as Protein Data Bank (PDB). There are databases providing literature sources such as PubMed. Databases exist with different quality of data (curated on a regular base or not). Databases can store a variety of information along with chemical structures. For example, the ChEMBL database includes the structure of compounds accompanied by bioactivity data collected from the literature.

For many cheminformatics applications, virtual collections of small molecules are needed. Representing molecules in computers involves scientific software and storing of the chemical data produced. A number of software (molecular editor/ structure sketcher) and the most commonly used formats for storing in cheminformatics are presented next. The representation of molecules with the use of computers and the creation of specific file formats for data exchange are vital for the construction of a virtual database [2–8].

Machine-readable formats for digital representations of chemical structures of small molecules: Figure 10.1 presents the formats commonly used in cheminformatics for the construction of a database consisting of small in size compounds.

1. ***SMILES (Simplified Molecular Input Line Entry System)***: A one-dimensional (1D) string (a line notation) which represents the two-dimensional (2D) chemical structure of a molecule. For example, the Canonical SMILES for aspirin is {CC(=O) OC1=CC=CC=C1C(=O)O} as provided from PubChem website (PubChem CID = 2244). For further reading, DAYLIGHT webpage provides information including uses of SMILES, rules concerning canonicalization (for a unique identifier),

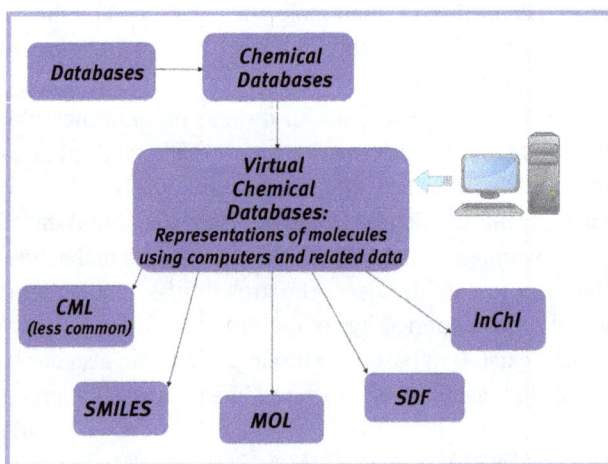

Figure 10.1: Common file formats used in Cheminformatics.

specification rules such as the representation of atoms and bonds. **SMARTS** is an extension of SMILES used for substructure searching [9–12].

2. **InChI (and InChIKey)**: IUPAC's (International Union of Pure and Applied Chemistry) chemical structure identifier standard. InChI is a unique string (a line notation) of symbols which a computer can "read" and, as a result of this, can represent a defined chemical structure. An InChI can be produced by drawing structures on a computer screen with the use of a specific software. InChIKey is the compressed version of InChI consisting of 27 characters. The creation of InChIKey intended for use on the internet or for a database searching, not for the representation of molecules. For example, aspirin's InChI is {1S/C9H8O4/c1-6(10)13-8-5-3-2-4-7(8)9(11)12/h2-5H,1H3,(H,11,12)} and InChIKey for aspirin is {BSYNRYMUTXBXSQ-UHFFFAOYSA-N} as found on PubChem website (Figure 10.2) [12–14].

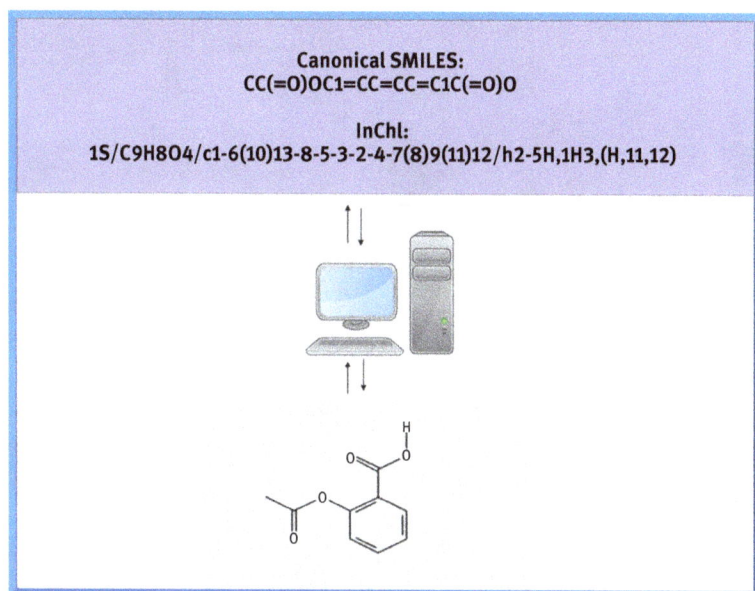

Canonical SMILES:
CC(=O)OC1=CC=CC=C1C(=O)O

InChI:
1S/C9H8O4/c1-6(10)13-8-5-3-2-4-7(8)9(11)12/h2-5H,1H3,(H,11,12)

Figure 10.2: Digital representation of a molecule.

3. **MOL**: IUPAC's Gold Book defines the MOL file format as a file format which encodes structures or substructures or conformations in the form of text-based connection tables (connection tables specify how atoms are connected). It is used for 2D and 3D structures [13, 15, 16].

4. **SDF (Structure-Data File)**: This format provides information for 2D or 3D structure (e.g., connectivity of atoms) and data annotation. It also uses the form of text-based connection tables as MOL file format. It is useful for more complicated tasks in cheminformatics. Figure 10.3 presents sections of the SDF file format for

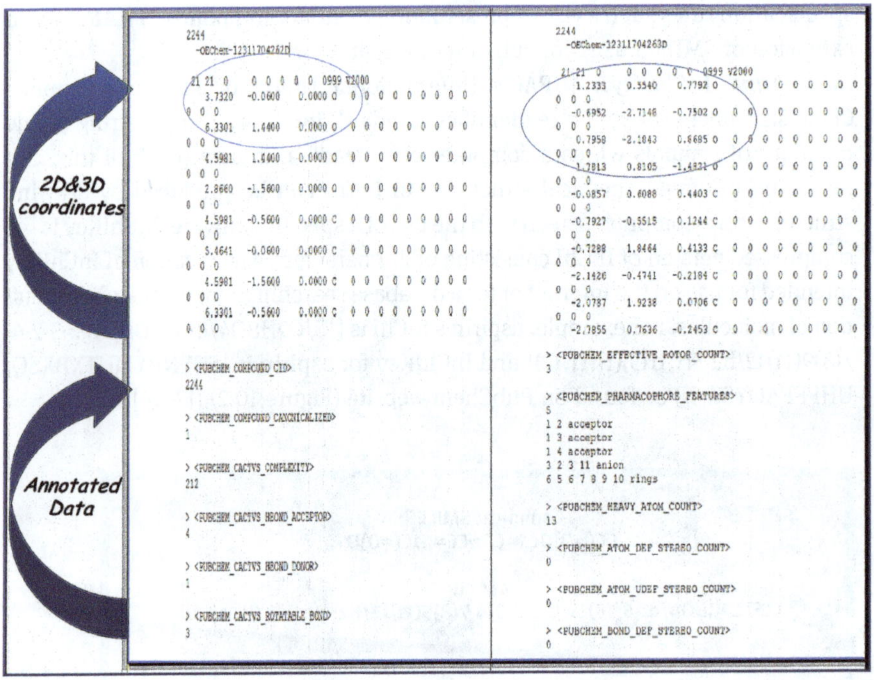

Figure 10.3: SDF file format for (2D)/left & (3D)/right structure of aspirin (PubChem).

the 2D structure of aspirin and the corresponding file for 3D structure of the molecule, extracted from PubChem, as text.

5. **Chemical Markup Language (CML)**: A less common format, which is an XML-based representation of chemical structures and can provide the structures of small molecules and more complex structures [14].

Tip: The users are able to view the file formats as a text by using a basic word processor such as a text editor (e.g. MS WordPad). The text reveals all the information included in the file, such as coordinates, pharmacophore features or molecular formula. This capability allows the users to make corrections e.g. during the curation of a database or interconversion of file formats.

In some computational techniques, such as Structure-Based Virtual Screening (SBVS) or target fishing, file formats for the representation of macromolecules are also used. PDB file format, XML, mmCIF, and FASTA are used in Protein Data Bank which is for the repositioning of proteins (targets) and other important macromolecules for the scientific research. An online guide is provided for "Understanding PDB Data". Here, the interested user can find a plethora of information, including how to view a PDB file which consists of coordinates of all the atoms and experimental observations such as missing coordinates or temperature factor (flexibility of macromolecules). This is useful for the preparation of a protein (corrections) for reliable results. It is also useful for understanding how file formats are constructed in general [2, 17–19].

Next, few examples of software used for drawing on-screen chemical structures are presented briefly.

– *Avogadro*: Used for editing and visualizing molecules. The user can download from databases such as PubChem. It is compatible with several operating systems. Free for use [20].
– *ChemDraw*: A drawing program with additional applications such as property calculators [21].
– *Maestro*: The user can draw chemical structures with Maestro and visualize them [22].
– *ChemSketch*: The user can draw chemical structures and calculate molecular properties [23].
– *MarvinSketch*: A drawing program for simple and complicated chemical structures [24].

There are two categories of structure sketchers: 3D sketchers for the creation of 3D chemical structures and 2D sketchers for the design of 2D structures, e.g. 2D similarity searches in a database [3].

Furthermore, there are web sketchers for supporting search of online databases. For example, there is the JSME molecular editor which offers editing of structures, even the complex ones, and reactions [25, 26]. PubChem has its own editor for searching the database by similarity or by structure/substructure [12].

A problem has resulted from the need for data exchange. The interconversion of molecular structures between different formats. Different file formats of data included in various databases, different file formats needed from different scientific software are points for which a computational chemist should be aware of in order to design experiments with a reliable outcome. During the process of interconversion, there is a possibility of alteration or loss of important information such as chirality [18]. The Open Babel software is a well-known free software for these complicated tasks with the capability to interconvert more than (110) file formats. These formats include those commonly used in applications of cheminformatics, formats used by various software for docking experiments, formats used by different software for visualization. Moreover, additional information can be retrieved using Open Babel and specific file formats. Finally, 2D or 3D structure of a molecule can be generated from a SMILES string by Open Babel [27].

Another example of a software with the capability of interconversion is the Chemistry Development Kit (CDK). This is an open-source application for use in chemoinformatics. It provides various functionalities such as the exchange of chemical data between different scientific programs, the identification of identical structures and fingerprint methods for similarity search procedures [28, 29].

Furthermore, there is the RDKit tool which handles various types of file formats such as SMILES, SDF and PDB. Beyond this function, RDKit also provides other capabilities including substructure searching and generation of fingerprints [30].

Lastly, an outline for the construction of a chemical database includes these basic points: Purpose of the database, sources of chemical information, data annotation (information that should be included), scientific software to create the collection, file format to store data, a database for internal use or a public repository, curation [31, 32].

10.2 Guidelines when preparing a compound database

The preparation of a compound database is crucial and affects the next steps of a drug discovery process. Here, we present some issues that should be taken into account during the construction of a virtual compound collection.

10.2.1 The purpose of the compound database

The purpose of a database is one major factor. Here are some interesting cases:
–   Diversity of a database should exist when the goal is the identification of a novel lead compound. For lead optimization, a library consisting of compounds with similar structures is advisable.
–   A target-focused library: these are compounds with known bioactivity against a single target.
–   A focused library includes compounds against a family of targets (e.g. GPCRs: G protein-coupled receptors).
–   A general-purpose database. Usually, a database of this kind combines with the Lipinski Rule of Five ("drug-like").
–   A database for 2D similarity search needs the 2D representations of molecules. A database for 3D similarity search or other computational techniques such as pharmacophore-based VS needs the 3D representations of molecules. In these techniques, a 3D database including conformers of each molecule is even needed.
–   If there is a time limitation, purchasable compounds should be included in a database.
–   Data annotation. The computational chemist must decide which kind of information to comprise in the database, apart from chemical structures e.g. literature references or descriptors needed for a QSAR study.
–   A specific purpose database (e.g. a collection of natural products).
–   A database for in-house use or an online database for public use [18, 33–38].

10.2.2 Molecular properties

Virtual databases, as it has already been mentioned, include chemical structures along with relevant data concerning molecular properties and bioactivity. These properties are expressed as arithmetical values and can be calculated from 2D or/and 3D chemical structures [36, 37]. These values are known as molecular descriptors and can be used for the filtering of a database in order to obtain a subset with specific characteristics (e.g. drug-like), exclude unwanted molecules or be used in computational methods such as QSAR [37, 39, 40]. A point that is worthy of attention is that there may be errors in chemical structures within a database, so attention must be paid for correct representations before calculating the desirable descriptors [4, 41]. Properties which can be provided within a collection of molecules are: logP (lipophilicity), number of hydrogen bond acceptors and donors, polarity, molecular weight, number of rotatable bonds, total polar surface area (TPSA), ADMET (Absorption/ Distribution/ Metabolism/ Excretion/ Toxicity) properties and more [40, 42]. Fingerprints can also be included in order to be used for similarity searches (*Fingerprint*: a bit-string where the presence of a feature is represented by one bit) [10]. There are various tools for the prediction of these descriptors such as QikProp, Molinspiration, E-Dragon, Bioclipse, FAFDrugs4 [22, 43–47]. *Tip*: The values given for bioactivity are usually obtained from various assays, so there is an assay comparability matter [18]. *Tip2*: There are various free databases which provide as data annotation many properties such as ADMET properties [6, 31].

10.2.3 3D representations

3D representations of molecules are vital in various computational techniques such as pharmacophore-based and structured-based VS. They should be of high quality for obtaining reliable results. A correct chemical structure is necessary and should be taken into account that there are errors in various databases [4, 41]. Ensuring the correct chemical structure (with removal of salt and addition of hydrogens as well if needed), the next step is the generation of 3D representations for each molecule of the database with an appropriate software such as LigPrep or MOE software. For some cases, it is important to create a number of low energy conformations after the generation of the 3D structures. In this case, the result is a multiple conformer 3D database. Taking into account factors including pH and charge (formal charge/partial charge) is also advisable, especially for docking studies. According to pH conditions, the ligand could have charged groups or not. Correct stereochemistry is needed and creation of tautomers, too. *Tip*: Be careful of the pH where the target is acting. *Tip2*: Check always the resulted log file [18, 22, 31, 42, 48, 49].

10.2.4 Format for storing chemical data

Depending on the use of the virtual chemical library, a choice should be made about the format of storing data. For a local library, the format must be compatible with the scientific software to be used. For an online database, different formats are useful in order to satisfy all interested users and be used with the majority of scientific software. For example, ZINC database offers various file formats for downloading molecules. *Tip*: There is always the case of interconversion for both local and online libraries. The possibility of errors and inconsistency between different file formats should be taken into account. Things are a bit different and easier when a database is constructed from the sketch with the same software package. *Tip 2:* Preparation of a ready to use collection of molecules with the software to be used in order to resemble a collection from the sketch [18, 41, 50–52].

10.2.5 Compounds with problematic functional groups

There are compounds which must be excluded from a virtual library because of problematic functional groups. Their presence is not approved in a collection of molecules for drug discovery. These compounds present toxicophore or reactive (reactive: undesired, probably covalent, interactions with the target) parts, such as triflates, alkyl halides or acrylamide. Applying filters is a way to avoid this category of compounds, especially when using various databases such as free online repositories [37, 49, 51, 53].

10.2.6 Compounds likely to interfere in assays

The presence of compounds known as PAINS (pan-assay interference compounds) should be considered carefully during the construction of a database since a large percentage of this category of chemicals presents false positive biological activity in various tests. Compounds belonging to this category are candidates for further thorough investigations, when proposed as a promising result. In the article by Bisson et al. (2016), IMPs (Invalid Metabolic Panaceas) are introduced as the corresponding category of PAINS in natural products. The interested reader can learn more about PAINS in: (https://chembiohub.ox.ac.uk/blog/2015/03/10/pains.html). In the article by Baell and Holloway, an analysis of PAINS and the filters used specifically for recognizing them in screening libraries and excluding them is reported. There is Supplementary Material along with the publication providing the PAINS filters as a Sybyl line notation and commands for use on a Linux operating. In the article by Pouliot and Jeanmart, PAINS and other categories of problematic compounds are introduced, regarding especially the agrochemical sector. Furthermore, rules for avoiding undesired compounds during the lead identification stage are reported in this article. For example, SMARTs filters introduced by the University of Dundee, the ALARM NMR tool, the GSK filters and other rules are discussed which are implemented by the pharmaceutical industry as well as academia [18, 54–57].

10.2.7 Diversity

As it has already been mentioned, the diversity of a chemical database depends on the purpose of the experiment. For the process of the identification of a lead compound, a collection of molecules with a vast variety of chemical scaffolds is needed. On the other hand, lead optimization asks for libraries with a high degree of similarity. A general-purpose drug discovery process invests in libraries consisting of molecules with drug-like properties. When referring to diversity, the total number of chemical compounds with different structures that can be created is the matter. This is called chemical space. For a virtual chemical library, this is a much easier task due to the fact that a lot of combinations between different building blocks can exist (huge chemical space). On the contrary, there is an obvious difficulty in the synthesis of all these compounds, which can be created virtually. Additionally, the chemical space decreases because of ADMET properties. Drug-like compounds occupy specific parts of the chemical space as clusters of the same scaffold. Therefore, a virtual experiment should take into account the scope and the capability for synthesis (Figure 10.4) [18, 58].

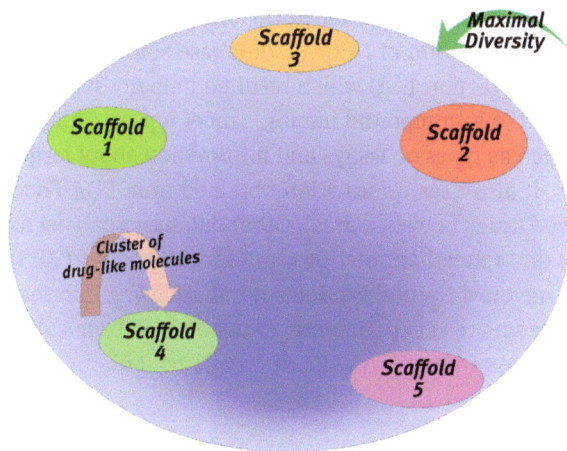

Figure 10.4: A basic conception for chemical diversity.

10.2.8 Tools for library design

Free and commercial software packages can be used for the design of a compound collection, corresponding to the specific needs of a drug discovery project. Here, a number of these tools are presented.

- *FAF-Drugs4* is a free web tool for filtering a compound library by PAINS, undesirable groups, toxicity, ADME properties and by Lipinski Rule of Five before using it in a virtual experiment. It also calculates molecular descriptors as already mentioned. Useful information is provided *via* the website. In the publication by Lagorce et al. (2015), a useful URL is also provided (www.vls3d.com, Villoutreix et al. Drug Discovery Today 2013, 18:1081–9) where various online tools are collected and introduced [44, 59].

– *e-LEA3D*: a web tool which combines three functions: the de novo drug design, filtering of libraries, and combinatorial library design [60].
– *ICM Chemist*, provided by Molsoft, can create focused libraries [61].
– *LipidMapsTools* for the construction of a virtual library consisting of lipids [62].
– *ilib diverse* is a tool provided by Inte:Ligand for the construction of a diverse library. The user creates chemical structures, even 3D, applies methods of filtering and exports the compounds in SDF file format or SMILES [63]. In the review article by Lionta et al. [32] with references therein, tools for library design are presented.

10.2.9 Sources

Literature is an essential source for the initial stage of the construction of a virtual database, in order to retrieve information about chemical structures, biological activities and more. For example, the creation of NPCARE database started with an exploration of PubMed (PubMed, see Section 10.1) which resulted in many scientific papers with the information needed (e.g. compound names, cancer types). ChEMBL database retrieves activity data such as types of assays and information about target molecules from scientific papers. It also uses information obtained from FDA-Food and Drug Administration approved drugs (a web source). Other databases can act as providers of chemical structures and related annotations. Gu and colleagues (2013) describe the construction of the Universal Natural Products Database using information from other databases. Another source is experimental data obtained from collaborative laboratories such as in the case of the p-ANAPL library [5, 64–66].

10.3 The use of virtual databases in computational chemistry

Virtual databases are of great importance for drug discovery process, especially for the initial stage where the use of computational chemistry is necessary and cost/time effective. These virtual collections are used during various in silico procedures, including pharmacophore modeling and QSAR, as a complementary stage. This stage is called VS and it is the virtual analogous procedure of High-Throughput Screening (HTS). Simply, VS means checking each molecule of the virtual library very quickly with an appropriate software using various queries. Here, their role in different computational techniques is discussed [31, 38, 67].

Pharmacophore-based Virtual Screening: The initial stage of this procedure is the generation of a pharmacophore model. With reference to IUPAC definition, a pharmacophore model compiles all the vital steric and electronic features (e.g., hydrophobic regions) needed for the optimal interactions between a molecule-target and a molecule-ligand. Two methods are used to construct a pharmacophore model. The Ligand-based and the Structure-Based methodology. In the second one, the use of a macromolecular repository such as PDB is essential. When an appropriate pharmacophore model is created, it can be used as a query for searching a 3D chemical database in order to find molecules presenting the features proposed by the pharmacophore model. The flexibility of the database's molecules must be taken into account before the screening procedure (generation of conformers) [2, 38, 67, 68].

Docking-based Virtual Screening (or *Structure-based Virtual Screening*): The purpose of a virtual molecular docking experiment is to investigate how a ligand is positioned in a macromolecule-ligand complex (interactions found between the ligand and the macromolecule, usually a protein) and to estimate the stability of this complex (the query for VS). The necessity of a collection consisting of 3D structures of macromolecules, such as PDB, is obvious for applying this method. Databases of small chemical molecules in 3D structure, then, can be screened with molecular docking method in order to discover compounds which also appear the ability to interact with the macromolecule and form a stable complex [32, 67].

(3D)-QSAR: QSAR methodology utilizes molecular descriptors such as the number of rotatable bonds in order to construct quantitative structure-activity relationship models. Since molecular descriptors can be calculated from structures using computer software, databases can be used in the initial stage of QSAR methodology as providers of chemical structures and of information regarding biological activities. Then, the models created can be used as a query for screening various virtual libraries [36, 67].

Similarity/substructure search (2D similarity-based or 3D similarity-based methods): Databases can be searched by similarity or by substructure using a query molecule (known ligand). If molecules are represented by fingerprints, the results from screening are based on the similarity between fingerprints (e.g. topological fingerprints are used for substructure search). Molecules of a database can also be represented by SMILES strings. In this case, the collection of molecules is searchable using the similarity between SMILES strings. Usually, the Tanimoto coefficient is used to compare the similarity between molecules (2D similarity search methods). 2D similarity-based methods are faster than 3D. 3D shape-based VS is a 3D similarity search approach based on the shape and size of molecules [10, 37, 69, 70].

Machine learning techniques: Another way to build a QSAR/QSPR (quantitative structure-activity or structure-property relationships) model is by using machine learning techniques such as random forests (RF) and partial least square (PLS) [36].

Chemical space analysis: Databases can be used for the analysis of chemical space (total possible number of descriptors/chemical diversity). A very common technique for this purpose is principal component analysis (PCA). In this way, clusters of compounds are created and can be used for further analysis with VS techniques [36, 71].

Target fishing: Another use of a virtual chemical database is for complementing the method called target fishing. The purpose of this method is to discover the interactions of bioactive, chemical compounds with different targets. In this way, possible side effects can be found or a different macromolecule-target for further investigation [19].

An example of the databases' use in computational chemistry, found in literature, is the discovery of a novel inhibitor for the cytosolic phospholipase $A_2\alpha$. In the article, the authors describe the generation of a 3D pharmacophore model based on the Ligand-based methodology. By using the virtual database provided by National Cancer Institute (NCI) with pharmacophore-based screening, they discovered a hit compound (hit: potential ligand) which was tested and presented biological activity [72].

The number of methods that use virtual chemical databases as a complementary part of their procedure, as described earlier, support the important role of virtual libraries during an in silico experiment (Figure 10.5). Emphasis must be placed on well-defined stereochemistry and structure of molecules included in chemical libraries as well as on preparation of these libraries by the user in accordance with the software to be used (e.g. file format), the purpose of the screening (Pharmacophore-based or Docking-based, 2D similarity search, target), the protonation states or tautomers needed before applying VS and more. In that way, all useful

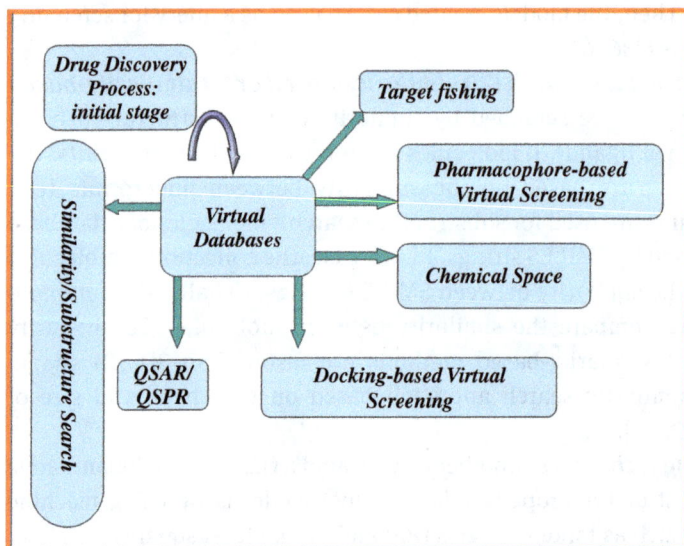

Figure 10.5: Use of virtual databases in computational chemistry.

information will be included in the virtual collection for a reliable VS procedure. In some cases, it is advisable to use filters in order to reduce the size of a database (e.g. Lipinski Rule of Five). As the result of the filters used, VS procedure becomes time effective and the performance of the computer is less demanding [18, 31, 32].

10.4 Tools for handling and managing database resources

Databases consist of various kinds of data such as chemical structures, physicochemical descriptors, literature references and more. This is a large amount of information which must be organized in ways that will help scientists to decide how they will proceed with a drug discovery project. The organization of the data is crucial and there are different tools to manipulate all the chemical information gathered from different resources.

First of all, a limitation can be introduced for the size of a database. Libraries exist in various sizes, consisting of a small number of molecules or a large number of compounds (e.g. ZINC database). The screening of a very large database is a task which needs time and computer resources. There are ways to reduce the size of a database by similarity and by filtering.

- 2D or 3D similarity search: especially useful when the goal is the creation of a target-focused library or a library for lead optimization [18, 37].
- Removal of compounds using filters. Compounds with undesirable functional groups and PAINS can be excluded [37]. "Lipinski Rule of Five" is, among others, a set of criteria which can be used to filter and retrieve molecules with desirable properties [31, 32, 35]. For example, there is *FAF-Drugs4* which has already been mentioned as a tool for filtering compounds. *ChemBioServer* also exists as a web tool for filtering compound libraries [73, 74].

Online databases are accessible via their website. Their web interface often provides the user with the ability to search using various filters or by similarity or substructure. For example, in NuBBE Database natural compounds can be filtered by molecular mass, logP, "Lipinski Rule of Five", biological, taxonomic, geographic, pharmacological information, etc. [75] ZINC12 (old website, see Section 10.5) database offers subsets filtered by various properties. ChEMBL database allows the interested user to draw a structure and search for compounds containing the structure or for similar compounds [76]. Online databases usually use various systems for managing databases in order to support their existence on the web such as *MySQL* and *CACTVS* [31, 77, 78].

For the management of local libraries, there are software packages such as *CORINA Symphony*: One of its features is the capability of excluding duplicates resulting from merging various databases. *Instant JChem (IJC)* by ChemAxon supports the management of databases. *ScreeningAssistant 2 (SA2)*: Among its basic functions is storing unique compounds and computing of descriptors [31, 47, 79–81]. *The RDKit database cartridge*: A free tool which also provides the capability to create databases with online guidance (see Section 10.1) [82].

There are also tools for the exchange of data in different formats. Open Babel example has already been discussed in Section 10.1. Moreover, standardization tools for unique entries based on the same rules for their representations have been developed such as *BIOVIA Chemical Registration* and *UniChem* for the comparison of structures between different databases. Finally, there are tools for data mining in order to retrieve information from various resources such as patents. *ChemCurator* by ChemAxon retrieves structures and related data from the literature [83–86].

10.5 Public and private databases

Due to the fact that the use of VS methods has increased, the number of virtual databases is increasing as well in order to contribute to a more effective procedure for drug discovery. There are two main categories of databases: commercial databases and free for users. An introduction of various commercial and open source libraries that exist for use in computer-based methodologies is presented here briefly. In Section 10.6, an introduction of databases of natural products is also provided. To begin with, there is an important number of freely available online databases which every interested user can take advantage of the data provided for his/her own research [6, 18, 31]. For example:

– *PubChem* which is described in the natural products section.
– *Commercial Compound Collection (CoCoCo)*: There are data about molecular structure for commercial compounds [87].
– *DrugBank database* includes information about drugs and targets [88].
– *BindingDB* which includes binding affinities between targets and small molecules [89].
– *ChemSpider*: A chemical structure database with annotated data [90].
– *ChemBank*: A database of small molecules with related data [91].
– *ChEBI* database with focus on small molecules (see also Section 10.6) [92].
– *Mother of All Databases (MOAD)*: a subset of Protein Data Bank with high-quality assemblies of proteins and ligands [93].
– *Inte: Ligand's* focused sample libraries for free [63].
– *eMolecules Plus Database*: Free version with structural data [94]. Next, three of the most commonly used free databases are presented with useful details.

ZINC: A curated database of purchasable compounds with 3D structures in various file formats and related data such as logP and number of hydrogen bond acceptors/donors. From the main menu, choice of "Substances" results in ways to search the database for desirable compounds by sketch, by SMILES, by ZINC id, by SMARTS or

InChI. By clicking on "Substances" and then "Subsets", a list of all subsets with a description, such as reactive and aggregators (reactive & aggregators subsets: useful for filters/similarity search to exclude these compounds from a database), is retrieved. An online guide provides help to interested users. The old website still exists at: http://zinc.docking.org/ (Accessed 7 January 2018). For natural products in ZINC, see Section 10.6 [50–52, 95].

ChEMBL: A curated database of compounds accompanied by information about bioactivity and ADMET properties. Search is available online for ligands by structure, keyword, ChEMBL id and SMILES. Search for targets is also provided. The interested user can browse drugs ("Browse Drugs" tab: drugs approved by FDA/compounds from USP dictionary) and drug targets as well. Icons are used for these compounds (drugs) in order to characterize them depending on the features they present. For example, clicking on the tab "Browse Drugs", a list of molecules is retrieved where icons reveal features for each compound such as drug type. The about tab and the FAQ section are online guidance. Compounds can be downloaded in various file formats [5, 76, 96].

Protein Data Bank (PDB): This curated database provides important information for 3D structures of macromolecules such as proteins and nucleic acids, which can be used in several computational techniques including structure-based VS and molecular docking. From the main menu "Search", there are options for advanced search, by sequences, by ligands (complex macromolecule-ligand), by drugs and drug targets and more. From the main menu "Learn", online guidance is offered for many interesting issues including information about biological assemblies, methods used for the determination of the 3D structures of macromolecules, the resolution matter and many other subjects. The website provides the capability of visualization of the 3D structures and a "Download" main menu option for downloading ligand files, sequences files and more [2, 17, 97].

Furthermore, there are various commercial databases such as GVKBIO databases and Reaxys collection of molecules (Reaxys, Section 10.6) [18, 98, 99].

10.6 Databases for natural products

Nature has always been an important source for the drug discovery process. Throughout the centuries, natural products (NPs) have been used for the treatment of various human diseases. For instance, in ancient Greece and in Ancient Egypt remedies were based on medicinal plants and animal products (e.g. Dioscorides, Ebers Papyrus). Nowadays, botanicals are used in Traditional Chinese Medicine (TCM) and in Ayurveda. Furthermore, many drugs are natural products, their derivatives, mimic natural products or contain their pharmacophore features. These drugs are produced or inspired by NPs coming from different sources such as bacteria or marine life. Despite the difficulties found in natural product-based drug discovery process, examples of drugs which have been developed from nature are increasingly found in the literature. Artemisinin is a medicinal agent derived from TCM. USA. FDA has approved compounds from plants as anticancer agents including cabazitaxel. Penicillin introduced the discovery of drugs from microbial sources. The use of natural products and derivatives in drug development has risen in the last years for many reasons. Among them are the structural diversity and the biological relevance natural products present. Furthermore, the number of synthetic drugs is declining and this fact led to the re-evaluation of the natural product-based drug discovery. Figures 10.6–10.8 present the chemical structure of two well-known compounds found in nature as well as few examples found in NuBBE database. The reader is advised to read the first chapter of this book for a thorough overview of natural product classes and scaffolds [67, 101, 102].

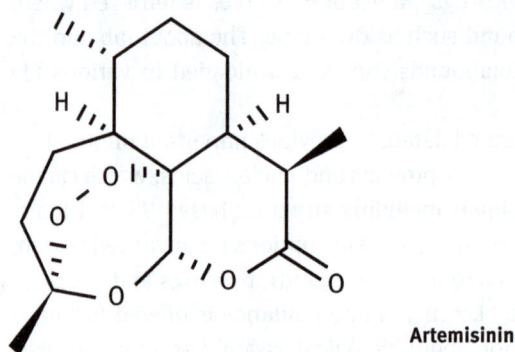

Artemisinin

Figure 10.6: The chemical structure of Artemisinin [67].

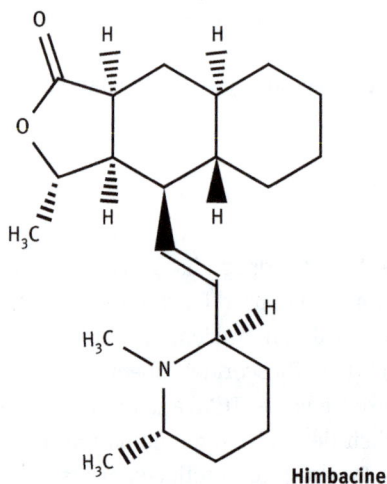

Himbacine

Figure 10.7: The chemical structure of Himbacine [100].

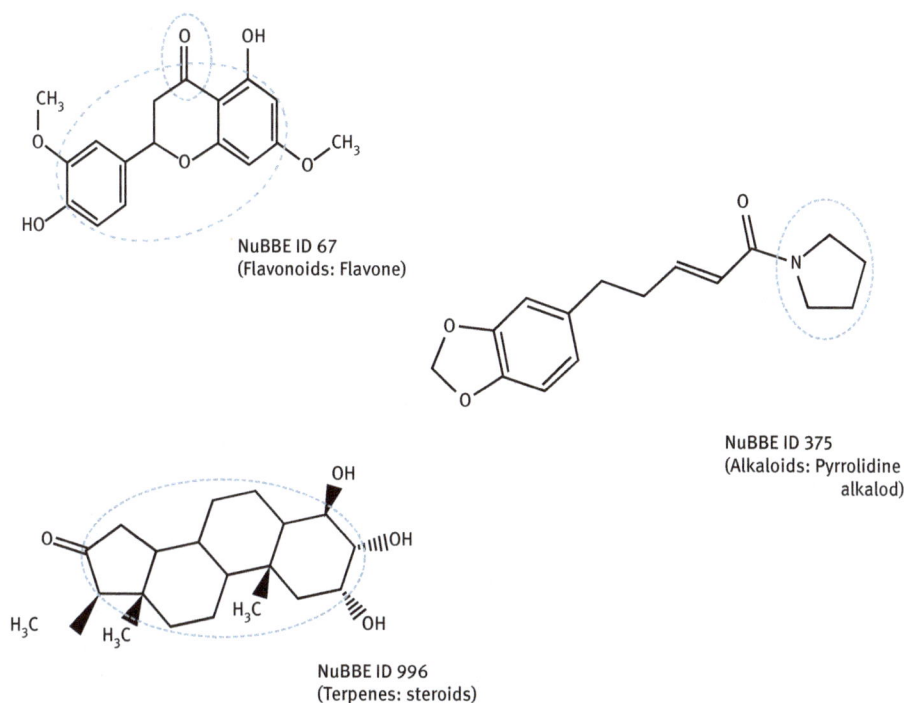

NuBBE ID 67
(Flavonoids: Flavone)

NuBBE ID 375
(Alkaloids: Pyrrolidine
alkalod)

NuBBE ID 996
(Terpenes: steroids)

Figure 10.8: Compounds found in NuBBE Database. Circled is the basic scaffold of their chemical class [75].

Due to the aforementioned facts, the need for natural products databases emerged for use in computational chemistry. Many virtual libraries of natural products exist for commercial use or for free. Here, a number of *virtual natural products collections* are presented for use in VS or another computer-based technique along with useful information for navigating each database.

– *Dictionary of Natural Products (DNP)*: A commercial database which provides the user with an online search tool with options such as a draw structure query combined with a substructure/exact match structure tool and an online detailed guide for assistance ("Tour of Dictionary of Natural Products"). The DNP intro- duction (pdf file for DNP on CD-ROM) informs the interested user about the characteristics of this database (e.g. classification of natural products in DNP, stereochemical conventions, numbering system). The Dictionary of Natural Products gives information for each entry on physicochemical properties (e.g. melting point and molecular weight), the biological source of the chemical substance, alternative names, bibliographic references and more [101, 103].
– *Dictionary of Marine Natural Products (DMNP)*: A subset of **DNP database** with online search capability and an online guide for assistance [101, 104].

- **CamMedNP**: The Cameroonian 3D structural natural products database with plant origins. For each compound in this database, the interested user can find the optimized 3D structure, known biological activities, literature resources (Supplementary Material - Additional file 1), calculated physicochemical properties and more. The 3D chemical structures are available in mdb file format provided with the Supplementary Material (Additional file 2) of the publication by Ntie-Kang et al. These 3D structures are those found in the literature. The user has the responsibility to treat the input structures for protonation and generation of tautomers [42].
- **Super Natural II**: The database includes an online search tool of the most common templates found as substructures, such as aromatics and fused rings ("Search a desired template" or main menu option "Templates"). For each result entry, there are the options of a "Start similarity search", "Go to Cluster of similar compounds" and "Show Pathways". From the main menu option "Compounds", the user is able to search a compound by properties, by name and by drawing a structure. Main menu option "MoA - Mechanism of Action Prediction" provides search by the target. The main menu option "Pathways" gives the opportunity for a compound-target relation into a cellular environment. Finally, "Clusters" main menu option enables the user to examine various clusters and their members and the "FAQ" main menu option is a detailed guide for a convenient usage of the database. This collection of molecules is free, contains a large number of substances from various resources, information on physicochemical properties (e.g. logP), SMILES and suppliers. A Mol file is provided for each molecule. Lastly, there is a toxicity prediction via ProTox. For more information, the user is advised to read the corresponding publication [77, 101, 105].
- **TCM Database@Taiwan**: A free database on TCM. The components are obtained from herbs, animal products, and minerals. Twenty-two different drug classes exist. Literature references are also provided for each compound as well as structural information. The "About TCM Database" menu option gives information about this collection of molecules. 3D optimized structures are ready to download in mol2 file format. 2D structures are also available in cdx and sdf file formats. A search of the database based on chemical names, molecular properties or molecular structures is possible. For example, the "Basic Search" menu option enables the user to search by chemical composition or by Chinese medicinal herb. The "Advanced Search" menu option allows for searching based on molecular properties such as polar surface area, molecular weight or SMILES [101, 106, 107].
- **AfroDb**: A free 3D structure collection of molecules from African medicinal plants with known bioactivities. The user can download the collection in different file formats from the supplementary electronic file of the publication (sdf and ldb). Other file formats are provided upon request to the authors of the article. The dataset includes computed physicochemical descriptors [101, 108, 109].

- *Northern African Natural Products Database (NANPDB)*: A free database. The compounds of this collection have been isolated from plants (mostly), animals, bacteria and fungi. Compounds can be searched online by name, compounds id, keywords, or by structure (by drawing a structure). The database includes information for each compound concerning SMILES, properties such as Lipinski violations or cLogP, the availability of source sample, literature references and more. The user can download the entire collection in different file formats (SMILES, SDF-2D, SDF-3D). A publication is provided as reference for further reading [101, 110, 111].
- *Traditional Chinese Medicines Integrated Database (TCMID)*: A free database. An online search tool is provided for search by prescription, by herb, by ingredient, by disease, by drug and by target. An online guide for the use of the database is found under the main menu option "Help". For an ingredient, information is given for SMILES, structure, related targets, related herbs. There are five files for downloading, each one with details for it's content [101, 112, 113].
- *Reaxys*: A commercial database with natural products from plants, animals, marine life, bacteria and other sources. There are literature references (e.g. journals, patents, reviews, conferences). A user interface is provided for searching compounds, reactions, citations and bioactivity data. The compounds can be exported in sdf file format or SMILES. There are useful videos on youtube offered by Reaxys Training for the use of this database [99, 101, 114].
- *Afrocancer*: The African anticancer natural products library. It includes compounds with anticancer activities. This free small data set can be downloaded from the supplementary file of the publication by Ntie-Kang et al. [101, 115, 116]
- *Chem-TCM*: A commercial digital database developed at King's College London, UK. The compounds are derived from plants used in the traditional Chinese herbal medicine. Chemical information is provided such as chirality, name, InChI Key. Botanical information is also provided. Predicted activities against important targets in Western medicine are included. A brief online guidance exists when the user chooses the menu option "Content View" and "Database". The sdf file format for the compounds is available [101, 117].
- *PubChem*: A free database. Users can choose the option "Databases" from the homepage, select "Compound" and type their query of interest (e.g. taxol). From the search results page, the user can select the compounds of interest and take action on the results by using the tool on the right named "Actions on your results". By clicking on the name of each compound, information is provided for 2D and 3D structure, computed descriptors such as SMILES and InChI, chemical and physical properties, chemical vendors, biological activity and more. With the option "Refine your results", the user can obtain a subset of the results. The PubChem Download Service offers various file formats such as sdf. Another way to search for natural products in PubChem is to choose "Databases" from the

homepage, select "Compound" and select "Chemical Structure Search" under the menu "PubChem tools". The user can choose to search by name/text, by drawing a structure and search for similar compounds and substructures, by molecular formula and by 3D conformer. Finally, the main menu on the homepage provides the option "help" (online guide) [12, 101, 118].

- **Taiwan indigenous plant database (TIPdb-3D)**: A free 3D structure database. With the use of a link provided or from the main menu "Download", the interested user can download two datasets of molecules (Structure Data Format). The first one is called "All chemicals". The second dataset is named "Drug-liked chemicals" and consists of chemicals filtered by Lipinski's rule of five. The database lists compounds from plants with anticancer, antiplatelet and antituberculosis properties. The main menu tool "Help" is an online guide for the right use of the database. Statistics concerning the database is given with the option "Statistics" (e.g. class/number of chemicals). With the tool "Browse", a search based on plants is achieved. With the function "Search", a search by part of a plant, by class of a chemical, by botanical name of a plant, by chemical name, by TIPID and by activity is feasible. By clicking on a chemical from the search result list (function "Search"), information about the compound is retrieved. The 2D structure is available as an image file. The 3D structure is available as mol file format. Details for properties such as XlogP, TPSA (Topological Polar Surface Area), plant sources with literature references and more are given. Due to the fact that the chemical structures and properties are manually curated, the database can be effective for QSAR modeling [101, 119–121].
- **South African Natural Compounds Database (SANCDB)**: A free online database with compounds from plants and marine life. The user can search compounds by name, by SMILES, by structure (draw a structure), by properties, by source organism, by classification, by use, by author, by reference and there is an option for advanced searching. For each compound the entry name is given, the 2D image, SMILES, literature references, classifications, other names, the source organism and compound uses. Also, each compound can be downloaded as mol2, pdb, pdb (minimized), sdf and SMILES [101, 122, 123].
- **Carotenoids Database**: This free collection of molecules provides 1172 natural carotenoids from various organisms, a complete list of which can be found by clicking on "699 source organisms". For each organism, information is introduced about the description of the organism, its carotenoid profile and literature references. More references can be found by google scholar provided as a link. By clicking on "1172 natural carotenoids" a complete list of the carotenoids of the database is retrieved. For each entry, there is a classification, names including IUPAC name, structure which can be obtained as a mol file, InChI and InChIKey, TPSA (Topological Polar Surface Area), source organisms, bibliographic references and more. There is also the google scholar tool. On the homepage, from the

menu on the left, classification of the molecules is provided based on C number, on chemical modifications, and by using carotenoid database chemical finger-prints. The database can be searched by keyword and by "Tools" including similarity search by using carotenoid database chemical fingerprints and searching for similar profiled organisms. There is a tool in order to predict the biological functions of carotenoids and a tool for searching possible biosynthetic routes of carotenoids in theory. Finally, the user is able to download fingerprints. By choosing "About" on the homepage, a pdf can be downloaded explaining the chemical fingerprints used. The "Help" function is an online guide for the database. At last, a "Biological Functions" and a "Statistics" option exist on the homepage [101, 124, 125].

– *NuBBE database (NuBBEDB)*: A free virtual 3D database from the biodiversity of Brazil. Brazil has an extremely rich biodiversity, accounting for approximately 20 % of all known living species, found in several important biomes, such as the Amazonian and the Atlantic forest regions [126]. The modern natural products research involving plant species from Brazilian biodiversity had a major boost in the studies carried out by the researchers Otto Gottlieb and Walter Mors in the 1950s. Since then, the phytochemistry of Amazonia plants has been recognized worldwide with research on Lauraceae and Myristicaceae [127–129] with an emphasis for lignans and neolignans [130, 131]. Thenceforth, more than 32,000 scientific papers were published on natural products studies in Brazil, showing the great potential of the molecular value of Brazilian biodiversity. There are several research groups in Brazil that focus on exploring this rich biodiversity rationally. One of these is the Nuclei of Bioassays, Biosynthesis and Ecophysiology of Natural Products (NuBBE, UNESP – Araraquara/Brazil) research group, which has been involved in the latest advances in natural product chemistry, including new analytical methods for metabolomics, proteo-mics, biosynthesis and medicinal chemistry. One of the focus of the group is the search for biologically active compounds from Brazilian plant species, with an emphasis on Cerrado and the Atlantic forest, endophytic fungi and marine species [132, 133]. Beside secondary metabolites, NuBBE has invested in a new topic, and recently has been dedicated to the studies on orbitides and cyclotides from Brazilian plant species [134–136].

Thus, NuBBE has accumulated valuable information on natural products from Brazilian biodiversity collected over the years. Thousands of natural products, belonging to several classes, and complex molecular features. However, these fantastic chemical data were scattered in many articles fragmented, and thus, very difficult to access readily. It could be defined as oceans of information and drops of organized information. With the objective to contribute for the organization of a database of all natural products already identified/isolated from Brazilian biodiversity, the first natural product library from Brazilian biodiversity (NuBBE Database) was created in 2013, at that time, comprising of 640 compounds isolated in NuBBE Lab [137, 138]. The NuBBE$_{DB}$ is the result of an effective collaborative project between the NuBBE group and the Laboratory of Computational and Medicinal Chemistry (LQMC, USP - São Carlos/Brazil). Currently, more than 2000 compounds are compiled, an estimate of approximately 5% of the information published. The database is an ongoing project and has already been used as an important resource for several areas of studies [139–144].

The scientific information becomes easier to access when standardized, certified and organized in a database. NuBBE Database is freely accessible online (Figure 10.9, http://nubbe.iq.unesp.br/portal/nubbedb.html) and provides valuable and integrative data such as chemical, spectroscopic, biological, taxonomic, geographic and pharmacological information. Statistical analysis currently shows that 78% of the compounds were isolated from plants, 15% are semi-synthetic products, 5% were obtained from microorganisms, 1.5% are biotransformation products, and the

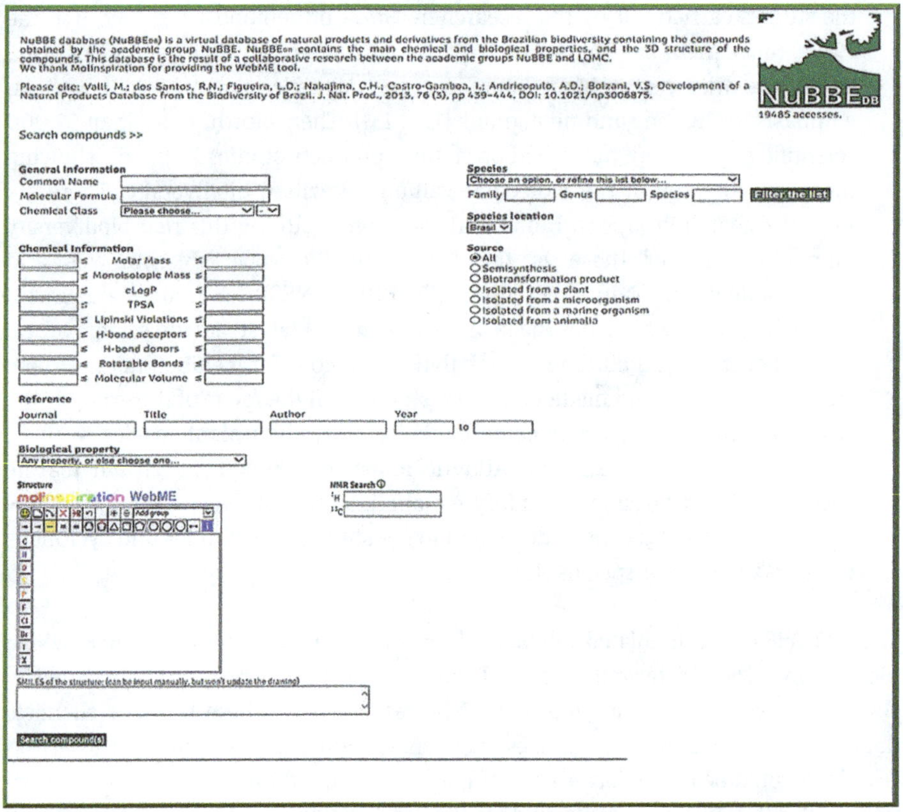

Figure 10.9: NuBBE Database search website. Compounds can be retrieved by property, chemical structure, any information shown, or a combination of criteria.

remaining from other environments. These statistical data show that plants remain the main source of the natural products studied in Brazil. Terrestrial microorganisms, such as plant endophytes, fungi, bacteria, plant rhizosphere microorganisms and marine organisms, have only recently been the object of study and reflecting a world trend in the area of natural products chemistry.

The database includes compounds identified in species from all six Brazilian biomes. The chemical diversity is considerably rich, including compounds from a variety of molecular classes, such as flavonoids, alkaloids, terpenes, iridoids, and lignans. This database could be very useful not only to scientific purposes, especially computational studies, and medicinal chemistry but also to assist the decisions and support public policies and private investments, preservation and sustainable use of the Brazilian biodiversity. General information such as molecular formula and chemical class can be used. Additional search criteria are species, sources such as plants or marine organism, chemical information including molecular volume, rotatable bonds, Lipinski violations, literature references, biological property and more. All the search results can be downloaded in mol2 file format. For each result entry structure, general information including common name, IUPAC name, InchI, InchIKey, SMILES is provided. Chemical information is also retrieved as well as literature references, biological properties, and spectral data [75, 101, 137].

Other virtual databases of natural products in fewer details are:

- **The Herb Ingredients' Targets (HIT) database** includes information for natural products' interactions with protein targets. Search is available by keyword (Chinese name, Latin name, chemical formula, etc.) or by similarity via compound structure or protein sequence. For further reading, the publication by Ye et al. is provided at the "References" section of this chapter [101, 145].

- **ZINC database** (free) includes a subset of natural products. From the main menu tool "Substances", the user selects "Subsets" and the webpage introduces a catalog with various subsets with a description for each one. Selection of the subset named "natural- products" retrieves a collection of these compounds. The user is able to download all the collection by clicking "Get Total". Various file formats are available from the tool next to "Get Total" option [95].

- **The African Antimalarial Natural Products Library (AfroMalariaDB)** lists compounds with plant origins. The database can be downloaded from the electronic supplementary file of the publication as sdf file format [101, 146, 147].

- **The Universal Natural Products Database (UNPD)**: A non-commercial database of compounds from plants, animals, and microorganisms. 3D structures are provided with defined stereochemistry [65, 101].

- **The Herbal Ingredients in-vivo Metabolism database (HIM)**: A free database for academic use which connects herbal active ingredients and their in-vivo metabolism. A structure similarity search and a substructure search function are provided [101, 148].

- **The Natural Products Database of the UEFS** (State University of Feira de Santana, Brazil): Provided by ZINC database (zinc.docking.org/catalogs/uefsnp, accessed: 21/12/2017) [101, 149].
- **Naturally Occurring Plant-based Anticancerous Compound-Activity-Target database (NPACT)**: A database which consists of compounds from plants with anti-cancerous activity. There is an online tool for search. 3D structures can be downloaded. From the function "Help" on the right, an online guide for the use of the database is provided [101, 150, 151].
- **Natural Products for Cancer Regulation (NPCARE)**: A free database. Natural products from plants, marine life, fungi and bacteria with anticancer activity [64, 101].
- **StreptomeDB**: There is an online guide to the database. The collection lists natural compounds produced by bacteria of the genus Streptomyces. The user can download all the collection in SD format [101, 152, 153].
- **Antibase database**: Commercially available collection from microorganisms and fungi [101, 154].
- **The MarinLit database**: An online guide exists with the option "Get Started with MarinLit". It is dedicated to marine natural products. The database is searchable by text and structure [101, 155].
- **Phytochemica**: A collection of phytochemicals [156, 157].
- **Alkamid®**: A database for N-alkylamides from plants [158, 159].
- **3DMET**: A database of 3D structures of natural metabolites [160, 161].
- **ChEBI**: A subset of natural products exist in this free database. Chemical structures are provided in sdf format and Molfile [92, 162].
- **NAPRALERT**: A database which consists of literature resources about natural products [163].
- **MPD3** is a free online database with 3D structures of phytochemicals in MDL MOL format [164].

In addition to virtual natural compounds collections, there are physical natural products databases from vendors which can be useful for VS and other cheminformatics applications. Usually, these collections include structures available in a file format. The computed structure representation is a good starting point for the construction of a virtual database using a variety of tools described previously in this chapter.

Next, a number of **physical NPs libraries** are presented:
- **AnalytiCon discovery**: Four collections of molecules. MEGx: Purified natural products. NATx: Synthetic compounds based on natural products. FRGx: Fragments from nature. MACROx: Macrocyclic compounds [101, 165].
- **The NCI Natural Products Repository**: From the introductory webpage with the option "Databases & Tools", the user can select "DTP Bulk Data for Download"

and then "Compound Sets". There are three sets of natural products with structures in SD file format [101, 166].

- **TimTec NPL-Natural Product Library**: Compounds are mostly from plants. Other sources are fungus, bacteria, and animals. From homepage (see references), by clicking on "Gossypol and its derivatives", the user can retrieve a pdf file with a list of Gossypol derivatives and there is also an option for SDF File Download via login. Secondly, by clicking on "NDL-3000", the NDL-3000 Natural Derivatives Library is presented. Here, by selecting the option "FL-500", the user has the opportunity for obtaining a pdf with a list of the flavonoid derivatives of the FL-500 collection. The database is searchable by structure or substructure and there is a similarity search tool too [101, 167].

- **The NPDI collection**: Sources are plants, fungal extracts and actinomycetous [168].

- **INDOFINE Chemical Company database**: From the main menu "Products", the selection of "Product Categories" and then "Natural Products" introduces a list of natural products. By selecting each entry, information is available about the entry compound. From the main menu "Products", selection of "Product Catalogs" presents catalogs in pdf format about natural products, flavonoids and coumarins, herbal standards and herbal and nutritional products. Finally, from the main menu "Products" the selection of "Download SDF Structures" enables the interested user to download the category of interest (e.g. Natural-Products-sdf.zip) [101, 169].

- **Ambinter collection of natural products**: Searching for compounds online is available (Text search and Structure search). From the section "Discover our Products & Services", selection of "Libraries" introduces a list of collections. Among them is the **Greenpharma** Natural Compound Library. An electronic file of the library exists with structures and calculated physicochemical descriptors [101, 170].

- **Interbioscreen collections**: Compounds can be searched online (Text search, Structure search, Advanced search). There are two collections consisting of natural products. The first one is called "The Natural Compound (NC) collection". The compounds listed here derive from various sources such as plants and microorganisms. The most representative categories of compounds are terpenoids, alkaloids, flavonoids, coumarins. The second library is called "Building Blocks" and contains among others natural building blocks [101, 171].

- **TargetMol Natural Compound Library**: The selection of "Libraries" and the category "Natural Compound Library" from homepage introduces the user to vendor's collection of natural products. By selecting menu "Libraries", the user can also be informed for the "Polyphenolic Natural Compound Library" and "Selected Plant-sourced compound Library" [101, 172].

- **PI Chemicals Collection**: Selection of the category "Natural Products" from the option "Products" (in the middle of the webpage), retrieves a list of the NPs that the company offers. Clicking on each entry, information about the compound is provided with an image of the 2D structure. For some entries, SMILES and InchIKey are also provided [101, 173].
- **Selleckchem Natural Product Library**: Selection of main menu option "Inhibitors" and then "By product type"-"Natural Products" a list of natural products is acquired. Their origins are from plants, marine life, and microbiological sources. They are presented in categories such as flavones, phenols, and xanthones [174].

There are more physical libraries containing natural products and natural products derivatives. Some of them are presenting here by name:
- **The Natural Products Library Initiative (NPLI)** at The Scripps Research Institute [175].
- **Quality Phytochemicals LLC Library** of diverse phytochemicals [176].
- **p-ANAPL database** which also exists as a virtual library provided as a supplementary electronic file of the corresponding publication by Ntie-Kang et al. [66, 101].
- **AK Scientific natural product library** [101, 177].
- **BioAustralis:** A collection of metabolites from microbes [178].
- **MedChem Express** natural products [101, 179].
- **Specs** natural products [18, 101, 180].
- **Chimiothèque Nationale (CN)**: Registered users can be provided with an SD file with all the structures [18, 181].

Apart from the aforementioned databases, a number of collections of natural products also exist which have not been referred here.

A conclusion that can be extracted from navigating some of the databases previously discussed is that the compounds included are mainly of plant origins. There are also compounds from other sources such as bacteria, marine life and fungi. At least, a chemical structure is provided by these collections and many of them also provide calculated molecular properties such as molecular weight and number of hydrogen bond acceptors and donors. Furthemore, there are databases which are focused on specific categories of compounds stemming from nature. For example, MarinLit is dedicated to compounds from marine life and StreptomeDB deals with compounds from streptomycetes.

In the article by Chen et al. (2017), a comparison between free virtual databases of NPs and database DNP reveals the existence of a great overlap. An analogous conclusion can be found concerning molecular scaffolds. There are categories of NPs that seem to predominate such as flavonoids and coumarins [101].

Table 10.1: Databases for Natural Products discussed in the text.

Virtual Databases of Natural Products	*Dictionary of Natural Products (DNP)*	*Dictionary of Marine Natural Products (DMNP)*
	CamMedNP	*Super Natural II*
	TCM Database@Taiwan	*AfroDb*
	Northern African Natural Products Database (NANPDB)	*Traditional Chinese Medicines Integrated Database (TCMID)*
	Reaxys	*AfroCancer*
	Chem-TCM	*PubChem*
	Taiwan indigenous plant database (TIPdb-3D)	*South African Natural Compounds Database (SANCDB)*
	Carotenoids Database	*NuBBE database (NuBBE$_{DB}$)*
	The Herb Ingredients' Targets (HIT) database	*ZINC database (subset Natural Products)*
	The African Antimalarial Natural Products Library (AfroMalariaDB)	*The Universal Natural Products Database (UNPD)*
	The Herbal Ingredients in-vivo Metabolism database (HIM)	*The Natural Products Database of the UEFS*
	Naturally Occurring Plant-based Anti-cancerous Compound-Activity-Target database (NPACT)	*Natural Products for Cancer Regulation (NPCARE)*
	StreptomeDB	*Antibase database*
	The MarinLit database	*Phytochemica*
	Alkamid®	*3DMET*
	ChEBI	*NAPRALERT*
	MPD3	*- - -*
Physical Natural Products Databases	*AnalytiCon discovery*	*The NCI Natural Products Repository*
	TimTec NPL-Natural Product Library	*The NPDI collection*
	INDOFINE Chemical Company database	*Ambinter collection of natural products (Greenpharma)*
	Interbioscreen collections	*TargetMol Natural Compound Library*
	PI Chemicals Collection	*Selleckchem Natural Product Library*
	The Natural Products Library Initiative (NPLI) (Scripps Research Institute)	*Quality Phytochemicals LLC Library*
	p-ANAPL database	*AK Scientific natural product library*
	BioAustralis	*MedChem Express*
	Specs	*Chimiothèque Nationale (CN)*

10.7 The purchasable chemical space of currently available natural product libraries

The chemical space that a library of molecules encloses is an important factor in a drug discovery process, especially when searching for novel lead compounds. Chen and colleagues compared free virtual NPs databases with commercial NPs libraries using unique InChIs (*Tip*: avoiding inconsistencies and duplicates with InChI representations, see Heller et al. 2013) and resulted in the fact that a great degree of overlapping exists between them. Additionally, they observed that diversity of structures is achieved by free virtual NPs collections, a fact that the creators of the UNPD database also noticed when they examined their collection with the application of PCA [65, 101]. However, drug discovery process is a time-consuming procedure and thus ways of reducing the cost of time are essential and should be taken into account especially in the complex field of natural products. One way is to search for compounds which are already purchasable or can be easily synthesized from purchasable fragments. So, the issue of the purchasable chemical space of NPs libraries shows up too.

ZINC is a good example of the estimated size of the purchasable chemical space. The number of purchasable compounds, that ZINC includes, increases with high rates. For example, concerning the subsets (home/substances/subsets) of compounds, including in ZINC15, the total number and the number of the purchasable ones are given. Natural products are included as a category with a total number presented and a number of commercial NPs given. In the article by Chen et al., it is reported that the number of purchasable natural products in the ZINC database represents the 10% of the total number of the NPs found in virtual libraries. It is also mentioned that a considerable number of NPs included in other databases can be prepared on demand. The ready-to-use NPs are usually small compounds making them suitable for use in fragment-based methodologies in a virtual environment or for an easier modification during the process of synthesis in the laboratory. Noticeably, macrocycles are becoming popular in drug design offering more opportunities for purchasable natural compounds [95, 101, 182]. A study by Lucas et al. could be consulted regarding the purchasable space of NPs as it describes the analysis of a very large number of commercially available compounds excluding those belonging to PAINS, toxic compounds, and other undesirable categories. The analysis contains purchasable natural products among other compounds [183].

10.8 A watchword when using natural product databases

Natural products are a promising resource for the discovery of lead compounds or drugs such as paclitaxel or artemisinin. However, there are some points that should be taken into account when using a natural product database due to the different properties that natural products present compared to synthetic compounds.

First of all, natural products are not restricted to drug or lead likeness rules such as the Lipinski's Rule of Five [65, 101]. In the review article by Doak et al. (2014), an analysis of drugs and clinical candidates showed that natural products are important for drug discovery, even with a molecular weight (MW) larger than those proposed by the rules. In the review article by Pascolutti and Quinn (2014), the creation of NP-lead-like libraries is presented with a natural product as a starting point with various values of molecular weight (MW). Considering the aforementioned, attention should be paid during the filtering process of virtual NPs libraries of the limits applied in order to obtain compounds with desirable values of physicochemical properties, especially in the process of lead identification [184, 185]. When filtering a natural products database for undesirable functional groups, a computational chemist should take into account the fact that NPs usually have reactive or other undesirable groups. Moreover, natural products are considered privileged structures which can present bioactivity for many targets. Databases of natural products should be diverse in order to obtain a promising result, especially for lead identification [101].

Furthermore, structural complexity is another difficulty that the computational chemist confronts. The digital representation of a natural product is, in many cases, difficult and an accurate structure is needed for methods such as molecular docking. The users of a natural product database should always check for errors especially for precise stereochemistry as the structures of NPs contain many chiral centers and databases do not always define stereochemistry. In such cases, it is advisable to create all possible configurations. Structural complexity is also a key factor for a synthetic route since a great amount of NPs is not purchasable [65, 101, 186].

Additionally, difficulties may exist with the experimental test for bioactivity caused by difficulties in the isolation of a natural compound and its concentration. For example, regarding natural products derived from plants, the availability of satisfactory amounts of a plant is an issue that creates problems to the whole procedure of drug discovery as well as the difficulties observed during the isolation of the compound. Usually, extracts are used for bioactivity tests. In this case, interference of various compounds in the extracts may occur resulting in false positives or negatives experimental observations about the bioactivity of the compound examined [67, 101, 186].

The careful choice of the molecular descriptors should be made depending on the goal of the research procedure. For example, Gu and colleagues calculated various descriptors, such as molecular weight and number of rings, in order to compare universal natural products database (UNPD) and FDA-approved drugs in DrugBank using PCA. In another study by Pereira et al., CDK and semi-empirical quantum-chemical descriptors were used with machine learning techniques for the creation of QSAR models aiming at the prediction of bioactivities of NPs. ADMET-related molecular descriptors can be calculated and used for ADMET predictions [65, 147, 187].

The choice of the scientific software to be used is another issue that should concern a computational chemist since scientific software packages were mostly created for the study of small, synthetic molecules. NPs structures may be more complicated in contrast to a less complex, small molecule. Nevertheless, a number of successful stories using natural products and computational chemistry have been reported in the literature. Considering the improvements applied in scientific software, there is an expectation for more successful paradigms in the future [101, 186].

Finally, there are issues concerning the patentability of natural products of plant origins [67].

10.9 Conclusion

Databases play an important role in drug discovery as a complementary tool in many computational techniques including molecular docking and pharmacophore modeling. A careful use of these collections of molecules can lead to the identification of promising compounds for use in medicinal chemistry. Natural products (NPs) have always been a resource of great value for agents against human diseases. In the last years, an increasing number of drugs from or inspired by nature has occurred. An interesting and extended analysis carried out by Pye et al. on NPs, stemming from marine life and microorganisms, indicated that the discovery of novel structures from nature continues and research on bioactivities of NPs has not revealed everything yet. Special characteristics of NPs, such as a variety of structures, result in their considerable role in drug discovery process. Databases containing natural products have been constructed or are being constructed (physical/virtual databases) in order to assist the scientific community in the identification of promising compounds from nature, despite the difficulties that scientists come across due to the complexity of these unusual chemical structures [101, 188, 189].

Acknowledgements: FNK acknowledges a Georg Forster return fellowship from the Alexander von Humboldt Foundation, Germany. MV and VSB gratefully acknowledge financial support from FAPESP (Fundação de Amparo à Pesquisa do Estado de São Paulo [grant number 2010/52327-5], [grant number 2013/07600-3]; CNPq (Conselho Nacional de DesenvolvimentoCientífico e Tecnológico) and CAPES (Coordenação de Aperfeiçoamento de Pessoal de Nível Superior). MV acknowledges from Finatec [grant number 120/2017]. The authors gratefully acknowledge the helpful comments and suggestions of the reviewers, which have improved the article. The authors are also grateful to Mr Conrad V. Simoben for helping in a number of articles. Financial support for this work is acknowledged from a ChemJets fellowship from the Ministry of Education, Youth and Sports of the Czech Republic awarded to FNK.

Abbreviations

VS	Virtual screening
0D/1D/2D/3D	Zero dimensional/one dimensional/two dimensional/three dimensional
IUPAC	International Union of Pure and Applied Chemistry
QSAR	Quantitative structure-activity relationship
RF	Random forests

PLS	Partial least square
PCA	Principal component analysis
NCI	National Cancer Institute
NPs	Natural products
HTS	High-throughput screening
TPSA	Topological polar surface area
FAQ	Frequently asked questions
logP	Lipophilicity
ADMET	Absorption/distribution/metabolism/excretion/toxicity
MW	Molecular weight
PAINS	Pan-Assay INterference compoundS
CDK	Chemistry development kit

References

[1] Lusher SJ, McGuire R, Schaik van RC, Nicholson CD, Vlieg de J. Data-driven medicinal chemistry in the era of big data. Drug Discov. Today. 2014;19:859–68.

[2] Berman HM, Westbrook J, Feng Z, Gilliland G, Bhat TN, Weissig H, et al. The protein data bank. Nucleic Acids Res. 2000;28:235–42.

[3] Ertl P. Molecular structure input on the web. J Cheminform. 2010;2:1.

[4] Gasteiger J. Chemoinformatics: achievements and challenges, a personal view. Molecules. 2016;21:151.

[5] Gaulton A, Bellis LJ, Bento AP, Chambers J, Davies M, Hersey A, et al. ChEMBL: a large-scale bioactivity database for drug discovery. Nucleic Acids Res. 2012;40:D1100–7.

[6] Jónsdóttir SO, Jørgensen FS, Brunak S. Prediction methods and databases within chemoinformatics: emphasis on drugs and drug candidates. Bioinformatics. 2005;21:2145–60.

[7] NCI Open Database website. Available at: https://cactus.nci.nih.gov/download/nci/. Accessed: 6 Oct 2018.

[8] PubMed. Available at: https://www.ncbi.nlm.nih.gov/pubmed/. Accessed: 30 Dec 2017.

[9] DAYLIGHT. Available at: http://www.daylight.com/dayhtml/doc/theory/theory.smiles.html. Accessed: 29 Dec 2017.

[10] Kristensen TG, Nielsen J, Pedersen CNS. Methods for similarity-based virtual screening. *Comput Struct Biotechnol J*. 2013;5:201302009.

[11] O'Boyle NM. Towards a Universal SMILES representation–A standard method to generate canonical SMILES based on the InChI. J Cheminform. 2012;4:22.

[12] PubChem. Available at: https://pubchem.ncbi.nlm.nih.gov/. Accessed: 20 Dec 2017.

[13] Heller S, McNaught A, Stein S, Tchekhovskoi D, Pletnev I. InChI-the worldwide chemical structure identifier standard. J Cheminform. 2013;5:7.

[14] Chepelev LL, Dumontier M. Chemical entity semantic specification: knowledge representation for efficient semantic cheminformatics and facile data integration. J Cheminform. 2011;3:20.

[15] IUPAC Gold Book. Available at: https://goldbook.iupac.org/html/M/MT06966.html. Accessed: 30 Dec 2017.

[16] IUPAC. Compendium of chemical terminology, 2nd ed. (the "Gold Book"). Compiled by A. D. McNaught and A. Wilkinson. Blackwell Scientific Publications, Oxford (1997). XML on-line corrected version: http://goldbook.iupac.org (2006-) created by M. Nic, J. Jirat, B. Kosata; updates compiled by A. Jenkins. ISBN 0-9678550-9-8.

[17] PDB Guide website. Available at: http://pdb101.rcsb.org/learn/guide-to-understanding-pdb-data/introduction. Accessed: 31 Dec 2017.

[18] Scior T, Bender A, Tresadern G, Medina-Franco JL, Martínez-Mayorga K, Langer T, et al. Recognizing pitfalls in virtual screening: a critical review. J Chem Inf Model. 2012;52:867–81.

[19] Wang L, Xie XQ. Computational target fishing: what should chemogenomics researchers expect for the future of in silico drug design and discovery?. Future Med Chem. 2014;6:247–9.

[20] Hanwell MD, Curtis DE, Lonie DC, Vandermeersch T, Zurek E, Hutchison GR. Avogadro: an advanced semantic chemical editor, visualization, and analysis platform. J Cheminform. 2012;4:17.

[21] ChemDraw/Cambridgesoft. Available at: http://www.cambridgesoft.com/software/overview. aspx. Accessed: 31 Dec 2017.

[22] Schrodinger website. Available at: https://www.schrodinger.com/maestro. Accessed: 31 Dec 2017.

[23] ChemSketch//ACD/Labs. Available at: http://www.acdlabs.com/resources/freeware/chems ketch/. Accessed: 31 Dec 2017.

[24] MarvinSketch/ChemAxon. Available at: https://www.chemaxon.com/products/marvin. Accessed: 31 Dec 2017.

[25] Bienfait B, Ertl P. JSME: a free molecule editor in JavaScript. J Cheminform. 2013;5:24.

[26] JSME. NA. Available at: http://peter-ertl.com/jsme/JSME_2017-02-26/JSME.html. Accessed: 1 Jan 2017.

[27] O'Boyle NM, Banck M, James CA, Morley C, Vandermeersch T, Hutchison GR. Open Babel: an open chemical toolbox. J Cheminform. 2011;3:33.

[28] Steinbeck C, Han Y, Kuhn S, Horlacher O, Luttmann E, Willighagen E. The chemistry development kit (CDK): an open-source java library for chemo- and bioinformatics. J Chem Inf Comput Sci. 2003;43:493–500.

[29] Willighagen EL, Mayfield JW, Alvarsson J, Berg A, Carlsson L, Jeliazkova N. The chemistry development kit (CDK) v2.0: atom typing, depiction, molecular formulas, and substructure searching. J Cheminform. 2017;9:33.

[30] RDKit Documentation. Available at: https://www.rdkit.org/RDKit_Docs.current.pdf, Greg Landrum, Release 2018.03.1/. Accessed: 2 Oct 2018.

[31] Moura Barbosa AJ, Del Rio A. Freely accessible databases of commercial compounds for high-throughput virtual screenings. Curr Top Med Chem. 2012;12:866–77.

[32] Lionta E, Spyrou G, Vassilatis DK, Cournia Z. Structure-based virtual screening for drug discovery: principles, applications and recent advances. Curr Top Med Chem. 2014;14:1923–38.

[33] Isberg V, Mordalski S, Munk C, Rataj K, Harpsøe K, Hauser AS, et al. GPCRdb: an information system for G protein-coupled receptors. Nucleic Acids Res. 2016;44:D356–64.

[34] Isberg V, Mordalski S, Munk C, Rataj K, Harpsøe K, Hauser AS, et al. Corrigendum: GPCRdb: an information system for G protein-coupled receptors. Nucleic Acids Res. 2017;45:2936.

[35] Lipinski CA, Lombardo F, Dominy BW, Feeney PJ. Experimental and computational approaches to estimate solubility and permeability in drug discovery and development settings. Adv Drug Deliv Rev. 1997;23:3–25.

[36] Prachayasittikul V, Worachartcheewan A, Shoombuatong W, Songtawee N, Simeon S, Prachayasittikul V, et al. Computer-aided drug design of bioactive natural products. Curr Top Med Chem. 2015;15:1780–800.

[37] Walters WP, Stahl MT, Murcko MA. Virtual screening-an overview. Drug Discov Today. 1998;3:160–78.

[38] Yang SY. Pharmacophore modeling and applications in drug discovery: challenges and recent advances. Drug Discov Today. 2010;15:444–50.

[39] Lee CH, Huang HC, Juan HF. Reviewing ligand-based rational drug design: the search for an atp synthase inhibitor. Int J Mol Sci. 2011;12:5304–18.

[40] Sebastián-Pérez V, Roca C, Awale M, Reymond JL, Martinez A, Gil C, et al. Medicinal and biological chemistry (MBC) library: an efficient source of new hits. J Chem Inf Model. 2017;57:2143–51.

[41] Akhondi SA, Kors JA, Muresan S. Consistency of systematic chemical identifiers within and between small-molecule databases. J Chem. 2012;4:35.

[42] Ntie-Kang F, Mbah JA, Mbaze LM, Lifongo LL, Scharfe M, Hanna JN, et al. CamMedNP: building the Cameroonian 3D structural natural products database for virtual screening. BMC Complement Alterna Med. 2013;13:88.

[43] Bioclipse. Available at: http://www.bioclipse.net/. Accessed: 5 Jan 2018.

[44] FAFDrugs4. Available at: http://fafdrugs3.mti.univ-paris-diderot.fr/. Accessed: 5 Jan 2018.

[45] E-Dragon. Available at: http://www.vcclab.org/lab/edragon/start.html. Accessed: 5 Jan 2018.

[46] Molinspiration. Available at: http://www.molinspiration.com/. Accessed: 5 Jan 2018.

[47] VLS3D.COM. Available at: http://www.vls3d.com/. Accessed: 9 Jan 2018.

[48] MOE (Chemical Computing Group). Available at: http://www.chemcomp.com/index.htm. Accessed: 4 Jan 2018.

[49] Forli S. Charting a path to success in virtual screening. Molecules. 2015;20:18732–58.

[50] Irwin JJ, Shoichet BK. ZINC - a free database of commercially available compounds for virtual screening. J Chem Inf Model. 2005;45:177–82.

[51] Irwin JJ, Sterling T, Mysinger MM, Bolstad ES, Coleman RG. ZINC: A free tool to discover chemistry for biology. J Chem Inf Model. 2012;52:1757–68.

[52] Sterling T, Irwin JJ. ZINC 15–ligand discovery for everyone. J Chem Inf Model. 2015;55:2324–37.

[53] Bruns RF, Watson IA. Rules for identifying potentially reactive or promiscuous compounds. J Med Chem. 2012;55:9763–72.

[54] Baell JB, Holloway GA. New substructure filters for removal of pan assay interference compounds (PAINS) from screening libraries and for their exclusion in bioassays. J Med Chem. 2010;53:2719–40.

[55] Bisson J, McAlpine JB, Friesen JB, Chen SN, Graham J, Pauli GF. Can invalid bioactives undermine natural product-based drug discovery? J Med Chem. 2016;59:1671–90.

[56] Pouliot M, Jeanmart S. Pan assay interference compounds (PAINS) and other promiscuous compounds in antifungal research. J Med Chem. 2016;59:497–503.

[57] Saubern S, Guha R, Baell JB. KNIME workflow to assess PAINS filters in SMARTS format. Comparison of RDKit and Indigo cheminformatics libraries. Mol Inf. 2011;30:847–50.

[58] Lipinski CA. Drug-like properties and the causes of poor solubility and poor permeability. J Pharmacol Toxicol Methods. 2000;44:235–49.

[59] Lagorce D, Sperandio O, Baell JB, Miteva MA, Villoutreix BO. FAF-Drugs3: a web server for compound property calculation and chemical library design. Nucleic Acids Res. 2015;43: W200–7.

[60] Douguet D. e-LEA3D: a computational-aided drug design web server. Nucleic Acids Res. 2010;38:W615–21.

[61] Molsoft. Available at: http://www.molsoft.com/chemical-library.html. Accessed: 8 Jan 2018.

[62] Sud M, Fahy E, Subramaniam S. Template-based combinatorial enumeration of virtual compound libraries for lipids. J Cheminform. 2012;4:23.

[63] Inte:Ligand. Available at: http://www.inteligand.com/. Accessed: 7 Jan 2018.

[64] Choi H, Cho SY, Pak HJ, Kim Y, Choi JY, Lee YJ, et al. NPCARE: database of natural products and fractional extracts for cancer regulation. J Cheminform. 2017;9:2.

[65] Gu J, Gui Y, Chen L, Yuan G, Lu HZ, Xu X. Use of natural products as chemical library for drug discovery and network pharmacology. PLoS ONE. 2013;8:e62839.

[66] Ntie-Kang F, Amoa Onguéné P, Fotso GW, Andrae-Marobela K, Bezabih M, Ndom JC, et al. Virtualizing the p-ANAPL library: a step towards drug discovery from African medicinal plants. PLoS ONE. 2014;9:e90655.

[67]   Atanasov AG, Waltenberger B, Pferschy-Wenzig EM, Linder T, Wawrosch C, Uhrin P, et al. Discovery and resupply of pharmacologically active plant-derived natural products: A review. Biotechnol Adv. 2015;33:1582–614.

[68]   Wermuth CG, Ganellin CR, Lindberg P, Mitscher LA. Glossary of terms used in medicinal chemistry (IUPAC Recommendations 1998). Pure Appl Chem. 1998;70:1129–43. https://www. iupac.org/publications/pac/pdf/1998/pdf/7005x1129.pdf. Accessed: 23 Jan 2018.

[69]   Ekins S, Mestres J, Testa B. In silico pharmacology for drug discovery: methods for virtual ligand screening and profiling. Br J Pharmacol. 2007;152:9–20.

[70]   Kaserer T, Temml V, Kutil Z, Vanek T, Landa P, Schuster D. Prospective performance evaluation of selected common virtual screening tools. Case study: cyclooxygenase (COX) 1 and 2. Eur J Med Chem. 2015;96:445–57.

[71]   Klenner A, Hähnke V, Geppert T, Schneider P, Zettl H, Haller S, et al. From virtual screening to bioactive compounds by visualizing and clustering of chemical space. Mol Inf. 2012;31:21–26.

[72]   Noha SM, Jassar B, Kuehnl S, Rollinger JM, Stuppner H, Schaible AM, et al. Pharmacophore-based discovery of a novel cytosolic phospholipase A2α inhibitor. Bioorg Med Chem Lett. 2012;22:1202–7.

[73]   ChemBioServer. Available at: http://bioserver-3.bioacademy.gr/Bioserver/ChemBioServer/. Accessed: 9 Jan 2018.

[74]   Athanasiadis E, Cournia Z, Spyrou G. ChemBioServer: a web-based pipeline for filtering, clustering and visualization of chemical compounds used in drug discovery. Bioinformatics. 2012;28:3002–3.

[75]   NuBBE database (NuBBEDB). Available at: http://nubbe.iq.unesp.br/portal/nubbedb.html. Accessed: 21 Dec 2017.

[76]   ChEMBL. Available at: https://www.ebi.ac.uk/chembl/. Accessed: 7 Jan 2018.

[77]   Banerjee P, Erehman J, Gohlke BO, Wilhelm T, Preissner R, Dunkel M. Super Natural II-a database of natural products. Nucleic Acids Res. 2015;43:D935–9.

[78]   CACTVS. Available at: http://xemistry.com/. Accessed: 9 Jan 2018.

[79]   Instant JChem (IJC). Available at: https://chemaxon.com/products/instant-jchem. Accessed: 9 Jan 2018.

[80]   CORINA Symphony. Available at: https://www.mn-am.com/products/corinasymphony. Accessed: 9 Jan 2018.

[81]   Screening Assistant 2. Available at: http://sa2.sourceforge.net/. Accessed: 9 Jan 2018.

[82]   The RDKit Database Cartridge: Open-source Cheminformatics and Machine Learning. Available at: http://www.rdkit.org/docs/Cartridge.html. Accessed: 10 Jan 2018.

[83]   BIOVIA Chemical Registration. Available at: http://accelrys.com/products/collaborative-science/biovia-registration/chemical-registration.html. Accessed: 10 Jan 2018.

[84]   ChemCurator. Available at: https://chemaxon.com/products/chemcurator. Accessed: 10 Jan 2018.

[85]   Hersey A, Chambers J, Bellis L, Bento AP, Gaulton A, Overington JP. Chemical databases: curation or integration by user-defined equivalence?. Drug Discov Today. 2015;14:17–24.

[86]   UniChem. Available at: https://www.ebi.ac.uk/unichem/widesearch/widesearch. Accessed: 10 Jan 2018.

[87]   CoCoCo. Available at: http://cococo.isof.cnr.it/cococo. Accessed: 6 Jan 2018.

[88]   DrugBank. Available at: https://www.drugbank.ca/. Accessed: 6 Jan 2018.

[89]   BindingDB. Available at: https://www.bindingdb.org/bind/index.jsp. Accessed: 6 Jan 2018.

[90]   ChemSpider. Available at: http://www.chemspider.com/Default.aspx. Accessed: 6 Jan 2018.

[91]   ChemBank. Available at: http://chembank.broadinstitute.org/. Accessed: 6 Jan 2018.

[92]   ChEBI. Available at: http://www.ebi.ac.uk/chebi/. Accessed: 25 Dec 2017.

[93]   Binding MOAD. Available at: http://www.bindingmoad.org/. Accessed: 6 Jan 2018.

[94] eMolecules. Available at: https://www.emolecules.com/info/plus/download-database. Accessed: 10 Jan 2018.

[95] ZINC15 database. Available at: http://zinc15.docking.org/. Accessed: 20 Dec 2017.

[96] Bento AP, Gaulton A, Hersey A, Bellis LJ, Chambers J, Davies M, et al. The ChEMBL bioactivity database: an update. Nucleic Acids Res. 2014;42:D1083–90.

[97] Rose PW, Prlić A, Altunkaya A, Bi C, Bradley AR, Christie CH, et al. The RCSB protein data bank: integrative view of protein, gene and 3D structural information. Nucleic Acids Res. 2017;45: D271–81.

[98] GVKBIO. Available at: https://www.gvkbio.com/. Accessed: 7 Jan 2018.

[99] Reaxys-Fact Sheet. Available at: https://www.elsevier.com/solutions/reaxys. Accessed: 19 Dec 2017.

[100] Chackalamannil S, Davies RJ, Asberom T, Doller D, Leone D. A highly efficient total synthesis of (+)-Himbacine. J Am Chem Soc. 1996;118:9812–3.

[101] Chen Y, Kops de Bruyn C, Kirchmair J. Data resources for the computer-guided discovery of bioactive natural products. J Chem Inf Model. 2017;57:2099–111.

[102] *INsPiRE workshop: Cell cycle and natural products, book of abstracts: Kinghorn AD. Discovery of anticancer agents of diverse natural origin*. Athens, Greece: 2014 May 8–9.

[103] Dictionary of Natural Products (DNP). Available at: http://dnp.chemnetbase.com. Accessed: 17 Dec 2017.

[104] Dictionary of Marine Natural Products (DMNP). Available at: http://dmnp.chemnetbase.com. Accessed: 18 Dec 2017.

[105] Super Natural II. Available at: http://bioinf-applied.charite.de/supernatural_new. Accessed: 18 Dec 2017.

[106] Chen CY-C. TCM Database@Taiwan: the world's largest traditional Chinese medicine database for drug screening in silico. PLoS ONE. 2011;6:15939.

[107] TCM Database@Taiwan. Available at: http://tcm.cmu.edu.tw/. Accessed: 18 Dec 2017.

[108] AfroDb. Available at: http://african-compounds.org/about/afrodb/. Accessed 19 Dec 2017.

[109] Ntie-Kang F, Zofou D, Babiaka SB, Meudom R, Scharfe M, Lifongo LL, et al. AfroDb: a select highly potent and diverse natural product library from African medicinal plants. PLoS ONE. 2013;8:e78085.

[110] Northern African Natural Products Database. Available at: http://african-compounds.org/nanpdb/. Accessed: 20 Dec 2017.

[111] Ntie-Kang F, Telukunta KK, Döring K, Simoben CV, Moumbock AFA, Malange YI, et al. NANPDB: A resource for natural products from Northern African sources. J Nat Prod. 2017;80:2067–76.

[112] Xue R, Fang Z, Zhang M, Yi Z, Wen C, Shi T. TCMID: traditional Chinese medicine integrative database for herb molecular mechanism analysis. Nucleic Acids Res. 2013;41:D1089–95.

[113] TCMID-Traditional Chinese Medicines Integrated Database. Available at: www.megabionet. org/tcmid. Accessed: 19 Dec 2017.

[114] Reaxys Training, Natural Products in Reaxys (Christine Flemming). Available at: https://www. youtube.com/watch?v=vJKXsDDhRyk Accessed: 19 Dec 2017.

[115] AfroCancer. Available at: http://african-compounds.org/about/afrocancer/. Accessed: 19 Dec 2017.

[116] Ntie-Kang F, Nwodo JN, Ibezim A, Simoben CV, Karaman B, Ngwa VF, et al. Molecular modeling of potential anticancer agents from African medicinal plants. J Chem Inf Model. 2014;54: 2433–50.

[117] Chem-TCM: Chemical Database of Traditional Chinese Medicine. Available at: http://www. chemtcm.com/. Accessed: 19 Dec 2017.

[118] Hao M, Cheng T, Wang Y, Bryant SH. Web search and data mining of natural products and their bioactivities in PubChem. Sci China Chem. 2013;56:1424–35.

[119] Lin YC, Wang CC, Chen IS, Jheng JL, Li JH, Tung CW. TIPdb: a database of anticancer, antiplatelet, and antituberculosis phytochemicals from indigenous plants in Taiwan. Sci World J. 2013; Article ID 736386.

[120] TIPdb-The Taiwan indigenous plant database. Available at: http://cwtung.kmu.edu.tw/tipdb/. Accessed: 20 Dec 2017.

[121] Tung CW, Lin YC, Chang HS, Wang CC, Chen IS, Jheng JL, et al. TIPdb-3D: the three-dimensional structure database of phytochemicals from Taiwan indigenous plants. Database. 2014;2014: Article ID bau055.

[122] Hatherley R, Brown DK, Musyoka TM, Penkler DL, Faya N, Lobb KA, et al. SANCDB: a South African natural compound database. J Cheminform. 2015;7:29.

[123] South African Natural Compounds Database (SANCDB). Available at: https://sancdb.rubi.ru.ac. za/. Accessed: 20 Dec 2017.

[124] Carotenoids Database. Available at: www.carotenoiddb.jp. Accessed: 21 Dec 2017.

[125] Yabuzaki J. Carotenoids database: structures, chemical fingerprints and distribution among organisms. Database. 2017;2017:Article ID bax004.

[126] Bolzani VS, Castro-Gamboa I, Silva DHS. In comprehensive natural products II chemistry and biology; Verpoorte R, Editor. Oxford-UK: Elsevier. Vol. 3, Chapter 3.05, pp. 95–133 2010.

[127] Pagotto CLAC, Barros JRT, Borin MRMB, Gottlieb OR. Quantitative chemical biology. II. Chemical mapping of Lauraceae. Anais da Academia Brasileira de Ciências. 1998;70: 705–9.

[128] Silva DHS, Cavalheiro AJ, Yoshida M, Gottlieb OR. The chemistry of Brazilian Myristicaceae. Xxxvii. Flavonolignoids from the fruits of *Iryanthera grandis*. Phytochemistry. 1995;38:1013–6.

[129] Kato MJ, Yoshida M, Gottlieb OR. The chemistry of Brazilian Myristicaceae.34. Flavones and lignans in flowers, fruits and seedlings of *Virola venenosa*. Phytochemistry. 1992;31:283–7.

[130] Cabral MMO, Azambuja P, Gottlieb OR, Garcia ES. Effects of some lignans and neolignans on the development and excretion of *Rhodnius prolixus*. Fitoterapia. 2000;71:1–9.

[131] Marques MOM, Yoshida M, Gottlieb OR, Maia JGS. The chemistry of Brazilian Lauraceae.97. Neolignans from *Licaria aurea*. Phytochemistry. 1992;31:360–1.

[132] Medina RP, Silva AD, Andersen RJ, Araújo AR, Silva DHS. Botryane sesquiterpenes and binaphthalene tetrols from endophytic fungi associated to the marine red algae *Asparagopsis taxiformis*. Planta Medica. 2016;82:S1–S381.

[133] Jasandrade T, Somensi A, Lopes MN, Araújo AR, Jaspars M, Silva DH. Citrinadin A derivatives from *Penicillium citrinum*, an endophyte from the marine red alga *Dichotomaria marginata*. Planta Med. 2014;80:776–76.

[134] Pinto MEF, Najas JZG, Magalhães LG, Bobey AF, Mendonça JN, Lopes NP, et al. Inhibition of breast cancer cell migration by cyclotides isolated from *Pombalia calceolaria*. J Nat Prod. 2018;81:1203–8.

[135] Pinto MEF, Batista JM, Koehbach J, Gaur P, Sharma A, Nakabashi M, et al. Ribifolin, an orbitide from *Jatropha ribifolia*, and its potential antimalarial activity. J Nat Prod. 2015;78:374–80.

[136] Ramalho SD, Pinto MEF, Ferreira D, Bolzani VS. Biologically active orbitides from the Euphorbiaceae family. Planta Med. 2018;84:558–67.

[137] Valli M, dos Santos RN, Figueira LD, Nakajima CH, Castro-Gamboa I, Andricopulo AD, et al. Development of a natural products database from the biodiversity of Brazil. J Nat Prod. 2013;76:439–44.

[138] Pilon AC, Valli M, Dametto AC, Pinto MEF, Freire RT, Castro-Gamboa I, et al. NuBBEDB: an updated database to uncover chemical and biological information from Brazilian biodiversity. Sci Rep. 2017;7:7215.

[139] Villoutreix BO, Lagorce D, Labbé CM, Sperandio O, Miteva MA. One hundred thousand mouse clicks down the road: selected online resources supporting drug discovery collected over a decade. Drug Discov Today. 2013;18:1081–9.

[140] Harvey AL, Edrada-Ebel RA, Quinn RJ. The re-emergence of natural products for drug discovery in the genomics era. Nat Rev Drug Discov. 2015;14:111–29.

[141] Kuenemann MA, Labbé CM, Cerdan AH, Sperandio O. Imbalance in chemical space: how to facilitate the identification of protein-protein interaction inhibitors. Sci Rep. 2016;6:23815.

[142] Tietz JI, Mitchell DA. Using genomics for natural product structure elucidation. Curr Top Med Chem. 2016;16:1645–94.

[143] Mohamed A, Nguyen CH, Mamitsuka H. Current status and prospects of computational resources for natural product dereplication: a review. Brief Bioinform. 2016;17:309–21.

[144] Valli M, Altei W, Santos RN, Lucca Jr. EC, Dessoy MA, Pioli RM, et al. Synthetic analogue of the natural product piperlongumine as a potent inhibitor of breast cancer cell line migration. J Braz Chem Soc. 2017;28:475–84.

[145] Ye H, Ye L, Kang H, Zhang D, Tao L, Tang K, et al. HIT: linking herbal active ingredients to targets. Nucleic Acids Res. 2011;39:D1055–59.

[146] AfroMalariaDB: African Antimalarial Natural Products Library. Available at: http://african-compounds.org/about/afromalariadb/. Accessed: 24 Dec 2017.

[147] Onguéné PA, Ntie-Kang F, Mbah JA, Lifongo LL, Ndom JC, Sippl W, et al. The potential of anti-malarial compounds derived from African medicinal plants, part III: an in silico evaluation of drug metabolism and pharmacokinetics profiling. Org Med Chem Lett. 2014;4:6.

[148] Kang H, Tang K, Liu Q, Sun Y, Huang Q, Zhu R, et al. HIM-herbal ingredients in-vivo metabolism database. J Cheminform. 2013;5:28.

[149] UEFS Natural Products Database. Available at: http://zinc.docking.org/catalogs/uefsnp. Accessed: 21 Dec 2017.

[150] Mangal M, Sagar P, Singh H, Raghava GPS, Agarwal SM. NPACT: naturally occurring plant-based anti-cancer compound-activity-target database. Nucleic Acids Res. 2013;41:D1124–9.

[151] NPACT. Available at: http://crdd.osdd.net/raghava/npact/. Accessed: 21 Dec 2017.

[152] Klementz D, Döring K, Lucas X, Telukunta KK, Erxleben A, Deubel D, et al. StreptomeDB 2.0-an extended resource of natural products produced by streptomycetes. Nucleic Acids Res. 2016;44:D509–14.

[153] StreptomeDB 2.0. Available at: http://132.230.56.4/streptomedb2/. Acceessed: 21 Dec 2017.

[154] AntiBase. Available at: https://application.wiley-vch.de/stmdata/antibase.php. Accessed: 21 Dec 2017.

[155] MarinLit database. Available at: http://pubs.rsc.org/marinlit/. Accessed: 21 Dec 2017.

[156] Pathania S, Ramakrishnan SM, Bagler G. Phytochemica: a platform to explore phytochemicals of medicinal plants. Database. 2015;2015:Article ID bav075.

[157] Phytochemica. Available at: faculty.iiitd.ac.in/~bagler/webservers/Phytochemica/index.php. Accessed: 21 Dec 2017.

[158] Alkamid®. Available at: http://alkamid.ugent.be/. Accessed: 25 Dec 2017.

[159] Boonen J, Bronselaer A, Nielandt J, Veryser L, De Tre´ G, De Spiegeleer B. Alkamid database: chemistry, occurrence and functionality of plant N-alkylamides. J Ethnopharmacol. 2012;142:563–90.

[160] 3DMET. Available at: http://www.3dmet.dna.affrc.go.jp/. Accessed: 25 Dec 2017.

[161] Maeda MH, Kondo K. Three-dimensional structure database of natural metabolites (3DMET): a novel database of curated 3D structures. J Chem Inf Model. 2013;53:527–33.

[162] Hastings J, de Matos P, Dekker A, Ennis M, Harsha B, Kale N, et al. The ChEBI reference database and ontology for biologically relevant chemistry: enhancements for 2013. Nucleic Acids Res. 2013;41:D456–63.

[163] NAPRALERT. Available at: https://www.napralert.org/. Accessed: 5 Jan 2018.

[164] MPD3. Available at: http://bioinform.info/. Accessed: 14 Jan 2018.

[165] AnalytiCon discovery. Available at: www.ac-discovery.com. Accessed: 22 Dec 2017.

[166] NCI Natural Products Repository. Available at: https://dtp.cancer.gov/organization/npb/introduction.htm. Accesssed: 22 Dec 2017.

[167] TimTec NPL. Available at: http://www.timtec.net/natural-compound-library.html. Accessed: 22 Dec 2017.

[168] NPDI. Available at: http://www.npdi-us.org/collection/. Accessed: 22 Dec 2017.

[169] INDOFINE Chemical Company. Available at: www.indofinechemical.com. Accessed: 22 Dec 2017.

[170] Ambinter. Available at: www.ambinter.com. Accessed: 22 Dec 2017.

[171] Interbioscreen. Available at: www.ibscreen.com. Accessed: 22 Dec 2017.

[172] TargetMol. Available at: www.targetmol.com. Accessed: 22 Dec 2017.

[173] PI Chemicals. Available at: www.pipharm.com. Accessed: 22 Dec 2017.

[174] Selleckchem. Available at: http://www.selleckchem.com. Accessed: 23 Dec 2017.

[175] NPLI. Available at: http://www.scripps.edu/shen/NPLI/npliattsri.html. Accessed: 25 Dec 2017.

[176] Quality Phytochemicals. Available at: http://www.qualityphytochemicals.com/. Accessed: 25 Dec 2017.

[177] AK Scientific. Available at: www.aksci.com. Accessed: 22 Dec 2017.

[178] BioAustralis. Available at: http://www.bioaustralis.com/. Accessed: 25 Dec 2017.

[179] MedChem Express. Available at: http://www.medchemexpress.com/. Accessed: 2 Dec 2017.

[180] Specs. NA. Available at: http://www.specs.net/page.php?pageid=2004111115353984&smenu=2008111411133023. Accessed: 5 Jan 2018.

[181] Chimiothèque Nationale (CN). Available at: http://chimiotheque-nationale.cn.cnrs.fr/?Presentation,18. Accessed: 7 Jan 2018.

[182] Irwin JJ, Gaskins G, Sterling T, Mysinger MM, Keiser MJ. Predicted biological activity of purchasable chemical space. J Chem Inf Model. 2018;58:148–64.

[183] Lucas X, Grüning BA, Bleher S, Günther S. The purchasable chemical space: a detailed picture. J Chem Inf Model. 2015;55:915–24.

[184] Doak BC, Over B, Giordanetto F, Kihlberg J. Oral druggable space beyond the rule of 5: insights from drugs and clinical candidates. Chem Biol. 2014;21:1115–42.

[185] Pascolutti M, Quinn RJ. Natural products as lead structures: chemical transformations to create lead-like libraries. Drug Discov Today. 2014;19:215–21.

[186] Ma DL, Chan DSH, Leung CH. Molecular docking for virtual screening of natural product databases. Chem Sci. 2011;2:1656–65.

[187] Pereira F, Latino DARS, Gaudêncio SP. QSAR-assisted virtual screening of lead-like molecules from marine and microbial natural sources for antitumor and antibiotic drug discovery. Molecules. 2015;20:4848–73.

[188] Pye CR, Bertin MJ, Lokey RS, Gerwick WH, Linington RG. Retrospective analysis of natural products provides insights for future discovery trends. Proc Natl Acad Sci USA. 2017;114:5601–6.

[189] Shen B. A new golden age of natural products drug discovery. Cell. 2015;163:1297–300.

Further Reading

Baell JB. Broad coverage of commercially available lead-like screening space with fewer than 350,000 compounds. J Chem Inf Model. 2013;53:39–55.

Bajusz D, Rácz A, Héberger K. Chemical data formats, fingerprints, and other molecular descriptions for database analysis and searching. In: Chackalamannil S, Rotella DP, Ward SE, editors. Comprehensive medicinal chemistry III, vol. 3. Oxford: Elsevier, 2017: 329–78.

Dias DA, Urban S, Roessner U. A historical overview of natural products in drug discovery. Metabolites. 2012;2:303–36.

Finn PW, Morris GM. Shape-based similarity searching in chemical databases. Wiley Interdiscip Rev Comput Mol Sci. 2013;3:226–41.

Lipinski CA. Rule of five in 2015 and beyond: Target and ligand structural limitations, ligand chemistry structure and drug discovery project decisions. Adv Drug Deliv Rev. 2016;101:34–41.

Mitchell JBO. Machine learning methods in chemoinformatics. Wiley Interdiscip Rev Comput Mol Sci. 2014;4:468–81.

Stratton CF, Newman DJ, Tan DS. Cheminformatic comparison of approved drugs from natural product versus synthetic origins. Bioorg Med Chem Lett. 2015;25:4802–7.

Tao L, Zhu F, Qin C, Zhang C, Chen S, Zhang P, Zhang C, Tan C, Gao C, Chen Z, Jiang Y, Chen YZ. Clustered distribution of natural product leads of drugs in the chemical space as influenced by the privileged target-sites. Sci Rep. 2015;5:9325.

Rita C. Guedes and Tiago Rodrigues

11 Drug target prediction using chem- and bioinformatics

Abstract: The biological pre-validation of natural products (NPs) and their underlying frameworks ensures an unrivaled source of inspiration for chemical probe and drug design. However, the poor knowledge of their drug target counterparts critically hinders the broader exploration of NPs in chemical biology and molecular medicine. Cutting-edge algorithms now provide powerful means for the target deconvolution of phenotypic screen hits and generate motivated research hypotheses. Herein, we present recent progress in artificial intelligence applied to target identification that may accelerate future NP-inspired molecular medicine.

Keywords: machine intelligence, cheminformatics, chemical biology, target identification, natural products

11.1 Introduction

Successful future chemical biology and drug discovery relies on the identification and validation of small molecule effectors [1, 2], despite the advent of biologics [3]. Indeed, drug target binding and engagement remains a hallmark for modulation of (patho)physiological events. It is now widely accepted that small molecules rarely are selective but engage dozens of related or unrelated targets [4], through what has been coined as polypharmacology or network pharmacology [5, 6]. For further reading on the topic please refer to the articles titled "Computer-based Techniques for lead Identification and Optimization I: Basics" [7] and "Computer-based Techniques for Lead Identification and Optimization II: Advanced Search methods" [8].

Natural products (NPs) have traditionally offered biologically motivated prototypes for optimization given the pre-validation of their underlying molecular architectures over millions of years of evolutionary pressure [9, 10]. In spite of their limited supply in nature, constant advances in synthetic chemistry are now meeting ends with regards to making amounts of chemical matter suitable for biological, phenotypic testing [11]. However, if test NPs are identified as hits in any given cell-based assay, researchers are faced with a riddle known as drug target deconvolution [12]. This is in fact true not only for NPs but also for phenotypic

This article has previously been published in the journal *Physical Sciences Reviews*. Please cite as: Guedes, R., Rodrigues, T. Drug target prediction using chem- and bioinformatics. *Physical Sciences Reviews* [Online] **2018** DOI: 10.1515/psr-2018-0112.

https://doi.org/10.1515/9783110579352-012

screen hits in general. Arguably the identification of drug targets remains the major bottleneck of modern drug discovery, as the recognition of ligand–drug target correlations still cannot be streamlined. Despite the current technical challenges for the full mapping of pharmacology networks, important advances are being made recently. Crucially, such knowledge may bring benefits to the design of leads with lower probability of attrition and ultimately afford efficacious disease modulators [1].

Chemoproteomics approaches remain the gold-standard method for the identification of macromolecular counterparts for small bioactive molecules. For example, using chemical proteomics, Cravatt and co-workers have recently identified the mitochondrial carnitine–acylcarnitine translocase SLC25A20 as a functional target of the diterpenoid ester ingerol mebutate that is used as a first-in-class treatment for actinic keratosis [13]. However, the method itself presents a multitude of drawbacks that have precluded this approach of becoming commonplace, particularly in academia. For example, the need for chemical tagging of the ligand of interest may inadvertently disrupt the binding affinity towards relevant on- and off-targets, the method is laborious, usually time consuming, may require expensive equipment and in the end only vouches for motivated research hypotheses that need to be verified in follow-up studies.

In silico tools have been typically employed to virtually screen an enumerated fraction of chemical space that is, ideally, physically accessible either through custom synthesis or vendors libraries [14]. By doing so, researchers expect to prioritize for testing potential bioactive molecules and, consequently, avoid the screening of thousands of entities against a target of interest that could be both time consuming and expensive. Here, the focus is the drug target for which ligands are sought after, and only then correlate engagement of the aforementioned target with modulation of disease or secondary pharmacology. Molecular docking and pharmacophore-based screening, either ligand- or target-based, are two of the most popular approaches to virtual screening, but they too suffer from important limitations (*vide infra*). Most importantly, the exponential growth of available biological and chemical data renders these methods suboptimal for most large-scale discovery programs. Fortunately, the advent of "big data" [15] is parallelized with more powerful computing capabilities and increased storage capacity that may facilitate the integration and analysis of unstructured, sparse and/or incomplete information. Finally, advanced algorithms for machine learning are also emerging as workhorse of modern bio and cheminformatics [16], without which artificial intelligence as a whole would be intractable.

Cheminformatics provides an array of tools to scrutinize NPs [17, 18]. Herein we provide an overview of select *in silico* methods for drug target deconvolution and deorphanization of biologically active NPs, with special focus on recent machine learning applications. We highlight strengths and weaknesses of different approaches in an attempt to rationalize method selection and as an alternative and/or complement to chemical proteomics.

11.2 Pharmacophore screening

In light of Wermuth's definition [19],

> a pharmacophore is the ensemble of steric and electronic features that is necessary to ensure the optimal supramolecular interactions with a specific biological target structure and to trigger (or to block) its biological response. A pharmacophore does not represent a real molecule or a real association of functional groups, but a purely abstract concept that accounts for the common molecular interaction capacities of a group of compounds toward their target structure. The pharmacophore can be considered the largest common denominator shared by a set of active molecules. This definition discards a misuse often found in the medicinal chemistry literature, which consists of naming as pharmacophores simple chemical functionalities such as guanidines, sulfonamides or dihydroimidazoles (formerly imidazolines), or typical structural skeletons such as flavones, phenothiazines, prostaglandins or steroids.

From the medicinal chemistry practice vantage point, a pharmacophore encodes the relevant features for ligand recognition by a receptor. Hence, its 3D projection can be used to relate ligands without explicit structure comparison and, therefore, assess the potential complementarity between small molecules and drug targets [20]. 3D Pharmacophore models describe molecules in abstract fashion, i. e. different chemical motifs can afford identical models [21]. Thus, one may consider that such models are primed for "scaffold hopping", and have been extensively used in virtual screening campaigns [22]. Because the successful use of 3D pharmacophores relies on the identification of bioactive molecular geometries, proper conformational sampling of the reference and search molecules is key. However, one may argue that finding reasonable low energy conformers for small molecules is a daunting task [23, 24]. This is especially true for NPs, whose high frequency of stereogenic centres can lead to particularly inaccurate and/or irrelevant conformers [25]. Nevertheless, irrespective of the intuitive importance of having accurate 3D data to start a virtual screening campaign, it has been suggested that the impact of the bioactive conformation on the overall database enrichment is limited [26–28]. Different programs, e. g. Corina [23, 29], Catalyst [30], MOE [31], employ different approaches, such as genetic algorithms [32] and other stochastic methods [33, 34], and molecular dynamics [35] simulations to generate conformers. Likewise, the generation of 3D pharmacophore models differs between software packages [36]. In general, each feature is given as a sphere corresponding to a tolerance zone that can be occupied by chemical motifs capable of exerting a given type of interaction (Figure 11.1). Using the resulting ensemble of features as fuzzy molecular representation, one can swiftly mine databases to identify similar molecules and infer identical biological effects. Among a manifold of tools, Catalyst [37], LigandScout [38], Phase [39] and MOE have been frequently used tools in 3D pharmacophore-based virtual screening.

   On occasion, it is advisable to generate several models reflecting the pharmacophoric features of structurally related chemotypes, which are more likely to present

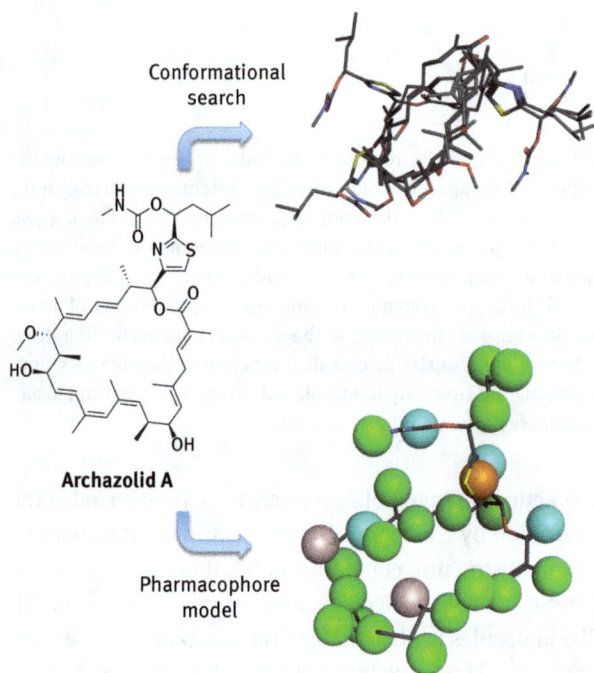

Figure 11.1: Conformational search and pharmacophore model exemplified for the macrocyclic natural product archazolid A (computed with MOE, Chemical Computting Group, Canada). The conformational search generates several plausible conformations with little agreement between each other, as observed by a rigid alignment, which exemplifies the difficulty of selecting meaningful 3D molecular models to derive a pharmacophore hypothesis. The pharmacophore model is generated automatically and can be manually pruned. Tolerance spheres are then used for virtual screening purposes. Green: hydrophobic feature; Orange: aromatic feature; Cyan: hydrogen bond donor feature; Salmon: hydrogen bond acceptor feature.

similar binding sites and ligand binding modes [40]. Multiple ligand alignments assume identical binding modes, which is often not realistic. Thus, making wrong assumptions in the model-building step increases the likelihood of failure in the downstream screening process. Typically, a pharmacophore-based screening workflow starts with the selection and superimposition of potential bioactive conformations of a training set, along with the generation of conformers for the query molecules [41]. Then, the software swiftly identifies conformers in the query database that match the previously constructed pharmacophore model(s) [41]. 3D Pharmacophore models are among the most commonly used virtual library screening techniques, including for the identification of drug targets for NPs.

Using this concept, Rollinger and colleagues have identified several NPs and NP-inspired entities as 11β-hydroxysteroid dehydrogenase modulators [42] and depside/depsidones as potent inhibitors of microsomal prostaglandin E2 synthase-1 (Figure 11.2)

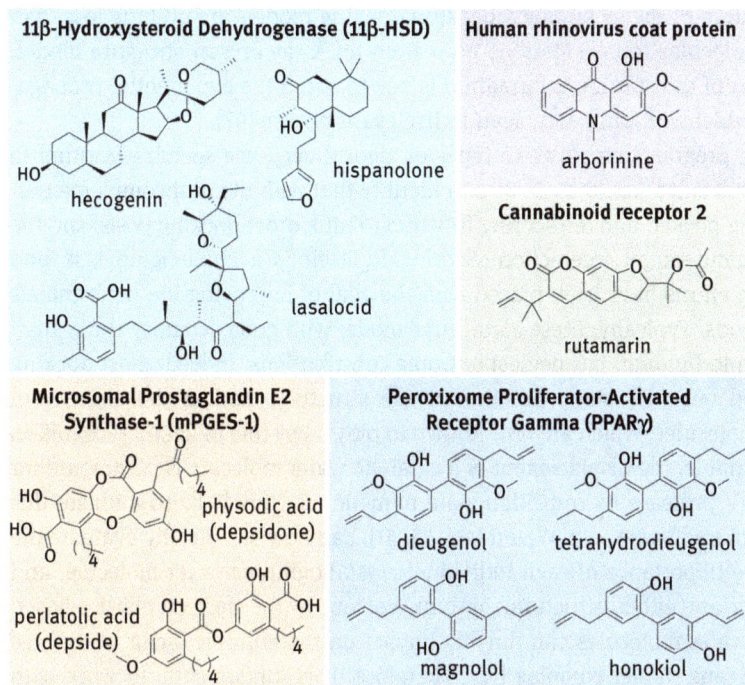

Figure 11.2: Structure of natural products and their targets, as identified by pharmacophore-based virtual screening.

[43]. Moreover, several metabolites from *Ruta graveolans* were interrogated with more than 2,000 3D pharmacophore models to suggest, and then experimentally confirm inhibition of the human rhinovirus coat protein and cannabinoid-2 receptor by arbor-inine and rutamarin, respectively [44]. Finally, peroxixome proliferator-activated receptor gamma (PPARγ) was found as target for biphenyl-based NPs, such as dieugenol, tetrahydrodieugenol, magnolol and honokiol [45, 46].

11.3 Molecular docking

Molecular docking has frequently been employed to screen *in silico* libraries of thousands to millions of potential ligands against a certain target of interest. Molecular docking has thus become one of the most popular techniques in computational medicinal chemistry given the power of predicting the so-called "docking pose" at a user-defined binding site and, therefore, rationalize the interactions between ligand and target, i.e. molecular recognition. Naturally, a good docking model can empower a skilled medicinal chemist with valuable information for the design of subsequent lead compound series, but several caveats must be kept in mind to allow a correct exploitation of the technology. Because docking models are nothing less than hypothetical molecular representations built from an X-ray crystal structure model, careful selection of the starting information is key to avoid the exponential propagation of errors, which will ultimately lead to irrelevant results [47].

All docking programs employ two types of algorithms – the search algorithm to sample small molecules' conformations and identify that with the best complementarity to the binding pocket, and the scoring function to rank order docking poses and the hypothetical ligand–target interaction. Scoring in itself is a challenging task and several research efforts have been placed with the goal of improving the performance of scoring methods. Typically, these algorithms model with good accuracy the enthalpic contribution to binding, but neglect entropic contributions. Indeed, most docking runs are carried out with static protein/enzyme structures and in the absence of explicit water molecules, which are now known to play a key role in ligand recognition [48, 49]. For example, the displacement of a catalytic water molecule is paramount for inhibition of HIV protease by transition state mimetic effectors [50]. To mitigate this limitation bespoke software, e. g. WaterMap [48, 51], can now statistically evaluate the position and the importance of each individual crystallographic water molecule, and estimate if they are either structural or bulk solvent – the correct physics-based modulation of water molecules can directly impact on the entropic factor of the free energy of the ligand–target complex [52]. Therefore, it is fundamental to work with high-resolution crystal structures to enable the detection of density attributable to water molecules and, if available, apply ancillary software for molecular dynamics simulations to predict a more global picture of the protein behaviour in solution.

As a result of the well-established limitations of molecular docking, state-of-the-art software tools predict the incorrect binding pose all-too-frequently. Indeed, when only a single fixed receptor conformation is used, the docking pose output is incorrect for up to 70 % of the screened ligands [53]. Still, the relative ease with which a docking run can be set up and the wealth of information it can provide makes it an attractive tool to deploy with NPs in an attempt to unveil putative on-/off-targets and modes of action or biochemical liabilities. For example, by relying on inverse molecular docking, i.e. docking of several proteins against a ligand of interest [54], cyclooxygenase-2 and PPARγ were identified as targets of meranzin, whose potency is comparable to that of indomethacin and rosiglitazone (Figure 11.3) [55].

11.4 Similarity-based searches

While 3D methods have been applied successfully to uncover putative targets for NPs, one may still argue that the need to compute meaningful small molecule conformations and search binding poses is computationally expensive and not high-throughput friendly. Topological (two-dimensional) methods offer alternatives that are several folds faster and surprisingly accurate, when compared to its 3D

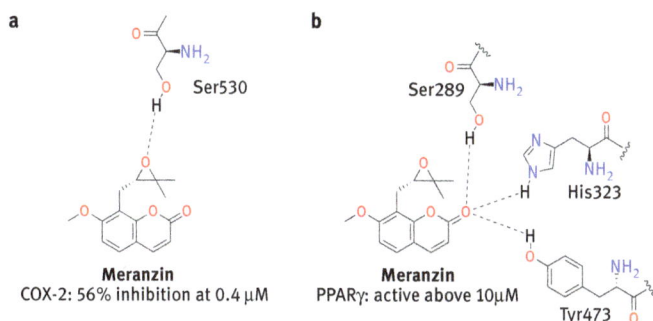

Figure 11.3: Predicted interactions between the natural product meranzin and COX-2 (panel a) and the peroxisome proliferator-activated receptor gamma (PPAR γ, panel b) as identified by inverse molecular docking. Meranzin displayed concentration-dependent effects against both targets, as observed experimentally, validating the docking protocol.

counterpart methods [56]. Indeed, the applicability of said topological methods is strongly grounded on the chemogenomics molecular similarity principle, wherein similar ligands are likely to bind to identical targets [57]. To this end, one must then calculate descriptors (e. g. physicochemical, MACCS keys, ECFP fingerprints) that will encode molecular structures into computable units and assess the pair-wise similarity by one of several available metrics, e. g. Tanimoto-Jacquard coefficient or index (eq. (11.1)) Dice similarity, Euclidean or Manhattan distances [58, 59]. The Tanimoto coefficient is one of the most widely employed metrics and it computes a value between zero and one, which measures the similarity of the fingerprints of molecules A and B. Whereas a value of zero indicates complete dissimilarity between the fingerprints of A and B, a value of one indicates full fingerprint identity, and by consequence similar (or identical) ligands.

$$T = \frac{c}{a + b - c} \tag{11.1}$$

where T is the Tanimoto index, a is the number of bits set in fingerprint A, b is the number of bits set in fingerprint B and c the number of common bits set in both fingerprints.

As a rule of thumb, high affinity ligands should be used as reference molecules for similarity-based virtual searches, given that the goal is to identify chemical matter active against a ligandable target. It is also established that the performance of the screening campaign is heavily dependent on the molecular descriptors used [60]. This is easily comprehensible if one considers the extreme case of encoding two molecules on separate occasions with the same fingerprint type but using two different lengths (e. g. fingerprint 1 with length 2 $vs.$ fingerprint 1 with length 2048). Naturally, the fingerprint length, which is a user-defined parameter, influences the similarity measure between both molecules. The same comparison is valid when considering different fingerprints. Computed similarities are stored in a database and sorted according to decreasing similarity in comparison to the query structure/fingerprint. The top molecule in the rank list, also known as the "nearest neighbour", is deemed the most similar to the query.

The similarity ensemble approach (SEA, webserver: http://sea.bkslab.org) [57] leverages the abovementioned concepts coupled to probabilistic models to identify and prioritize drug targets for screening with speed vastly unmatched by 3D-pharmacophore and molecular docking screens. While the method was developed primarily with the aim enabling the identification of drug targets for small molecules of synthetic origin, it has also been successfully applied to unveil the antiplasmodial effects of physalins B, D, F and G (Figure 11.4) [61].

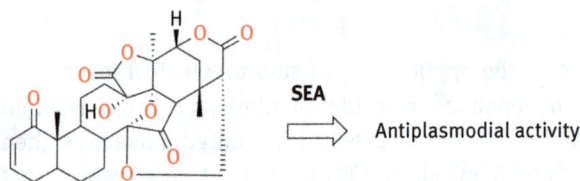

Physalin B: Δ^5
Physalin D: 5α-OH, 6β-OH
Physalin F: 5β,6β-epoxy
Physalin G: Δ^4, 6α-OH

Figure 11.4: Application of the Similarity Ensemble Approach (SEA) to identify the antiplasmodial activity of physalins.

Other tools have equally proven utility for the identification of bioactivities for NPs with minimal computational effort. PASS, which like SEA is available as a webserver (http://www.pharmaexpert.ru/passonline/) [62, 63] uses topological fragment structure descriptors [64] as a means of identifying similarity and infer on the biological activities of the studied chemotypes. Albeit lacking experimental validation of the generated research hypotheses, the application of PASS to > 90 marine sponge alkaloids suggested anti-tumoural activity for the majority (80 %) of the queried structures [65].

Recognizing the need to curate the scattered bioactivity information of NPs, the same authors have compiled NPASS (Natural Product Activity and Species Source), which annotates NP bioactivities with the aim to optimize *in silico* algorithms for advanced NPs applications [66]. One may consider that substructural descriptors are not ideal when analysing NPs, given the well-known dissimilarity between their intricate frameworks and the scaffolds entailed in the reference databases, which are usually of synthetic origin. Thus, the algorithms discussed in this section were primarily designed for synthetic small molecules and may, on average, provide less accurate predictions than software tailored for NPs. The later are detailed in the following section together with select applications.

11.5 Statistical learning and machine intelligence

Machine learning is an emerging technology in drug discovery that leverages significant amounts of data, otherwise intractable, for pattern recognition and correlation analyses [67]. Indeed, statistical learning methodologies have seen a vast array of applications and promise to reshape how both fundamental and applied science are carried out [16]. For example, regression and classification models have been successfully employed for a multitude of tasks, including *de novo* design of small molecules [68], prediction of ADME properties [69], prediction of drug-likeness of traditional medicines [70], conformational sampling [25], prediction of synthetic routes [71], among others, in an attempt to speed up discovery chemistry. Hence, machine intelligence has tremendous potential in early drug discovery as it may streamline processes by cutting out inefficiencies and feed development pipelines in sustainable fashion. The same premises are true for the identification of drug targets for bioactive matter, including NPs, where different methods of machine intelligence (e.g. deep learning, random forests, k-nearest neighbours support vector machines) may help prioritizing targets for experimental validation and help eliminate the execution of unmotivated assays [72].

To mitigate some of the limitations of software tools previously discussed in this chapter and associated descriptors, Schneider and co-workers reported in 2014 the Self-Organizing Maps (SOM)-based prediction of drug equivalence (SPiDER) software [4] that uses unsupervised learning, i.e. it only takes into account the structure of the input data and runs until convergence or for a user-defined number of epochs, without knowing the "correct" output. SOMs employ heuristics inspired in neural networks that lead to a weighted projection of the chemical feature space, while preserving neighbourhoods between data points, which constitutes an important advantage for interpretability. The method is thus appropriate for data clustering and visualization, with the interpretation of both being intuitive for the user. To ensure that the applicability domain of SPiDER could be extended to NPs, two sets of "fuzzy" molecular descriptors were employed: (i) 2D physicochemical properties as computed by the MOE software (Chemical Computing Group, Canada) and (ii) type-scaled CATS2 topological pharmacophore descriptors [73], which computes a vector of pairwise feature correlations (hydrogen bond donor/acceptor, positive/negative charge, aromatic, hydrophobic) up to a distance of 10 bonds. Alongside a custom-built, manually curated reference ligand database [74], these constitute major differences compared to other tools, which result in a broad domain of applicability including *de novo* designed molecules and NPs. By using both descriptor sets independently, SPiDER tessellates chemical space from distinct, yet ideally complementary, vantage points to afford predictions via the nearest neighbour concept. Through arithmetic combination of both prediction outputs, and analyses of background ligand distances, a consensus prediction score is afforded together with a *p*-like value that reflects its significance (Figure 11.5) [4].

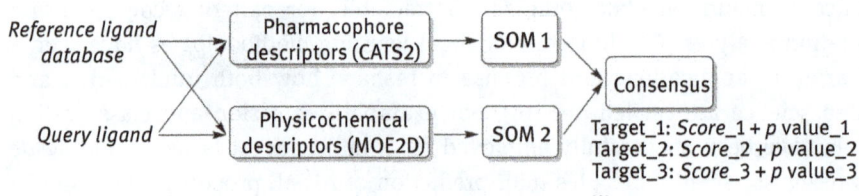

Figure 11.5: Schematics of the SPiDER software [4] workflow. Two independent self-organizing maps (SOMs) are generated from different descriptor sets. A consensus prediction is obtained by arithmetic combination of both outputs. To assess the significance of the prediction a *p*-like value is calculated on the basis of background ligand distances.

SPiDER has been extensively validated for deorphanizing and clarifying the polypharmacology traits of distinct NPs, including fragment-like [2, 75] and macrocyclic entities [76]. In its first report in 2014, the software was applied to pharmacophoricaly-motivated archazolid A fragments, acting as a means of encoding bioactivity fingerprints [76]. The resulting predictions were then inferred back to the parent NP. The approach was motivated by the fact that archazolid A differs from the reference ligands in SPiDER, which led to lower confidence predictions when compared to its derived fragments. Indeed, different fragments provided overlapping confident predictions, which further motivated the biochemical tests against a range of different targets, which included cyclooxygenase-2 (COX-2), PPARγ, glucocorticoid receptor (GR), mPGES-1 among others. Albeit not confirmed experimentally, modulation of these targets may contribute for the already known anticancer activity (Figure 11.6) [76]. Likewise, the anticancer macrocycle doliculide was deorphanized as a nanomolar-potent prostanoid receptor 3 antagonist which may be involved in cancer progression [77]. More recently, targets were identified for the fragment-like NPs (-)-sparteine [2], isomacroin and graveolinin [75], suggesting that optimization of their molecular frameworks may provide potent modulators of disease-relevant targets. Finally, (-)-englerin A was unveiled as a moderate voltage-gated calcium $Ca_v1.2$ channel antagonist [78]. Although the finding is not relevant to explain the anticancer activity of (-)-englerin A, the study suggests that this NP presents moieties that can be grafted onto molecules tailored for $Ca_v1.2$ inhibition and that machine learning may be a suitable technology for systems pharmacology studies.

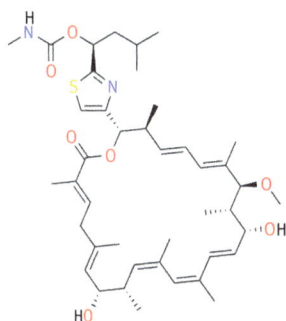

Archazolid A

Target_1: mPGES-1 ($p = 0.002$)
Target_2: GR ($p = 0.012$)
Target_3: COX-2 ($p = 0.081$)
Target_4: FXR ($p = 0.012$)
Target_5: PPARγ ($p = 0.014$)
Target_6: 5-LO ($p = 0.066$)

EC$_{50}$ (agonist) = 8 μM
22% effect at 10 μM
24% effect at 10 μM
EC$_{50}$ (agonist) = 0.2 μM
EC$_{50}$ (agonist) = 8 μM
EC$_{50}$ (antagonist) = 11 μM

Doliculide

Target: EP3 ($p < 0.05$)   IC$_{50}$ = 16 nM

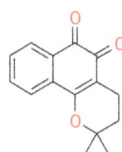

β-Lapachone

Target_1: 5-LO ($p = 0.010$)   IC$_{50}$ = 240 nM
Target_2: EP3 ($p = 0.017$)   IC$_{50}$ = 20 μM

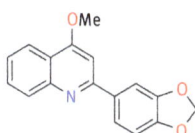

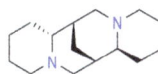

Graveolinine

Target_1: COX-2 ($p = 0.0072$)   79% effect at 150 μM
Target_2: 5-HT2B (p = 0.0096)   IC$_{50}$ = 12 μM

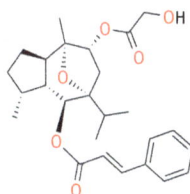

(-)-Sparteine

Target: Kappa opioid ($p < 0.05$)   EC$_{50}$ = 245 μM

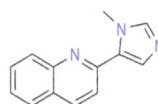

(-)-Englerin A

Target: Ca^{2+} channel ($p = 0.007$) K_i = 6 μM

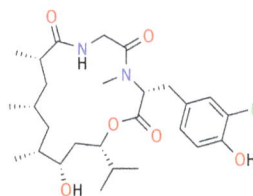

Isomacroin

Target: PDGFRα ($p = 0.0025$)   IC$_{50}$ = 25 μM

Figure 11.6: Examples of the successful application of SPiDER to identify drug targets for diverse natural products.

In similar fashion to SPiDER, the Target Inference Generator (TIGER) also leverages a consensus of two SOMs to afford target predictions. Built with the ability of inferring up to 331 qualitative ligand–target relationships, it has been productively applied to the marine NP (±)-marinopyrrole A to unveil modulation of orexin-1/2 and glucocorticoid receptors, as well as cholecystokinin 2 (Figure 11.7) [79]. In a second case study

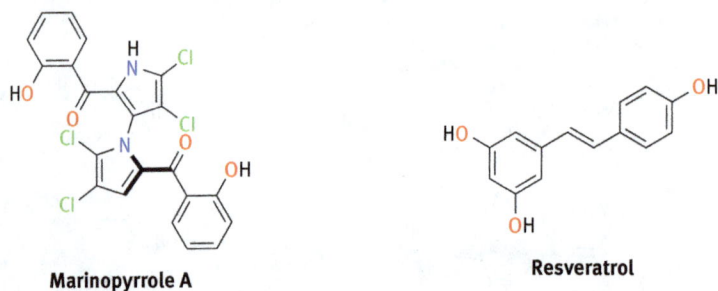

Marinopyrrole A

K_B (Glucocorticoid receptor) $= 0.7 \,\mu M$
K_B (Cholecystokinin 2) $= 1 \,\mu M$
K_B (Orexin-1) $= 0.3 \,\mu M$
K_B (Orexin-2) $= 0.6 \,\mu M$

Resveratrol

K_i (ERα) $= 4 \,\mu M$
K_i (ERβ) $= 0.4 \,\mu M$

Figure 11.7: Structures of (±)-marinopyrrole A and resveratrol, and targets unveiled by TIGER.

using TIGER, resveratrol was found to modulate the estrogen receptor β (ERβ, $K_i = 0.4$ μM) with a reasonable degree of selectivity over its α counterpart (ERα, $K_i = 4 \,\mu M$) [80].

More recently, Rodrigues and Bernardes reported the Drug–Target Relationship Predictor (DEcRyPT) [81] software that uses the regressor random forest technology as an orthogonal machine intelligence approach to SPiDER. While SPiDER affords a binary answer (bind or does not bind) for each target, DEcRyPT predicts an affinity value by using an ensemble of decision trees built from CATS2 descriptors. Each tree analyses part of the data to afford a weak estimator. However, when put together, the decision trees afford a prediction that ideally tends to balance out variance and bias and avoid the over-fitting of a single decision tree. To that end, the authors collected, curated and transformed bioactivity data from the ChEMBL22 database prior to model building. Using DEcRyPT as a method to further scrutinize SPiDER predictions, 5-lipoxygenase (5-LO) was confidently predicted as a binding counterpart for the clinical stage, naphthoquinone NP β-Lapachone. Cell-free assays confirmed that the hydroquinone form of β-Lapachone acts as a nanomolar inhibitor of 5-LO (IC$_{50}$ $= 0.24 \,\mu M$) and that modulation was largely independent of the presence of Triton X-100, which ruled out unspecific nuisance behaviour. Moreover, equipotent 5-LO inhibition was confirmed on a whole cell assay system. In a rather unexpected event, β-Lapachone also proved to have selectivity for 5-LO over 15- and 12-LO, which suggested binding to an allosteric pocket. In fact, β-Lapachone did not compete with arachidonic acid in co-incubation functional assays but its potency was significantly affected by increasing concentrations of phosphatidylcholine, which binds at the interface of the catalytic and C2-like domains. Of importance, inhibition of 5-LO by β-Lapachone correlated well with its anticancer effects, which shows the power of machine intelligence for deconvoluting complex phenotypes (Figure 11.6 and Figure 11.8).

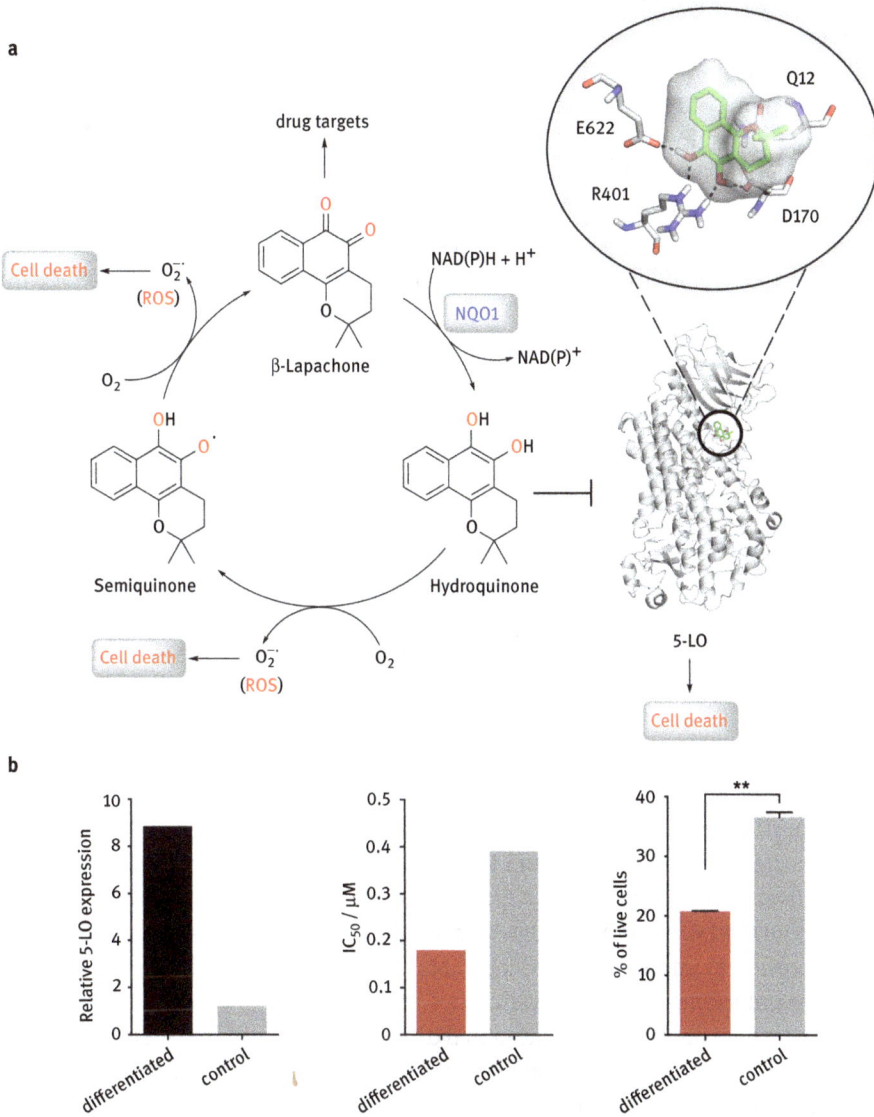

Figure 11.8: Mechanism of anticancer activity of β-Lapachone. a) β-Lapachone acts as a redox cycler, generating reactive oxygen species (ROS) that may partly contribute to the cytotoxicity. The hydroquinone form of β-Lapachone is a potent, reversible, allosteric inhibitor of 5-lipoxygenase (5-LO). b) Differentiated HL-60 cell lines overexpress 5-LO (left) and are more sensitive to β-Lapachone (middle and right). IC_{50} (differentiated) = 0.18 μM; IC_{50} (control) = 0.39 μM (middle). Percentage of live HL-60 cells in the differentiated and control groups when treated with 0.5 μM of β-Lapachone. Statistics: two-tailed t-Student test; **$p < 0.005$.

In another case study, DEcRyPT was extended to include bioactivity data on > 1000 liganded targets, and used to unveil secondary pharmacology by DMP-1, a synthetic analogue of militarinone A (Figure 11.9). Potent modulation of the cannabinoid receptor 1 (CB1) was predicted with high confidence (predicted affinity = 0.16 µM) using this random forest routine. As proof-of-concept, the usefulness and accuracy of the method were both confirmed experimentally, by determining functional antagonistic effects with a K_B value of 0.32 µM and displacement of a radiolabelled ligand with a K_i value of 3.2 µM. However, given the magnitude of the functional effects it is likely that targets other than CB1 mediate the inhibition of autophagy by DMP-1 [82].

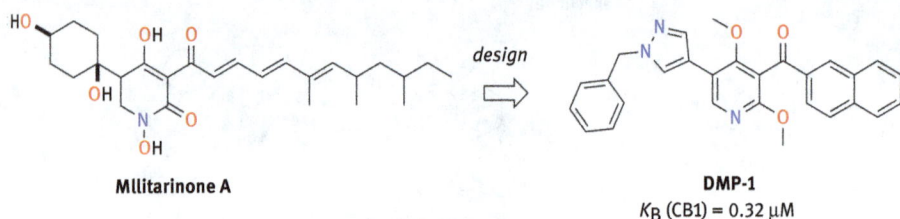

Mllitarinone A

DMP-1
K_B (CB1) = 0.32 µM

Figure 11.9: The militarinone A analogue DMP-1 is a potent cannabinoid receptor 1 antagonist, as predicted by the machine intelligence software DEcRyPT.

11.6 Outlook

It is unquestionable that NPs still inspire medicinal chemists for molecular design and that a significant percentage of FDA-approved drugs are either unmodified NPs or derivatives/analogues thereof [83]. The efficient navigation of chemical space is currently pursued with the aid of algorithms. Indeed, software tools are possibly even more important to unravel the pharmacology of "dark matter" [84] of natural origin, which can unlock new research avenues.

We have here presented and discussed different methods for the identification of drug targets of NPs. Those are however universally applicable to phenotypic hits, irrespective of its origin. Nevertheless, we do caution the reader that, like for any other *in silico* method, only research hypotheses are generated that ought to be confirmed experimentally. Also, any approach, including those discussed here, is certain to fail at least occasionally. As a rule of thumb, any method is only as good as its start-up information package.

Molecular docking and 3D-pharmacophore modelling have traditionally been workhorses in molecular informatics, despite being computationally expensive and processing only a limited amount of data at a time. Thus, with the ever-increasing research expenditure, new, fast and economical technologies are in high demand for sustainable drug discovery. Machine learning algorithms are primed to crunch numbers in the arising "big data" era, but sound knowledge of its workings is required to avoid using such methods as black boxes and evade pitfalls. For example, while artificial neural networks require scaling of the descriptor set, the same is not true for decision trees. Also, different machine learning methods require different computational resources for an efficient prospective use. Our experience suggests that machine intelligence methods are not a panacea, but have given us rather high success rates when applied to target discovery in the NP realm. As more information becomes available, with the development of efficient algorithms and computational resources, success rates only tend to improve. We thus foresee that target identification can finally be automated in future drug discovery or, at the bare minimum, artificially rationalized.

Acknowledgements: Rita Guedes thanks the funding received from the European Structural & Investment Funds through the COMPETE Programme and from National Funds through FCT under the Programme grant SAICTPAC/0019/2015, PDTC/QEQ-MED/7042/2014 and UID/DTP/04138/2013. Tiago Rodrigues thanks generous support by the Marie Curie Actions (IF grant 743640 and TWINN-2017 ACORN, Grant 807281) and FCT/FEDER (02/SAICT/2017, Grant 28333).

References

[1]   Rodrigues T. Harnessing the potential of natural products in drug discovery from a cheminformatics vantage point. Org Biomol Chem. 2017;15:9275–82.

[2]   Rodrigues T, Reker D, Schneider P, Schneider G. Counting on natural products for drug design. Nat Chem. 2016;8:531–41.

[3]   Mullard A. 2017 FDA drug approvals. Nat Rev Drug Discov. 2018;17:81–5.

[4]   Reker D, Rodrigues T, Schneider P, Schneider G. Identifying the macromolecular targets of de novo-designed chemical entities through self-organizing map consensus. Proc Natl Acad Sci USA. 2014;111:4067–72.

[5]   Hopkins AL. Network pharmacology: the next paradigm in drug discovery. Nat Chem Biol. 2008;4:682–90.

[6]   Hopkins AL. Network pharmacology. Nat Biotechnol. 2007;25:1110–1.

[7]   Maruca A, Ambrosio FA, Lupia A, Romeo I, Rocca R, Moraca F, et al. Computer-based techniques for lead identification and optimization I: Basics. Phys Sci Rev 2018. DOI: 10.1515/psr-2018-0113. Forthcoming.

[8]   Lupia A, Moraca F, Maruca A, Ambrosio FA, Catalano R, Romeo I, et al. Computer-based techniques for lead identification and optimization II: advanced search methods. Phys Sci Rev 2018. DOI: 10.1515/psr-2018-0114. Forthcoming.

[9]   Wetzel S, Bon RS, Kumar K, Waldmann H. Biology-oriented synthesis. Angew Chem Int Ed. 2011;50:10800–26.

[10] Van Hattum H, Waldmann H. Biology-oriented synthesis: harnessing the power of evolution. J Am Chem Soc. 2014;136:11853–9.

[11] Baran PS. Natural product total synthesis: as exciting as ever and here to stay. J Am Chem Soc. 2018;140:4751–5.

[12] Laraia L, Waldmann H. Natural product inspired compound collections: evolutionary principle, chemical synthesis, phenotypic screening, and target identification. Drug Discov Today Technol. 2017;23:75–82.

[13] Parker CG, Kuttruff CG, Galmozzi A, Jorgensen L, Yen CH, Hermanson DJ, et al. Chemical proteomics identifies SLC25A20 as a functional target of the ingenol class of actinic keratosis drugs. ACS Cent Sci. 2017;3:1276–85

[14] Schneider G. Virtual screening: an endless staircase? Nat Rev Drug Discov. 2010;9:273–6.

[15] Singh G, Schulthess D, Hughes N, Vannieuwenhuyse B, Kalra D. Real world big data for clinical research and drug development. Drug Discov Today. 2018;23:652–60.

[16] Lavecchia A. Machine-learning approaches in drug discovery: methods and applications. Drug Discov Today. 2015;20:318–31.

[17] Henkel T, Brunne RM, Muller H, Reichel F. Statistical investigation into the structural complementarity of natural products and synthetic compounds. Angew Chem Int Ed. 1999;38:643–7.

[18] Stahura FL, Godden JW, Xue L, Bajorath J. Distinguishing between natural products and synthetic molecules by descriptor Shannon entropy analysis and binary QSAR calculations. J Chem Inf Comput Sci. 2000;40:1245–52.

[19] Wermuth C-G, Ganellin CR, Lindberg P, Mitscher LA. Glossary of terms used in medicinal chemistry (IUPAC recommendations 1997). Annu Rep Med Chem. 1998;33:385–95.

[20] Seidel T, Ibis G, Bendix F, Wolber G. Strategies for 3D pharmacophore-based virtual screening. Drug Discov Today Technol. 2010;7:e203–70.

[21] Langer T, Wolber G. Pharmacophore definition and 3D searches. Drug Discov Today Technol. 2004;1:203–7.

[22] Schneider G, Schneider P, Renner S. Scaffold-hopping: how far can you jump? QSAR Comb Sci. 2006;25:1162–71.

[23] Gasteiger J, Rudolph C, Sadowski J. Automatic generation of 3D-atomic coordinates for organic molecules. Tetrahedron Comput Method. 1990;3:536–47.

[24] Chen I-J, Foloppe N. Conformational sampling of drug like molecules with MOE and catalyst: implications for pharmacophore modeling and virtual screening. J Chem Inf Model. 2008;48:1773–91.

[25] Rupp M, Bauer MR, Wilcken R, Lange A, Reutlinger M, Boeckler FM, et al. Machine learning estimates of natural product conformational energies. PLoS Comput Biol. 2014;10:1003400.

[26] Renner S, Schwab CH, Gasteiger J, Schneider G. Impact of conformational flexibility on three-dimensional similarity searching using correlation vectors. J Chem Inf Model. 2006;46:2324–32.

[27] Zhang Q, Muegge I. Scaffold hopping through virtual screening using 2D and 3D similarity descriptors: ranking, voting, and consensus scoring. J Med Chem. 2006;49:1536–48.

[28] Hawkins PC, Skillman AG, Nicholls A. Comparison of shapematching and docking as virtual screening tools. J Med Chem. 2007;50:74–82.

[29] Sadowski J, Gasteiger J, Klebe G. Comparison of automatic threedimensional model builders using 639 X-ray structures. J Chem Inf Comput Sci. 1994;34:1000–8.

[30] Güner O, Clement O, Kurogi Y. Pharmacophore modeling and three dimensional database searching for drug design using catalyst: recent advances. Curr Med Chem. 2004;11:2991–3005.

[31] Molecular Operating Environment (MOE) 2012.10 (2012).

[32] McGarrah DB, Judson RS. Analysis of the genetic algorithm method of molecular conformation determination. J Comput Chem. 1993;14:1385–95.

[33] Saunders M. Stochastic exploration of molecular mechanics energy surfaces. Hunting for the global minimum. J Am Chem Soc. 1987;109:3150–2.

[34] Saunders M. Stochastic search for the conformations of bicyclic hydrocarbons. J Comput Chem. 1989;10:203–8.

[35] Chang G, Guida WC, Still WC. An internal coordinate Monte Carlo method for searching conformational space. J Am Chem Soc. 1989;111:4379–86.

[36] Wolber G, Seidel T, Bendix F, Langer T. Molecule-pharmacophore superpositioning and pattern matching in computational drug design. Drug Discov Today. 2008;13:23–9.

[37] Catalyst, Accelrys Inc., http://accelrys.com

[38] Wolber G, Langer T. LigandScout: 3-D pharmacophores derived from protein-bound ligands and their use as virtual screening fi lters. J Chem Inf Model. 2005;45:160–9.

[39] Dixon SL, Smondyrev AM, Rao SR. PHASE: a novel approach to pharmacophore modeling and 3D database searching. Chem Biol Drug Des. 2006;67:370–2.

[40] Lower M, Geppert T, Schneider P, Hoy B, Wessler S, Schneider G. Inhibitors of helicobacter pylori protease HtrA found by 'virtual ligand' screening combat bacterial invasion of epithelia. PLoS One. 2011;6:17986

[41] Rodrigues T, Moreira R, Gut J, Rosenthal PJ, O'Neill PM, Biagini GA, et al. Identification of new antimalarial leads by use of virtual screening against cytochrome bc1. Bioorg Med Chem. 2011;19:6302–8

[42] Vuorinen A, Nashev LG, Odermatt A, Rollinger JM, Schuster D. Pharmacophore model refinement for 11beta-hydroxysteroid dehydrogenase inhibitors: search for modulators of intracellular glucocorticoid concentrations. Mol Inf. 2014;33:15–25.

[43] Bauer J, Waltenberger B, Noha SM, Schuster D, Rollinger JM, Boustie J, et al. Discovery of depsides and depsidones from lichen as potent inhibitors of microsomal prostaglandin E2 synthase-1 using pharmacophore models. ChemMedChem. 2012;7:2077–81

[44] Rollinger JM, Schuster D, Danzl B, Schwaiger S, Markt P, Schmidtke M, et al. In silico target fishing for rationalized ligand discovery exemplified on constituents of Ruta graveolens. Planta Med. 2009;75:195–204

[45] Fakhrudin N, Ladurner A, Atanasov AG, Heiss EH, Baumgarter L, Markt P, et al. Computer-aided discovery, validation, and mechanistic characterization of novel neolignan activators of peroxisome proliferator-activated receptor gamma. Mol Pharmacol. 2010;77:559–66

[46] Atanasov AG, Wang JN, Gu SP, Bu J, Kramer MP, Baumgarter L, et al. Honokiol: a non-adipogenic PPARgamma agonist from nature. Biochim Biophys Acta. 2013;1830:4813–9

[47] Brooijmans N, Kuntz ID. Molecular recognition and docking algorithms. Annu Rev Biophys Biomol Struct. 2003;32:335–73.

[48] Bertoldo JB, Rodrigues T, Dunsmore L, Aprile FA, Marques MC, Rosado LA, et al. A water-bridged cysteine-cysteine redox regulation mechanism in bacterial protein tyrosine phosphatases. Chem. 2017;3:665–77

[49] Filippakopoulos P, Qi J, Picaud S, Shen Y, Smith WB, Fedorov O, et al. Selective inhibition of BET bromodomains. Nature. 2010;468:1067–73

[50] Brik A, Wong CH. HIV-1 protease: mechanism and drug discovery. Org Biomol Chem. 2003;1:5–14.

[51] Cappel D, Sherman W, Beuming T. Calculating water thermodynamics in the binding site of proteins - applications of WaterMap to drug discovery. Curr Top Med Chem. 2017;17:2586–98.

[52] Ladbury JE. Just add water! The effect of water on the specificity of protein-ligand binding sites and its potential application to drug design. Chem Biol. 1996;3:973–80.

[53] Totrov M, Abagyan R. Flexible ligand docking to multiple receptor conformations: a practical alternative. Curr Opin Struct Biol. 2008;18:178–84.

[54] Chen YZ, Zhi DG. Ligand-protein inverse docking and its potential use in the computer search of protein targets of a small molecule. Proteins. 2001;43:217–26.

[55] Do QT, Lamy C, Renimel I, Sauvan N, Andre P, Himbert F, et al. Reverse pharmacognosy: identifying biological properties for plants by means of their molecule constituents: application to meranzin. Planta Med. 2007;73:1235–40

[56] Brown RD, Martin YC. The information content of 2D and 3D structural descriptors relevant to ligand-receptor binding. J Chem Inf Comp Sci. 1997;37:1–9.

[57] Keiser MJ, Roth BL, Armbruster BN, Ernsberger P, Irwin JJ, Shoichet BK, et al. Relating protein pharmacology by ligand chemistry. Nat Biotechnol. 2007;25:197–206

[58] Bajusz D, Racz A, Heberger K. Why is Tanimoto index an appropriate choice for fingerprint-based similarity calculations? J Cheminf. 2015;7:20.

[59] Sheridan RP, Kearsley SK. Why do we need so many chemical similarity search methods? Drug Discov Today. 2002;7:903–11.

[60] Bender A, Jenkins JL, Scheiber J, Sukuru SC, Glick M, Davies JW. How similar are similarity searching methods? A principal component analysis of molecular descriptor space. J Chem Inf Model. 2009;49:108–19

[61] Sa MS, de Menezes MN, Krettli AU, Ribeiro IM, Tomassini TC, Ribeiro dos Santos R, et al. Antimalarial activity of physalins B, D, F, and G. J Nat Prod. 2011;74:2269–72

[62] Poroikov V, Filimonov D, Lagunin A, Gloriozova T, Zakharov A. PASS: identification of probable targets and mechanisms of toxicity. SAR QSAR Environ Res. 2007;18:101–10.

[63] Lagunin A, Stepanchikova A, Filimonov D, Poroikov V. PASS: prediction of activity spectra for biologically active substances. Bioinformatics. 2000;16:747–8.

[64] Filimonov D, Poroikov V, Borodina Y, Gloriozova T. Chemical similarity assessment through multilevel neighborhoods of atoms: definition and comparison with the other descriptors. J Chem Inf Comp Sci. 1999;39:666–70.

[65] Lagunin A, Filimonov D, Poroikov V. Multi-targeted natural products evaluation based on biological activity prediction with PASS. Curr Pharm Des. 2010;16:1703–17.

[66] Zeng X, Zhang P, He W, Qin C, Chen S, Tao L, et al. NPASS: natural product activity and species source database for natural product research, discovery and tool development. Nucleic Acids Res. 2018;46:D1217–D1222

[67] Butler KT, Davies DW, Cartwright H, Isayev O, Walsh A. Machine learning for molecular and materials science. Nature. 2018;559:547–55.

[68] Merk D, Friedrich L, Grisoni F, Schneider G. De novo design of bioactive small molecules by artificial intelligence. Mol Inf. 2018;37:1700153.

[69] Kirchmair J, Williamson MJ, Afzal AM, Tyzack JD, Choy AP, Howlett A, et al. FAst MEtabolizer (FAME): A rapid and accurate predictor of sites of metabolism in multiple species by endogenous enzymes. J Chem Inf Model. 2013;53:2896–907

[70] Tian S, Wang J, Li Y, Xu X, Hou T. Drug-likeness analysis of traditional Chinese medicines: prediction of drug-likeness using machine learning approaches. Mol Pharm. 2012;9:2875–86.

[71] Segler MH, Preuss M, Waller MP. Planning chemical syntheses with deep neural networks and symbolic AI. Nature. 2018;555:604–10.

[72] Mayr A, Klaumbauer G, Unterthiner T, Steijaert M, Wegner JK, Cuelemans H, et al. Large-scale comparison of machine learning methods for drug target prediction on ChEMBL. Chem Sci. 2018;9:5441–51

[73] Reutlinger M, Koch CP, Reker D, Todoroff N, Schneider P, Rodrigues T, et al. Chemically advanced template search (CATS) for scaffold-hopping and prospective target prediction for 'orphan' molecules. Mol Inf. 2013;32:133–8

[74] Schneider P, Schneider G. Collection of bioactive reference compounds for focused library design. QSAR Comb Sci. 2003;22:713–8.

[75] Rodrigues T, Reker D, Kunze J, Schneider P, Schneider G. Revealing the macromolecular targets of fragment-like natural products. Angew Chem Int Ed. 2015;54:10516–20.

[76] Reker D, Perna AM, Rodrigues T, Schneider P, Reutlinger M, Monch B, et al. Revealing the macromolecular targets of complex natural products. Nat Chem. 2014;6:1072–8

[77] Schneider G, Reker D, Chen T, Hauenstein K, Schneider P, Altmann KH, et al. Deorphaning the macromolecular targets of the natural anticancer compound doliculide. Angew Chem Int Ed. 2016;55:12408–11

[78] Rodrigues T, Sieglitz F, Somovilla VJ, Cal PM, Galione A, Corzana F, et al. Unveiling (-)-englerin A as a modulator of L-type calcium channels. Angew Chem Int Ed. 2016;55:11077–81

[79] Schneider P, Schneider G. De-orphaning the marine natural product (+/-)-marinopyrrole A by computational target prediction and biochemical validation. Chem Commun. 2017;53:2272–4.

[80] Schneider P, Schneider G. A computational method for unveiling the target promiscuity of pharmacologically active compounds. Angew Chem Int Ed. 2017;56:11520–4.

[81] Rodrigues T, Werner M, Roth J, da Cruz EHG, Marques MC, Akkapeddi P, et al. Machine intelligence decrypts β-lapachone as an allosteric 5- lipoxygenase inhibitor. Chem Sci. 2018;9:6899–903.

[82] Robke L, Rodrigues T, Schroder P, Foley DJ, Bernardes GJL, Laraia L, et al. Discovery of 2,4-dimethoxypyridines as novel autophagy inhibitors. Tetrahedron. 2018;74:4531–7. DOI: 10.1016/j.tet.2018.1007.1021

[83] Patridge E, Gareiss P, Kinch MS, Hoyer D. An analysis of FDA-approved drugs: natural products and their derivatives. Drug Discov Today. 2016;21:204–7.

[84] Wassermann AM, Lounkine E, Hoepfner D, King FJ, Studer C, Peltier JM, et al. Dark chemical matter as a promising starting point for drug lead discovery. Nat Chem Biol. 2015;11:958–66

Annalisa Maruca, Francesca Alessandra Ambrosio, Antonio Lupia, Isabella Romeo, Roberta Rocca, Federica Moraca, Carmine Talarico, Donatella Bagetta, Raffaella Catalano, Giosuè Costa, Anna Artese and Stefano Alcaro

12 Computer-based techniques for lead identification and optimization I: Basics

Abstract: This chapter focuses on computational techniques for identifying and optimizing lead molecules, with a special emphasis on natural compounds. A number of case studies have been specifically discussed, such as the case of the naphthyridine scaffold, discovered through a structure-based virtual screening (SBVS) and proposed as the starting point for further lead optimization process, to enhance its telomeric RNA selectivity. Another example is the case of Liphagal, a tetracyclic meroterpenoid extracted from *Aka coralliphaga*, known as PI3Kα inhibitor, provide an evidence for the design of new active congeners against PI3Kα using molecular dynamics (MD) simulations. These are only two of the numerous examples of the computational techniques' powerful in drug design and drug discovery fields. Finally, the design of drugs that can simultaneously interact with multiple targets as a promising approach for treating complicated diseases has been reported. An example of polypharmacological agents are the compounds extracted from mushrooms identified by means of molecular docking experiments. This chapter may be a useful manual of molecular modeling techniques used in the lead-optimization and lead identification processes.

Keywords: drug discovery, drug-like properties, virtual screening, pharmacophore models, molecular recognition, molecular dynamics

12.1 Introduction

Drug discovery is the result of an expensive and time-consuming process, since 800 million to 1 billion dollars and a gap of about 15 years are needed for each new marketed compound. In this context, natural products play a critical role in the discovery of bioactive molecules for the treatment of human diseases. In details, for the treatment of cancer and infectious disease, about 70% of new drugs arise from natural sources [1]. The increasing economic pressure on the pharmaceutical industry in the development of new drugs has led to the introduction of computer-aided

This article has previously been published in the journal *Physical Sciences Reviews*. Please cite as: Maruca, A., Ambrosio, F.A., Lupia, A., Romeo, I., Rocca, R., Moraca, F., Talarico, C., Bagetta, D., Catalano, R., Costa, G., Artese, A., Alcaro, S. Computer-based Techniques for Lead Identification and Optimization I: Basics. *Physical Sciences Reviews* [Online] **2019** DOI: 10.1515/psr-2018-0113.

https://doi.org/10.1515/9783110579352-013

molecular design methods in order to ensure a rapid lead structure discovery [2]. Computational techniques represent useful and powerful tools to enhance the rational drug design in the medicinal chemistry field (Figure 12.1). The combination of *in silico* approaches with advances in purification, structure investigation and the streamlining of the screening process has resulted in the reduced timeline associated with the natural drugs, with the aim to improve the treatment of human illnesses [3, 4].

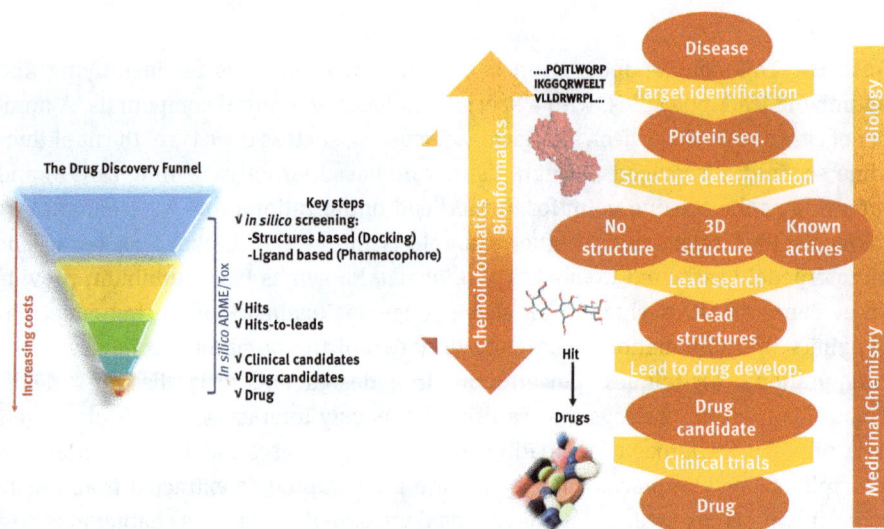

Figure 12.1: The drug design process for lead identification and optimization.

12.2 Virtual screening: A tool for lead identification

Virtual screening (VS) is a powerful technique for the identification of hit molecules. It uses computer-based methods to predict new potential biological binders, with the goal to reduce enormous virtual chemical databases of organic molecules, thus prioritizing compounds with the better chemical features for the biological activity (Figure 12.2). Generally, VS approaches can be distinguished into SBVS [5], ligand-based VS (LBVS) [6] and inverse VS [7]. While SBVS exploits the target structures information as screening templates, LBVS focuses on reference compounds with known activity with the aim to select structurally different molecules which have a similar activity. Meanwhile, in inverse VS, the compounds are tested on a set of targets in order to identify a specific pharmacological activity (set of anticancer, antibacterial, and antiviral targets) [8]. Thus, using several methods, it is possible to concentrate biological tests on just a few promising compounds selected by the drug-likeness properties, such as oral bioavailability, low toxicity,

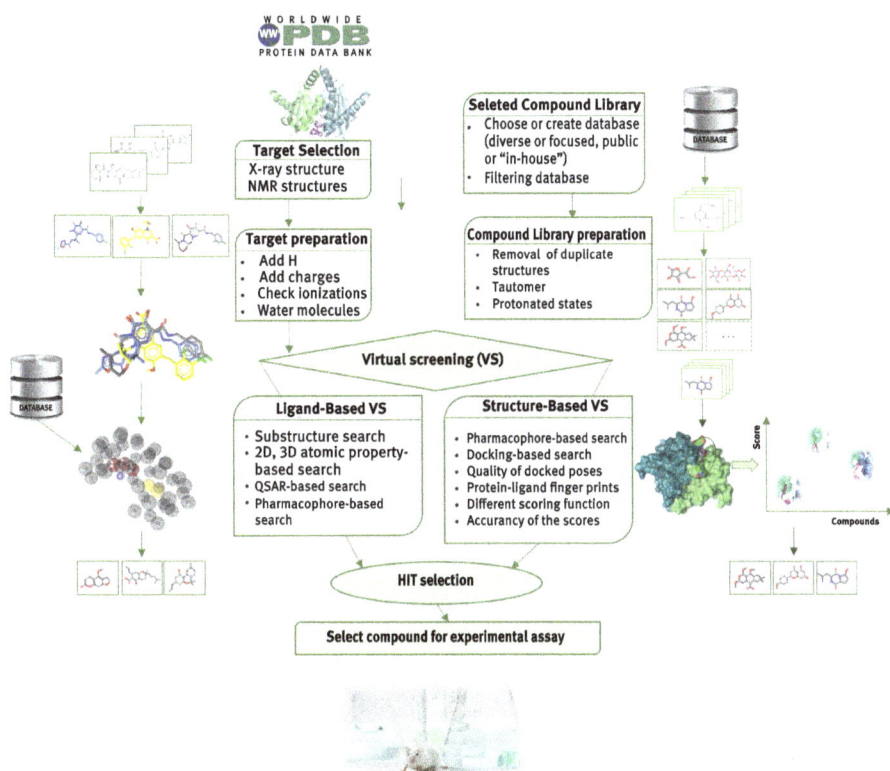

Figure 12.2: Workflow of structure- and ligand-based VS.

membrane permeability, metabolic stability, and to select only those that exhibit favorable ADMET (Absorption, Distribution, Metabolism, Excretion and Toxicity) parameters [9].

12.2.1 Drug-like properties prediction of natural compounds

Drug-like properties comprise structural, physiochemical, biochemical, pharmacokinetic and toxicity characteristics of a compound and provide a powerful source for selecting the "*hits*" that are suitable starting points for research of new candidate in the drug discovery process. The concepts of 'drug-like' and 'lead-like' chemical properties are having a major influence on the selection of compounds for high-throughput screening and in the design of lead generation libraries. Several rules have been applied to ensure 'drug-likeness', with the most popular being the "Lipinski Rule of Five" [10], and the "Jorgensen Rule-of-Three". An entire chapter in this book is dedicated to "drug-likeness". In the last years, a number of computational software able to predict a range of ADMET properties have been released (Table 12.1). An increasing number of *in silico*

Table 12.1: Popular commercial software available for predicting ADME and ADME-related properties (BBB: Blood-brain barrier permeability; Bio: Oral bioavailability; C2: Caco-2 permeability; Mtb: Metabolism; PPB: Plasma-protein binding; Sol: Solubility) [11].

Software	Company	pK$_a$	logP	logD	Sol	C2	BBB	Bio	Mtb	PPB
Cerius2	Accelrys				✓		✓		✓	✓
Qikprop	Schrödinger		✓		✓	✓	✓		✓	✓
ACD/LABS	ACD Labs	✓	✓	✓	✓					
Volsurf	Tripos				✓	✓	✓		✓	✓
ADME Boxer	Pharma Algorithms	✓	✓	✓	✓	✓		✓		✓
ADME Predictor	Simulations Plus	✓	✓	✓	✓		✓			✓
KnowItAll	Bio-Rad Lab	✓	✓	✓	✓		✓	✓		✓
NorayMet ADME	Noraybio	✓	✓	✓	✓	✓			✓	✓
Idea ADME	LION Bioscience				✓	✓	✓	✓	✓	
PreADMET	BMDRC		✓		✓	✓	✓			✓
ADME Collection	Scitegic				✓		✓		✓	✓
Jchem	ChemAxon	✓	✓	✓						
StarDrop	Biofocus	✓	✓	✓	✓		✓		✓	✓
ADME-WORKS	Fujitsu		✓		✓		✓		✓	
Meta	Multicase								✓	
MetaSite	Molecular Discovery								✓	

models for these purposes have been reported, including solubility, molecular weight, logP, Caco-2 permeability, human intestinal absorption, oral bioavailability, blood–brain partitioning, P-glycoprotein-mediated transport, plasma-protein binding, metabolism, volume of distribution, clearance and even half-life [11].

12.2.2 Drug-likeness of commercially available compound libraries for drug discovery

Many current drug discovery and development projects rely on libraries of small molecules to find novel active compounds against various extracellular and intracellular molecular targets. In recent years, various high-throughput technologies for combinatorial chemistry and parallel chemical synthesis helped to increase the size and structural diversity of such compound collections. Similarly, high-throughput screening (HTS) technologies helped to increase the number of compounds which can be screened against a given target, from just 50–200 thousand compounds before 2000, to over 1.5 million compounds in current HTS campaigns. More drug discovery programs today originate by selection of "*hit*" molecules resulting from assays against large compound screening libraries. As mentioned above, most commercial compounds in chemical libraries have a higher molecular weight and greater hydrophobicity compared to orally available drugs. This shift to higher hydrophobicity and lower solubility is also reflected in compounds with a significant biological activity found in the ChEMBL database, indicating a trend toward larger and more hydrophobic compounds. In particular, drugs that have been recently approved were found to possess much lower median logP values than those previously suggested as optimal [12]. It must be noted that the higher the hydrophorbicity or lipophility of a compound, the higher their logP value. It has been demonstrated that the docking of huge compound libraries, such as PubChem (30 million compounds) and ChemSpider (26 M compounds), is time-consuming and computationally demanding, and results in redundant possibilities with a great burden for compound selection. In addition to filtering for lead-like properties, one should exclude known toxicophores or metabolically liable moieties [13]. Moreover, several novel filters have been recently developed, aiming at the quality enhancement of database content. The Pan Assay Interference Compounds (PAINS) [14] and the ALARM-NMR [15] filters contain compounds found to be chemically reactive and assay interfering, appear as frequent hitters, and are not identified by toxicophoric filters. Therefore, a combination of filtering for desired pharmacological and ADMET properties is advisable early in the drug design process.

12.2.3 Software tools for pre-filtering of the compounds libraries

To filter in an efficient manner a library of compounds against such criteria, several online tools have been developed. Chembioserver is a publicly available online application specializing in filtering and selection of small molecules [16]. The objective of this application is to facilitate compound preparation prior to (or after) VS computations, by using (i) basic search, (ii) filtering (steric clashes and toxicity), (iii) advanced filtering based on custom chosen physicochemical properties, (iv) clustering (according to structure and compound physicochemical properties providing representative compounds for each cluster), (v) customized pipeline and (vi) visualization of compounds properties through property graphs and thus, increase the efficiency and the quality of compounds that proceed to *in vitro* assaying.

FAF-Drugs2 server is a public tool for computationally filtering compounds by considering ADMET and physicochemical properties and identifying key functional or undesirable moieties, by choosing from 23 physicochemical and 204 substructure-searching rules [17]. Finally, yet importantly, this tool provides numerous distribution diagrams of the properties of the filtered compounds in a web-server version [18].

12.2.4 Compound libraries preparation

Selected compound libraries should be also pre-processed in realistic 3D representations. Therefore, the compound set to be used for SBVS should have realistic bond lengths and angles as these may not change during docking. Furthermore, it must be free from counter-ions, metals and solvent molecules. Thus, all compounds need to have assigned bond order and filled valences, partial charges, an appropriate proto-nation state at physiological pH or at the pH of the interest and proper tautomeric states [19, 20]. However, it was shown that consideration of all possible tautomers (stable and unstable) yields slightly poorer results than including only the most stable form in water [21]. Docking software usually include ligand preparation sections, such as AutoDock Tools [22], LigPrep within the Schrödinger suite [23], MOE [24] and the MAPS platform [25]. Moreover, stand-alone programs exist for ligand preparation, such as DISI [26], Pipeline pilot [27], or Hyperchem [28]. Grinter et al. have created a workflow through custom-made script files for preparing large libraries for docking, including scripts for ligand preparation, freely offered to the academic community [29].

12.2.5 Pharmacophore

The identification of common chemical features of molecules was the starting point for the introduction of the pharmacophore concept in 1900s by Paul Ehrlich, namely the "father of drug discovery". Later in the 1970s, Camille G. Wermuth *et al.* intro-duced the actual definition of the term: *"A pharmacophore is the ensemble of steric and electronic features that ensure the optimal supramolecular interactions with a specific biological target, triggering or blocking the relative biological response"* [30]. Therefore, by using a 3D pharmacophore, it is possible to describe the common chemical features of active ligands or the interaction keys between a ligand and its biological target, aligned in three-dimensional space. These interactions can be summarized as follows: hydrogen-bond donor and acceptors (HBD, HBA); hydro-phobic (HYD); negative and positive charges (POS/NEG CHARGEs); negative and positive ionizable (NI, PI); ring aromatic (RA). The basic pharmacophore features are shown in Figure 12.3. After generating the pharmacophore model, the quality of the hypothesis may be validated by means of different methods [32].

12.2.5.1 Assessing the performance of pharmacophore models
Pharmacophores are very useful in VS for hit identification. However, a big issue is that in many cases a few percentages of the selected virtual hits are bioactive, and the screening results bear a higher "false negative" rate. False negatives are bioactive compounds which are not identified on the hit list during VS. To address this problem, a comprehensive validation and optimization to the pharmacophore model are required.

A series of descriptors, capable to estimate the enrichment of active molecules from a database containing active molecules and decoys, for which activity is not known (Figure 12.3). This seems to be a useful method to evaluate the VS perfor-mance. Some of these measures are:

Hydrogen Bond Donor		**Aromatic Ring**
Hydrogen Bond Acceptor		**Iron Binding Location**
Positive Ionizable Area		**Zinc Binding Location**
Negative Ionizable Area		**Magnesium Binding Location**
Hydrophobic Interactions		

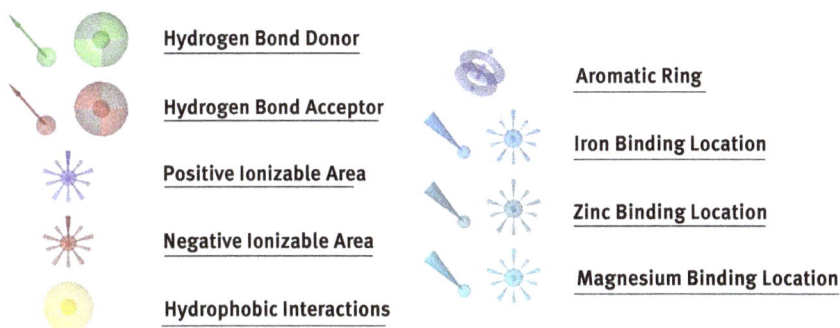

Figure 12.3: Basic pharmacophore features of LigandScout [31]. Figure reproduced from LigandScout Software Manual (http://www.inteligand.com/ligandscout) with permission to republish from InteLigand GmbH, Austria.

- **Sensitivity (Se):** as reported in eq. (12.1), it describes the ratio of the active molecules number found by VS method to the number of all active database compounds,

$$Se = \frac{N \text{ selected actives}}{N \text{ total actives}} = \frac{TP}{TP + FN} \tag{12.1}$$

where TP is the number of true positive compounds and FN are the false negatives. TP are bioactive compounds which appear on the VS hit list.

- **Specificity (Sp):** represents the *ratio* between number of inactive compounds not selected by the VS protocol, and the number of all inactive molecules included into the chemical database eq. (12.2) [33],

$$Sp = \frac{N \text{ discarded inactives}}{N \text{ total inactives}} = \frac{TN}{TN + FP} \tag{12.2}$$

where TN and FP are true negatives and false positive, respectively. FP represent decoys (false positive compounds) which appear on the hit list.

Specificity ranges from 0 to 1 and denotes the percentage of truly inactive compounds. $Sp = 0$ defines the worst-case scenario where all inactives are selected by error as actives, whereas $Sp = 1$ means that all inactive compounds have been correctly rejected during the screening process.

- **Enrichment Factor (EF):** calculates the fraction of retrieved known actives in the top x % percent of the ranked list, compared to the *ratio* between binders and decoys in the entire database (Figure 12.4). EF is defined in eq. (12.3):

$$EF = \frac{TP/n}{A/N} \tag{12.3}$$

where *TP* is the number of true positive, *n* is number of molecules from a database containing *N* entries and *A* are the active molecules.

A disadvantage of the *EF* is its high dependency on the *ratio* of active compounds of the screened database; furthermore, enrichment curves highlight only one of the two aspects depicted by the ROC plots, which is sensitivity (Figure 12.4).

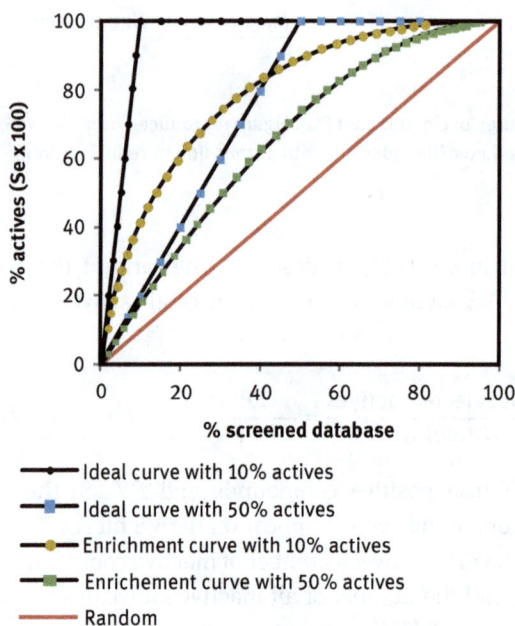

Figure 12.4: Theoretical enrichment curves of the same test for two different ratios of actives. This graph illustrates the difficulty of comparing test performances by taking into account enrichment curves [34].

- **ROC Curve (Receiver Operating Characteristic):** represents ideal distributions, where no overlap between the scores of active molecules and decoys exists (Figure 12.5); it proceeds from the origin to the upper-left corner until all the actives are retrieved and *Se* reaches the value of 1 [34].
- **AUC (Area Under the Curve)** is the sum of all rectangles formed by the S_e and $1-S_p$ values for the different thresholds. Threshold S_i is the score of the ith active molecule eq. (12.4).

$$AUC = \sum_i [(Se_{i+1})(Sp_{i+1} - Sp_i)] \tag{12.4}$$

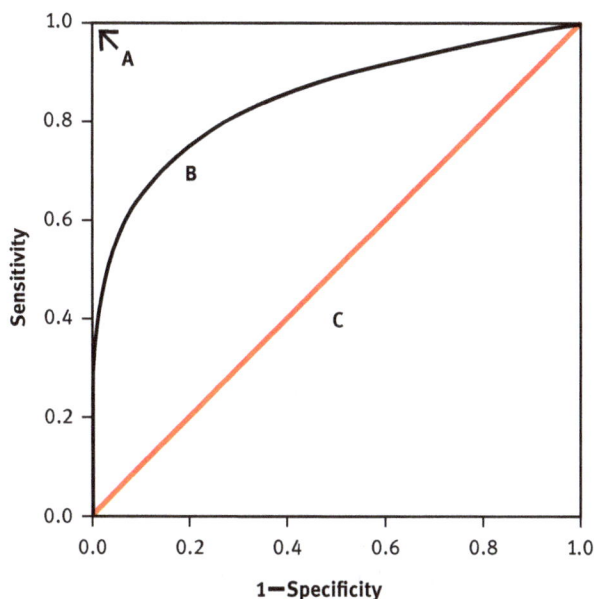

Figure 12.5: Three hypothetical ROC curves. The diagnostic accuracy of the gold standard (lines A; AUC=1), a typical ROC curve (curve B; AUC=0.85), and a diagonal line corresponding to random chance (line C; AUC=0.5) [35].

AUC values between 0.5 and 1.0 means that the used VS workflow is able to operate a good discrimination between actives and *decoys* [34].

- *RIE* **(Robust initial enhancement)** provides a descriptor that does not suffer from large value variations if only a small number of actives are investigated. For these reasons, the rank of the ith active molecule is correlated to the number of scored compounds investigated a (eq. (12.5)).

$$S = \sum_{i=1}^{\text{actives}} \exp(-\text{rank}(i)/a) \tag{12.5}$$

If the RIE values is greater than 1, then the VS method is capable to score active compounds higher than a random distribution [36].

- *BEDROC* **(Boltzmann-enhanced discrimination of ROC)** is a novel descriptor developed in 2007 by Truchon and collaborators. It's a generalized AUC descriptor that includes a decreasing exponential weighting function that focuses on active molecules ranked at the beginning of the ordered list eq. (12.6) [32].

$$BEDROC = RIE \times \frac{R_a \sinh\left(\frac{\alpha}{2}\right)}{\cosh\left(\frac{\alpha}{2}\right) - \cosh\left(\frac{\alpha}{2} - \alpha R_a\right)} + \frac{1}{1 - e^{\alpha(1-Ra)}} \approx \frac{RIE}{\alpha} + \frac{1}{1 - e^{\alpha}}, \quad (12.6)$$

In order to obtain comparable BEDROC values from VS workflows with different underlying distributions of actives and *decoys*, αR_a should be smaller than 1, so that the BEDROC descriptor is independent from the *ratio* of actives. It is recommended a value of 20 for α if the BEDROC descriptor is used for the evaluation of VS methods. This means that the first 8 % of the relative rank contributes to 80 % of the BEDROC value [32].

12.2.5.2 Application methods of the pharmacophore models

There are two main approaches for developing pharmacophore models: Ligand-based and Structure-based design [37]. The ligand-based approach aims at identifying common chemical features of a group of well-known molecules in order to know ligands that are likely to interact with the target [38]. The fundamental step of the ligand-based approach is the conformational analysis of ligands and the molecular alignment. For this purpose, there are a considerable number of advanced algorithms to sample the conformational search, for example, ConfGen, MonteCarlo, OMEGA, Corina [39]. The structure-based approach requires the existence of a 3D structure of the target. The aim of this approach is to exploit the molecular recognition between a ligand and a biological structure, to select out chemical entities that strongly bind to the active sites of biologically relevant targets for which the 3D structures are known or inferred. In such an approach, docking methods assume a primary role. Some of the most widely used docking programs in virtual high-throughput searches are DOCK [40], FlexX [41], GOLD [42], Glide [43], ICM [44], FRED [45] and AutoDock [22]. One of the advantages of these approaches is the possibility to carry out the VS without ligand information, but considering the characteristics (shape, size, chemical and electrostatic features) of the binding pocket. It could be possible to find additional sites whose targeting could produce a desired biological response. Among the software that allow identifying and characterizing binding sites, we can cite SiteMap [46] and FLAPsite [47]. SiteMap provides quantitative and graphical information with the aim to improve the pockets' analysis and help to select virtual binders in a lead discovery application. FLAPsite uses similar approaches but different metrics to evaluate the quality of a putative binding site. The algorithm uses the GRID force field to calculate molecular interaction fields (MIFs) [48], which are then used to identify energetically favorable binding sites on a target for probes and use these molecular maps to design optimized ligands. With the time machine evolution, a variety of automated tools for pharmacophore modeling emerged, and thanks to these software a huge amount of *in silico* studies were carried out using Ligand-based (LB) and Structure-based (SB) approaches. There are several software, such as

LigandScout [31], FLAPpharm [49], Catalyst [50] and MOE [24] which can automatically generate potential pharmacophores from a list of known active molecules.

12.2.6 Successful applications of VS

In the following paragraphs, we will discuss successful applications of VS methods used to identify new potential compounds. Here we report a short overview, based on recent publications, highlighting interesting aspects of VS methods applied to identify bioactive natural compounds. For more details, we refer to the original literature.

12.2.6.1 From nature to virtual multi-target activity

Recently, the scientific interest in natural compounds extracted from the fungal species has increased because these compounds are also known to have pharmacological/biological activities. The reported case study is a clear example of how an *"in-house"* chemoinformatics database should be built, starting from a deep knowledge of natural mushroom extracts, and how it can be applied for a VS campaign. All data were collected from the available literature at the National Center for Biotechnology Information (NCBI) website [51], by obtaining a database of fungal extracts with more than 1103 natural compounds present in 162 different fungal species. The 2D chemical structures of the identified fungal extracts were downloaded from free access websites such as PubChem [52], ChEMBL [53] and ChemSpider [54].

Missing chemical structures were integrated using MarvinSketch [55] InstantJchem [56] was used for computing physico-chemical properties and for adding CAS number, IUPAC names, doi, and references. All ligands were optimized and, subsequently, PAINS were removed in order to reduce the number of false positive hits (Figure 12.6). The results obtained by means of molecular docking experiments highlighted those compounds with a good theoretical binding affinity against targets involved in several complicated diseases. Indeed, chemical scaffolds from mushrooms provide realistic opportunities for finding new lead structures for medicinal chemistry possessing. The identified compounds showed theoretical anti-cancer, anti-inflammatory and anti-neurodegenerative effects, while some others could be good starting points for the treatment of metabolic disease. Among these molecules, some fungal hits were able to non-selectively interact with most of the targets with a good theoretical binding affinity. These non-selective molecules could be considered as an example of compounds with off-target activity. Indeed, VS methods can be helpful to identify sources of off-target drug effects, investigating their adverse or desirable side effects, obtained as a result of modulation of other targets. The construction of the chemoinformatic database, which includes new chemical entities and new experimental/theoretical activities data linked to known compounds, has led to deeper knowledge about species

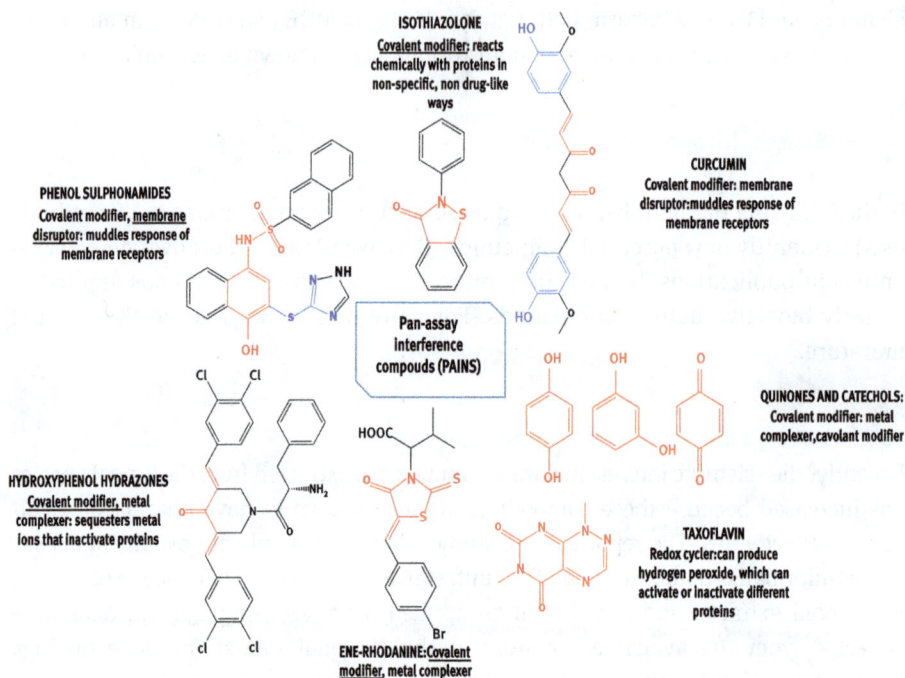

Figure 12.6: The chemical substructures of PAINS (reactive portions are shown in blue and red) [4].

that are still poorly explored or unknown. The chemical structures of these compounds that could have a second chance will be made available online on a publicly accessible website [57–59].

12.2.6.2 Natural compounds active on nucleic acids

Progress in computer sciences applied to molecular drug discovery fields allowed to carry out a large number of studies on different kinds of molecular targets, including both nucleic acids and proteins.

In a recent publication, the authors reported some innovative scaffolds able to selectively bind specific DNA and RNA G-quadruplex isoforms [60]. A SBVS workflow was applied to telomeric tetraplex forms of DNA (Tel$_2$) and RNA (TERRA$_2$), involved in fundamental biological processes, thus representing a very interesting anti-cancer target. 3D Structure-based pharmacophores were developed by means of the software, LigandScout best-ranked docked pose of 9 actives obtained against the previously validated ensemble receptor. Each of the validated pharmacophore models was used to perform VS separately using a multi conformational database of natural compounds. About 257,000 natural compounds were screened and, after a careful visual inspection analysis of the best docking poses and the evaluation of their commercial availability, the best 20 final hits were purchased and submitted to biological tests.

It is clear how by combining molecular modeling and biophysical studies, novel promising stabilizers against DNA and RNA telomeric G-quadruplex forms can be retrieved (Table 12.2). In particular, a naphthyridine scaffold was proposed as the most interesting one against $TERRA_2$ and could be the starting point for a further lead optimization to enhance its telomeric RNA selectivity. On the other hand, two derivatives with a furochromen and a benzofuran scaffolds preferentially recognized Tel_2, thus their affinity profile could be investigated also against other G4s topologies.

Table 12.2: In bold, the 20 hits selected from molecular modelling studies and submitted to biophysical tests [60].

$TERRA_2$		Tel_2	
Hit #	ΔT_m (°C)	Hit #	ΔT_m (°C)
1	<1.0	7	8.3 ± 0.3
2	<1.0	8	<1.0
3	<1.0	9	<1.0
4	<1.0	10	<1.0
5	<1.0	11	<1.0
6	<1.0	12	<1.0
7	6.8 ± 0.4	13	<1.0
		14	<1.0
		15	9.0 ± 0.3
		16	<1.0
		17	9.1 ± 0.2
		18	1.8 ± 0.2
		19	<1.0
		20	<1.0

ΔT_m of $TERRA_2$ RNA G4 (4 µM single strand concentration) and Tel_2 DNA G4 (8 µM single strand concentration) in 100 mM K^+, in the presence of the compounds (8 µM and 16 µM, respectively), measured by melting CD spectroscopy. ΔT_m values were reported in Celsius degrees (°C).

12.3 Molecular mechanics simulations

Molecular mechanics simulations are pivotal tools aimed to understand the physical basis of the structure and function of biological targets. The hinge bending modes in the opened and closed conformations of the active sites, the flexibility of tRNA, the impact of configurational entropy in proteins and nucleic acids and the fluctuations requested for ligand entry and exit in the heme proteins represent just a few examples [61–64]. It is important to bear in mind that the comparisons of simulation and experimental data help to test the accuracy of the calculated results and to give parameters in order to improve the methodology. Three types of simulation methods are applied for a long time: the first is characterized by simulation simply with a view to sampling configuration space, by using molecular dynamics with simulated annealing protocols in order to refine structures with data obtained from experiments; the second is involved in the investigation of the system at equilibrium, comprising structural and motional features and the values of thermodynamic parameters, analyzing the conformational space with appropriate Boltzmann factor; the third type uses simulations with the aim to examine the actual dynamics and must be done so that it properly represents the development of the model over time. Monte Carlo simulations and Molecular Dynamics are useful for the first two types, meanwhile for the third only the Molecular Dynamics can be used due to the importance of the structural motion over time [65].

12.3.1 Monte Carlo simulations

The aim of the Monte Carlo (MC) simulation is to collect an ensemble of representative configurations within particular thermodynamics conditions for a macromolecular system, not considering the time evolution, but evaluating the probability distributions [66].

Among the key elements of the MC simulations, there are the partition functions skilled to identify the number of microstates associated with a macromolecule, the probability of occurrence of a specific conformation and the ensemble averages of observables. The number of the microstates is related to the entropy and the Boltzmann constant. As a result, the ensemble average of an observable is derived from the weight of the various achievements of the observable by their related probability. Afterwards, with the MC integration process is possible to efficiently calculate the multidimensional integrals related to the probability and the partition function. In agreement with Markovian process, the integral is approximated by the average of the corresponding sample states. Thus, each state is defined from the previous one. However, from the experimental point of view, the micro-canonical partition function is not realistic. For this reason, the simulations are performed at constant volume and temperature (canonical ensemble, NVT) or at constant pressure and temperature (isobaric-isothermal ensemble, NPT).

In certain cases, in order to activate the process of biomolecular systems, there is a high-energy barrier to overcome the initial and final states [67]. To accomplish this, the replica exchange is used although this is computationally expensive [68]. In details, the replicas represent a certain number of non-interacting simulations, performed in parallel and with its own temperature: the local minima are explored by low temperature simulations, meanwhile with high temperature simulations is possible to overcome energy barriers and move in between local minima. Also, in the replica exchange, the replicas are periodically swapped considering the acceptance probability, with the difference that each state is marked by its own temperature. At the end of this process, the simulations resume as long as another exchange is performed, thus obtaining a better sampling of the macromolecular states. In the scientific literature there are several cases reporting the importance of the conformational analysis to gain insight on the relation between the role of conformational flexibility of drugs and their activity. Larsen *et al.* in a review summarized the relevance of the MC conformational search of polyketides, an important class of natural secondary metabolites characterized by a wide range of structural features that play a key role in defining their conformational preference through fundamental steric, electronic and electrostatic interactions [69]. In particular, in this review different examples wherein conformational analysis complements classic structure-activity relationship in the design of bioactive natural products analogues are reported, such as bryostatin, laulimalide etc.

12.3.2 Molecular Dynamics: A classical tool driving drug discovery

MD is a method to compute the equilibrium and transport properties of a classical many-body system, considering the laws of classical mechanics for the nuclear motion of the constituent particles [70]. This method investigates the temporal evolution of the coordinate and the state of a selected macromolecular target, creating the trajectory by solving Newton's equation of motion ($F = ma$). The interaction between non-bonded atoms are established from standard potentials, electrostatic (Coulomb) and Lennard-Jones potential (Van der Waals), bonded-atoms interactions are considered as springs, using harmonic potentials with equilibrium distances and angles. These parameters, in addition to charges and force constants, are grouped in a functional form known as force-field. The reader is invited to consult specialized textbooks on force-fields and MD.

The most popular potentials are CHARMM, AMBER and OPLS applied to analyze proteins, lipids, ethers and carbohydrates [71].

In order to refine experimentally determined structures, the local optimization is advisable to relax the system, which presents distorted bond length and bond angles. The most used algorithms are the steepest descent algorithm [72] and the conjugate gradient algorithm [73], based on the gradient and the Newton-Raphson method established on the Hessian [74].

Instead, when the global optimization is needed, the simulated annealing is the most appropriate MC method in which the position of atoms undergoes little random displacements [75]. At first, the increased temperature leads to the transitions from lower to higher energy with no negligible probability to escape local minima. After, the decreased temperature allows for the reduction of the occurrence of such a transition. In an MD simulation, the solvation of the whole system represents an essential condition and can be implicit or explicit.

As concerns the implicit solvation, the water molecules are replaced by a potential that defines their average action, meanwhile regarding the explicit solvation, the system is surrounded by a water box, by choosing a range of 10–20 Å of water buffer between the outline of the structure and the box size [76]. Afterwards, with the purpose to neutralize the net charge of the system, counterions are also added. For reproducing the physiological conditions, the next step is to add the thermostat and barostat (constant temperature and pressure, respectively). For this reason, consistent with the kinetic theory of gases, the initial velocities of atoms are produced randomly using a Maxwell-Boltzmann distribution at the initial low temperature. Firstly, it is recommended to use the NVT ensemble to gradually heat the system, thus avoiding spikes in the temperature, which for example could lead to the inversion of chiral centers. After reaching the production temperature, the barostat is added and the pre-equilibration dynamics is carried out. Due to the finite dimensions of the system, the solvation box presents unnatural boundary effects, such as the periodic boundary conditions (PBC). In this way, the system box is virtually embedded in each direction by copies of itself in order that the atoms can interact with atoms in another PBC image without the containment effects. In order to better calculate the electrostatic interaction, the *Particle Mesh Ewald* (PME) method has been developed, splitting the *cut-off* for the calculation of non-bonded interactions in short and long-range regions [77].

12.3.3 MD-derived successful stories of natural compounds

In the following paragraph, we will discuss MD-based approaches helpful to investigate binding modes, providing thermodynamic data being crucial for the lead optimization step and further drug discovery process. Herein, we reported short case studies, based on recent publications, highlighting interesting aspects of MD methods applied to natural compounds. For more details, we refer to the original literature [78–87].

The study of molecular interactions of natural compounds by means of MD simulations is a potential tool to get new insights of the driving forces for the observed activity. In particular, MD runs are performed after docking in order to evaluate the *in silico* predicted binding modes of the top-ranking compounds as a final filter to guide lead optimization [78]. In 2014, in order to identify new potential natural anticancer compounds, a high-throughput in silico screening of commercially available alkaloid derivative databases by means of a structure-based approach based on docking and MD simulations against the human telomeric and the *c-myc* promoter sequence G-quadruplex structures was performed [79]. Among the best hits, a berberine derivative, already known to remarkably inhibit telomerase activity, was identified. With the aim to explain the binding stability of this compound and to clarify the role of its side chain, MD simulations were performed using NAMD program [80]. In particular, MD confirmed the stabilization of both targets bound to the berberine derivative and the better affinity of this compound for *c-myc*, associated to the van der Waal's term as the major contribution, accordingly to its sandwiched binding mode [79].

Another attractive anticancer target is the phosphatidylinositol 3-kinase α (PI3Kα). Natural products are, in fact, widely used as chemical probes to unfold the biological aspects of PI3K signaling [81, 82]. In particular, liphagal, a tetracyclic meroterpenoid extracted from *Aka coralliphaga*, resulted as a potent PI3Kα inhibitor, with tenfold selectivity over PI3Kγ [83, 84]. Furthermore, the dynamic behavior of PI3K binding with liphagal was explored, leading to the identification of the most favorable binding mode among the alternate orientations [85]. Finally, the key interactions for the affinity of liphagal to PI3K were determined, providing an evidence for the design of new active congeners of liphagal against PI3Kα [86].

Interestingly, MD simulations can be also useful for investigating the structural manifestations of membrane-bound protein-ligand complexes at the atomic level. For example, by means of the MD simulation of bryostatin 1/protein kinase C (PKC) complex, Ryckbosch *et al.* proposed a method to elucidate the ligand-induced positioning of ligand-protein complexes in the membrane. Bryostatin 1 is a marine macrolide lactone and binds to the cysteine-rich domains of PKC, resulting in its activation and translocation to the cell membrane. To explain how the ligand-exposed surface might influence the dynamics of PKC-ligand complexes within the membrane, a long-timescale all-atom MD simulation of PKC-bryostatin 1 membrane complex was performed. This MD run elucidated the positioning of the complex within the membrane, linked to interactions with water network and membrane lipid head groups, which were impossible to address without MD simulations [87]. These findings might enable new structural approaches for the design of more effective and simpler analogs of known natural PKC analogues.

12.4 Conclusions

Natural drug discovery was particularly significant as a strategic source for the discovery and the development of new medicines. As described in this chapter, the synergy between experimental studies and computational approaches is increasing and crucial to improve the treatment of human illnesses. In particular, VS approaches coupled to molecular docking and dynamics are very helpful to perform a systematic search of bioactive natural compounds towards their molecular target of interest. In addition, the increasing number and diversity of new natural databases certainly ensures that natural compounds will remain an attractive starting point for discovery and design of optimized lead molecules, which may advance to approved drugs. Till now, drugs are thought as magic bullets characterized by high selectivity and potency toward a single target of interest, but it is clear now that complex diseases, such as cancer, may require a multi-target therapeutic approach. For this purpose, natural compounds offer higher potential efficacy because of their pre-validated polypharmacological profile with respect to synthetic drugs. In this chapter, we have shown a growing interest in predictive basic computational approaches to facilitate research on natural drugs to unravel unmet medical needs. However, the discovery of novel natural (as well as synthetic) drugs has improved, in recent years, thanks to the development of more advanced computational techniques, known as enhancing sampling methods. Their utility in the drug discovery process, especially when the protein-ligand binding free energy needs to be accurately characterized, has been demonstrated by several literature data [88–90]. A detailed discussion about such enhanced sampling techniques and their applications will be addressed in the next chapter.

Acknowledgements: This work was partially supported by Prof. Francesco Ortuso. The authors also gratefully acknowledge the helpful comments and suggestions of the book editor and reviewers, which have improved the presentation.

References

[1]   Newman DJ, Cragg GM. Natural products as sources of new drugs from 1981 to 2014. J Nat Prod. 2016;79:629–61.

[2]   Langer T, Hoffmann RD. Virtual screening: an effective tool for lead structure discovery? Curr Pharm Des. 2001;7:509–27.

[3]   Artese A, Alcaro S, Moraca F, Reina R, Ventura M, Costantino G, et al. State-of-the-art and dissemination of computational tools for drug-design purposes: a survey among Italian academics and industrial institutions. Future Med Chem. 2013;5:907–27.

[4]   Maruca A, Moraca F, Rocca R, Molisani F, Alcaro F, Gidaro MC, et al. Chemoinformatic database building and in silico hit-identification of potential multi-targeting bioactive compounds extracted from mushroom species. Molecules. 2017;22:1571.

[5]   Lionta E, Spyrou G, Vassilatis DK, Cournia Z. Structure-based virtual screening for drug discovery: principles, applications and recent advances. Curr Top Med Chem. 2014;14:1923–38.

[6]   Ripphausen P, Nisius B, Bajorath J. State-of-the-art in ligand-based virtual screening. Drug Discov Today. 2011;16:372–6.

[7]   Chen YZ, Zhi DG. Ligand-protein inverse docking and its potential use in the computer search of protein targets of a small molecule. Proteins. 2001;43:217–26.

[8]   Lauro G, Masullo M, Piacente S, Riccio R, Bifulco G. Inverse Virtual Screening allows the discovery of the biological activity of natural compounds. Bioorg Med Chem. 2012;20:3596–602.

[9]   Vyas V, Jain A, Jain A, Gupta A. Virtual screening: a fast tool for drug design. SciPharm. 2008;76:333–60.

[10] Lipinski CA, Lombardo F, Dominy BW, Feeney PJ. Experimental and computational approaches to estimate solubility and permeability in drug discovery and development settings. Adv Drug Deliv Rev. 2001;46:3–26.

[11] Cheng F, Li W, Liu G, Tang Y. In silico ADMET prediction: recent advances, current challenges and future trends. Curr Top Med Chem. 2013;13:1273–89.

[12] Zuegg J, Cooper MA. Drug-likeness and increased hydrophobicity of commercially available compound libraries for drug screening. Curr Top Med Chem. 2012;12:1500–13.

[13] Blagg J. Structure–activity relationships for in vitro and in vivo toxicity. Annu Rep Med Chem. 2006;41:353–68.

[14] Baell JB, Holloway GA. New substructure filters for removal of pan assay interference compounds (PAINS) from screening libraries and for their exclusion in bioassays. J Med Chem. 2010;53:2719–40.

[15] Metz JT, Huth JR, Hajduk PJ. Enhancement of chemical rules for predicting compound reactivity towards protein thiol groups. J Comput Aided Mol Des. 2007;21:139–44.

[16] Athanasiadis E, Cournia Z, Spyrou G. ChemBioServer: a web-based pipeline for filtering, clustering and visualization of chemical compounds used in drug discovery. Bioinformatics. 2012;28:3002–3.

[17] Lagorce D, Sperandio O, Galons H, Miteva MA, Villoutreix BO. FAF-Drugs2: free ADME/tox filtering tool to assist drug discovery and chemical biology projects. BMC Bioinf. 2008;9:396.

[18] Lagorce D, Maupetit J, Baell J, Sperandio O, Tufféry P, Miteva MA, et al. The FAF-Drugs2 server: a multistep engine to prepare electronic chemical compound collections. Bioinformatics. 2011;27:2018–20.

[19] Kalliokoski T, Salo HS, Lahtela-Kakkonen M, Poso A. The effect of ligand-based tautomer and protomer prediction on structure-based virtual screening. J Chem Inf Model. 2009;49:2742–8.

[20] Sadowski J, Rudolph C, Gasteiger J. The generation of 3D-models of host-guest. Anal Chim Acta. 1992;265:233–41.

[21] Milletti F, Vulpetti A. Tautomer preference in PDB complexes and its impact on structure-based drug discovery. J Chem Inf Model. 2010;50:1062–74.

[22] Morris GM, Huey R, Lindstrom W, Sanner MF, Belew RK, Goodsell DS, et al. AutoDock4 and AutoDockTools4: automated docking with selective receptor flexibility. J Comput Chem. 2009;30:2785–91.

[23] LigPrep. Version 3.9. New York (USA): Schrödinger LLC, 2016.

[24] Canada, H3A 2R7: Montreal, QC. Molecular Operating Environment (MOE), St West, Suite #910, 2013.

[25] MAPS. Version 3.4. Paris, France: Scienomics SARL, 2014.

[26] http://wiki.uoft.bkslab.org/index.php/Preparing_the_ligand.

[27] http://www.accelrys.com.

[28] Hyperchem. Gainesville, FL: HyperCube.

[29] Grinter SZ, Yan C, Huang SY, Jiang L, Zou X. Automated large-scale file preparation, docking, and scoring: evaluation of ITScore and STScore using the 2012 community structure-activity resource benchmark. J Chem Inf Model. 2013;53:1905–14.

[30] Wermuth CG, Ganellin CR, Lindberg P, Mitscher LA. Glossary of terms used in medicinal chemistry (IUPAC Recommendations 1998).

[31] Wolber G, Langer T. LigandScout: 3-D pharmacophores derived from protein-bound ligands and their use as virtual screening filters. J Chem Inf Model. 2005;45:160–9.

[32] Kirchmair J, Markt P, Distinto S, Wolber G, Langer T. Evaluation of the performance of 3D virtual screening protocols: RMSD comparisons, enrichment assessments, and decoy selection—what can we learn from earlier mistakes? J Comput Aided Mol Des. 2008;22:213–28.

[33] Braga RC, Andrade CH. Assessing the performance of 3D pharmacophore models in virtual screening: how good are they? Curr Top Med Chem. 2013;13:1127–38.

[34] Triballeau N, Acher F, Brabet I, Pin JP, Bertrand HO. Virtual screening workflow development guided by the "receiver operating characteristic" curve approach. Application to high-throughput docking on metabotropic glutamate receptor subtype 4. J Med Chem. 2005;48:2534–47.

[35] Zou KH, O'Malley AJ, Mauri L. Receiver-operating characteristic analysis for evaluating diagnostic tests and predictive models. Circulation. 2007;115:654–7.

[36] Sheridan RP, Singh SB, Fluder EM, Kearsley SK. Protocols for bridging the peptide to nonpeptide gap in topological similarity searches. J Chem Inf Comput Sci. 2001;41:1395–406.

[37] McInnes C. Virtual screening strategies in drug discovery. Curr Opin Chem Biol. 2007;11:494–502.

[38] Sun H. Pharmacophore-based virtual screening. Curr Med Chem. 2008;15:1018–24.

[39] Yang SY. Pharmacophore modeling and applications in drug discovery: challenges and recent advances. Drug Discov Today. 2010;15:444–50.

[40] Ewing TJA, Kuntz ID. Critical evaluation of search algorithms for automated molecular docking and database screening. J Comput Chem. 1997;18:1175–89.

[41] Rarey M, Kramer B, Lengauer T, Klebe G. A fast flexible docking method using an incremental construction algorithm. J Mol Biol. 1996;261:470–89.

[42] Jones G, Willett P, Glen RC, Leach AR, Taylor R. Development and validation of a genetic algorithm for flexible docking. J Mol Biol. 1997;267:727–48.

[43] Friesner RA, Banks JL, Murphy RB, Halgren TA, Klicic JJ, Mainz DT, et al. Glide: a new approach for rapid, accurate docking and scoring. Method and assessment of docking accuracy. J Med Chem. 2004;47:1739–49.

[44] Abagyan R, Totrov M. Biased probability Monte Carlo conformational searches and electrostatic calculations for peptides and proteins. J Mol Biol. 1994;235:983–1002.

[45] McGann M. FRED and HYBRID docking performance on standardized datasets. J Comput Aided Mol Des. 2012;26:897–906.

[46] Halgren T. New method for fast and accurate binding-site identification and analysis. Chem Biol Drug Des. 2007;69:146–8.

[47] Baroni M, Cruciani G, Sciabola S, Perruccio F, Mason JS. A common reference framework for analyzing/comparing proteins and ligands. Fingerprints for ligands and proteins (FLAP): theory and application. J Chem Inf Model. 2007;47:279–94.

[48] Goodford PJ. A computational procedure for determining energetically favorable binding sites on biologically important macromolecules. J Med Chem. 1985;28:849–57.

[49] Cross S, Baroni M, Goracci L, Cruciani G. GRID-based three-dimensional pharmacophores I: fLAPpharm, a novel approach for pharmacophore elucidation. J Chem Inf Model. 2012;52:2587–98.

[50] Catalyst. Version 4.11. San Diego, CA, USA: Accelry's Inc., 2007.

[51] https://www.ncbi.nlm.nih.gov/.

[52] Kim S, Thiessen PA, Bolton EE, Chen J, Fu G, Gindulyte A, et al. PubChem substance and compound databases. Nucleic Acids Res. 2016;44:D1202–13.

[53] Bento AP, Gaulton A, Hersey A, Bellis LJ, Chambers J, Davies M, et al. The ChEMBL bioactivity database: an update. Nucleic Acids Res. 2014;42:D1083–90.

[54] http://www.chemspider.com.

[55] https://chemaxon.com/products/marvin.

[56] https://chemaxon.com/products/instant-jchem.

[57] http://www.mutalig.eu/.

[58] http://www.unicz.chemotheca.it.

[59] Ortuso F, Bagetta D, Maruca A, Talarico C, Bolognesi ML, Haider N, et al. The Mu.Ta.Lig. Chemotheca: a community-populated molecular database for multi-target ligands identification and compound-repurposing. Front Chem. 2018;6:130.

[60] Rocca R, Moraca F, Costa G, Nadai M, Scalabrin M, Talarico C, et al. Identification of G-quadruplex DNA/RNA binders: structure-based virtual screening and biophysical characterization. Biochim Biophys Acta Gen Subj. 2017;1861:1329–40.

[61] Brooks B, Karplus M. Harmonic dynamics of proteins: normal modes and fluctuations in bovine pancreatic trypsin inhibitor. Proc Natl Acad Sci USA. 1983;80:6571–5.

[62] Case DA, Karplus M. Dynamics of ligand binding to heme proteins. J Mol Biol. 1979;132:343–68.

[63] Colonna-Cesari F, Perahia D, Karplus M, Eklund H, Bräden CI, Tapia O. Interdomain motion in liver alcohol dehydrogenase. Structural and energetic analysis of the hinge bending mode. J Biol Chem. 1986;261:15273–80.

[64] Harvey SC, Prabhakaran M, Mao B, McCammon JA. Phenylalanine transfer RNA: molecular dynamics simulation. Science. 1984;223:1189–91.

[65] Karplus M, McCammon JA. Molecular dynamics simulations of biomolecules. Nat Struct Biol. 2002;9:646–52.

[66] Fichthorn KA, Weinberg WH. Theoretical foundations of dynamical Monte Carlo simulations. J Chem Phys. 1991;95:1090–6.

[67] Minary P, Levitt M. Probing protein fold space with a simplified model. J Mol Biol. 2008;375:920–33.

[68] Huber T, van Gunsteren WF. SWARM-MD: searching conformational space by cooperative molecular dynamics. J Phys Chem A. 1998;102:5937-43.

[69] Larsen EM, Wilson MR, Taylor RE. Conformation-activity relationships of polyketide natural products. Nat Prod Rep. 2015;32:1183–206.

[70] Allen MP. Computer simulation of liquids. Oxford: Oxford University Press, 2007.

[71] Vanommeslaeghe K, Hatcher E, Acharya C, Kundu S, Zhong S, Shim J, et al. CHARMM general force field: a force field for drug-like molecules compatible with the CHARMM all-atom additive biological force fields. J Comput Chem. 2010;31:671–90.

[72] Arfken G. "The method of steepest descents." § 7.4. Mathematical methods for physicists, 3rd ed. Orlando, FL: Academic Press, 1985:428–36.

[73] Hestenes MR, Stiefel E. Methods of conjugate gradients for solving linear systems. J Res Natl Bur Stand. 1952;49:409–36.

[74] Nocedal J, Wright SJ. Numerical optimization Vol. 35. New York: Springer, 2006.

[75] Kirkpatrick S, Gelatt CD, Vecchi MP. Optimization by simulated annealing. Science. 1983;220:671–80.

[76] Kohn W, Sham LJ. Self-consistent equations including exchange and correlation effects. PhysRev. 1965;140:A1133.

[77] Darden T, York D, Pedersen L. Particle mesh Ewald: an $N \cdot \log(N)$ method for Ewald sums in large systems. J Chem Phys. 1993;98:10089–92.

[78] Zhao H, Caflisch A. Molecular dynamics in drug design. Eur J Med Chem. 2015;91:4–14.

[79] Rocca R, Moraca F, Costa G, Alcaro S, Distinto S, Maccioni E, et al. Structure-based virtual screening of novel natural alkaloid derivatives as potential binders of h-telo and c-myc DNA G-quadruplex conformations. Molecules. 2014;20:206–23.

[80] Phillips JC, Braun R, Wang W, Gumbart J, Tajkhorshid E, Villa E, et al. Scalable molecular dynamics with NAMD. J Comput Chem. 2005;26:1781–802.

[81] Harris SJ, Parry RV, Westwick J, Ward SG. Phosphoinositide lipid phosphatases: natural regulators of phosphoinositide 3-kinase signaling in T lymphocytes. J Biol Chem. 2008;283:2465–9.

[82] Wymann MP, Zvelebil M, Laffargue M. Phosphoinositide 3-kinase signalling which way to target? Trends Pharmacol Sci. 2003;24:366–76.

[83] Sundstrom TJ, Anderson AC, Wright DL. Inhibitors of phosphoinositide-3-kinase: a structure-based approach to understanding potency and selectivity. Org Biomol Chem. 2009;7:840–50.

[84] Marion F, Williams DE, Patrick BO, Hollander I, Mallon R, Kim SC, et al. Liphagal, a Selective inhibitor of PI3 kinase alpha isolated from the sponge akacoralliphaga: structure elucidation and biomimetic synthesis. Org Lett. 2006;8:321–4.

[85] Li T, Wang G. Computer-aided targeting of the PI3K/Akt/mTOR pathway: toxicity reduction and ther-apeutic opportunities. Int J Mol Sci. 2014;15:18856–91.

[86] Gao Y, Ma Y, Yang G, Li Y. Molecular dynamics simulations to investigate the binding mode of the natural product liphagal with phosphoinositide 3-Kinase α. Molecules. 2016;21:857.

[87] Ryckbosch SM, Wender PA, Pande VS. Molecular dynamics simulations reveal ligand-controlled positioning of a peripheral protein complex in membranes. Nat Commun. 2017;8:6.

[88] Limongelli V, Bonomi M, Marinelli L, Gervasio FL, Cavalli A, Novellino E, et al. Molecular basis of cyclooxygenase enzymes (COXs) selective inhibition. Proc Natl Acad Sci USA. 2010;107:5411–6.

[89] Troussicot L, Guillière F, Limongelli V, Walker O, Lancelin JM. Funnel-metadynamics and solution NMR to estimate protein–ligand affinities. J Am Chem Soc. 2015;137:1273–81.

[90] Yuan X, Raniolo S, Limongelli V, Xu Y. The molecular mechanism underlying ligand binding to the membrane-embedded site of a g-protein-coupled receptor. J Chem Theory Comput. 2018;14:2761–70.

Antonio Lupia, Federica Moraca, Donatella Bagetta,
Annalisa Maruca, Francesca Alessandra Ambrosio, Roberta Rocca,
Raffaella Catalano, Isabella Romeo, Carmine Talarico,
Francesco Ortuso, Anna Artese and Stefano Alcaro

13 Computer-based techniques for lead identification and optimization II: Advanced search methods

Abstract: This paper focuses on advanced computational techniques for identifying and optimizing lead molecules, such as metadynamics and a novel dynamic 3D pharmacophore analysis method called *Dynophores*. In this paper, the first application of the funnel metadynamics of the Berberine binding to G-quadruplex DNA is depicted, disclosing hints for drug design, in particular clarifying water's role and suggesting the design of derivatives able to replace the solvent-mediated interactions between ligand and DNA to achieve more potent and selective activity. Secondly, the novel dynamic pharmacophore approach is an extension of the classic 3D pharmacophores, with statistical and sequential information about the conformational flexibility of a molecular system derived from molecular dynamics (MD) simulations.

Keywords: advanced computational techniques, drug discovery, *Dynophores*, funnel metadynamics, pharmacophore models

13.1 Introduction

Understanding the molecular interactions between ligands and macromolecular targets, as well as the accurate prediction of their binding affinity, is of remarkable relevance for the success of a drug discovery process, especially when referring to structure-based drug design and lead optimization. To this aim, a variety of computational methods have been proposed over the years. We will focus our attention on metadynamics and molecular dynamics (MD) combined with pharmacophores *Dynophore*. Metadynamics is one of the most powerful techniques for studying protein-ligand binding, especially when the exact ligand binding pose is unknown. In such cases, in fact, many docking algorithms are widely adopted to predict a large number of ligand binding poses, but the accuracy of their binding affinities is

This article has previously been published in the journal *Physical Sciences Reviews*. Please cite as: Lupia, A., Moraca, F., Bagetta, D., Maruca, A., Ambrosio, F.A., Rocca, R., Catalano, R., Romeo, I., Talarico, C., Ortuso, F., Artese, A., Alcaro, S. Computer-based Techniques for Lead Identification and Optimization II: Advanced Search Methods. *Physical Sciences Reviews* [Online] **2019** DOI: 10.1515/psr-2018-0114.

https://doi.org/10.1515/9783110579352-014

neglected. To overcome such limitations, more accurate methods are needed. In this paper, after a detailed description of the advanced computational algorithm, we will highlight the application of funnel metadynamics (FM), a specialized method to compute the absolute protein-ligand binding free energy. Specifically, we have used FM to study the binding and pre-binding processes of the natural alkaloid Berberine. The estimation of its absolute binding free energy was in good agreement with experimental data. The second approach, *Dynophore*, has been recently developed to improve on the classical concept pharmacophore model providing a more evolved method for molecular design. Classical pharmacophore models, in fact, allow only a static view on a single ligand-target conformation. The aim of *Dynophores* is to develop pharmacophore models taking into account the target flexibility that derives from MDs. Pharmacophore features detected by the MD trajectory are grouped into so-called superfeatures according to feature type, and these are statistically and sequentially analyzed in terms of their occurrence. *Dynophores* can be extremely useful for medicinal chemistry and structural biology, especially to improve dynamic information in pharmacophore-based virtual screening (VS) workflow [1]

13.2 Metadynamics: An enhanced sampling method to accelerate rare events

Although MDs is currently a precious instrument for understanding the mechanisms underlying complex processes, many relevant conformational regions remain unexplored [2]. Indeed, many systems are characterized by several metastable states separated by high energetic barriers, the transition between one state to the other can occur on long timescales, and the whole process is referred to as a "rare event" [3]. Most phenomena of biological interest, including conformational changes in solution, protein folding, chemical reactions as well as ligand–binding, belong to the so-called rare events due to the fact that they take place on large time scales ranging from microseconds to seconds. Therefore, an exhaustive exploration of the conformational space is not accessible within the standard MD simulation time (Figure 13.1).

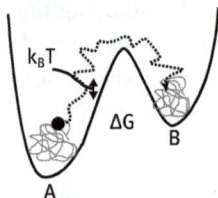

Figure 13.1: Example of *rare events*. Two conformational minima separated by a high energetic barrier. With MD simulations, the probability to overcome such a barrier, in order to visit other minima (B) in a finite reasonable computational time, is very low.

To deal with this problem, a wide variety of algorithms, also called *enhanced sampling* methods, have been proposed over the course of the last few years. Here, we will limit our discussion to metadynamics (MetaD) [4], successfully applied in different fields. During a MetaD simulation, the dynamics of the system is accelerated by adding a history-dependent bias potential (V_G) built as a sum of Gaussian functions (eq. (13.1)) centered on a selected number of degrees of freedom (S), referred to as collective variables (CV) that function as microscopic coordinates of the system (R):

$$V_{(s,t)} = \sum_{i=1}^{t/\tau_G} w_G exp\left[-\sum_{\alpha=1}^{N_{CV}} \frac{(S_i(R) - (S_i(R(\tau_G))^2}{2\sigma_i^2} \right]$$

(13.1)

where w_G and σ represent, respectively, the Gaussian height and width, while τ_G is the frequency at which Gaussians are added.

The accuracy and efficiency of free energy reconstruction depend on these three parameters. In fact, excessive Gaussians width (σ) leads to a quick exploration of the CV space, but the free energy estimation will be affected by heavy errors. On the contrary, when Gaussians are small or placed infrequently, the reconstruction will be accurate, but it will take a longer time. Moreover, it is worth mentioning that w_G and τ_G in eq. (13.1) are not independent and what really matters is their rate $\omega = w_G/\tau_G$, also known as the *deposition rate*.

Following eq. (13.1), the history-dependent potential iteratively compensates the underlying free energy and the system tends to escape from any free energy minimum via the lowest free energy saddle point (Figure 13.2).

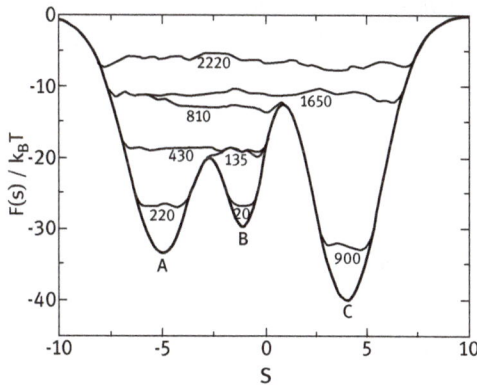

Figure 13.2: Example of a Metadynamics simulation in a one-dimensional model potential. The time *t* is measured by counting the number of Gaussians deposited. The schematic representation of the progressive filling of the underlying potential (thick line) by means of the Gaussians deposited along the trajectory. The sum of the underlying potential and of the Metadynamics bias is shown at different times (thin lines).

After a sufficiently long time, $V_G(S, t)$ provides an estimation of the underlying free energy:

$$\lim_{t \to \infty} V_G(S, t) \sim - F(S) + C \tag{13.2}$$

where C is an additive constant.

Equation (13.2) establishes that an equilibrium quantity, the free energy, may be estimated by a non-equilibrium dynamic in which the potential below is changed each time a new Gaussian is added [5].

13.2.1 The well-tempered metadynamics and the reweighting procedure

Two main drawbacks are related to standard MetaD:
1. The statistical error on the reconstructed profile grows with the deposition rate (ω);
2. V_G does not converge on the free energy but oscillates around it.

A solution to these two problems is provided by an improved variant of standard MetaD, namely Well-tempered Metadynamics (WT-MetaD) [6]. In WT-MetaD, the initial bias deposition rate (ω) decreases to $1/t$, by rescaling the Gaussian height (wG):

$$w_G = \omega \tau_G e^{-\frac{V_G(S, t)}{K_B \Delta T}} \tag{13.3}$$

where τ_G is the Gaussian deposition stride, ΔT is an user-defined constant parameter with the dimension of a temperature that regulates the extent of the exploration of the free energy landscape [7] and $V_G(S,t)$ is the bias potential accumulated in S over time t. At variance with standard MetaD, the bias potential does not fully compensate the free energy, but it converges as follows:

$$V_G(S, t \to \infty) = - \frac{\Delta T}{T + \Delta T} F(S) \tag{13.4}$$

thus resulting in the sampling of CVs at an effectively higher temperature $T + \Delta T$, where T is the temperature of the system. This leads to the following distribution of S:

$$P(S, t \to \infty) \propto e^{-\frac{F(S)}{k_B(T + \Delta T)}} \tag{13.5}$$

By tuning ΔT, one can increase barrier crossing and facilitate the exploration in the CVs space. Furthermore, using a finite value of ΔT, one automatically limits the exploration of the free energy region to an energy range of the order $T + \Delta T$. Hence, the exploration of free energy can be limited to the physically interesting regions of S. The risk of overfilling is avoided, and optimal use is made of the computer time [8]. In conclusion, WT-MetaD solves the convergence problems of MetaD and allows the computational effort to be focused on the physically relevant regions of the conformational space [6]. It was further demonstrated that WT-MetaD converges asymptotically to the final state [9].

The introduction of a history-dependent potential alters the probability distribution. Although from eq. (13.5) the probability distribution for the CVs can be easily reconstructed, for the other degrees of freedom, it is distorted in a nontrivial way. This problem has been solved by introducing a reweighting algorithm for recovering the unbiased probability distribution of any variable (Boltzmann distribution) from a WT-MetaD simulation [7]:

$$P_B(R) \propto e^{+\beta V_G(S(R),\, t)} P(R, t) \tag{13.6}$$

where $P_B(R)$ is the Boltzmann distribution.

Therefore, one does not need to use CVs directly related to measurable quantities to make quantitatively contact with experiments [6]. However, it must be pointed out that, in WT-MetaD, convergence is estimated by measuring the free energy difference of various points of the free energy surface (FES), without solving the problem of estimating local convergence. This aspect has been completed by constructing a time-independent free energy estimator, that enables not only the evaluation of the local quality of the convergence in different parts of the FES, but it also allows a direct comparison of the FES coming from multiple well-tempered and standard Metadynamics run with different simulation parameters thanks to the introduction of a time-independent estimator of the free energy [10]. The new algorithm also simplifies the previous reweighting procedure of Bonomi et al. [7]. It is worth mentioning that later Bonomi et al. [8] showed that it is possible to perform WT-MetaD simulation using the potential energy of the system as CV, namely a "well-tempered ensemble", which can be used in practice to enhance the efficiency of parallel tempering simulations, thus reducing the number of replicas [11].

13.2.2 Funnel Metadynamics: a valuable method for an accurate estimation of the ligand–binding free energy

A detailed description of the ligand/target events and an accurate estimation of the drug affinity to its target is of great help in speeding up drug discovery strategies because it facilitates many intricate aspects, such as structure-based drug design and lead optimization. Over the years, a variety of methods that are able to describe ligand/protein interactions in a more accurate way with respect to docking and VS simulations have been proposed. Among them, metadynamics, which has been widely used to study long-timescale processes particularly in complicated ligand/protein binding cases, offering a qualitative estimation of the protein-ligand binding free energy [12–14]. This is because, once the ligand leaves the binding pocket, it starts exploring all of the possible solvated states and rarely rebinds leading to a poorly converged free-energy calculation. In order to overcome this limitation and with the aim of providing an accurate estimation of the absolute target–ligand binding free energy, a metadynamics-based method, called FM, has been developed [15]. In FM, a funnel-shaped restraint potential is applied to the system, reducing the space to explore in the unbound state (Figure 13.3) process.

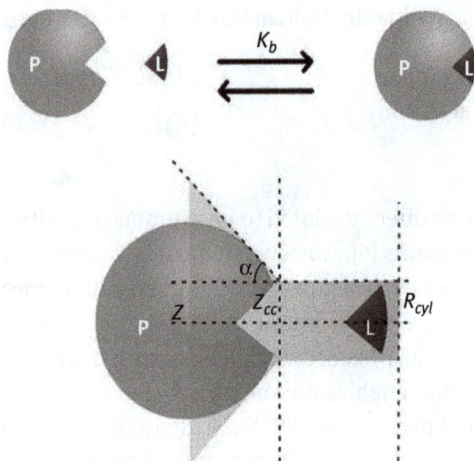

Figure 13.3: Schematic representation of the ligand/protein binding process and the funnel restraint potential used in FM calculations. The shape of the funnel can be customized on the target by setting a few parameters. In particular, given z, the axis defining the exit-binding path of the ligand, z_{cc} is the distance where the restraint potential switches from a cone shape into a cylinder. The α-angle defines the amplitude of the cone, and R_{cyl} is the radius of the cylindrical section.

In such a way, the sampling of ligand-bound and ligand-unbound states is highly enhanced, thus leading to an accurate estimation of the binding FES within an affordable simulation time. Using FM, one can compute the absolute protein-ligand binding free energy AG_O typically expressed by the following formula:

$$\Delta G_b^0 = -\frac{1}{\beta} \ln\left(C^0 K_b\right)$$
(13.7)

where $C^0 = 1/1.660$ Å^{-3} is the standard concentration and K_b is the equilibrium binding constant computed as follows [16]:

$$K_b = C^0 \pi R_{cyl}^2 \int_{site} dz e^{-\beta[W(z) - Wref]}$$
(13.8)

where C^0 is the standard concentration of 1 M and is equal to $1/1.660$ Å^{-3}, πR^2_{cyl} is the surface of the cylinder used as restraint potential that accounts for the entropic restraint applied to the system in the unbound state, $W(z)$ is the potential value in the unbound state and Wref can be derived from potential of mean force (PMF).

FM ensures a number of recrossing events between the bound and unbound states during the simulation, leading to a quantitatively well-characterized free-energy profile and a converged estimation of ΔG. FM has been successfully applied in a number of studies [17–20], but in the following paragraph, we will present only the case study of FM applied to the natural alkaloid Berberine [16].

13.2.1.1 Funnel metadynamics: A successful case study on the natural alkaloid berberine

Alkaloids that possess an isoquinoline moiety constitute one of the largest groups of natural substances. They are not a structurally homogenous group, however, based on their intramolecular rearrangements and on the presence of additional rings connected to the main system, they may be divided into several subgroups including the proto-berberines, which represent the largest group since they constitute 25 % of all eluci-dated structures of isoquinoline alkaloids [21]. One of the most representative alkaloids of the protoberberine group having a benzylisoquinoline moiety is Berberine, obtained from many plants like *Hydrastis Canadensis, Rhizoma Coptidis* and *Berberis Vulgaris*. This alkaloid exerts several pharmacologic effects like anti-inflammatory and anti-diabetic [22] and it has been proposed also as a promising anti-cancer agent [23]. In particular, concerning its anti-cancer effects, it has been found that one of the possible mechanisms of action is due to the stabilization of G-quadruplex (G4) DNA through the π–π stacking interactions between aromatic moiety of N^+-containing Berberine with the G-quartet [24] and several studies have addressed this subject [25]. In order to give a detailed atomistic description of the binding of Berberine to the human telomeric G4 DNA and calculate its absolute binding free energy, an FM study was conducted on the X-Ray of Berberine/G4 complex (PDB code: 3R6R).

In particular, a funnel-shaped restraint potential has been built at the 3'-end binding site of the Berberine/G4 complex, so that the conic region of the funnel included the whole 3'-end binding site, while cylindrical restraint was directed toward the bulk water region (Figure 13.4). This was achieved by properly setting the funnel parameters: the angle α, the distance z_{cc} and the radius of the cylinder R_{cyl}. When the ligand visits states inside the funnel volume, the potential felt by the system is null and, as the ligand reaches the edge of the funnel, a repulsive bias is applied to the system in order to prevent the ligand from visiting bulk regions outside the funnel. At the same time, the ligand sampling inside the funnel is not affected by the potential restraint. In addition to the funnel-shaped potential, since a key step in metadynamics is the *a priori* identification of the right set of CVs, the Berberine distance (d) and the torsion (π) CVs with respect to the G-quartet of the DNA where chosen. In this way, the ligand binding,

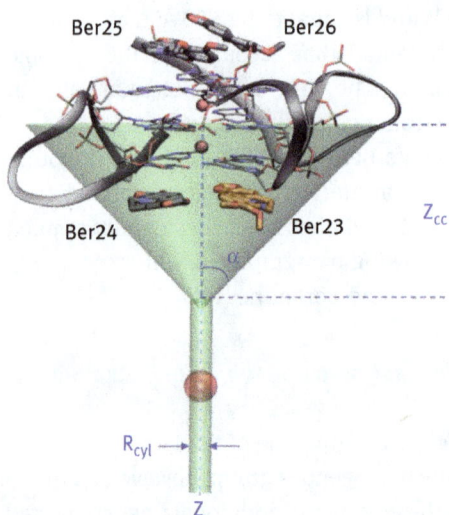

Figure 13.4: The funnel shape potential restraint applied at the 3'-end binding site of the Berberine/G4 complex. Distance and torsion CVs were applied only on the Ber23 residue (depicted as yellow stick). The funnel parameters: angle (α), distance (zcc) and the radius of the cylinder (R_{cyl}) are also highlighted.

which is typically a long-timescale event, is accelerated under FM and several forth-and-back events between the bound and unbound states are observed (Figure 13.5A), leading to the convergence in the estimation of the Berberine/G4 binding free energy (-10.3 ± 0.5 kcal/mol) (Figure 13.5B), found to be in good agreement with the experimental value calculated by steady-state fluorescence experiments [7].

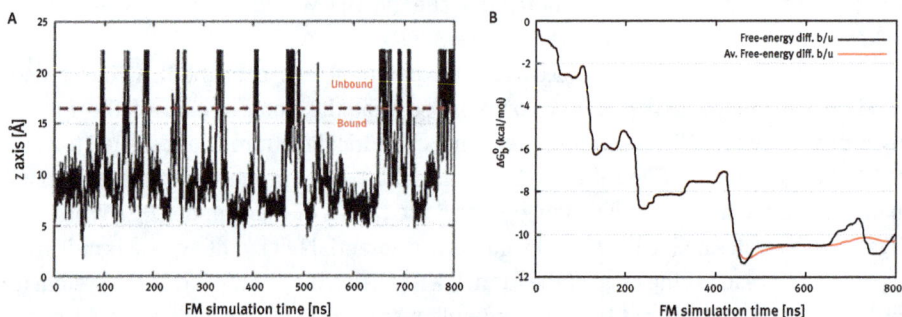

Figure 13.5: (A) Recrossing events between the Berberine bound and unbound states are sampled during 800 ns of FM. (B) The plot of the free-energy difference between the bound and unbound state as a function of the simulation time calculated by applying the analytical correction. Red line shows the weighted average estimation of the absolute Berberine/G4 binding free energy equal to -10.3 ± 0.5 kcal/mol in good agreement with the experimental value.

The computational protocol adopted in this study led us to obtain, at the end of ~0.8 µs simulation, a quantitatively well-characterized FES, and an accurate Berberine/G4 absolute binding free-energy estimate calculated as equation 13. In addition, mechanistic details of the Berberine binding to G4 have been provided using the reweighting algorithm (Figure 13.6) [7].

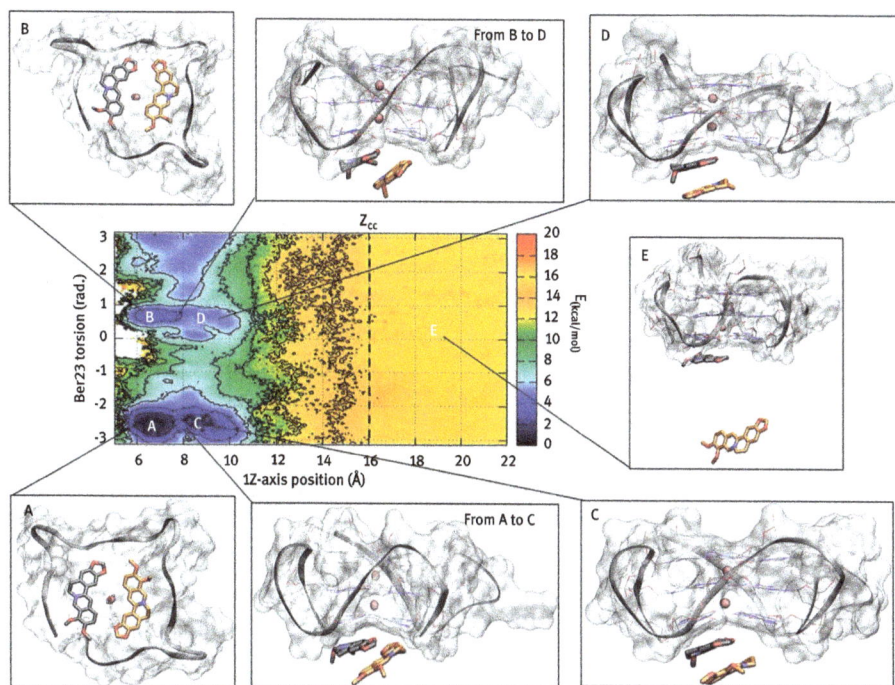

Figure 13.6: The binding FES calculated using the reweighting algorithm as a function of Berberine torsion (π) CV and the funnel z-axis position. The pre-binding states of Berberine were found in basins C and D.

Another interesting aspect of the work was to elucidate the water solvent role. Indeed, in the last decades, computational methods have provided strong support to deeply understand the role of aqueous solvent in a large number of chemical systems using *ab initio* MD simulations [26], [27]. Several efforts have been undertaken to clarify also the role of the hydrogen-bond network of water combining ab initio and parallel tempering with the well-tempered ensemble metadynamics (PTWTE) [28], [29] and the implications of hydrogen-bond network on biological processes such as the protein fibril formation [29] and enzymatic functions [30]. Other scientific works discuss the role of water networks in numerous other contexts like ion solvation and diffusion [31], [32] or proton transfer [33], [34].

The role of the water has been investigated also in a wide number of works focusing on fundamental biophysical processes, such as protein folding or solvent-protein interactions. For instance, Jong et al. investigated complexity of the water network surrounding the hydrated peptide tri-alanine [35]; while the electronic properties of three hydrated zwitterionic amino acids forming the collagen were analyzed using ab initio MD simulations [36]. Further studies about the investigation of the water role can be found in the reviews of Laage et al. and Kim et al. [37], [38]. Given the importance of water molecules in most biological systems, the solvent contribution in DNA-ligand binding [16] was investigated by adopting two approaches: the first was useful to assess the energetic contribution of the water solvent; reconstructing the FES as a function of Berberine torsion (π) and the water bridge CVs (*watCV*) using the reweighting algorithm (Figure 13.7A). In particular, looking at the small energetic difference between basins Aa and Ab (~1.9 kcal/mol), it can be observed that the presence of water at the interface between the ligands and the 3′-end G-quartet is possible, but not necessary for ligand binding. The second approach has been performed generating an unbiased GBPM (Grid-Based Pharmacophore Model) [39] in order to rationalize the solvent role. The GBPM analysis has been performed in the presence of the bridging water

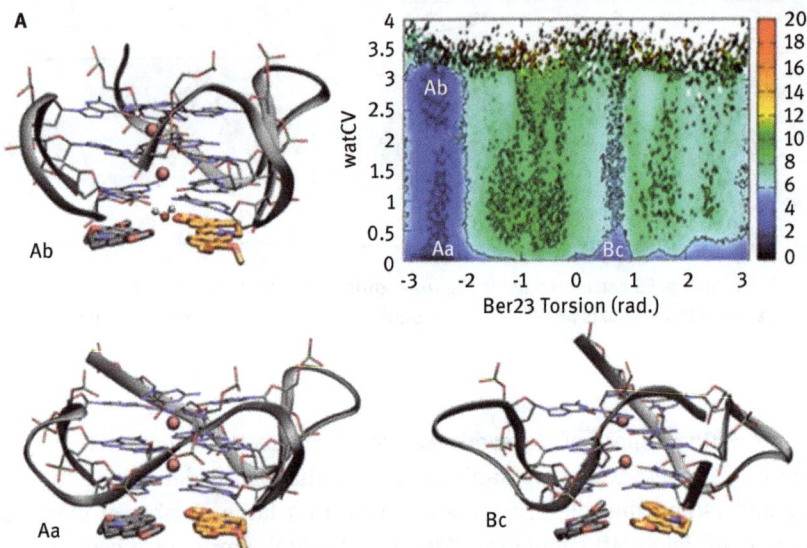

Figure 13.7: (A) The reweighted FES as a function of the Berberine torsion (π) and the interfacial water CVs (*watCV*), with the two energetic minima Aa and Ab. (B–C) Top and lateral views of the GBPM applied at the 3′-end of the Berberine/G4 X-Ray pose. Regions with favorable interactions explored by the DRY, N1, and O probes are shown as yellow, blue and red contour maps, respectively, whereas their most relevant interaction points, identified with the Minim utility, are depicted as yellow, blue, and red spheres, respectively. K^+ ions are represented as pink spheres.

molecule; forming H-bond bridge interactions between the nucleobases at the 3′-end. This water was considered part of the target. The utility of this method consists in proposing a possible role of the water-mediated H-bond network on the Ber23 binding affinity and also to rationally guide the structure-based design of more potent derivatives. The DRY probe, which mimics the aromatic and aliphatic carbon atoms, was used to detect the target regions with potential hydrophobic interaction, whereas the O and the N1 probes were chosen to mimic, respectively, the sp^2 carbonyl oxygen and the amide NH group as hydrogen-bond acceptor and donor, respectively. The GBPM analysis reveals the important role of water in mediating favorable H-bond interactions with the four O6 atoms of the guanine residues G10:G16:G4:G22 (blue and red contour maps), whereas the hydrophobic interactions (yellow contour maps) involve the isoquinoline core of Berberine (Figure 13.7B and Figure 13.7C).

13.3 *Dynophores*: The use of molecular dynamics in pharmacophore modeling

In medicinal chemistry, 3D-pharmacophore modeling has evolved more and more over the last few years, becoming a well-established method for molecular design and for the description of molecules and biological systems. In chapter 16, we focus on the pharmacophore concept and its applicability. Here, we introduce a new method of analysis based on the pharmacophore model but more evolved than the classic 3D-pharmacophore. A 3D-pharmacophore model derived from X-ray or NMR complex (structure-based design), which contains a set of information about the chemical interactions of a ligand with a binding pocket. When information about the ligand co-crystallized in the binding site is not available, it is common to use the best docking pose as the starting model for structure-based design. This could be a valid alternative to the more common method that involves active/inactive compounds in the ligand-based design. This docking approach allows scoring of several possible interaction modes for different alignments. However, in both of the aforementioned cases, the limit is represented by the static conformation of the binding site allowing only a static view on single ligand-target conformation for the classical pharmacophore models [40]. To investigate in detail the interactions involved in the stabilization of the ligands into the binding site, a dynamic pharmacophores analysis method was recently developed. The goal of this method is to extrapolate the ligand conformational interactions from the binding site during the MD simulations and to include this information in one single dynamic pharmacophore model, termed *Dynophore*.

For *Dynophore* generation [41], with each snapshot of an MD simulation, pharmacophores are automatically generated by using the program DynophoreApp [42], which is based on the API of the ilib/LigandScout framework [1], [41]. For each frame of the trajectories of an MD simulation, each feature is grouped into *Dynophore superfeatures* and the frequency of its occurrence in the trajectory is collected and illustrated in bar codes, while distance distributions are represented in histograms. If a feature of the same type and the same involved atoms on the ligand-site reoccurs over time, these features will build one group of *superfeatures* to monitor its behavior throughout the whole trajectory. These *superfeatures* contain information about the feature's atom type. *Dynophores* represent a novel approach in molecular design which includes conformational flexibility of a biological system derived from MD simulation (MDs) that gives more information based on the interaction pattern analysis.

13.3.1 *Dynophores*: A case study on the catalytic site modulation of arginase

In 2017, Mortier *et al.* applied *Dynophores* with the goal of better understanding the intermolecular interactions involved in the stabilization of the ligands in a binding site [43]. This study concerned the metalloenzyme Arginase and, in particular, a careful computer-aided structural investigation of this enzyme was performed using MD simulations and *Dynophores*, detecting new possibilities of modulation in the catalytic site. The novelty of this method was the construction of a single 3D-model comprising all ligand-enzyme interactions occurring throughout a complete MD trajectory. Furthermore, it was possible to highlight the plasticity of the size of the Arginase active site and the loop flexibility conformational changes, leading to the hypothesis that larger ligands could enter the cavity of Arginase. At first, all co-crystallized inhibitors available from the Protein Data Bank (PDB) were collected and 3D-pharmacophore models were created for each of them. Furthermore, the last pharmacophore model that shared features from all models describing ligand-enzyme interactions was constructed and used to screen a database of commercial libraries comprising over 106 *drug-like* compounds (from providers Asinex, Vitas M, Life Chemicals, Chembridge, Maybridge, Specs, Enamine, and Prestwick). By using Ligand Scout, the best *hits* were ranked based on the quality of their 3D superposition with the pharmacophore model and subsequently tested. Considering that significant activity was not observed in all the best hits, the authors chose to analyze the mechanisms of Arginase inhibition particularly focusing on the Arginase pocket size. In this regard, the Pocket Volume Measurer (POVME) algorithm has been adopted as a measuring tool for characterizing pocket volume, highlighting that the presence of the ligand could induce an effect on the binding pocket volume. Moreover, even the best hits selected with 5- and 6-membered rings as linkers, indicating that the geometry of the pocket could allow the entry of more cumbersome chemical groups. For this reason, the active site of Arg1 was investigated using MD simulations with three different systems: (a) the *apo* structure of Arg1, (b) the Ornithine–Arg1 complex and (c) the ABH–Arg1 complex (Figure 13.8). The obtained results showed highest movements by average deviation of Cα's from the initial conformation (RMSD) for the apoenzyme with a value reaching a plateau around 3 Å, while deviations observed with two other complexes were between

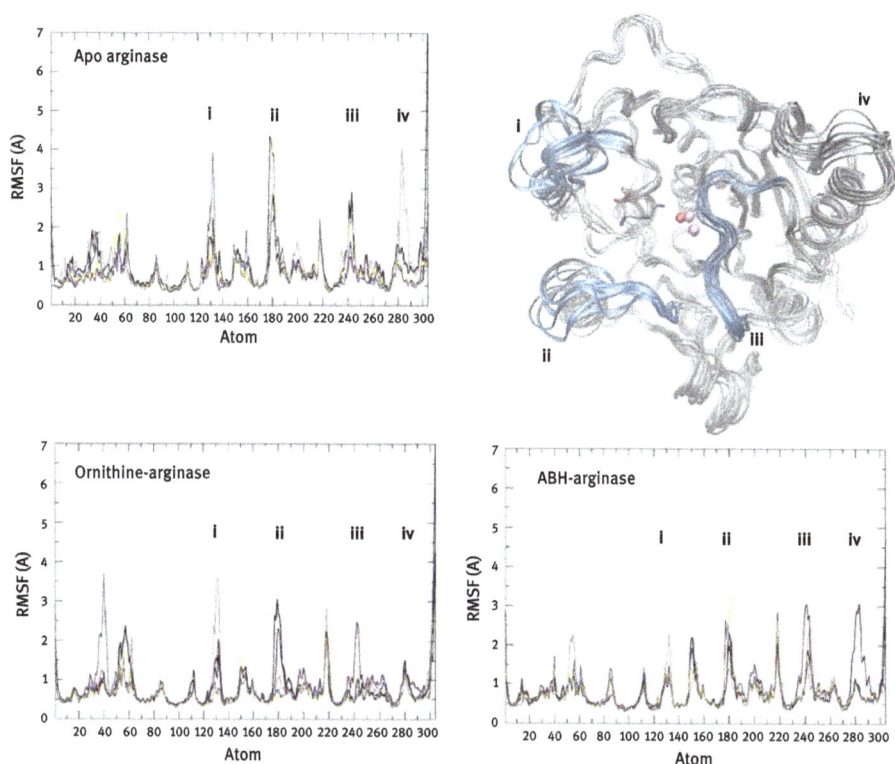

Figure 13.8: Root mean square fluctuations (RMSF) overlay for the five 200 ns repeats conducted with three different systems (left) and superposition of 10 protein conformations extracted from 200ns simulation of the Ornithine–Arginase 1 complex (right). The three flexible loops responsible for the opening of the binding pocket are highlighted in blue in the 3D view. Loops surrounding the active site cavity are labeled with roman numbers (i–iv). Mn atoms are represented in mauve and, the hydroxide group, in red. RMSF plots can be found in a bigger format in the supporting information.

1 Å and 3 Å. The trajectories observed in ABH-Arginase and the Ornithine–Arginase complex showed movements of the loops in the area of the binding site. For the first complex, a closure of the cavity around the inhibitor was observed, while relaxation of the active site was observed in the second. These major differences in the RMSD between the ABH-Arginase and the Ornithine–Arginase complexes were evidenced by movements of the loop at the surface of the protein and confirmed by the analysis of the pocket volume. In fact, the highest volumes (>750 Å³) were measured for three out of five 200 ns trajectories, while the volume of the same pocket remained below 500 Å³ with the *apo* form of Arginase (Figure 13.9).

In the second part of the work, two *Dynophores* were built from all interactions detected in the MD simulations of ARG1–Ornithine, and ABH–Arginase complexes (Figure 13.10). For both complexes, all features represented in the *Dynophores* were

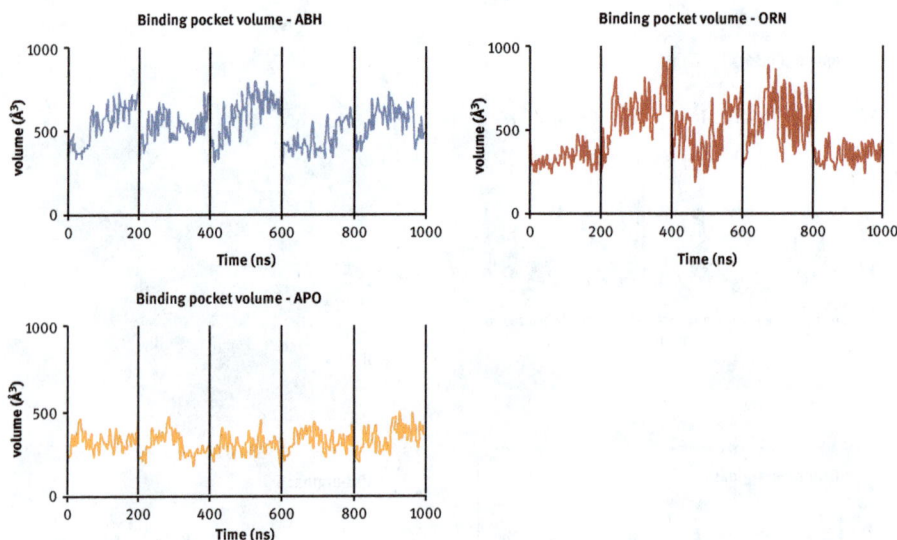

Figure 13.9: The binding pocket volume of three systems ABH, ORN and *apo*. The 200-ns repeats are separated by a vertical line.

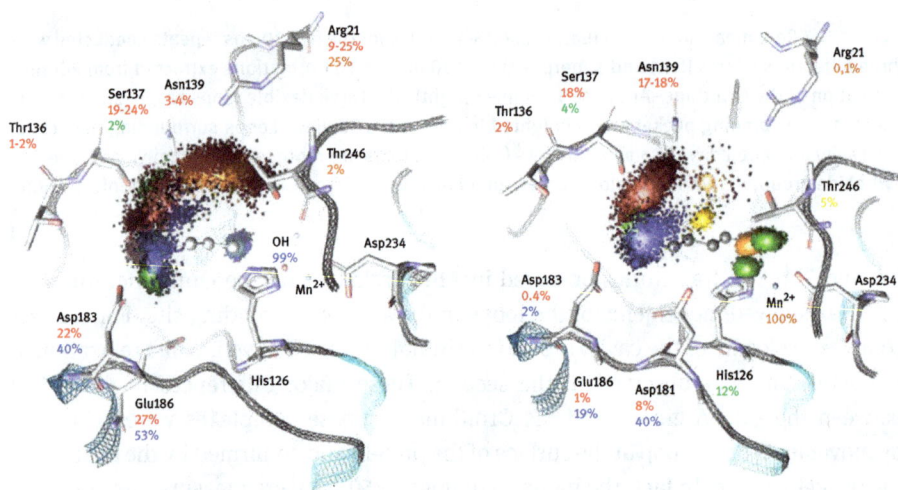

Figure 13.10: *Dynophores* constructed with frames from 1-µs MD simulation with enzymatic product ornithine (left) and inhibitor ABH (right). Color code: red for the H-bond acceptor, orange for the negative ionizable feature, green for the H-bond donor, blue for the positive ionizable moiety and yellow for hydrophobic contact. The manganese ions are represented by lavender spheres and the hydroxide group is in red.

extracted from 1 µs (5000 frames) of MD simulations trajectory (frames were recorded every 5 ns). The interactions of the same type detected for one single ligand atom were grouped into super-features, namely graphical representations as 3D points cloud, statistically described by their frequency occurrence. The frequency of each interaction in the trajectory was illustrated in bar codes and was simply calculated relative to the sum of all considered frames. Additionally, for estimating feature quality, the distance distributions were represented as histograms for each reported interaction (interaction distance histogram). Here, the ARG1–Ornithine complex was selected as a single example of this type of descriptor. In particular, highlighting the Ornithine atoms involved in detected interactions with positive ionizable (PI) and hydrogen-bond donor (HBD) super-features (Figure 13.11).

In the first case (Ornithine–Arginase complex), the dynamic 3D model showed local flexibility in the Arginase binding site where H-bonds and Coulomb interactions were detected during the time simulation between the hydroxide (with surrounding

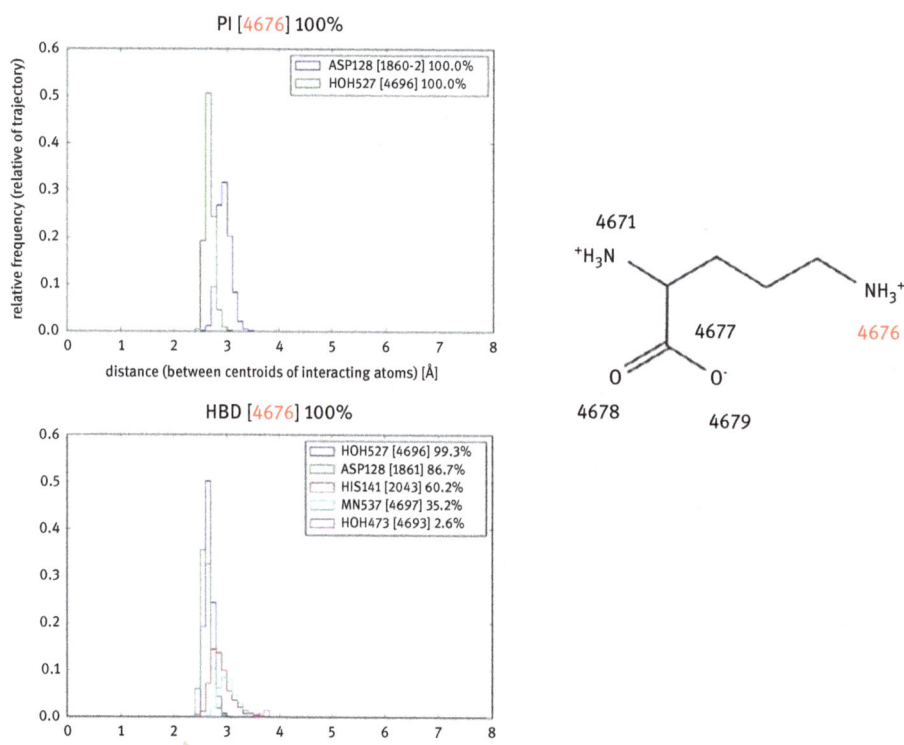

Figure 13.11: On the right of the figure, the 2D representation of the Ornithine with the heavy atom numbers is shown. On the left of the figure, the distance distributions for the interactions detected between the positive ionizable (PI) and hydrogen-bond donor (HBD) super-features, and the heavy atom 4676 of the ornithine (red) are shown. The y-axis represents the relative occurrence frequency.

residues) and the primary amine on the side chain of L-Ornithine. Other interactions were detected and shown as percentages of interaction. The carboxylate group of Ornithine interacted through electrostatic interactions (24.7 %) as well as H-bonds (64.4 % and 62.9 % for each oxygen, respectively). Thr246 and Arg21 became the main interaction partners when the active site was opened. In the second case (ABH–arginase complex), two main interacting groups were distinguished: the amino-acid moiety and the boronate. The *Dynophore* indicates the presence of H-bonds between the carboxylate of ABH and residues 130–139 during 36.8 % to 36.9 % of the simulation time. Others H-bonds interactions involved Ser137, Asn130, Thr135, Thr136, and Asn139. Thr246 and Arg21 were almost never reached by the COO– group (<1 %). Instead, the carboxylate of ABH was held in the region of loop 130–139, allowing the ligand to conserve an ideal orientation for **complexation of the ion cluster by the boronate. This great difference between the dynamic binding of ABH and Ornithine was illustrated by an interaction cloud more concentrated around this particular loop of ABH. Secondly, the amino moiety interacted as a positive ionizable feature. Charge interactions were detected in 26.5 % of simulation time with residues Glu186, Asp183 and Asp181. H-bonds stabilizing the amino group were detected less frequently (7.6 % of the analyzed frames, mostly with Ser137). Finally, the boronate was depicted in the *Dynophore* as 3 H-bond donors at the level of the 3-hydroxyl groups as well as one H-bond acceptor and one negative ionizable feature on the boron atom.

In the last part of the work, in order to consolidate the finding that larger molecules can enter the catalytic site of Arginase, a fragment-based approach was followed. Among all commercially available compounds identified, hydrophobic fragments with a larger volume than ABH were selected. Subsequently, the inhibitory potency of all compounds was experimentally tested against Arginase 1. Out of 30 small fragments, two compounds inhibited more than 50 % of Arginase activity at a concentration of 1 mM (Figure 13.12). Residual activity of Arginase reached 35 ± 3 % in the presence of the 4-Sulfamoylbenzeneboronic acid (BA-11), while Cyclohexylboronic acid (BA-25) showed the highest potency with 25 ± 4 % residual activity.

This new insight of ARG1 enzymatic cavity showed a rigid architecture at the level of the ions cluster, and more flexibility in the region interacting with the amino acid moiety. Contrary to static Arginase–ligand complexes, the development of dynamic pharmacophore (*Dynophore*) allowed to better understand the interactions in the Arginase binding site.

13.4 Pharmacophore models deriving from MD simulation and virtual screening

Pharmacophore approaches, successfully and extensively applied in VS, are one of the most useful tools in drug discovery. In the following paragraphs, the screening performance of structure-based pharmacophore models deriving from MD

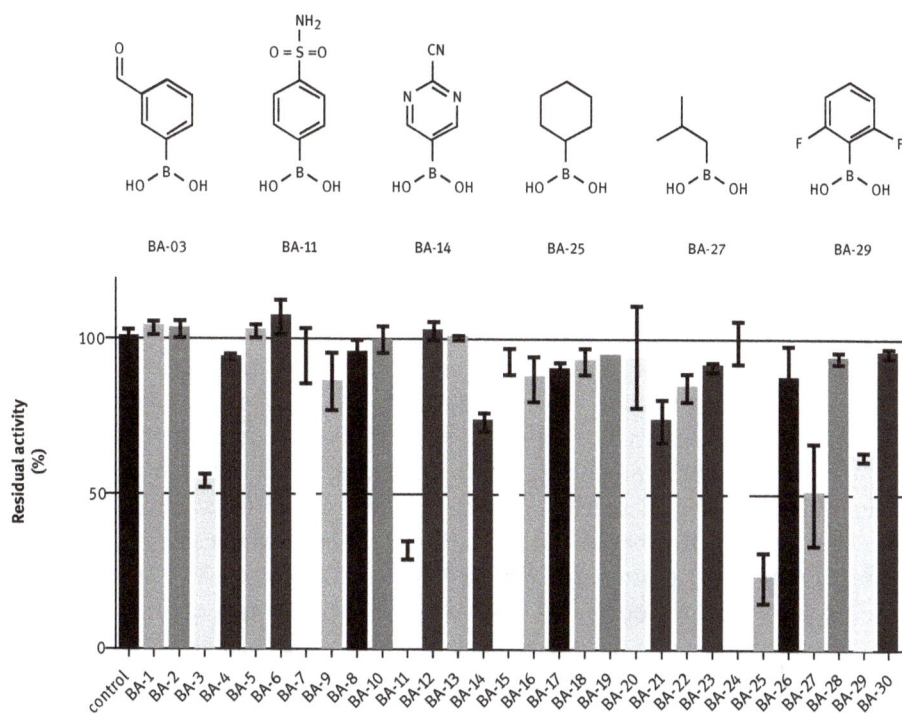

Figure 13.12: Residual arginase activity in presence of Boronic acid fragments (concentration 1 mM) and chemical structure of the most potent compounds.

simulations has been evaluated using two different approaches. In the first case, the screening performance of the pharmacophore model deriving from the crystal structure has been compared with the pharmacophore model derived from the last frame of a MD simulation. In fact, as a structure-based method, it depends on the correct interpretation of protein-ligand interactions. Therefore, there are reasonable concerns about the accuracy of the bound ligand and about non-physiological interactions with parts of the crystal as well as the solvent effects that influence the protein structure. The final structure of an MD simulation represents a possible way to refine the structure of a protein-ligand system. The second reported approach uses the multiple coordinate sets saved during the MD simulations and, for each frame, it generates a pharmacophore model. Pharmacophore models with the same features are polled in order to reduce the high number of resulting pharmacophores to only representative models. Each representative model is screened, and the screening results are combined and re-scored to generate a single *hit*-list. The score for a particular molecule is calculated based on the number of representative pharmacophore models (RPMs) that classify it as active. Hence, such a method has been named "Common Hit Approach" (CHA).

13.4.1 Virtual screening performance and comparison of pharmacophore models

Wieder *et al.* compared the pharmacophore model obtained from the crystal structure of a protein-ligand complex with the pharmacophore model derived from the last frame of a 20-ns MD simulation using Ligand Scout software [41], [44]. To analyze the differences in their ability to discriminate between active and decoy compounds, the initial pharmacophore model and the MD-refined model were first visually inspected, then characterized by different features. For analysis, six different systems with PDB codes 1J4H, 3BQD, 2HZI, 3L3M, 1UYG, and 3EL8 were chosen from the DUD-E database. This database provides known actives and decoys that are calculated using similar 1D physicochemical properties as the actives (e.g. molecular weight, calculated LogP) but dissimilar 2D topology (based on ECFP4 fingerprints) [45]. As shown in Figure 13.13 and Figure 13.14, for all analyzed systems, the pharmacophore model retrieved from the X-ray structure was different with respect to that obtained from the last frame of MD simulation. Focusing on specific features, it is interesting to note that (with the exception of the PDB model 1J4H) the hydrophobic features remained unchanged; none of the aromatic features appear in the MD-refined pharmacophore model; hydrogen-bond acceptors and donors are the most accountable features of variability.

Regarding the screening results (Table 13.1 and Table 13.2), except for the PDB model 1J4H, in which pharmacophores performed poorly, both the refined pharmacophore model and the initial pharmacophore model were able to favor active compounds over inactive ones—in some cases, e.g. 3BQD both were able to distinguish between the groups. Considering the enrichment factor at 100 %, the MD-refined pharmacophore model performed better for 1J4H (resulting performance was still low), as well as on average for 1UYG, 2HZI, and 3BQD. In the investigated cases, the overall enrichment factor reflected the results obtained from the early enrichment outcomes. It could be supposed that the enhanced performance of the MD-refined pharmacophore model of 1UYG was the result of the structural movement of the ligand, but Wieder *et al.* observed that there was no evident correlation between ligand or protein RMSD and the screening performance. The findings reported in this study suggest that the refinement of pharmacophore models using MD simulations is advantageous in more than 50 % of the cases to achieve improved screening results since additional protein-ligand interaction information can be revealed. However, for some systems, MD refinement did not yield better results; in these cases, additional operations are required to increase the pharmacophore model efficiency.

13.4.2 Common hit approach

Wieder *at al.* developed the new Common Hits Approach (CHA) that incorporates flexibility based on extensive MD simulations of protein-ligand complexes into

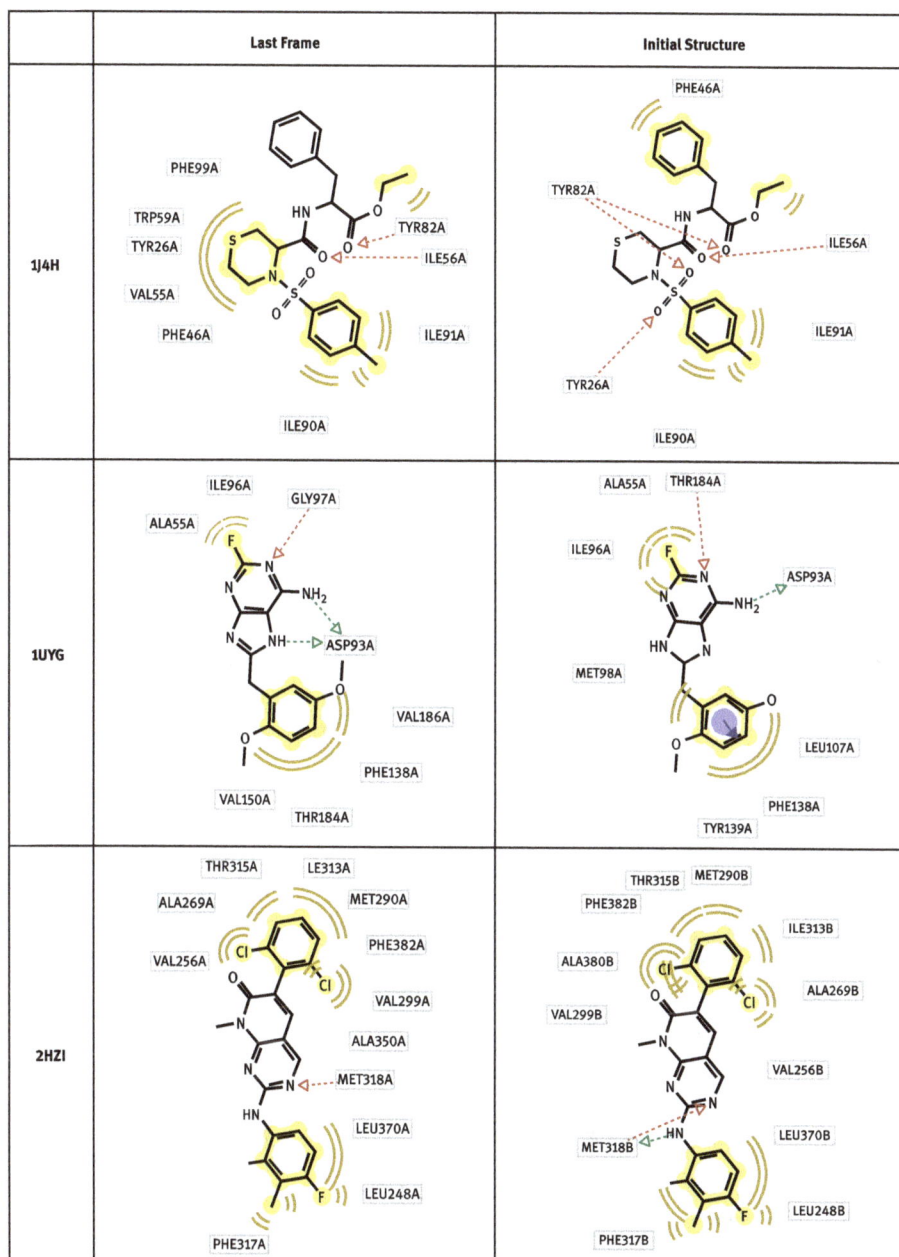

Figure 13.13: Comparison of the initial pharmacophore model and the MD-refined one for the 1J4H, 1UYG, and 2HZI PDB models. The features in yellow indicate hydrophobic features, the vector features in green indicate hydrogen-bond donors, the features spheres in blue with associated vectors indicate aromatic features and the features in blue with multiple lines associated indicate salt bridges.

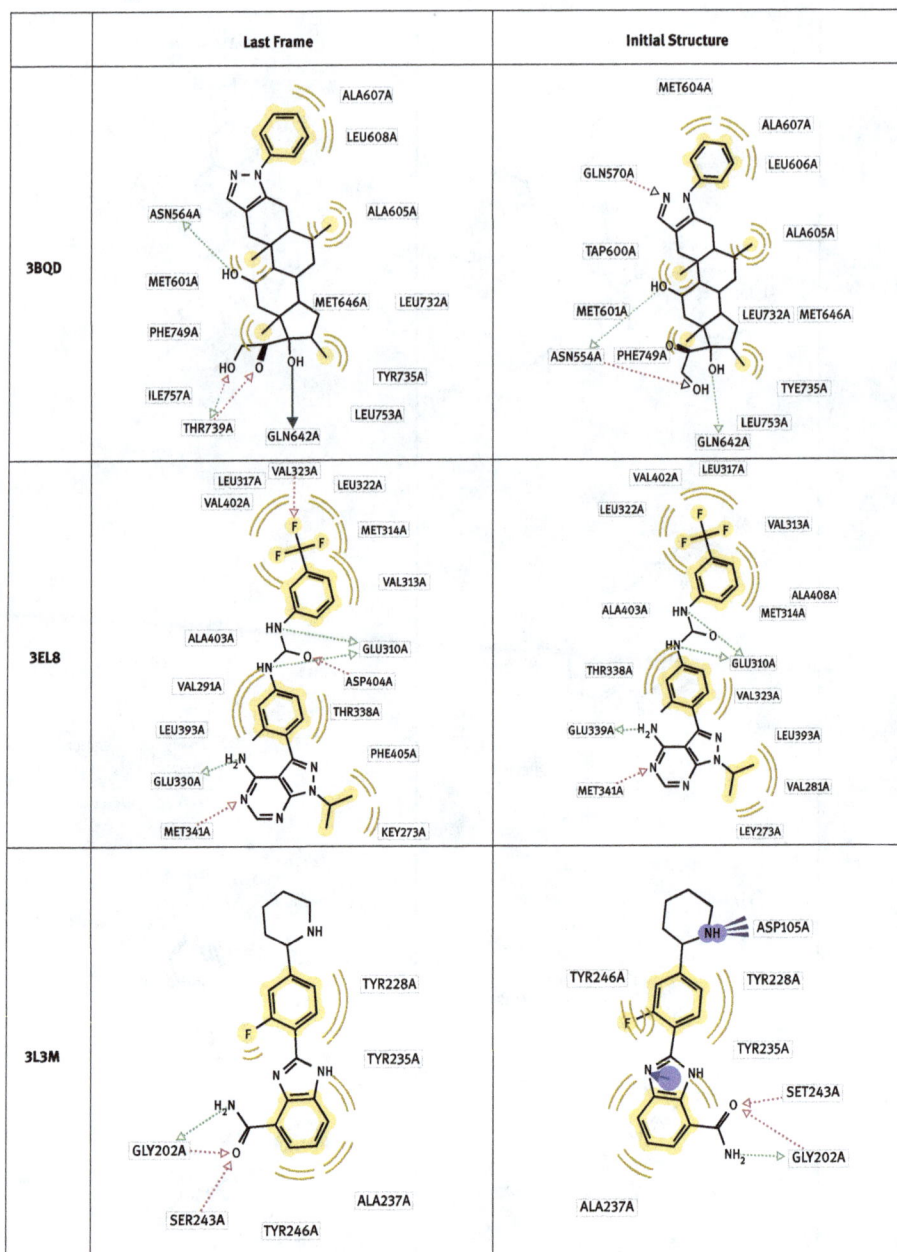

Figure 13.14: Comparison of the initial pharmacophore model and the MD-refined one for the 3BQD, 3EL8 and 3L3M PDB models. The features in yellow indicate hydrophobic feature, the vector features in green indicate hydrogen-bond donors, the features spheres in blue with associated vectors indicate aromatic features and the features in blue with multiple lines associated indicate salt bridges.

Table 13.1: For each PDB model the results obtained by the screening of pharmacophore model deriving from the last frame are shown.

PDB	Omitted features	No. of features	No. of hits	AUC				EF			
				1%	5%	10%	100%	1%	5%	10%	100%
1J4H	0	6	290	0.00	0.00	0.25	0.48	0.00	0.00	0.00	0.00
	1	5	2,922	0.00	0.19	0.20	0.32	0.00	0.40	0.40	0.30
1UYG	0	5	379	0.42	0.86	0.83	0.53	4.10	1.90	1.60	1.60
	1	4	3,259	0.82	0.86	0.83	0.62	3.30	2.60	2.40	1.00
2HZI	0	7	9	0.64	0.53	0.51	0.50	4.20	4.20	4.20	4.20
	1	6	314	0.89	0.54	0.50	0.50	2.10	0.90	0.90	0.90
	2	5	2,626	0.87	0.89	0.87	0.54	4.10	2.5	1.80	1.20
3BQD	0	10	0	0.00	0.00	0.00	0.00	0.00	0.00	0.00	0.00
	1	9	1	0.50	0.50	0.50	0.50	27.90	27.90	27.90	27.90
	2	8	47	0.99	1.00	0.80	0.53	20.20	20.20	20.20	20.20
3EL8	0	10	0.0	0.00	0.00	0.00	0.00	0.00	0.00	0.00	0.00
	1	9	4	0.56	0.51	0.50	0.50	10.80	10.80	10.80	10.80
	2	8	53	0.69	0.54	0.52	0.50	3.30	3.30	3.30	3.30
3L3M	0	6	0	0.00	0.00	0.00	0.00	0.00	0.00	0.00	0.00
	1	5	14,149	0.43	0.78	0.84	0.63	2.20	4.20	3.60	1.30

Table 13.2: For each PDB model the results obtained by the screening of pharmacophore model deriving from the initial structure are shown.

PDB	Omitted features	No. of features	No. of hits	AUC				EF			
				1%	5%	10%	100%	1%	5%	10%	100%
1J4H	0	8	0	0.00	0.00	0.00	0.00	0.00	0.00	0.00	0.00
	1	7	22	0.31	0.46	0.48	0.50	0.00	0.00	0.00	0.00
1UYG	0	5	389	0.61	0.71	0.48	0.49	2.40	1.00	0.60	0.60
	1	4	3,290	0.00	0.64	0.81	0.63	0.00	5.30	3.20	1.10
2HZI	0	8	0	0.00	0.00	0.00	0.00	0.00	0.00	0.00	0.00
	1	7	50	0.98	0.60	0.55	0.50	3.10	3.10	3.10	3.10
	2	6	664	0.94	0.95	0.79	0.51	5.50	1.60	1.40	1.40
3BQD	0	9	0	0.00	0.00	0.00	0.00	0.00	0.00	0.00	0.00
	1	8	3	0.68	0.53	0.52	0.50	18.60	18.60	18.60	18.60
	2	7	63	0.98	0.75	0.62	0.51	6.60	6.60	6.60	6.60
3EL8	0	8	8	0.49	0.50	0.50	0.50	0.00	0.00	0.00	0.00
	1	7	72	0.47	0.50	0.50	0.50	0.60	0.60	0.60	0.60
	2	6	825	0.46	0.66	0.58	0.51	1.70	1.70	1.70	1.70
3L3M	0	8	0	0.00	0.00	0.00	0.00	0.00	0.00	0.00	0.00
	1	7	117	1.00	1.00	0.94	0.54	23.70	23.70	23.70	23.70

structure-based pharmacophore modeling and VS [46]. For this study, 43 different protein-ligand complexes have been selected from the RCSB PDB considering the size of the system, the presence only of a single ligand, the absence of metal ions in the binding pocket and the availability of active and decoy compounds in the DUD-E database. Each protein-ligand complex has been submitted to 10 independent MD simulations starting from the same initial coordinates but considering different and randomly seeded velocities. This resulted in 20,000 coordinate sets per system and for each of them, the pharmacophore model was generated by Ligand Scout. A feature vector, which is a specific combination of pharmacophore features, was generated for each pharmacophore model; those derived from MD simulations and those from the PDB model and represented as a bit string (Figure 13.15). Every dimension of the vector represents a unique pharmacophore feature of the system. Wieber *et al.* considered pharmacophore features unique if they differed in both pharmacophore feature type and in the ligand atoms

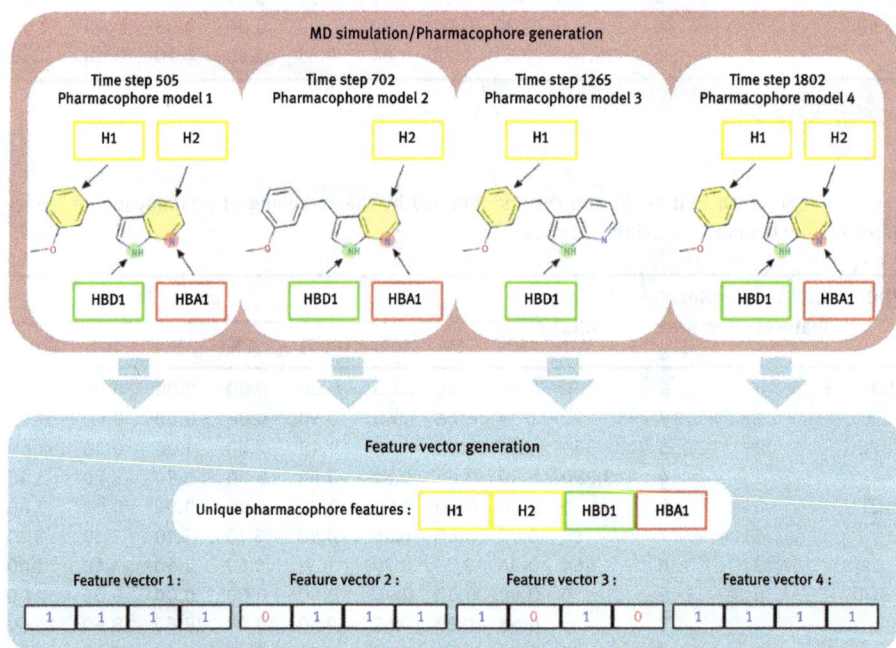

Figure 13.15: Generation of distinct feature vectors: four pharmacophore models obtained at different time steps during the MD simulation are reduced to four feature vectors. The four entries in the feature vector corresponding to the four different pharmacophore features (H1, H2, HBD1, HBA1). The yellow spheres indicate hydrophobic features (H1 and H2), the green small spheres show hydrogen-bond donors (HBD1) and the small red spheres show hydrogen-bond acceptors (HBA1). In this example, there are only three distinct feature vectors since feature vector 1 and feature vector 4 have the same combination of pharmacophore features.

involved. The entries of the feature vectors were initialized with zero; if one of the unique features appeared in a pharmacophore model, the value of the corresponding entry was set to 1.

Then, the authors counted how often a particular feature vector appears (appearance count) during the MD simulations. Thus, instead of using 20,000 individual feature vectors, they obtained a smaller number of distinct feature vectors, which have been observed one or more times during the MD simulations. All feature vectors detected just once were rejected because they were evaluated as random artifacts; therefore, only distinct feature vectors with an appearance count ≥2 were considered in the following step. Since the feature vectors are deficient in structural information required for VS, the conformational energy of the ligand for all structures belonging to a distinct feature vector was calculated using the OpenBabel [47] implementation of the MMFF94 force field [48]. On the basis of the resulting distribution of energies, only the 3D pharmacophore structure corresponding to the ligand conformation with the median energy was taken into account for VS with the aim to avoid extreme ligand conformations (Figure 13.16).

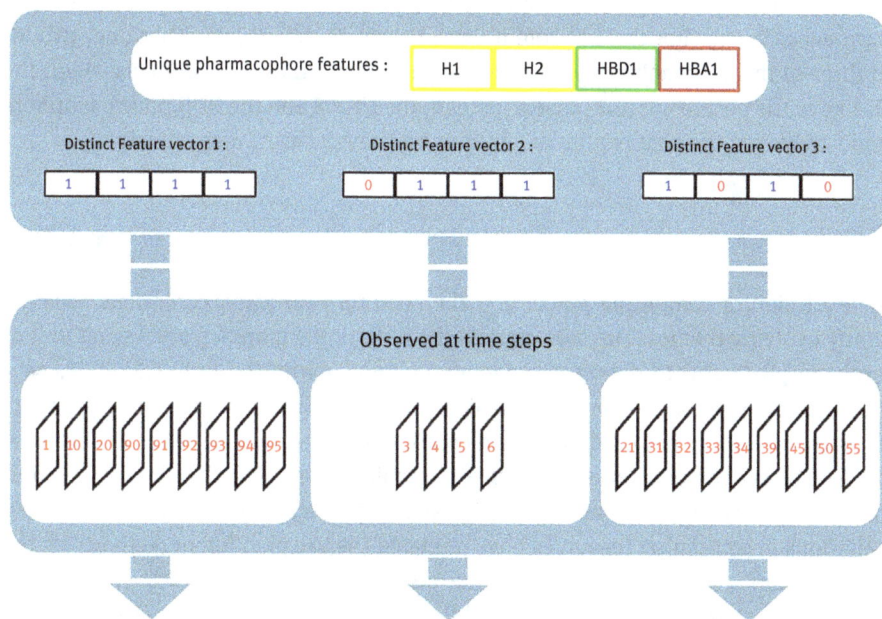

Figure 13.16: Workflow from distinct feature vector to representative pharmacophore model (RPM): the conformational energy of all ligand coordinates having the same distinct feature vector and appearing more than once during the MD simulations was computed with the MMFF94 force field. The coordinate set having the median energy was identified and chosen to build the RPM corresponding to this distinct particular feature vector.

The term RPM has been chosen to denote the combination of the distinct feature vector and representative 3D structural information. For each protein-ligand complex, the RPMs, the pharmacophore with the highest appearance (HRPM) and the PDB pharmacophore model were all screened. In particular, the information contained in the results of the individual VS of RPM was merged and estimated by the CHA. This method operated on the multiple *hit*-lists and combined them into a merged *hit*-list consisting only of unique compounds. VS provided a ranked list (*hit*-list) of compounds; a successful screening ranks active molecules higher than inactive ones.

Finally, the performance of the PDB pharmacophore, of the HRPM, and of the CHA against the (median) performance of the individual RPM was compared. The CHA with respect to the two other approaches was able to obtain remarkable results. In fact, in 80 % of the examined systems, the CHA returns a final *hit*-list, which performed better than a random classifier and better than RPM (median). Considering the HRPM, the number of systems with ROC-AUC >0.5 was 27 out of 40 (compared to 24 for the PDB pharmacophore-based approach), and for the CHA the success rate increased to 32 out of 40. The CHA reduced the number of systems for which VS failed to give usable results from ≈40 % for the PDB pharmacophore model to just 20 %. In addition, the ROC-AUC values obtained with the CHA outperformed the other approaches by far. For the 34 protein-ligand systems, for which either one, two, or all three of the approaches were able to perform better than a random classifier, the CHA gave the best result in 23 cases. Instead, the HRPM and the PDB pharmacophore model achieved the best results in just four and seven cases, respectively.

13.5 Progress in computational technology: The role of the GPU

The evolution of computing power and the need for ever faster computers are constantly evolving themes. The '80s marked the birth of the graphics processing unit or graphical processor, to date, more commonly known with the term GPU [49]. As a crucial component in the computational calculation, GPUs have become a premium necessity in order to help chemistry and computational biology researchers overcome the limitations of previous techniques and obtain more detailed information about complex systems and with more extensive simulation times. Indeed, for many years, limitations in computing resources have rendered the use of different methods for the binding affinity between a drug and the target protein, such as MD simulations, molecular mechanics (MM) or alchemical free energy methods as low performing.

In this regard, the use of modern GPUs have improved the performance simulation of the computational physical methods, such as thermodynamic integration (TI) and free energy perturbation (FEP), and common MD simulations such as quantum chemistry and/or docking simulations, to the point that they approach quantitative predictive accuracy. In fact, compared to CPUs, GPUs have increased throughput from 2 to 5 times faster and software packages such as NAMD [50], AMBER [51], OpenMM [52], GROMACS [53], and CHARMM [54] have been implemented.

Lee et al. described a GPU-accelerated TI implementation of the AMBER 16 pmemd program (pmemdGTI), highlighting that this code can be used to explore simulation time scales up to 100 times longer than a single CPU core for the calculation of ligand-protein binding affinities [55]. The pmemdGTI code performs TI at the speed of 70 % of running an MD simulation with the fast SDFP precision mode, like the ratios seen in CPU versions.

On the same line of research, in 2018 Giese et al., developed and tested a parameter-interpolated thermodynamic integration (PI-TI) method for TI free energy simulations on GPUs [56]. The TI-PI method is suitable for different types of free energy application fields, including solvation, metal ion solvation, QM/MM charge-corrections, point mutations and finally pKa shift prediction. Based on the experimental support, the authors calculated the pKa values of small double-stranded RNA sequences, performing the simulations with the GPU-accelerated version of PMEMD. Although they pushed up to 160 ns, only approximately 100 ns of statistics were necessary, to obtain converged and reproducible results to within 0.25 pKa units.

Based on these scientific observations, the GPU-accelerated alchemical free energy methods represent a powerful tool for new and emerging drug discovery applications.

13.6 Conclusions

The application of enhanced computational techniques for identifying and optimizing lead molecules is currently expanding, but it is still restrained by simulated time scales. However, in the near future, the time scale may improve due to the development of more powerful hardware.

The story of Berberine represents a successful example in this field. Elucidating the structural and energetic requisites underlying the recognition process between Berberine and its molecular target, the G-quadruplex DNA, through Metadynamics simulations, is of paramount relevance to lead efficacious drug design strategies. In this paper, we also described the *Dynophore*, a dynamic pharmacophore that takes into account the conformational flexibility of both the ligands and the targets derived from MD simulations. Although we do not yet have examples of its application on natural compounds, *Dynophore* works as a powerful analysis and validation tool for the evaluation of pharmacophores, ligand-target interaction patterns, and MD simulations. In fact, ligand binding often involves dynamic conformational transitions that may not be evident from a single and static structure. Furthermore, molecular geometry and chemical characteristics of the binding pocket in protein molecules certainly play a crucial role in ligand binding.

In conclusion, the most important aspect of any advanced computational method is its accuracy, Metadynamics has a privileged position amongst enhanced sampling techniques, while *Dynophore* will potentially enhance the predictive power of 3D pharmacophores.

Acknowledgements: This work was partially supported by Dr. Giosuè Costa. The authors also gratefully acknowledge the helpful comments and suggestions of the reviewers, which have improved the presentation.

References

[1] Nizami B, Sydow D, Wolber G, Honarparvar B. Molecular insight on the binding of NNRTI to K103N mutated HIV-1 RT: molecular dynamics simulations and dynamic pharmacophore analysis. Mol Biosyst. 2016;12:3385–95.

[2] Abrams C, Bussi G. Enhanced sampling in molecular dynamics using metadynamics, replica-exchange, and temperature-acceleration. Entropy. 2014;16:163.

[3] Valsson O, Tiwary P, Parrinello M. Enhancing important fluctuations: rare events and metadynamics from a conceptual viewpoint. Annu Rev Phys Chem. 2016;67:159–84.

[4] Laio A, Parrinello M. Escaping free-energy minima. Proc Natl Acad Sci USA. 2002;99:12562–6.

[5] Laio A, Gervasio FL. Metadynamics: a method to simulate rare events and reconstruct the free energy in biophysics, chemistry and material science. Rep Prog Phys. 2008;71:126601.

[6] Barducci A, Bussi G, Parrinello M. Well-tempered metadynamics: a smoothly converging and tunable free-energy method. Phys Rev Lett. 2008;100:020603.

[7] Bonomi M, Barducci A, Parrinello M. Reconstructing the equilibrium Boltzmann distribution from well-tempered metadynamics. J Comput Chem. 2009;30:1615–21.

[8] Bonomi M, Parrinello M. Enhanced sampling in the well-tempered ensemble. Phys Rev Lett. 2010;104:190601.

[9] Dama JF, Parrinello M, Voth GA. Well-tempered metadynamics converges asymptotically. Phys Rev Lett. 2014;112:240602.

[10] Tiwary P, Parrinello M. A time-independent free energy estimator for metadynamics. J Phys Chem B. 2015;119:736–42.

[11] Lipkowitz KB, Cundari TR, Gillet VJ, Boyd DB. Reviews in computational chemistry. Weinheim, Germany: Wiley, 1995. ISSN: 1934-5372. DOI: 10.1002/SERIES6143.

[12] Grazioso G, Limongelli V, Branduardi D, Novellino E, De Micheli C, Cavalli A, et al. Investigating the mechanism of substrate uptake and release in the glutamate transporter homologue Glt (Ph) through metadynamics simulations. J Am Chem Soc. 2012;134:453–63.

[13] Limongelli V, Bonomi M, Marinelli L, Gervasio FL, Cavalli A, Novellino E. Molecular basis of cyclooxygenase enzymes (COXs) selective inhibition. Proc Natl Acad Sci USA. 2010;107:5411–16.

[14] Limongelli V, Marinelli L, Cosconati S, La Motta C, Sartini S, Mugnaini L, et al. Sampling protein motion and solvent effect during ligand binding. Proc Natl Acad Sci USA. 2012;109:1467–72.

[15] Limongelli V, Bonomi M, Parrinello M. Funnel metadynamics as accurate binding free-energy method. Proc Natl Acad Sci USA. 2013;110:201303186.

[16] Moraca F, Amato J, Ortuso F, Artese A, Pagano B, Novellino E, et al. Ligand binding to telomeric G-quadruplex DNA investigated by funnel-metadynamics simulations. Proc Natl Acad Sci USA. 2017;114:E2136–E45.

[17] Bruno A, Scrima M, Novellino E, D'errico G, D'ursi AM, Limongelli V, et al. The glycan role in the glycopeptide immunogenicity revealed by atomistic simulations and spectroscopic experiments on the multiple sclerosis biomarker CSF114(Glc). Sci Rep. 2015;5:9200.

[18] Troussicot L, Guillière F, Limongelli V, Walker O, Lancelin JM, et al. Funnel-metadynamics and solution NMR to estimate protein-ligand affinities. J Am Chem Soc. 2015;137:1273–81.

[19] Comitani F, Limongelli V, Molteni C. The free energy landscape of GABA binding to a pentameric ligand-gated ion channel and its disruption by mutations. J Chem Theory Comput. 2016;12:3398–406.

[20] Yuan X, Raniolo S, Limongelli V, Xu Y. The molecular mechanism underlying ligand binding to the membrane-embedded site of a G-protein-coupled receptor. J Chem Theory Comput. 2018;14:2761–70.

[21] McCreath BS, Badal S, Delgoda R. Pharmacognosy: fundamentals, applications and strategies. Academic Press: 2017.

[22] Yin J, Xing H, Ye J. Efficacy of berberine in patients with type 2 diabetes mellitus. Metabolism. 2008;57:712–17.

[23] Sun Y, Xun K, Wang Y, Chen X. A systematic review of the anticancer properties of berberine, a natural product from Chinese herbs. Anticancer Drugs. 2009;20:757–69.

[24] Bazzicalupi C, Ferraroni M, Bilia AR, Scheggi F, Gratteri P. The crystal structure of human telomeric DNA complexed with berberine: an interesting case of stacked ligand to G-tetrad ratio higher than 1: 1. Nucleic Acids Res. 2012;41:632–8.

[25] Bessi I, Bazzicalupi C, Richter C, Jonker HR, Saxena K, Sissi C, et al. Spectroscopic, molecular modeling, and NMR-spectroscopic investigation of the binding mode of the natural alkaloids berberine and sanguinarine to human telomeric G-quadruplex DNA. ACS Chem Biol. 2012;7:1109–19.

[26] Hassanali AA, Zhong D, Singer SJ. An AIMD study of CPD repair mechanism in water: role of solvent in ring splitting. J Phys Chem B. 2011;115:3860–71.

[27] Kao Y-T, Guo X, Yang Y, Liu Z, Hassanali A, Song QH, et al. Ultrafast dynamics of nonequilibrium electron transfer in photoinduced redox cycle: solvent mediation and conformation flexibility. J Phys Chem B. 2012;116:9130–40.

[28] Gasparotto P, Hassanali AA, Ceriotti M. Probing defects and correlations in the hydrogen-bond network of ab initio water. J Chem Theory Comput. 2016;12:1953–64.

[29] Jong K, Grisanti L, Hassanali A. Hydrogen bond networks and hydrophobic effects in the amyloid β30–35 chain in water: A molecular dynamics study. J Chem Inf Model. 2017;57:1548–62.

[30] Wang L, Fried SD, Markland TE. Proton network flexibility enables robustness and large electric fields in the ketosteroid isomerase active site. J Phys Chem B. 2017;121:9807–15.

[31] Crespo Y, Hassanali A. Unveiling the Janus-like properties of OH–. J Phys Chem Lett. 2015;6:272–8.

[32] Giberti F, Hassanali AA. The excess proton at the air-water interface: the role of instantaneous liquid interfaces. J Chem Phys. 2017;146:244703.

[33] Hassanali AA, Giberti F, Sosso GC, Parrinello M. The role of the umbrella inversion mode in proton diffusion. Chem Phys Lett. 2014;599:133–8.

[34] Cuny J, Hassanali AA. Ab initio molecular dynamics study of the mechanism of proton recombination with a weak base. J Phys Chem B. 2014;118:13903–12.

[35] Jong K, Hassanali AA. A data science approach to understanding water networks around biomolecules: the case of tri-alanine in liquid water. J Phys Chem B. 2018;122:7895–906.

[36] Ulman K, Busch S, Hassanali AA. Quantum mechanical effects in zwitterionic amino acids: the case of proline, hydroxyproline, and alanine in water. J Chem Phys. 2018;148:222826.

[37] Laage D, Elsaesser T, Hynes JT. Water dynamics in the hydration shells of biomolecules. Chem Rev. 2017;117:10694–725.

[38]   Kim S, Peterson AM, Holten-Andersen N. Enhanced water retention maintains energy dissipation in dehydrated metal-coordinate polymer networks: another role for Fe-catechol crosslinks? Chem Mater. 2018;30:3648–3655.

[39]   Ortuso F, Langer T, Alcaro S. GBPM: GRID-based pharmacophore model: concept and application studies to protein-protein recognition. Bioinformatics. 2006;22:1449–55.

[40]   Mobley DL, Dill KA. Binding of small-molecule ligands to proteins: "what you see" is not always "what you get". Structure. 2009;17:489–98.

[41]   Wolber G, Langer T. LigandScout: 3-D pharmacophores derived from protein-bound ligands and their use as virtual screening filters. J Chem Inf Model. 2005;45:160–9.

[42]   Sydow D. Dynophores: novel dynamic pharmacophores. Humboldt-Universität zu Berlin, Lebenswissenschaftliche Fakultät, 2015.

[43]   Mortier J, Prévost Jrc, Sydow D, Teuchert S, Omieczynski C, Bermudez M, et al. Arginase structure and inhibition: catalytic site plasticity reveals new modulation possibilities. Sci Rep. 2017;7:13616.

[44]   Wieder M, Perricone U, Seidel T, Boresch S, Langer T. Comparing pharmacophore models derived from crystal structures and from molecular dynamics simulations. Monatshefte Für Chemie-Chemical Monthly. 2016;147:553–63.

[45]   Mysinger MM, Carchia M, Irwin JJ, Shoichet BK. Directory of useful decoys, enhanced (DUD-E): better ligands and decoys for better benchmarking. J Med Chem. 2012;55:6582–94.

[46]   Wieder M, Garon A, Perricone U, Boresch S, Seidel T, Almerico AM, et al. Common hits approach: combining pharmacophore modeling and molecular dynamics simulations. J Chem Inf Model. 2017;57:365–85.

[47]   O'Boyle NM, Banck M, James CA, Morley C, Vandermeersch T, Hutchison GR. Open Babel: an open chemical toolbox. J Cheminform. 2011;3:33.

[48]   Halgren TA. Merck molecular force field. II. MMFF94 van der Waals and electrostatic parameters for intermolecular interactions. J Comput Chem. 1996;17:520–52.

[49]   Hopgood FR, Hubbold RJ, Duce D. Advances in computer graphics II. Springer Science & Business Media: 1986.

[50]   Phillips JC, Braun R, Wang W, Gumbart J, Tajkhorshid E, Villa E, et al. Scalable molecular dynamics with NAMD. J Comput Chem. 2005;26:1781–802.

[51]   Salomon-Ferrer R, Case DA, Walker RC. An overview of the Amber biomolecular simulation package. Wiley Interdiscip Rev: Comput Mol Sci. 2013;3:198–210.

[52]   Eastman P, Pande V. OpenMM: a hardware-independent framework for molecular simulations. Comput Sci Eng. 2010;12:34–9.

[53]   Pronk S, Páll S, Schulz R, Larsson P, Bjelkmar P, Apostolov R, et al. GROMACS 4.5: a high-throughput and highly parallel open source molecular simulation toolkit. Bioinformatics. 2013;29:845–54.

[54]   Brooks BR, Brooks III CL, Mackerell AD, Nilsson L, Petrella RJ, Roux B, et al. CHARMM: the biomolecular simulation program. J Comput Chem. 2009;30:1545–614.

[55]   Lee TS, Hu Y, Sherborne B, Guo Z, York DM. Toward fast and accurate binding affinity prediction with pmemdGTI: an efficient implementation of GPU-accelerated thermodynamic integration. J Chem Theory Comput. 2017;13:3077–84.

[56]   Giese TJ, York DM. A GPU-accelerated parameter interpolation thermodynamic integration free energy method. J Chem Theory Comput. 2018;14:1564–82.

Ricardo Bruno Hernández-Alvarado, Abraham Madariaga-Mazón
and Karina Martinez-Mayorga

14 Prediction of toxicity of secondary metabolites

Abstract: The prediction of toxicological endpoints has gained broad acceptance; it is widely applied in early stages of drug discovery as well as for impurities obtained in the production of generic or equivalent products. In this work, we describe methodologies for the prediction of toxicological endpoints compounds, with a particular focus on secondary metabolites. Case studies include toxicity prediction of natural compound databases with anti-diabetic, anti-malaria and anti-HIV properties.

Keywords: computational toxicology, secondary metabolites, drug discovery, pesticides, predictive models, QSAR

14.1 Introduction

The prediction of toxicity endpoints relies heavily on QSAR models. The evolution of these models is described in the literature [1, 2]. Validated models are now accepted in several countries for the prediction of toxicity endpoints for regulatory purposes [3]. The OECD guidelines for the testing of chemicals are widely accepted for safety testing and assessment of pesticides, personal care products, industrial chemicals and even to aid in decision-making in emergency responses [4, 5]. In November 2004, the OECD member countries agreed on the "OECD Principles for the Validation, for Regulation Purposes, of (Quantitative) Structure-Activity Relationship Models" [6]. For a discussion on the comparison and complementarity of the steps of KDD and QSAR methodologies and how they fit in the OECD principles, the reader is referred to the literature [7]. On the preceding chapter (Toxicity of Secondary Metabolites), we present an overview of different sources of secondary metabolites and their toxicities. In addition, we elaborate on toxicity assessment in drug discovery and agrochemistry. Here, we provide an overview of tools and methods for the development of predictive models with an emphasis in regulatory settings, and we finalize with selected examples, including the prediction of toxicity of secondary metabolites.

14.2 Predictive models of toxicological endpoint

Toxicity prediction tools are becoming more useful in many areas. Food-chemistry and pharmaceutical sciences are implementing innovative methods in order to

This article has previously been published in the journal *Physical Sciences Reviews*. Please cite as: Hernandez-Alvarado, R.B., Madariaga-Mason, A., Martinez-Mayorga, K. Prediction of toxicity of secondary metabolites. *Physical Sciences Reviews* [Online] **2019** DOI: 10.1515/psr-2018-0107.

https://doi.org/10.1515/9783110579352-015

reduce costs, time and pollution. In addition, scientists are also developing new algorithms, including machine learning and deep learning methods, to improve the accuracy of the predictions. For example, eToxPred allows the estimation of toxicity of drug candidates based on Restricted Boltzmann Machine (RBM), Deep Belief Network (DBN) and Extremely Randomized Trees, or Extra Trees (ET) algorithm [8]. ToxiPred predicts aqueous toxicity of small chemical molecules in T. pyriformis [9]. Protox, based on a total of 33 machine-learning models, allows for the prediction of various toxicity endpoints such as acute toxicity, hepatotoxicity, cytotoxicity, carcinogenicity, mutagenicity, immunotoxicity, adverse outcomes (Tox21) pathways and toxicity targets [10] among others.

Consequently, the number of publications regarding toxicology predictions is increasing every year, as shown in Figure 14.1.

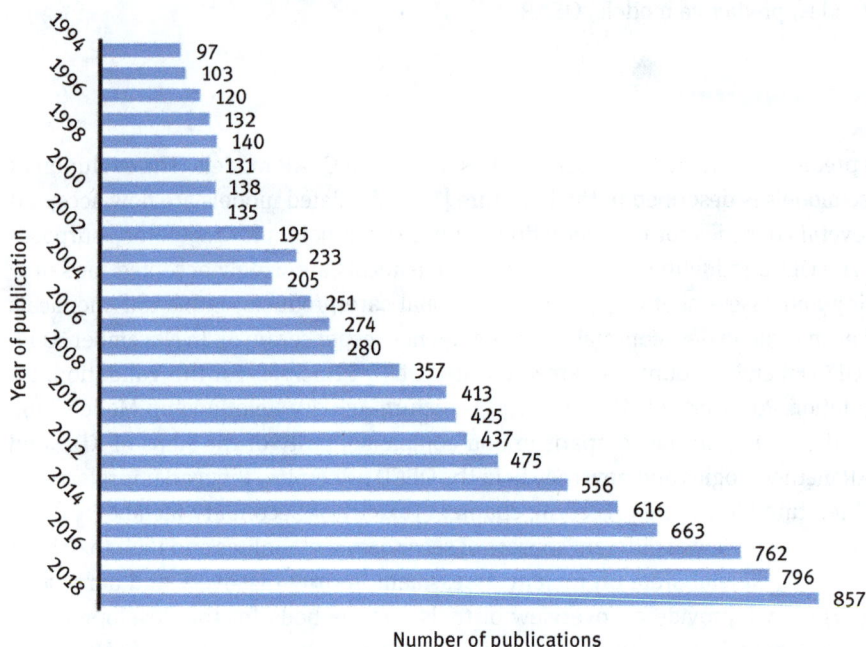

Figure 14.1: The number of publications on Web of Science, retrieved using the keyword "Toxicity Prediction", throughout the years.

14.2.1 Toxicological endpoints commonly predicted

The biological, toxicological and physicochemical endpoints commonly predicted highly depend on the regulation of each country. We have recently collected a representative list of endpoints [7] used in different countries. The evaluations for the testing of acute effects of chemicals are commonly called "6-pack" and include acute inhalation, acute oral, medial lethal dose (LD_{50}), acute dermal, eye irritation, skin irritation, dermal sensitization. Until recently, these evaluations were all performed experimentally in animal models. However, the use of *in vitro* and *in silico* evaluations is now accepted for regulatory purposes. These alternatives are cheaper and faster than *in vivo* full animal experiments, allowing to significantly reduce the use of experimental animals, decrease costs and time.

14.2.2 Databases of toxic compounds

Databases with biological and chemical information are essential in drug discovery campaigns. These databases contain information from experimental data published in scientific papers or specialized literature and they may also contain computational analyses. Being toxicity a key aspect in drug discovery and environmental sciences, there have been efforts to collect toxicological information in databases. Table 14.1 summarizes databases and servers with toxicological relevance.

Table 14.1: Resources with annotated toxicological information.

Name (reference)	Description
Toxnet (https://toxnet.nlm.nih.gov/)	Group of databases covering chemicals and drugs, diseases and the environment, environmental health, occupational safety and health, poisoning, risk assessment and regulations, and toxicology. Maintained by the National Library of Medicine (NLM) [11].
DSSTox (https://www.epa.gov/chemical-research/distributed-structure-searchable-toxicity-dsstox-database)	Distributed Structure-Searchable Toxicity (DSSTox) Database provides chemical information and annotations associated with toxicity data, to address mainly the needs for building quality models in predictive toxicology. Provides a structure browser [12].
Leadscope Toxicity Database (http://www.leadscope.com/toxicity_database/)	A database containing over 180,000 chemical structures with over 400,000 toxicity study results.
Akos (commercial) (http://www.akosgmbh.de/accelrys/data bases/toxicity.htm)	Database with over 170,000 chemical substances and comprehensive coverage of reported toxic properties.
ToxCast Dashboard (https://www.epa.gov/chemical-research/toxcast-dashboard)	An online platform with data on over 9,000 chemicals and more than 1,000 high-throughput assay endpoint components [13].

(continued)

Table 14.1 (continued)

Name (reference)	Description
AcTOR (http://actor.epa.gov/actor)	Aggregated Computational Toxicology Online Resource (AcTOR) from the Environmental Protection Agency (EPA) aggregates public sources of chemical toxicity from over 1,000 public sources on over 500,000 chemicals. Searchable by chemical name or structure [14].
CompTox (https://comptox.epa.gov/dashboard)	Computational Toxicology (CompTox) Chemicals Dashboard is an online tool that integrates chemical, toxicity and exposure information for over 760,000 chemicals [15].
ATSDR (https://www.atsdr.cdc.gov/substances/index.asp)	Agency for toxic substances & Disease Registry (ATSDR) is a web portal with information about toxic substances, toxicological information by health effects or chemical class and toxicological information by the audience (toxicologists, health care experts, etc.) [16].
ECOTOX (http://www.epa.gov/ecotox/)	The Eco-Toxicology knowledgebase is a comprehensive, publicly available database providing single chemical environmental toxicity data on aquatic life, terrestrial plants and wildlife. It integrates three existing USEPA datafiles, AQUIRE (aquatic organisms), Phytotox (terrestrial plants) and Terretox (wildlife species). Each record contains information about the chemical, organism, exposure condition and observed effect under which the toxicity test was conducted. The toxicological databases are accessible online.
IRIS (http://www.epa.gov/iris/index.htm)	The Integrated Risk Information System is an online database of toxicity information, providing quantitative human health carcinogenic/hazard data, ambient water quality criteria and maximum contaminant levels. The database is regularly updated and reviewed [17].
SuperToxic (http://bioinformatics.charite.de/super toxic/)	Compiles about 60,000 compounds and their structures, classified according to their toxicity, based on more than 2 million measurements. This information can be used to aid in the evaluation of the risks of newly designed compounds and an indication of biological interactions with the aid of similarity searches [18].
ArachnoServer (http://www.arachnoserver.org/)	Is a manually curated database containing information on the sequence, three-dimensional structure and biological activity of protein toxins derived from spider venom [19].

14.2.3 Methods and software to predict toxicity

In general, *in silico* prediction of toxicology follows an algorithm that consists in four main steps: (a) data collection and cleaning, (b) calculation and selection of molecular descriptors, (c) model generation, and (d) model evaluation. A variety of independent tools are able to accomplish each of these steps and can be executed through different software; a list of frequently used software is provided in Table 14.2. Some of these programs are commercially available and others open code. The differences in databases and algorithms used on each software frame their use and scope.

Table 14.2: Common software used in the development of models to predict toxicity.

Software		Brief description
Structural alerts		
Derek Nexus	https://www.lhasalimited.org/products/derek-nexus.htm	This software is a commercial offering from Lhasa Limited. Derek Nexus is an expert knowledge-based software which uses structure-activity relationships to predict the toxicity of novel compounds. It can assess toxicity for mammals and bacterium across 74 different endpoints including Mutagenicity, Skin Sensitisation, and Chromosome Damage.
Toxtree	http://toxtree.sourceforge.net	Open-source application based on tree decision toward estimation of toxic hazard. Toxtree was developed by the European Commission Joint Research Centre's European Chemicals Bureau. It was firstly designed for Cramer Classification, but the last versions include skin irritation and corrosion assessment [20].
ToxAlerts	http://ochem.eu/alerts	Open web-based platform, where users can upload new toxicophores and SAs. This web application is capable of virtually screen databases or compound libraries to identify certain alerts on chemicals [21].
Sahara Nexus	https://www.lhasalimited.org/products/sarah-nexus.htm	This is commercial software available from Lhasa Limited. Sarah Nexus is a statistical tool for predicting the mutagenic potential of compounds. It is commonly used alongside Derek Nexus to provide a prediction which can be used to meet the ICH M7 regulatory requirements.

(continued)

Table 14.2 (continued)

Software		Brief description
Physiologically based Toxicokinetic and Toxicodynamic models		
MeGen	https://megen.useconnect.co.uk	Web application, able to generate custom exportable codes to use in commercial software. It includes a physiologically based pharmacokinetics database [22].
WinNonLin	https://www.certara.com	Specialized in the non-compartmental analysis regarding PBTK models. It has the capacity to integrate built models and is compatible with statistical software [23].
Read-across		
OECD QSAR Toolbox	https://www.qsartoolbox.org/	Free downloadable software developed by OECD and released in 2008. It provides the ability to categorize chemicals by the mechanism of action. It is also possible to elucidate metabolic profiles. The toolbox is intended to be used in government affairs, so detailed reports could be generated by the software to support a certain prediction.
ToxRead	http://www.toxread.eu/	User-friendly tool programmed in Java (able to run on different operating systems). It provides a chemoinformatics package with useful features as similarity calculation, molecular depiction, among others [24].
QSAR		
T.E.S.T.	https://www.epa.gov	Free EPA software with precomputed models. It estimates different endpoints such as LD_{50} (quantitative) and Ames mutagenicity (qualitative), through different methods such as K-Nearest Neighbors and Hierarchical clustering. This software enables batch predictions and shows the final calculated model.
VEGA	http://www.caesar-project.eu/	VEGA provides access to different QSAR programs and models. The result can be outputted as an Excel file. Additionally, CAESAR software is compatible with VEGA as a module for diverse endpoints such as skin sensitization. CAESAR was firstly developed to produce QSAR models adapted to REACH legislation [25].

(continued)

Table 14.2 (continued)

Software		Brief description
QSARINS	http://www.qsar.it/	User-friendly platform intended to develop a build from scratch QSAR models. The license could be obtained by request. It is focused on external validation of QSAR models and implements a genetic algorithm to the selection of variables [26].
Online platforms		
Chembench	https://chembench.mml.unc.edu/	Public web portal for the development and distribution of QSAR models, as well as teaching [27].
QSARDB	https://qsardb.org/	A repository for (Q)SAR/QSPR models and datasets for the discovery, exploration, citing and predicting of models [28].

Additionally, statistical packages and data mining programs allow the construction of predictive models from scratch. Nevertheless, the best insight to choose relies on the characteristics of the problem at hand.

Methodologies commonly used for toxicity prediction are:

Structural Alerts (SAs). SAs relate the information on the potential toxicity of molecular fragments associated with high chemical reactivity, as shown in Figure 14.2. Additional examples are shown in Table 14.3. These alerts are based on information reported in the literature and the opinion of human experts plays a key role. The interest in SAs increased since the European Registration, Evaluation, Authorisation, and Restriction of Chemicals (REACH) regulation accepted the methodology in 2006. SAs

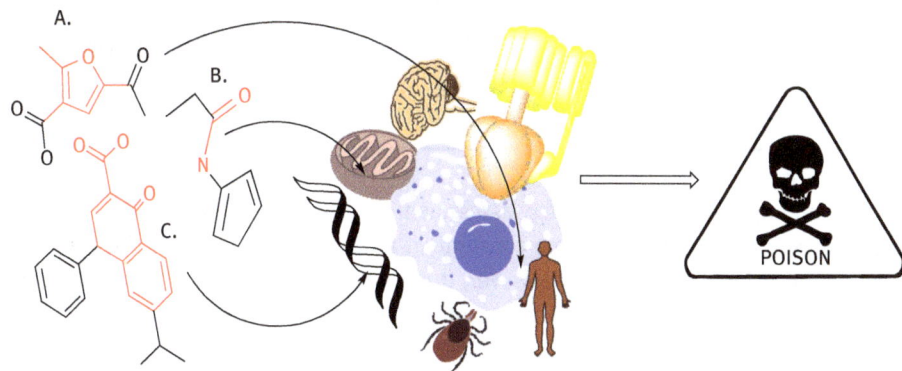

Figure 14.2: Mutagenic structural alerts (marked in red) of different molecular fragments such as furan (A), aromatic amides (B) and quinolones (C).

Table 14.3: Examples of known structural alerts. The different groups are characterized by their high reactivity.

Group	Fragment	Structure	Observations
Alkyl Groups	Aldehydes		Covalent bond formation with nucleophilic sites on DNA and enzymes [29].
	N-nitrosamines		Able to cause point mutations, *a priori* metabolic activation, leading to oncogene activation [30].
	Hydrazines		Classical model hepatotoxin, it has presented carcinogenicity and chronic toxicity in rats and mice, but molecular mechanisms remain elusive [31].
	Carbamates		Inhibition of Acetylcholinesterase, leading to encephalopathies, photophobia and irritability [32].
Alkyl halides	Alkyl halogen Allylic halogen		Functional moieties that have been related to DNA reactivity, i. e. mutagenic activity. These alerting structures are presented in common impurities in pharmaceutical products [33].
	Heteroatomic halogen		
Aromatic groups	Nitroaromatic		Well-recognized moieties for mutagenicity. The molecular mechanism is believed as the formation of superoxide anion radicals [34].
	N-Acylated aminoaryls		DNA altering groups. They were classified in Group 1 on the DNA-affecting functional groups by the Pharmaceutical Research and Manufacturing Association in 2004 [35].
	Aminoaryls		

became an important chemical safety assessment due to several factors such as simplicity, easy interpretation, short time consuming, among others. These studies provide information to predict toxicants. However, it does not include unknown toxicophores and could not make a broader prediction on less-studied cases.

Physiologically based Toxicokinetic and Toxicodynamic models (PBTK, PBTD). This recent approach involves anatomical and physiological descriptors that modify the Administration, Distribution, Metabolism, and Excretion (ADME) process of toxicants. Then, the relationship among the descriptors is transformed into a mathematical equation, in order to predict the concentration of the molecule of interest at a given time, on the site of action or in a specific organ. Similar to SAs, the models are constructed with previous observations and information from the literature. An advantage of PBTK and PBTD is the possibility of species extrapolation because it uses independent organ descriptors [36]. Due to the complexity of these studies, only a few PBTK software packages are available [23].

Read-Across. Read-Across uses chemical analogs to predict the toxicity of an unknown molecule [7]. For example, a compound with unknown LD_{50} must have similar toxicokinetics to the molecules with known the LD_{50}, from which the LD_{50} will be estimated. Read-across may give a qualitative or quantitative result, depending on the information available. The qualitative read-across involves binary answers, e. g. mutagenic or not mutagenic. The transparency and easy-interpretation of read-across have produced a particular interest in regulation purposes. Nevertheless, it has also limitations such as the dependence of accuracy and strength of the prediction to the number of analogs on the group. If the number of analogs is insufficient or they have complex toxicity profiles, it is preferred to use QSAR or another similar approach [37].

Quantitative structure-activity relationship (QSAR). QSAR is one of the most used tools for elicit fast and reliable predictions, not only on toxicology disciplines but also in medicinal chemistry, biology and in many other fields [38, 39]. For example, the number of hits obtained on Web of Science using the keyword "Toxicity Prediction", in combination with "QSAR", is ten times more frequent than when using the keywords "Structural alerts" and "Read Across". QSAR methodologies consist of the generation of mathematical models which are able to correlate physicochemical properties (represented by descriptors) or biological activities to chemical structures [40]. Since the origins of QSAR, back in 1962 [41], continuous efforts have been made to develop useful approaches based on the relationship between molecular structure and activity. Nowadays, it is possible to develop QSAR models to predict potential toxicological endpoints such as LD_{50}, developmental toxicity, etc. However, it should be noted that complex endpoints (such as *in vivo* toxicological ones) are more challenging than those based on *in vitro* assays or the prediction of physicochemical properties. The robustness of QSAR predictions in toxicity has led to its acceptance regulatory settings; this will be discussed in Section 14.5. Whenever possible, QSAR models should be interpretable to increase the clarity and transferability of the model.

14.3 Toxicity prediction in regulatory settings

The regulatory compliance of toxicity assessment is mostly handled by the OECD. A series of guides are available online and are continuously updated. The information provided also includes current and emerging challenges in public health. The OECD guidelines for the testing of chemicals is a collection of about 150 testing methods. The "Guidance Document on the Validation of (Quantitative) Structure-Activity

Relationship [(Q)SAR] Models" [6], provides a detailed description of the use of QSAR models for the prediction of toxicity endpoints.

The prediction of toxicity for regulatory purposes does not apply to any and all toxicity endpoints. Moreover, each country or geographic region (i. e. the European Union) establishes its own requirements. In terms of methodological protocols to perform and report QSAR models, there is a consensus among several countries to follow OECD principles. A list of endpoints requested by some countries is reported by Gomez, et al. [7]. Needless to say, that even in the case that a given endpoint is eligible for prediction of toxicity, additional limitations apply. For example, the compounds evaluated must fall within the applicability domain, and defined criteria for the statistics of the models should be met. As mentioned above, the prediction of toxicity for regulatory purposes is only allowable for impurities in the industrial chemical procedures, not for the active ingredients, which in fact is in a much higher concentration with respect to the impurities. Last but not least, risk assessment is performed taking into account the relative toxicity of the impurity with respect to the active ingredient, and its concentration. For pesticides, the calculation of worst-case-possible contribution by an impurity to the toxic hazards of the active ingredient is described in the "Manual on the development and use of FAO and WHO specifications for pesticides".

The amount, type and quality of the information available largely define the type of model to be developed, for a particular endpoint. While knowledge-based predictions can be used for small data sets, statistically-based predictions are more suited for large datasets. In general, it can fall into ligand-based or target-based models.

Ligand-based models are typically performed using SAR and/or QSAR methodologies. These models are based on the premise that similar compounds will behave similarly, in terms of biological and toxicological activities. However, it is also known that there are exceptions, called activity cliffs [42–44]. Ideally, after the identification and potential remission of activity cliffs, predictive tools would allow for the identification of potential hazards, such as physicochemical property or structural features that are associated with adverse effects. Predictive models can be used on a wide range of endpoints, from enzymatic assays to animal models, including those focused on pharmacokinetics and pharmacodynamics profiles.

Target-based models are particularly valuable when there is not enough information for the development of SAR or QSAR models or when the binding recognition process is particularly relevant. Analysis of potential off-target effects may aid to prevent potential safety issues. This approach requires a detailed analysis of the interaction of the ligands with the biological target under study. For example, following this structure-based approach, Szczepek et al. improved the safety profile of a set of compounds by reducing off-target cleavage of zinc-finger nucleases, associated with cytotoxicity [45].

The more complex endpoint is the more challenging the prediction becomes. Complex endpoints are characterized by having multiple mechanisms involved, either by the interaction of one ligand with multiple targets, by the interaction of multiple ligands to the same target, or a combination of both, which is observed for instance in the recognition of odors and fragrances. Complexity might also arise from the interaction at different binding sites. In other words, for the development of predictive models (particularly linear-based regression models), the set of molecules under study should follow the same kinetic mechanism (agonism, antagonism, etc.). It could even be observed that the effect of a given molecule could be dose-dependent, being an agonist at certain concentrations and an antagonist at a different concentration. The reason for this apparent controversy lies in the binding sites occupied. In this case, the binding site where the agonistic effect takes place is saturated, and the compound can bind to an allosteric site with an antagonistic effect. Statistics-based models are particularly sensitive to this type of divergent information. In turn, knowledge-based models are often interpretable and could even inform of adverse outcome pathways AOP.

Recently, Allen et al. [46] show an improvement on predictive toxicity models by incorporating three types of descriptors: chemical descriptors, protein target descriptors from protein-ligand interactions and descriptors derived from *in vitro* cell cytotoxicity dose-response data. The addition of the two later descriptors sets enhanced classification rates as well as interpretability of the models.

For an overview of the relevance and challenges of toxicity predictions in the context of regulatory settings, the reader is referred to the review article by Gleeson et al. [47] and references cited therein. As pointed out by the authors, low concordance has been observed between the toxicity on animal models and humans [48] which could explain the appearance of previously undetected toxicity issues that lead to the withdraw of approved drugs [49]. Relations of toxicity values across different species have been attempted using different approaches; although rough estimates can be proposed, prediction of toxicity values across species is rarely met.

Last but not least, the use of deep learning has made its appearance for the prediction of toxicity endpoints. In 2016, Mayr et al. won the Tox21 Data Challenge using deep learning [50]. For this data, the authors found that deep learning excelled in toxicity prediction and outperformed many other computational approaches like naive Bayes, support vector machines and random forests. Deep learning holds good promises, particularly for multi-objective problems. Gathering of further examples will frame the applicability scope of this method and will help avoid building false expectations.

14.4 Case studies

14.4.1 Toxicity prediction using different software

Concerning toxicity prediction of secondary plant metabolites, Glück et al. performed the *in silico* prediction of genotoxicity of 609 food-relevant phytochemicals. The authors found that the sensitivity and specificity were improved by the combined use software, such as VEGA models SARpy, KNN, ISS, CAESAR and LAZAR. They found that (-)-asimilobine, aloin, annoretine, chrysothrone, coptisine, elymoclavine and thalicminineare predicted to be genotoxic with high probability and that it is highly probable that pyrrolizidine alkaloids are carcinogenic [51].

Additional applications of toxicity prediction of secondary metabolites have focused on screening [52]; identification and elucidation of adverse outcome pathways [53]; or for the comparison with available experimental data [54].

The different software available can complement each other, for example, based on the chemical space coverage. However, if the compounds under study are within the applicability domain and the predictions have high confidence, the results obtained from different software coincide. Table 14.4 lists examples where T.E.S.T. and Derek correctly predict the mutagenicity of a series of secondary metabolites reported in the literature.

Table 14.4: Predicted and reported mutagenicity of selective secondary metabolites.

Source	Molecule	T.E.S.T.	Derek	Reported mutageni-city	Reference
Cinnamomum zeylanicum	Cinnamaldehyde	Positive	Plausible	Positive	Reported in EPA-T. E.S.T. dataset
Taxus brevifolia	Paclitaxel	Negative	Inactive	Negative	Reported in EPA-T . E.S.T. dataset
Cinchona spp.	Quinine	Negative	Inactive	Negative	Munzer [55]
Atropa belladona	Atropine	Negative	Inactive	Negative	Ramel [56]
Colchicum autum-nale L.	Colchicine	Negative	Inactive	Negative	Reported in EPA-T. E.S.T. dataset
Nicotiniana spp.	Nicotine	Negative	Inactive	Negative	Reported in EPA-T. E.S.T. dataset
Psilocybe spp.	Psilocybins	Negative	Inactive	Negative	Van Went [57]
Gyromitra esculenta	Gyromitrin	Negative	Inactive	Negative	Reported in EPA-T. E.S.T. dataset

Interestingly, in these examples, only the cinnamaldehyde is reported as mutagenic and is predicted as such by both programs used, giving confidence of future compounds of the same type.

14.4.2 Toxicity prediction of an entire database of natural products with antidiabetic activity

In a recent study, we reported a database named DiaNat-DB [58] (available at http://rdu. iquimica.unam.mx/handle/20.500.12214/1186). This database contains 336 compounds (secondary metabolites) with antidiabetic activity from plants used in traditional medicine in different countries. An estimation of the mutagenicity *in vitro*, evaluated in Sarah Nexus is summarized in Table 14.5. From the 336 compounds contained in DiaNat-DB, only 19 were out of the domain, and 31 resulted in equivocal classification (there is information against and in favor of a certain prediction). This shows that the chemical space covered, and information contained in Sarah Nexus allowed the evaluation of the 85 % of the secondary metabolites contained in DiaNat-DB. Interestingly, 245 of those compounds are predicted negative for mutagenicity *in vitro*.

Table 14.5: Mutagenicity *in vitro* evaluated in Sarah Nexus.

Classification	Number of molecules
Equivocal	31
Negative	245
Out of domain	19
Positive	39
Total	334

Further analysis of this data will be performed to explore the structural details of secondary metabolites that are predicted mutagenic. This information will be informative to focus on those compounds and revise their use in folk medicine. If required, the experimental evaluation for mutagenicity will place a warning on its use.

It should be noted that there is a fundamental difference between profiling a database for further development and the use of predictive methods in substitution of experimental evaluation. While in the first case the risk of false positives derives in costs; the second case would place a potential hazard to humans or the environment. Therefore, in regulatory settings, the use of toxicity prediction values as substitution of experimental evaluation is admissible only for impurities derived of manufacturing processes, not for the main components (active ingredient).

14.4.3 Toxicity prediction of natural compounds from the African flora

Another *in silico* toxicity assessment regarding secondary metabolites was that carried out by Onguéné et al. [59] In that study, databases of natural products (total of 806 chemical entries) from African flora (p-ANAPL, AfroMalariaDb, Afro-HIV) were compared for structural diversity using Principal Component Analysis and most common substructure analyses. In addition, structural alert analyses were performed using Lhasa Ltd. Briefly, the results showed that only a small portion of the libraries were predicted beyond the toxicity thresholds. This finding encourages the possibility to further analyze this database for additional applications [59].

Comparisons and further analyses of large compound collections as the ones described here provide a useful means for the discovery of new applications and structurally novel active compounds.

14.5 Conclusions

The assessment of toxicity values is valuable for the development of drugs, agrochemicals, cosmetics, food additives, etc. Toxicity values predicted using computational models are now accepted in regulatory settings as an alternative to animal testing if particular requirements are met. The OECD guidance for the development, use, and reporting of QSAR models are followed around the world and adherence of other countries is expected. The continuous update of databases; the development and implementation of improved algorithms; and novel strategies of external validation will contribute to further develop predictive models of toxicity. Furthermore, a large compound collection of natural products from Africa is available [60]. This comprehensive database contains more than 4,500 natural products and includes the botanical classification.

Acknowledgements: This work was supported by Instituto de Química-UNAM, and DGAPA-UNAM (PAPIIT IN210518). The authors thank ChemAxon and Lhasa Limited, for kindly providing academic licenses of their software.

References

[1]   Tropsha A, Golbraikh A. Predictive QSAR modeling workflow, model applicability domains, and virtual screening. Curr Pharm Des. 2007;13:3494–504.

[2]   Tropsha A, Wang SX. QSAR modeling of GPCR ligands: methodologies and examples of applications. In: Bourne H, Horuk R, Kuhnke J, Michel H, editors. GPCRs: from deorphanization to lead structure identification. Berlin: Springer, 2007:49–74.

[3]   Veith GD. On the nature, evolution and future of quantitative structure-activity relationships (QSAR) in toxicology. SAR QSAR Environ Res. 2004;15:323–30.

[4]   Demchuk E, Ruiz P, Chou S, Fowler BA. SAR/QSAR methods in public health practice. Toxicol Appl Pharmacol. 2011;254:192–7.

[5]   Ruiz P, Begluitti G, Tincher T, Wheeler J, Mumtaz M. Prediction of acute mammalian toxicity using QSAR methods: a case study of sulfur mustard and its breakdown products. Molecules. 2012;17:8982–9001.

[6]   Guidance Document on the Validation of (Quantitative) Structure-Activity Relationship [(Q)SAR] Models. OECD. 2014. DOI: 10.1787/9789264085442-en.

[7]   Gómez-Jiménez G, Gonzalez-Ponce K, Castillo-Pazos DJ, Madariaga-Mazón A, Barroso-Flores J, Cortés-Guzman F, et al. The OECD principles for (Q)SAR models in the context of knowledge

discovery in databases (KDD). In: Karabencheva-Christova T, Christov C, editor(s). Advances in protein chemistry and structural biology Vol. 113. Amsterdam: Elsevier, 2018:85–117.

[8] Pu L, Naderi M, Liu T, Wu H, Mukhopadhyay S, Brylinski M. eToxPred: a machine learning-based approach to estimate the toxicity of drug candidates. BMC Pharmacol Toxicol. 2019;20:2.

[9] Mishra N, Singla D, Agarwal S, Consortium OSDD, Raghava GPS. ToxiPred: a server for prediction of aqueous toxicity of small chemical molecules in T. Pyriformis. J Transl Toxicol. 2014;1:21–7.

[10] Drwal , Banerjee P, Dunkel M, Wettig MR, Preissner R. ProTox: a web server for the in silico prediction of rodent oral toxicity. Nucleic Acids Res. 2014;42:53–8.

[11] Wexler P. TOXNET: an evolving web resource for toxicology and environmental health information. Toxicol. 2001;157:3–10.

[12] Richard A, Williams C. DSSTox chemical-index files for exposure-related experiments in arrayexpress and gene expression omnibus: enabling toxico-chemogenomics data linkages. Bioinformatics. 2009;25:692–4.

[13] Richard A, Judson RS, Houck KA, Grulke CM, Volarath P, Thillainadarajah I, et al. ToxCast chemical landscape: paving the road to twenty-first century toxicology. Chem Res Toxicol. 2016;26:1225–51.

[14] Judson R, Richard A, Dix D, Houck K, Elloumi F, Martin M, et al. ACToR–aggregated computational toxicology resource. Toxicol Appl Pharmacol. 2008;233:7–13.

[15] Williams AJ, Grulke CM, Edwards J, McEachran AD, Mansouri K, Baker NC, et al. The CompTox chemistry dashboard: a community data resource for environmental chemistry. J Chem inform. 2017;9:61.

[16] Fay M, Donohue J, De Rosa C. ATSDR evaluation of health effects of chemicals. VI. Di(2-ethylhexyl)phthalate. Agency for Toxic Substances and Disease Registry. Toxicol Ind HealthToxicol. 1999;15:651–746.

[17] Review of EPA's Integrated Risk Information System (IRIS) Process. EPA 2014. DOI: 10.17226/18764.

[18] Schmidt U, Struck S, Gruening B, Hossbach J, Jaeger IS, Parol R, et al. SuperToxic: a comprehensive database of toxic compounds. Nucleic Acids Res. 2008;37:D295–9.

[19] Pineda S, Chaumeil P, Kunert A, Kaas Q, Thang M, Le L, et al. ArachnoServer 3.0: an online resource for automated discovery, analysis and annotation of spider toxins. Bioinformatics. 2018;34:1074–6. DOI: 10.1093/bioinformatics/btx661.

[20] Bhatia S, Schultz T, Roberts D, Shen J, Kromidas L, Api AM. Comparison of Cramer classification between Toxtree, the OECD QSAR Toolbox and expert judgment. Regul Toxicol Pharmacol. 2015;71:52–62.

[21] Sushko I, Salmina E, Potemkin VA, Poda G, Tetko IV. ToxAlerts: a web server of structural alerts for toxic chemicals and compounds with potential adverse reactions. J Chem Inf Model. 2012;52:2310–16.

[22] Loizou G, Hogg A. MEGen: a physiologically based pharmacokinetic model generator. Front Pharmacol. 2011;2:56.

[23] Bessems JG, Loizou G, Krishnan K, Clewell HJ, Bernasconi C, Bois F, et al. PBTK modelling platforms and parameter estimation tools to enable animal-free risk assessment. Regul Toxicol Pharmacol. 2014;68:119–39.

[24] Gini G, Franchi AM, Manganaro A, Golbamaki A, Benfenati E. ToxRead: a tool to assist in read across and its use to assess mutagenicity of chemicals. SAR QSAR Environ Res. 2014;25:999–1011.

[25] Chaudhry Q, Piclin N, Cotterill J, Pintore M, Price NR, Chrétien JR, et al. Global QSAR models of skin sensitisers for regulatory purposes. Chem Cent J. 2010;4:S5.

[26] Gramatica P, Cassani S, Chirico N. QSARINS-chem: insubria datasets and new QSAR/QSPR models for environmental pollutants in QSARINS. J Comput Chem. 2014;35:1036–44.

[27] Capuzzi SJ, Kim IS-J, Lam WI, Thornton TE, Muratov EN, Pozefsky D, et al. Chembench: a publicly accessible, integrated cheminformatics portal. J Chem Inf Model. 2017;57:105–8.

[28] Ruusmann V, Sild S, Maran U. QSAR DataBank repository: open and linked qualitative and quantitative structure–activity relationship models. J Cheminform. 2015;7:32.

[29] LoPachin RM, Gavin T. Molecular mechanisms of aldehyde toxicity: a chemical perspective. Chem Res Toxicol. 2014;27:1081–91.

[30] Mehta R, Schrader TJ. Carcinogenic substances in food: mechanisms. In: Caballero B, Finglas P, Toldra F, editor(s). Encyclopedia of food sciences and nutrition. USA: Academic Press, 2005:117.

[31] Matsumoto M, Kano H, Suzuki M, Katagiri T, Umeda Y, Fukushima S. Carcinogenicity and chronic toxicity of hydrazine monohydrate in rats and mice by two-year drinking water treatment. Regul Toxicol Pharmacol. 2016;76:63–73.

[32] Vale A, Lotti M. Organophosphorus and carbamate insecticide poisoning. In: Lotti M, Bleecker M, editor(s). Handbook of clinical neurology Vol. 131. Amsterdam: Elsevier, 2015:149–68.

[33] Müller L, Mauthe RJ, Riley CM, Andino MM, De Antonis D, Beels C, et al. A rationale for determining, testing, and controlling specific impurities in pharmaceuticals that possess potential for genotoxicity. Regul Toxicol Pharmacol. 2006;44:198–211.

[34] Kovacic P, Somanathan R. Nitroaromatic compounds: environmental toxicity, carcinogenicity, mutagenicity, therapy and mechanism. J Appl Toxicol. 2014;34:810–24.

[35] Reddy AV, Jaafar J, Umar K, Majid ZA, Aris AB, Talib J, et al. Identification, control strategies, and analytical approaches for the determination of potential genotoxic impurities in pharmaceuticals: a comprehensive review. J Sep Sci. 2015;38:764–79.

[36] Gehring R, Van Der Merwe D. Toxicokinetic-toxicodynamic modeling. In: Gupta R, editor. Biomarkers in toxicology. USA: Academic Press, 2014.

[37] Raies AB, Bajic VB. In silico toxicology: computational methods for the prediction of chemical toxicity. Wiley Interdiscip Rev Comput Mol Sci. 2016;6:147–72.

[38] Ding B, Hua C, Kepert CJ, D'Alessandro DM. Influence of structure–activity relationships on through-space intervalence charge transfer in metal–organic frameworks with cofacial redox-active units. Chem Sci. 2019;10:1392–400.

[39] Romero-Estudillo I, Viveros-Ceballos JL, Cazares-Carreño O, González-Morales A, Flores de Jesus B, López-Castillo M, et al. Synthesis of new α-aminophosphonates: evaluation as anti-inflammatory agents and QSAR studies. Bioorg Med Chem. 2018;27:2376–86.

[40] Begam BF, Kumar JS. Computer assisted QSAR/QSPR approaches – a review. Indian J Sci Technol. 2016;9:8.

[41] Yousefinejad S, Hemmateenejad B. Chemometrics tools in QSAR/QSPR studies: a historical perspective. Chemom Intell Lab Syst. 2015;149:177–204.

[42] Medina-Franco JL, Navarrete-Vázquez G, Méndez-Lucio O. Activity and property landscape modeling is at the interface of chemoinformatics and medicinal chemistry. Future Med Chem. 2015;7:1197–211.

[43] Maggiora GM. On outliers and activity cliffs – why QSAR often disappoints. J Chem Inf Model. 2006;46:1535.

[44] Dimova D, Bajorath J. Advances in activity cliff research. Mol Inform. 2016;35:181–91.

[45] Szczepek M, Brondani V, Büchel J, Serrano L, Segal DJ, Cathomen T. Structure-based redesign of the dimerization interface reduces the toxicity of zinc-finger nucleases. Nat Biotechnol. 2007;25:786–93.

[46] Allen CHG, Koutsoukas A, Cortés-Ciriano I, Murrell DS, Malliavin TE, Glen RC, et al. Improving the prediction of organism-level toxicity through integration of chemical, protein target and cytotoxicity qHTS data. Toxicol Res (Camb). 2016;5:883–94.

[47] Gleeson MP, Modi S, Bender A, Robinson RL, Kirchmair J, Promkatkaew M, et al. The challenges involved in modeling toxicity data in silico: a review. Curr Pharm Des. 2012;18:1266–91.

[48] Kalgutkar A, Didiuk M. Structural alerts, reactive metabolites, and protein covalent binding: how reliable are these attributes as predictors of drug toxicity? Chem Biodivers. 2009;6: 2115–37.

[49] Lasser KE, Allen PD, Woolhandler SJ, Himmelstein DU, Wolfe SM, Bor DH. Timing of new black box warnings and withdrawals for prescription medications. J Am Med Assoc. 2002;287:2215–20.

[50] Mayr A, Klambauer G, Unterthiner T, Hochreiter S. DeepTox: toxicity prediction using deep learning. Front Environ Sci. 2016;3:80.

[51] Glück J, Buhrke T, Frenzel F, Braeuning A, Lampen A. In silico genotoxicity and carcinogenicity prediction for food-relevant secondary plant metabolites. Food Chem Toxicol. 2018;116:298–306.

[52] Arvidson KB, Valerio LG, Diaz M, Chanderbhan RF. In silico toxicological screening of natural products. Toxicol Mech Methods. 2008;18:229–32.

[53] Ruiz-Rodríguez MA, Vedani A, Flores-Mireles AL, Cháirez-Ramírez MH, Gallegos-Infante JA, González-Laredo RF. In silico prediction of the toxic potential of lupeol. Chem Res Toxicol. 2017;30:1562–71.

[54] Valerio LG, Arvidson KB, Chanderbhan RF, Contrera JF. Prediction of rodent carcinogenic potential of naturally occurring chemicals in the human diet using high-throughput QSAR predictive modeling. Toxicol Appl Pharmacol. 2007;222:1–16.

[55] Münzner R, Renner HW. Mutagenicity testing of quinine with submammalian and mammalian systems. Toxicol. 1983;26:173–8.

[56] Ramel C, Alekperov UK, Ames BN, Kada T, Wattenberg LW. Inhibitors of mutagenesis and their relevance to carcinogenesis: report by ICPEMC expert group on antimutagens and desmutagens. Mutat Res Genet Toxicol. 1986;39:511–7.

[57] Van Went GF. Mutagenicity testing of 3 hallucinogens: LSD, psilocybin and 9-THC using the micronucleus test. Experientia. 1978;34:342–5.

[58] Noriega-Colima K, Martinez-Mayorga K, Madariaga-Mazon A. DiaNat-DB: una base de datos de agentes antidiabéticos de origen natural: generación y análisis d elas propiedades fisicoquímicas y estructurales. Available at: http://132.248.9.195/ptd2019/marzo/0786633/Index.html. 2018.

[59] Onguéné A, Simoben C, Fotso G, Andrae-Marobela K, Khalid S, Ngadjiu B, et al. In silico toxicity profiling of natural product compound libraries from African flora with anti-malarial and anti-HIV properties. Comput Biol Chem. 2018;72:136–49.

[60] Ntie-Kang A, Telukunta KK, Döring K, Simoben CV, Moumbock AFA, Malange YI, et al. NANPDB: a resource for natural products from Northern African sources. J Nat Prod. 2017;80:2067–76.

Part IV: **Case Studies**

Samuel Egieyeh, Sarel F. Malan and Alan Christoffels

15 Cheminformatics techniques in antimalarial drug discovery and development from natural products 1: basic concepts

Abstract: A large number of natural products, especially those used in ethnomedicine of malaria, have shown varying in vitro antiplasmodial activities. Facilitating antimalarial drug development from this wealth of natural products is an imperative and laudable mission to pursue. However, limited manpower, high research cost coupled with high failure rate during preclinical and clinical studies might militate against the pursuit of this mission. These limitations may be overcome with cheminformatic techniques. Cheminformatics involves the organization, integration, curation, standardization, simulation, mining and transformation of pharmacology data (compounds and bioactivity) into knowledge that can drive rational and viable drug development decisions. This chapter will review the application of cheminformatics techniques (including molecular diversity analysis, quantitative-structure activity/property relationships and Machine learning) to natural products with in vitro and in vivo antiplasmodial activities in order to facilitate their development into antimalarial drug candidates and design of new potential antimalarial compounds.

Keywords: natural products, antiplasmodial, cheminformatics, profiling, antimalarial, drug development

15.1 Introduction

15.1.1 Natural products and malaria

Nature has provided the medicinal agents used for centuries to treat injuries and diseases [1]. These natural medicinal agents are mostly metabolites and/or by-products from biological sources, e.g. microorganisms, plants or animals. With respect to malaria, natural products have been the most consistent and successful source of antimalarial drugs [2, 3]. This is evidenced by the discovery of quinine from *Cinchona succiruba* (Rubiaceae), which became the milestone in the history of modern medicine for malaria. The molecular framework of quinine (Figure 15.1) became a template for

This article has previously been published in the journal *Physical Sciences Reviews*. Please cite as: Egieyeh, S., Malan, S.F., Christoffels, A. Cheminformatics techniques in antimalarial drug discovery and development from natural products I: basic concepts. *Physical Sciences Reviews* [Online] **2019** DOI: 10.1515/psr-2018-0130.

https://doi.org/10.1515/9783110579352-016

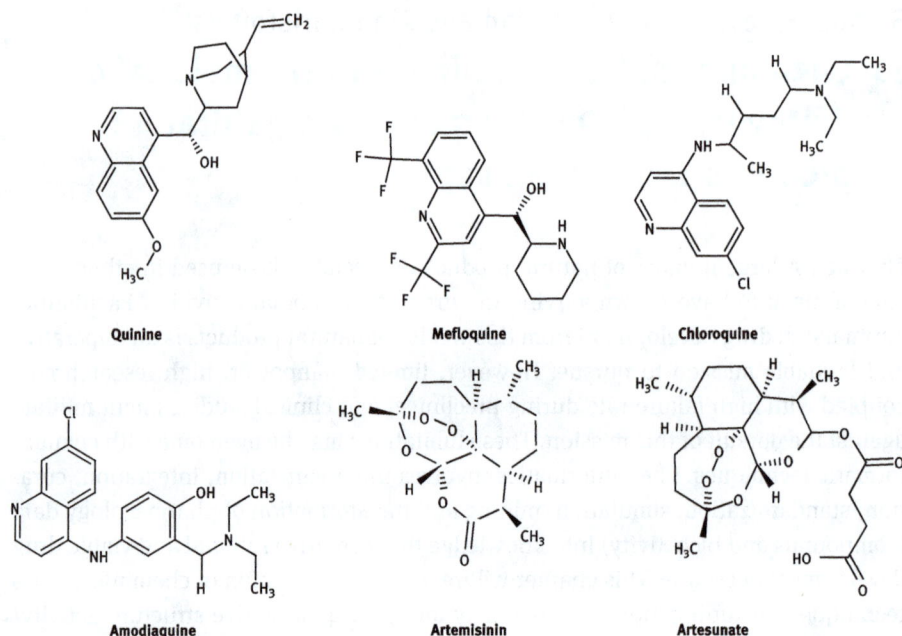

Figure 15.1: The chemical structures of quinine from Cinchona sp. and its synthetic derivatives. The aminoquinoline molecular framework of quinine was the inspiration for more active synthetic derivatives. Artemisinin from *Artemisia annua* and its derivatives.

the synthesis of aminoquinoline-based antimalarial drugs (chloroquine, amodiaquine, and mefloquine) shown in Figure 15.1 [4]. Recently, natural products have once again become the mainstay of chemotherapy for malaria. The isolation of artemisinin (qinghaosu) (Figure 15.1) from *Artemisia annua* [5] has led to the development of more active derivatives against malaria, such as artesunate (Figure 15.1) [6, 7]. These drugs have become a major part of the standard treatment guideline for malaria recommended by the World Health Organization (WHO) [8, 9]. A review of the literature revealed an increasing number of natural products with good in vitro and/or in vivo antiplasmodial activities [10–15].

15.1.2 Cheminformatics: definitions and background

The terms "cheminformatics," "chem[o]informatics," "chemical informatics" and "chemical information" have all been used, but "cheminformatics" is the most commonly used name [16]. Johann Gasteiger and Thomas Engel defined cheminformatics succinctly as "application of informatics methods to solve chemical problems" [17]. Cheminformatics involves the organization, integration, curation, standardization, simulation, mining and transformation of pharmacology data (compounds and bioactivity) into knowledge that can drive rational and viable drug development decisions [18]. One aspect that justifies the emergence of cheminformatics is the sheer magnitude of chemical information that must be processed. For example, Chemical Abstracts Service adds over three-quarters of a million new compounds to its database annually, for which large amounts of physical and chemical property data are available [19]. Cheminformatic techniques provide the ability to store and retrieve chemical information from such databases.

Cheminformatics may provide answers to the following typical questions: of the over 1000 or so natural compounds included in this study, how many could be predicted to lack drug-like properties (molecular weight too large? logP too high?)? Do the most active of these natural compounds share a set of molecular properties that are distinct from the less active compounds and the currently registered antimalarial drugs? How diverse are these natural compounds from currently registered antimalarial drugs? Given the structure-activity relationships (SAR) observed within these natural compounds, could one determine the chemical features necessary for antiplasmodial activity? Can one design a virtual compound library that may contain novel antimalarial compounds? Can a predictive equation or model relating molecular properties and in vitro bioassay activity for these natural compounds be built? Based on the predictive equation or model, can one identify potentially active compounds from a virtual compound library? These are representative questions facing the current drug design community and focus of cheminformatics.

15.2 Cheminformatics techniques in the development of antimalarial drugs from natural products

In view of limited resources, the high cost, low prospect and the higher cost of failure during preclinical and clinical drug development, it has become important to identify, as early as possible, compounds that are most likely to make it through the drug development pipeline. In addition, resistance by the causative organism, *Plasmodium*, to current chemotherapy necessitates identification or design of novel chemotypes that may have a unique mechanism of action (MOA). Therefore, the immediate logical steps to take given the collection of natural compounds that have shown antiplasmodial activities is to characterize these compounds with the aim to prioritize and select possible candidates for the next stage of drug development. Therefore, cheminformatics characterization of natural products with antiplasmodial activities is a crucial step to identifying and prioritizing potential antimalarial drugs from nature.

Cheminformatics strategies may be used to extract chemical information from these natural compounds. Various data mining techniques may then be used to relate the chemical information from the natural compounds to observed antiplasmodial activity and thus may transform data into information and information into knowledge for the intended purpose of making better decisions faster in the area of antimalarial drug design, lead identification, prioritization and organization [20, 21]. The forthcoming paragraphs give a brief insight into the concept of cheminformatics, the approaches used and its potential role in advancing antimalarial drug discovery/ design from nature.

Cheminformatics characterization or profiling of natural products with antiplasmodial activities may facilitate the identification of novel compounds (unique molecular scaffolds and chemical features) with potentially new MOA, and desirable pharmacokinetic profiles. Cheminformatics techniques can model and predict essential molecular properties required to characterize these natural compounds thus eliminating the cost of equipment and reagents to conduct in vitro or in vivo experiments to ascertain such properties. Compounds prioritized following the characterization or profiling of the natural products with antiplasmodial activities (referred to as "Hits") may then be taken into the next stage of preclinical drug development, "hit to lead" optimization [22]. Furthermore, aspects of cheminformatics like the prediction of drug-likeness [23], molecular scaffold analysis [24, 25], quantitative structure-activity relationship [26] and virtual library enumeration [27–29] may be applied on these natural products to mine information that may become valuable knowledge in antimalarial drug discovery/design.

Some cheminformatics techniques that have contributed to drug design include: chemical structure handling and classification [30], building and maintenance of chemical databases [31], calculation of molecular descriptors (MD) [32], ligand efficiency (LE) metrics [33], structure and substructure searching in two-dimension (2D) and three dimension (3D) [34], generation of 3D structures from 2D structures [35], molecular similarity search and analysis [36, 37], chemical structural activity relationships [38], prediction of absorption, distribution, metabolism, excretion and toxicity (ADMET) of hit compounds [39], bioactivity predictive models [40], virtual compound library enumerations [41] and data mining [42]. The description and usefulness of some of these cheminformatics techniques in drug discovery and development are discussed and highlighted below with relevant examples/case studies related to natural products with *in vitro* antiplasmodial activities (NAA).

15.3 Molecular descriptors

Molecular descriptors (MD) are numerical values that characterize properties of molecules e.g. physicochemical properties. Over the past years, many MD have been introduced, and new ones are continually being proposed [32, 43]. The reason for this proliferation is the need to find the most information rich and most suited MD to model chemical, biological and physical properties [43]. MD may represent the complete molecules, e.g. partition coefficient (LogP), molar refractivity or may be calculated from 2D graphs of the chemical structure, e.g. Topological Indexes and 2D fingerprints. However, some MD requires 3D representations of the chemical structure, e.g. pharmacophore-based descriptors. The Handbook of MD published in 2000 is the standard reference that gives an overview of the MD [32].

MD are pivotal to most cheminformatics processes in drug design. For example, there is usually a need to recognize or identify drug-like compounds from a compound library. This has been achieved by calculating MD of the compounds and scoring each compound based on predetermined values of MD that are observed in marketed drugs [44–46]. Another area where the calculation of MD has found relevance is in profiling compound libraries or hits from a bioassay or high throughput screening [47–49].

15.4 LE metrics

The use of LE metrics that normalize bioactivity values of ligands (hit compounds) with physicochemical properties, such as molecular weight or lipophilicity, has gained a significant attraction with many medicinal chemists in recent years [50, 51] and enabled the progression of a number of successful drug discovery projects [50–52]. These LE metrics (LE, ligand lipohilicity efficiency and ligand efficiency dependent lipophilicity (LELP)) are described below.

15.4.1 Ligand efficiency

LE assesses the contribution of heavy atoms in or molecular weight of a compound to potency or binding affinity of such a compound (i. e. potency or binding affinity per heavy atom/molecular weight, given by eq. (15.1)) [33].

$$LE = \Delta G / HA \qquad (15.1)$$

where $\Delta G = -RT \ln [IC_{50}/2]$, R = gas constant and T = absolute temperature, IC_{50} = half maximal reported antiplasmodial bioactivity. The unit of LE is kcal/mol/non-hydrogen atom or heavy atom.

The exemplary values for LE should be greater than 0.3 kcal per mole per heavy atom [53], which indicates potency at the right weight. It is particularly important to identify compounds with low weight and low potency because it has been reported that such compounds have "room" for optimization to increase potency and pharmacokinetic properties without the risk of losing LE [33, 54, 55]. A downside of LE is that it does not take lipophilicity, which is an important determinant of binding and/ or potency, into account in its estimation of efficiency of binding or potency [33]. Ligand lipophilicity efficiency (LLE) however provides a link between potency/binding affinity and lipophilicity.

15.4.2 Ligand lipophilicity efficiency (LLE)

LLE measures how efficiently a ligand/compound exploits its lipophilicity to bind to a target protein or create its potency (eq. (15.2)). In other words, it evaluates how well compounds improve potency while maintaining low lipophilicity [56].

$$LLE = pIC_{50} - clogP \ [or \ logD \ if \ the \ compound \ is \ ionisable] \qquad (15.2)$$

where IC_{50} = half maximal inhibitory concentration and clogP = partition coefficient.

The ideal values for LLE should be greater than five [56] (Figure 15.2). LLE values have been used as criteria to find compounds suitable as starting points for optimization and drug development [57]. This is because such compounds have big "lipophilicity room" that are generally "filled" during optimization towards improved potency. Monitoring the LLE of a compound collection during optimization will also allow medicinal chemists to track the efficiency of each lipophilic addition made towards improved potency. However, one limitation of LLE is that it does not account for molecular size (heavy atom or molecular weight). A binding efficiency metric that combines lipophilicity, molecular size and potency is the LELP index [50].

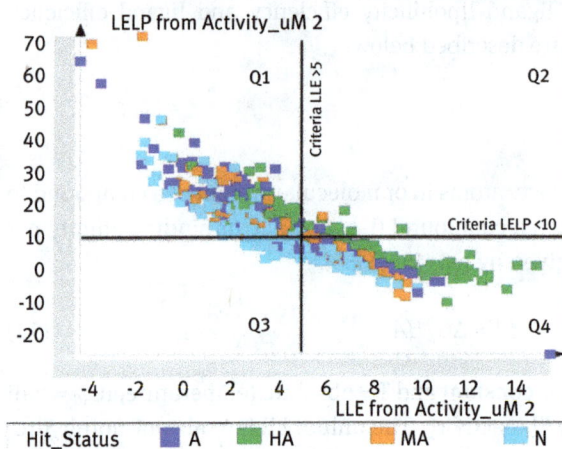

Figure 15.2: LELP versus LLE for a sample set of natural products with antiplasmodial activities [61]. Criteria for LELP and LLE was used to divide the plot area into four quadrants: Q1 (likely position for hits from bioassays and leads compounds), Q2 (no description), Q3 (likely position for successful leads) and Q4 (likely position for compounds in Phase 2 clinical trials and approved drugs).

15.4.3 Ligand efficiency dependent lipophilicity (LELP)

LELP, calculated using eq. (15.3), normalizes bioactivity or potency with lipophilicity and molecular size. LELP has been shown to reliably identify fragments, lead-like and drug-like compounds [33, 58]. Moreover, LELP was a better predictor of pharmacokinetic liabilities than LLE [33]. The ideal LELP values have been stated to be between −10 and 10 for acceptable leads [33].

$$LELP = logP/LE \qquad\qquad (15.3)$$

where logP = logarithm of partition coefficient, LE = Ligand efficiency.

It has been reported that compounds with LELP values outside the exemplar range may not proceed far in the drug development pipeline [59]. Therefore, lead optimization strategies should aim to increase LE or reduce logP in order to bring elevated LELP values within the desired range. Monitoring LELP will help to control essential physicochemical properties that will maintain desirable potency and pharmacokinetic profile during optimization [55, 60].

In light of the influence of logP and molecular size (heavy atoms or molecular weight) on potency and pharmacokinetic properties, the use of ligand-binding efficiency indices (LE, LLE and LELP) as a guiding criteria is important not only for hit selection, but also for lead generation and optimization [55, 60]. A plot of LELP against LLE (Figure 15.2), previously described by Tarcsay et al. [58], provides a visualization that may give medicinal chemists an idea of where the "hit" compounds are in terms of these parameters (LELP and LLE) and guide the optimization process to get the compounds to the desired region (quadrant Q4 in Figure 15.2) [61]. A key consideration is to be aware of the optimizing strategies that can increase potency and keep LE more or less constant or within exemplar limits.

15.5 Data mining

Data mining is the process of extracting trends or patterns from data to promote knowledge discovery [62, 63]. The data mining process takes the raw, experimental results and transforms it into useful and understandable information. Data mining is a vital process used in cheminformatics to assist in the discovery of novel bioactive molecules.

15.5.1 Data mining: molecular similarity and exploration of chemical space

Molecular similarity measures are sometimes referred to as molecular similarity coefficients or similarity indices [64]. Usually, similarity is subjective and relies upon relative judgments. There is no absolute standard of similarity, rather "like beauty, it is in the eye of the beholder" [64]. Therefore, it is difficult to develop methods for explicitly computing the similarities of large sets of molecules [64–66]. One of the most widely used molecular similarity measure is the Tanimoto similarity coefficient [S_{Tan}]. Tanimoto similarity coefficient [S_{Tan}] for objects A and B is given in set-theoretic language by eq. (15.4) below:

$$S_{Tan}[A, B] = \frac{|A \cap B|}{|A \cup B|} \tag{15.4}$$

By changing the form of the denominator:

$$S_{Tan}[A, B] = \frac{|A \cap B|}{[|A - B| + |B - A| + |A \cap B|]} \tag{15.5}$$

The equation can be simplified as

$$S_{Tan}[A, B] = \frac{a}{[a + b + c]} \tag{15.6}$$

where:
- a, Number of features in A but not in B
- b, Number of features in B but not in A
- c, Number of features common to A and B.

Concerning its usefulness, cheminformatics has also adopted molecular similarity for assessment of molecular diversity and similarity of a compound library before bioassay or hit compounds from a bioassay. Molecular recognition is essential for the interaction of small molecules with macromolecules in biological systems. These interactions produce the effect observed from such small molecules. It is assumed that similar molecules are more likely to interact with a given receptor site than molecules that differ dramatically in size, shape or electronic distribution [67]. This has led to the desire to compare molecules computationally before bioassay in order to prioritize them and test only those molecules most likely to have the desired activity. For example, in a random-screening program, a compound that generates the desired biological effect may be found. One would like to examine the compound library (which may exceed 1,000,000 compounds in most pharmaceutical companies), using molecular diversity and similarity assessment methods, to select a manageable number (say 2,000 compounds) that are similar to the "hit" compound and are most likely to show the similar activity. Hence, the concept of diversity and similarity is an important focus of cheminformatics.

15.6 Structure-activity landscape: identifying activity cliffs

Exploration of structure-activity landscape represents a core aspect of medicinal chemistry [68]. Activity cliff has been defined as pair of structurally similar compounds with a large difference in bioactivity/potency [69] and has been of interest to the medicinal and computational chemist for a long time [70, 71]. Structure-activity similarity analysis of a bioassay/bioactivity screening set using DataWarrior [72] can identify a pair of compounds that display activity cliffs.

Structure-activity landscape analysis, visualized with a self-organizing map (Figure 15.3), may identify pairs of compounds that display activity cliffs, i. e. structurally similar compounds with diverse bioactivity.

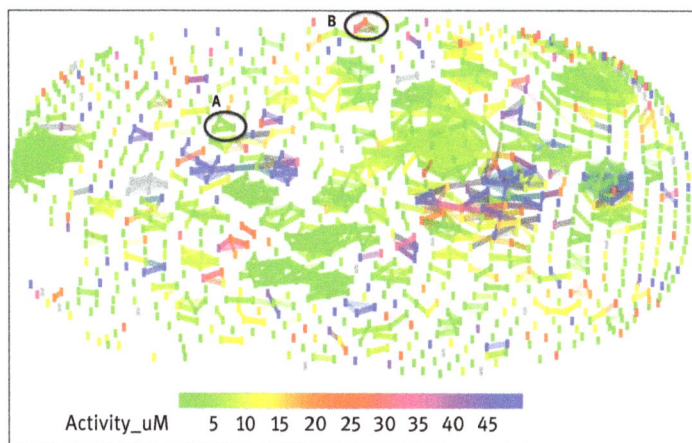

Figure 15.3: Self-organizing map depicting molecular similarity (based on SkelSpheres structural fingerprint) amongst natural products with antiplasmodial activities. Markers connected by lines represent similar compounds. Markers are coloured by activities (IC_{50}). Connected markers that have different colours (e. g. cluster marked B) represent similar compounds with different activity profile (activity cliff).

Although Maggiora [71] proposed that activity cliffs might be responsible for the inefficient performance of many QSAR models [73], activity cliffs help pinpoint regions of the activity landscape that contain maximum information for SAR studies. This is because it allows the medicinal chemist to identify the subtle molecular difference between a pair of compounds that is responsible for a dramatic shift in bioactivity (Figure 15.4), thus facilitate structural modification towards optimization of desired properties. The rich SAR information from activity cliffs has been used in many drug discovery studies [74–76].

15.7 Conclusions

In view of limited resources, the high cost, low prospect and the higher cost of failure during preclinical and clinical drug development, it has become important to identify, as early as possible, compounds that are most likely to make it through the drug development pipeline. In addition, resistance by the causative organism,

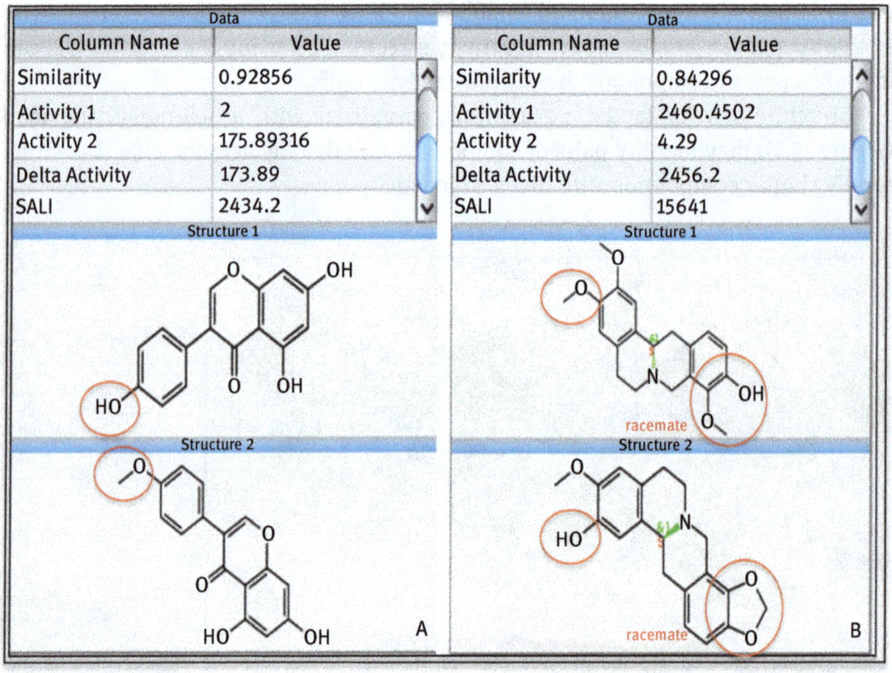

Figure 15.4: Examples of pair of natural products, with antiplasmodial activities, that represents activity cliffs. The value of molecular similarity [Similarity], activity of each compound [Activity 1 and Activity 2], the difference in activity [Delta activity] and Structure-Activity Landscape Index [SALI] are shown. The major structural differences between the compounds are highlighted with red circles. This provides valuable information about the structural features required for activity.

Plasmodium, to current chemotherapy necessitates identification or design of novel chemotypes that may have unique MOA. Therefore, chemoinformatic characterization of natural products with *in vitro* antiplasmodial activity (Figure 15.5) to prioritize hit compounds and identify optimization towards drug candidates is a logical and vital step in antimalarial drug development from nature.

In spite of the huge contributions that cheminformatics has made and can make to drug design, this relatively new field may encounter issues in the process of drug discovery/design. One of the challenges to cheminformatics in drug discovery is being able to effectively and efficiently extract knowledge from a big dataset, e. g. large-scale raw high-throughput screen data [77–79]. Therefore, data mining algorithms used in cheminformatics are constantly being improved to handle chemical information from big datasets. Cheminformatics approaches have had a huge impact on drug discovery and may be relevant in the quest to identify novel potential antimalarial drug candidates from natural products that have shown in vitro antiplasmodial activities (Figure 15.5) [21, 80, 81].

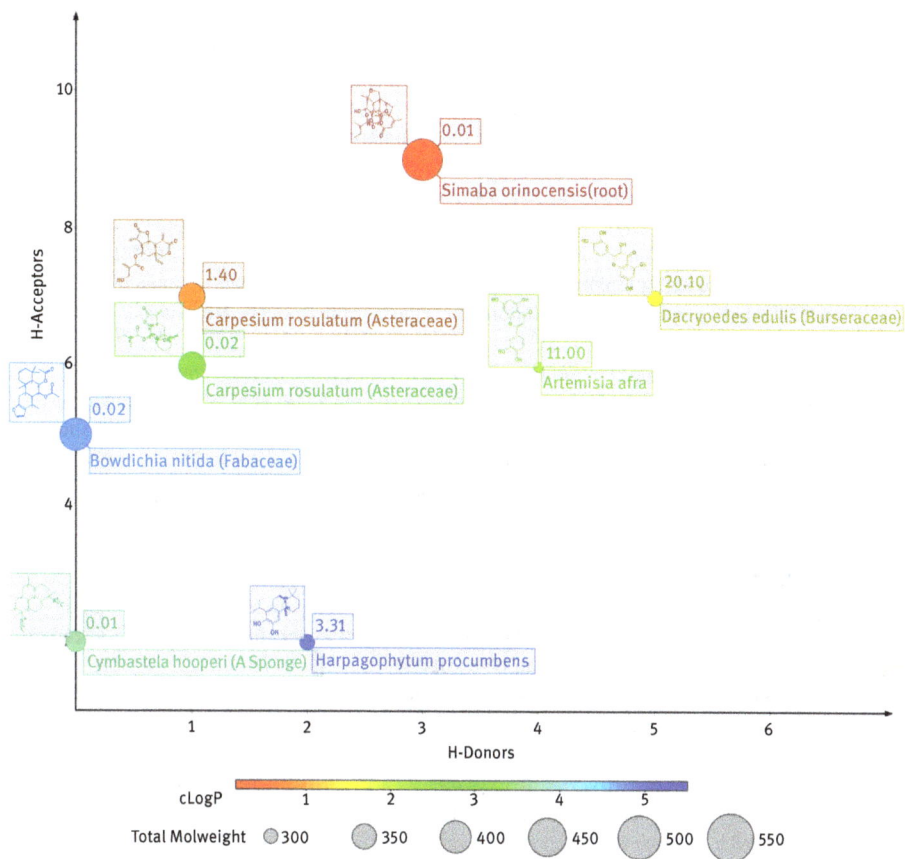

Figure 15.5: A few natural products with reported antiplasmodial activities. The natural source, IC_{50} and structures are shown within the Lipinski's rule of five space.

References

[1] Price M. History of ancient medicine in Mesopotamia and Iran. Iranchamber.com, October 2001.
[2] Willcox M, Bodeker G, Rasoanaivo P, Addae-Kyereme J. Traditional medicinal plants and malaria. Boca Raton, Florida 33431: CRC Press, 2004: ISBN 9780415301121.
[3] Yuan H, Ma Q, Ye L, Piao G. The traditional medicine and modern medicine from natural products. Molecules. 2016;21:559.
[4] O'Neill PM, Ward SA, Berry NG, Jeyadevan J, Biagini GA, Asadollaly E, et al. A medicinal chemistry perspective on 4-aminoquinoline antimalarial drugs. Curr Top Med Chem. 2006;6:479–507.
[5] ElSohly HN, Croom EM, Jr, El-Feraly FS, El-Sherei MM. A large-scale extraction technique of artemisinin from Artemisia annua. J Nat Prod. 1990;53:1560–4.
[6] De Vries PJ, Dien TK. Clinical pharmacology and therapeutic potential of artemisinin and its derivatives in the treatment of malaria. Drugs. 1996;52:818–36.

[7]   Price RN, Nosten F, Luxemburger C, Ter Kuile F, Paiphun L, Chongsuphajaisiddhi T, et al. Effects of artemisinin derivatives on malaria transmissibility. The Lancet. 1996;347:1654–8.

[8]   World Health Organization. Guidelines for the treatment of malaria. CH-1211 Geneva 27 Switzerland: World Health Organization, 2006.

[9]   World Health Organization. Guidelines for the treatment of malaria. CH-1211 Geneva 27 Switzerland: World Health Organization, 2015.

[10]   Laurent D, Pietra F. Antiplasmodial marine natural products in the perspective of current chemotherapy and prevention of malaria. A review. Mar Biotechnol. 2006;8:433–47.

[11]   Kaur K, Jain M, Kaur T, Jain R. Antimalarials from nature. Bioorg Med Chem. 2009;17:3229–56.

[12]   Batista R, De Jesus Silva Júnior A, De Oliveira AB. Plant-derived antimalarial agents: new leads and efficient phytomedicines. Part II. Non-alkaloidal natural products. Molecules. 2009;14:3037–72.

[13]   Mojab F. Antimalarial natural products: a review. Avicenna J Phytomedicine. 2012;2:52.

[14]   Nondo RS, Zofou D, Moshi MJ, Erasto P, Wanji S, Ngemenya MN, et al. Ethnobotanical survey and in vitro antiplasmodial activity of medicinal plants used to treat malaria in Kagera and Lindi regions, Tanzania. J Med Plants Res. 2015;9:179–92.

[15]   Harvey AL, Edrada-Ebel R, Quinn RJ. The re-emergence of natural products for drug discovery in the genomics era. Nat Rev Drug Discovery. 2015;14:111.

[16]   Jamal S, Grover A. Cheminformatics approaches in modern drug discovery. InDrug Design: principles and Applications. Singapore: Springer, 2017:135–48.

[17]   Gasteiger J. Handbook of chemoinformatics. Weinheim, Germany: Wiley-VCH, 2003:3–5.

[18]   Ertl P. Cheminformatics and its role in the modern drug discovery process. Basel, Switzerland: University of Strasbourg. Novartis. nd Web 2014:18.

[19]   Weisgerber DW. Chemical abstracts service chemical registry system: history, scope, and impacts. J Am Soc Inf Sci. 1997;48:349–60.

[20]   Covell DG. In: Tudor Oprea (University of New Mexico), editor. Chemoinformatics in drug discovery. Weinheim: Wiley-VCH, 2005:xxii 493. 17 × 25 cm. ISBN 3-527-30753-2.

[21]   Brown FK. Chemoinformatics: what is it and how does it impact drug discovery. Annu Rep Med Chem. 1998;33:375–84.

[22]   Ekins S, Freundlich JS, Hobrath JV, White EL, Reynolds RC. Combining computational methods for hit to lead optimization in Mycobacterium tuberculosis drug discovery. Pharm Res. 2014;31:414–35.

[23]   Tian S, Wang J, Li Y, Li D, Xu L, Hou T. The application of in silico drug-likeness predictions in pharmaceutical research. Adv Drug Deliv Rev. 2015;23:2–10.

[24]   Hu Y, Stumpfe D, Bajorath J. Lessons learned from molecular scaffold analysis. J Chem Inf Model. 2011;51:1742–53.

[25]   Yongye AB, Waddell J, Medina-Franco JL. Molecular scaffold analysis of natural products databases in the public domain. Chem Biol Drug Des. 2012;80:717–24.

[26]   Topliss J. Quantitative structure-activity relationships of drugs. Amsterdam, Netherlands: Elsevier, 2012.

[27]   Reymond J, Ruddigkeit L, Blum L, van Deursen R. The enumeration of chemical space. Wiley Interdiscip Rev: Comput Mol Sci. 2012;2:717–33.

[28]   Sud M, Fahy E, Subramaniam S. Template-based combinatorial enumeration of virtual compound libraries for lipids. J Cheminform. 2012;4:23.

[29]   Sanhueza CA, Cartmell J, El-Hawiet A, Szpacenko A, Kitova EN, Daneshfar R, et al. Evaluation of a focused virtual library of heterobifunctional ligands for Clostridium difficile toxins. Org Biomol Chem. 2015;13:283–98.

[30]   Feunang YD, Eisner R, Knox C, Chepelev L, Hastings J, Owen G, et al. ClassyFire: automated chemical classification with a comprehensive, computable taxonomy. J Cheminform. 2016;8:61.

[31] Ghahremanpour MM, Van Maaren PJ, Van Der Spoel D. The Alexandria library, a quantum-chemical database of molecular properties for force field development. Sci Data. 2018;5:180062.

[32] Todeschini R, Consonni V. Handbook of molecular descriptors. Weinheim: John Wiley & Sons, 2008.

[33] Planey SL, Kumar R. Lipophilicity indices for drug development. J Appl Biopharm Pharmacokinet. 2013;1:31–6.

[34] Kratochvíl M, Vondrášek J, Galgonek J. Sachem: a chemical cartridge for high-performance substructure search. J Cheminform. 2018;10:27.

[35] Evans DA. History of the Harvard ChemDraw project. Angew Chem Int Ed. 2014;53:11140–5.

[36] Backman TW, Cao Y, Girke T. ChemMine tools: an online service for analyzing and clustering small molecules. Nucleic Acids Res. 2011;39:W486–91.

[37] Skinnider MA, Dejong CA, Franczak BC, McNicholas PD, Magarvey NA. Comparative analysis of chemical similarity methods for modular natural products with a hypothetical structure enumeration algorithm. J Cheminform. 2017;9:46.

[38] Saldívar-González FI, Naveja JJ, Palomino-Hernández O, Medina-Franco JL. Getting SMARt in drug discovery: chemoinformatics approaches for mining structure–multiple activity relationships. RSC Adv. 2017;7:632–41.

[39] Daina A, Michielin O, Zoete V. SwissADME: a free web tool to evaluate pharmacokinetics, drug-likeness and medicinal chemistry friendliness of small molecules. Sci Rep. 2017;7:42717.

[40] Chen H, Engkvist O, Wang Y, Olivecrona M, Blaschke T. The rise of deep learning in drug discovery. Drug Discov Today. 2018;23:1241–50.

[41] Hessler G, Baringhaus K. Artificial intelligence in drug design. Molecules. 2018;23:2520.

[42] Ekins S, Clark AM, Dole K, Gregory K, Mcnutt AM, Spektor AC. Data mining and computational modeling of high-throughput screening datasets. Methods in Molecular Biology. Heidelberg: Springer, 2018:197–221.

[43] Grisoni F, Ballabio D, Todeschini R, Consonni V. Molecular descriptors for structure–activity applications: a hands-on approach. Computational toxicology. Heidelberg: Springer, 2018:3–53.

[44] Subramanian U, Sivapunniyam A, Pudukadu Munusamy A, Sundaram R. An in silico approach towards the prediction of druglikeness properties of inhibitors of plasminogen activator inhibitor1. Adv Bioinf. 2014;2014:385418.

[45] Tian S, Wang J, Li Y, Li D, Xu L, Hou T. The application of in silico drug-likeness predictions in pharmaceutical research. Adv Drug Deliv Rev. 2015;86:2–10.

[46] Wu Y, Hu M, Yang L, Li X, Bian J, Jiang F, et al. Novel natural-product-like caged xanthones with improved druglike properties and in vivo antitumor potency. Bioorg Med Chem Lett. 2015;15:2584–8.

[47] Wawer MJ, Li K, Gustafsdottir SM, Ljosa V, Bodycombe NE, Marton MA, et al. Toward performance-diverse small-molecule libraries for cell-based phenotypic screening using multiplexed high-dimensional profiling. Proceedings of the National Academy of Sciences. 2014;111:10911–16.

[48] Giroud C, Du Y, Marin M, Min Q, Jui NT, Fu H, et al. Screening and functional profiling of small-molecule HIV-1 entry and fusion inhibitors. Assay Drug Dev Technol. 2017;15:53–63.

[49] Xia J, Hu H, Xue W, Wang XS, Wu S. The discovery of novel HDAC3 inhibitors via virtual screening and in vitro bioassay. J Enzyme Inhib Med Chem. 2018;33:525–35.

[50] Scott JS, Waring MJ. Practical application of ligand efficiency metrics in lead optimisation. Bioorg Med Chem. 2018;26:3006–15.

[51] Heikal A, Nakatani Y, Jiao W, Wilson C, Rennison D, Weimar MR, et al. 'Tethering' fragment-based drug discovery to identify inhibitors of the essential respiratory membrane protein type II NADH dehydrogenase. Bioorg Med Chem Lett. 2018;28:2239–43.

[52] Stamford AW, Scott JD, Li SW, Babu S, Tadesse D, Hunter R, et al. Discovery of an orally available, brain penetrant BACE1 inhibitor that affords robust CNS Aβ reduction. ACS Med Chem Lett. 2012;3:897–902.

[53] Hopkins AL, Groom CR, Alex A. Ligand efficiency: a useful metric for lead selection. Drug Discov Today. 2004;9:430–1.

[54] Carr RA, Congreve M, Murray CW, Rees DC. Fragment-based lead discovery: leads by design. Drug Discov Today. 2005;10:987–92.

[55] Mortenson PN, Murray CW. Assessing the lipophilicity of fragments and early hits. J Comput Aided Mol Des. 2011;25:663–7.

[56] Leeson PD, Springthorpe B. The influence of drug-like concepts on decision-making in medicinal chemistry. Nat Rev Drug Discovery. 2007;6:881–90.

[57] Mowbray CE, Burt C, Corbau R, Gayton S, Hawes M, Perros M, et al. Pyrazole NNRTIs 4: selection of UK-453,061 (lersivirine) as a development candidate. Bioorg Med Chem Lett. 2009;19:5857–60.

[58] Tarcsay A, Nyíri K, Keserű GM. Impact of lipophilic efficiency on compound quality. J Med Chem. 2012;55:1252–60.

[59] Wager TT, Chandrasekaran RY, Hou X, Troutman MD, Verhoest PR, Villalobos A, et al. Defining desirable central nervous system drug space through the alignment of molecular properties, in vitro ADME, and safety attributes. ACS Chem Neurosci. 2010;1:420–34.

[60] Keserű GM, Makara GM. The influence of lead discovery strategies on the properties of drug candidates. Nat Rev Drug Discovery. 2009;8:203–12.

[61] Egieyeh SA, Syce J, Malan SF, Christoffels A. Prioritization of anti-malarial hits from nature: chemo-informatic profiling of natural products with in vitro antiplasmodial activities and currently registered anti-malarial drugs. Malar J. 2016;15:50.

[62] Larose DT. Discovering knowledge in data: an introduction to data mining. Hoboken, New Jersey.: John Wiley & Sons, 2014.

[63] Thomas MC, Zhu W, Romagnoli JA. Data mining and clustering in chemical process databases for monitoring and knowledge discovery. J Process Control. 2018;67:160–75.

[64] In: Bajorath J, Bajorath J, editor(s). Chemoinformatics and computational chemical biology. Totowa, New Jersey: Humana Press, 2011.

[65] Bajorath J. Chemoinformatics: concepts, methods, and tools for drug discovery. Heidelberg: Springer Science & Business Media, 2004.

[66] Sheridan RP, Kearsley SK. Why do we need so many chemical similarity search methods? Drug Discov Today. 2002;7:903–11.

[67] Varnek A, Tropsha A. Chemoinformatics approaches to virtual screening. Cambridge, United Kingdom: Royal Society of Chemistry, 2008.

[68] Guha R. Exploring structure–activity data using the landscape paradigm. Wiley Interdiscip Rev: Comput Mol Sci. 2012;2:829–41.

[69] Hu Y, Stumpfe D, Bajorath J. Advancing the activity cliff concept. F1000Research. 2013;2:199.

[70] Stumpfe D, Hu Y, Dimova D, Bajorath J. Recent progress in understanding activity cliffs and their utility in medicinal chemistry: miniperspective. J Med ChemJJ. 2013;57:18–28. DOI: 10.1021/jm401120g.

[71] Maggiora GM. On outliers and activity cliffs why QSAR often disappoints. J Chem Inf Model. 2006;46:1535–1535.

[72] Sander T, Freyss J, von Korff M, Rufener C. DataWarrior: an open-source program for chemistry aware data visualization and analysis. J Chem Inf Model. 2015;55:460–73.

[73] Golbraikh A, Tropsha A. 12 QSAR/QSPR revisited. In: Engel T, Gasteiger J, editor(s). Chemoinformatics: basic concepts and methods. Weinheim: Wiley-VCH, 2018:465.

[74] Dimova D, Stumpfe D, Bajorath J. Systematic assessment of coordinated activity cliffs formed by kinase inhibitors and detailed characterization of activity cliff clusters and associated SAR information. Eur J Med Chem. 2015;90:414–27.

[75] Naveja JJ, Medina-Franco JL. Activity landscape of DNA methyltransferase inhibitors bridges chemoinformatics with epigenetic drug discovery. Expert Opin Drug Discov. 2015;10:1059–70.

[76] Ojeda-Montes MJ, Gimeno A, Tomas-Hernández S, Cereto-Massagué A, Beltrán-Debón R, Valls C, et al. Activity and selectivity cliffs for DPP-IV inhibitors: lessons we can learn from SAR studies and their application to virtual screening. Med Res Rev. 2018;38:1874–915.

[77] Afendi FM, Ono N, Nakamura Y, Nakamura K, Darusman LK, Kibinge N, et al. Data mining methods for omics and knowledge of crude medicinal plants toward big data biology. Comput Struct Biotechnol J. 2013;4:1–14.

[78] Baumann K, Becker GF, Mestres J, Schneider G. Systems approaches and big data in molecular informatics. Mol Inform. 2015;34:2–2.

[79] Fourches D. Cheminformatics: at the crossroad of eras. Application of computational techniques in pharmacy and medicine. Heidelberg: Springer, 2014:539–46.

[80] Melagraki G, Afantitis A. Editorial [Thematic issue: advances in cheminformatics: drug discovery, computational toxicology and nanomaterials [Part I]]. Comb Chem High Throughput Screen. 2015;18:236–7.

[81] Xu J, Hagler A. Chemoinformatics and drug discovery. Molecules. 2002;7:566–600.

Samuel Egieyeh, Sarel F. Malan and Alan Christoffels

16 Cheminformatics techniques in antimalarial drug discovery and development from natural products 2: Molecular scaffold and machine learning approaches

Abstract: A large number of natural products, especially those used in ethnomedicine of malaria, have shown varying in-vitro antiplasmodial activities. Cheminformatics involves the organization, integration, curation, standardization, simulation, mining and transformation of pharmacology data (compounds and bioactivity) into knowledge that can drive rational and viable drug development decisions. This chapter will review the application of two cheminformatics techniques (including molecular scaffold analysis and bioactivity predictive modeling via Machine learning) to natural products with in-vitro and in-vivo antiplasmodial activities in order to facilitate their development into antimalarial drug candidates and design of new potential antimalarial compounds.

Keywords: natural products, cheminformatics, scaffold, predictive modeling, machine learning, antimalarial, drug development

16.1 Introduction

Malaria is a major health burden in Africa where 90% of the malaria deaths occur among children under the age of 5 years and pregnant women [1]. It is caused by infections with the blood parasite *Plasmodium*, transmitted by female mosquitoes (*Anopheles gambiae complex*) [2]. More than 100 different species of *Plasmodium* exist [3] but 4 species cause malaria in humans [4]. *Plasmodium falciparum* is responsible for most malaria deaths, especially in Africa [5]. *Plasmodium vivax,* the most geographically widespread of the species, produces less severe symptoms but relapses can occur for up to 3 years, and chronic disease is debilitating [6, 7]. *Plasmodium malariae* infections not only produce typical malaria symptoms but also can persist in the blood for very long periods, possibly decades, without ever producing symptoms [8]. *Plasmodium ovale* is rare, can cause relapses, and generally occurs in West Africa [9].

The discovery of quinine from *Cinchona succiruba* (Rubiaceae) and its subsequent development into a reliable antimalarial drug represents a milestone in the history of

This article has previously been published in the journal *Physical Sciences Reviews*. Please cite as: Egieyeh, S., Malan, S.F., Christoffels, A. Cheminformatics techniques in antimalarial drug discovery and development from natural products II: machine learning approaches. *Physical Sciences Reviews* [Online] **2019** DOI: 10.1515/psr-2019-0029.

https://doi.org/10.1515/9783110579352-017

modern medicine for malaria. Recently, natural products have once again become the mainstay of chemotherapy for malaria. The isolation of artemisinin (qinghaosu) from *Artemisia annua* [10] has led to the development of more active derivatives against malaria, such as artemether, arteether, dihydroartemisinin and artesunate [11, 12]. These drugs have become a major part of the standard treatment guideline for malaria recommended by the World Health Organization (WHO) [13, 14]. However, there have been reports of resistance to this recent treatment for malaria [15]. Thankfully, nature may have a lot more to offer in antimalarial therapy. A review of the literature revealed an increasing number of natural products, especially from the rich ethnobotanical heritage of malaria-endemic countries, with good *in vitro* and/or *in vivo* antiplasmodial activities [16–21]. However, a major limitation to the successful movement of these natural products through the antimalarial drug development process from this rich ethnobotanical resource may be the lack of skilled manpower, inadequate infrastructure, poor access to technological platforms to conduct drug discovery and development. This is compounded with the high cost, low prospect and the higher risk of failure during preclinical and clinical drug development.

Therefore, computer-aided drug design strategies, including cheminformatics techniques, have become important to simulate (model) some of the required preclinical studies for antimalarial drug development from natural products. Cheminformatics characterization of these natural products is a crucial step to identify, as early as possible, natural products that are most likely to make it through the antimalarial drug development pipeline.

In the previous chapter [22], we focused on the application of molecular descriptors, ligand efficiency metrics, molecular diversity analysis, and quantitative structure-activity relationships to the characterization of natural products with in vitro and in vivo antiplasmodial activities in order to facilitate their development into antimalarial drug candidates and design of new potential antimalarial compounds. In this chapter, the role of molecular scaffold analysis [23, 24] and bioactivity predictive modeling via machine learning (ML) for antimalarial drug development from natural products will be described.

16.2 Exploration of molecular scaffolds

It is believed that natural products are a good source of novel molecular scaffolds [25, 26] and scaffolds derived from natural compounds have preferable or privileged scaffold architectures [26, 27]. Identification of scaffolds that are present in natural antimalarial compounds may be the starting points for new classes of antimalarial drug candidates. Scaffold diversity assesses chemical diversity based on the scaffolds and ring systems in chemical structures [28–33]. The Bemis-Murcko framework (Figure 16.1) [34] is a suitable representation of molecular scaffolds that are used to analyze the structures of known drugs [34, 35], screening libraries and natural products [36].

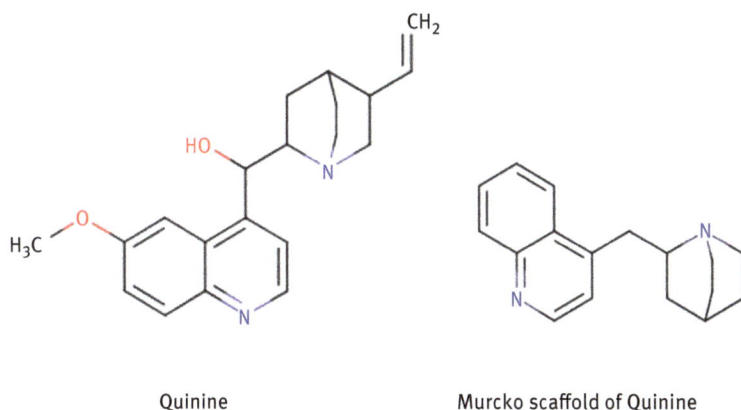

Figure 16.1: The Bemis-Murcko framework (scaffold) of quinine.

A Scaffold Tree (Figure 16.2) gives a visual hierarchical arrangement of scaffolds obtained by stepwise trimming of all terminal side chains of molecules [37]. Scaffold Hunter has been used to generate Scaffold Trees for large datasets and to identify "virtual scaffolds" that may have similar bioactivity with the parent molecule or child scaffolds [38]. The scaffold diversity of natural products with antiplasmodial activities may be evaluated and compared to current antimalarial drugs as well as synthetic compounds with antiplasmodial activities using the following parameters: the scaffold counts and cumulative scaffold frequency plot.

The scaffold counts are estimated from the ratios of scaffolds (N_s) to molecules (M), singleton scaffolds (N_{ss}) to a molecule (M), and singleton scaffolds (N_{ss}) to total scaffolds (N_s). Cumulative Scaffold Frequency Plot (CSFP) gives an indication of the distribution of molecules in a dataset over the molecular scaffolds. To generate a CSFP, first, the scaffolds should be sorted by their scaffold frequency (most frequent to least frequent). Then the cumulative percentage of scaffolds is plotted against the cumulative scaffold frequency as a percentage of total molecules in the dataset. A CSFP that shows a diagonal line (Figure 16.3) point to an equal distribution of molecules across the molecular scaffolds, while curves that are above the diagonal line i. e. with steeper gradients (Figure 16.3) represents compound datasets with low molecular scaffold diversity.

16.3 Bioactivity predictive models via machine-learning

A number of publications have reported the *in vitro* antiplasmodial activities of natural products from plants [39, 40], marine life forms [41–43] and synthetic compounds that are available in the public domain [44–46]. The availability of such data for drug discovery of malaria has motivated the development of predictive models

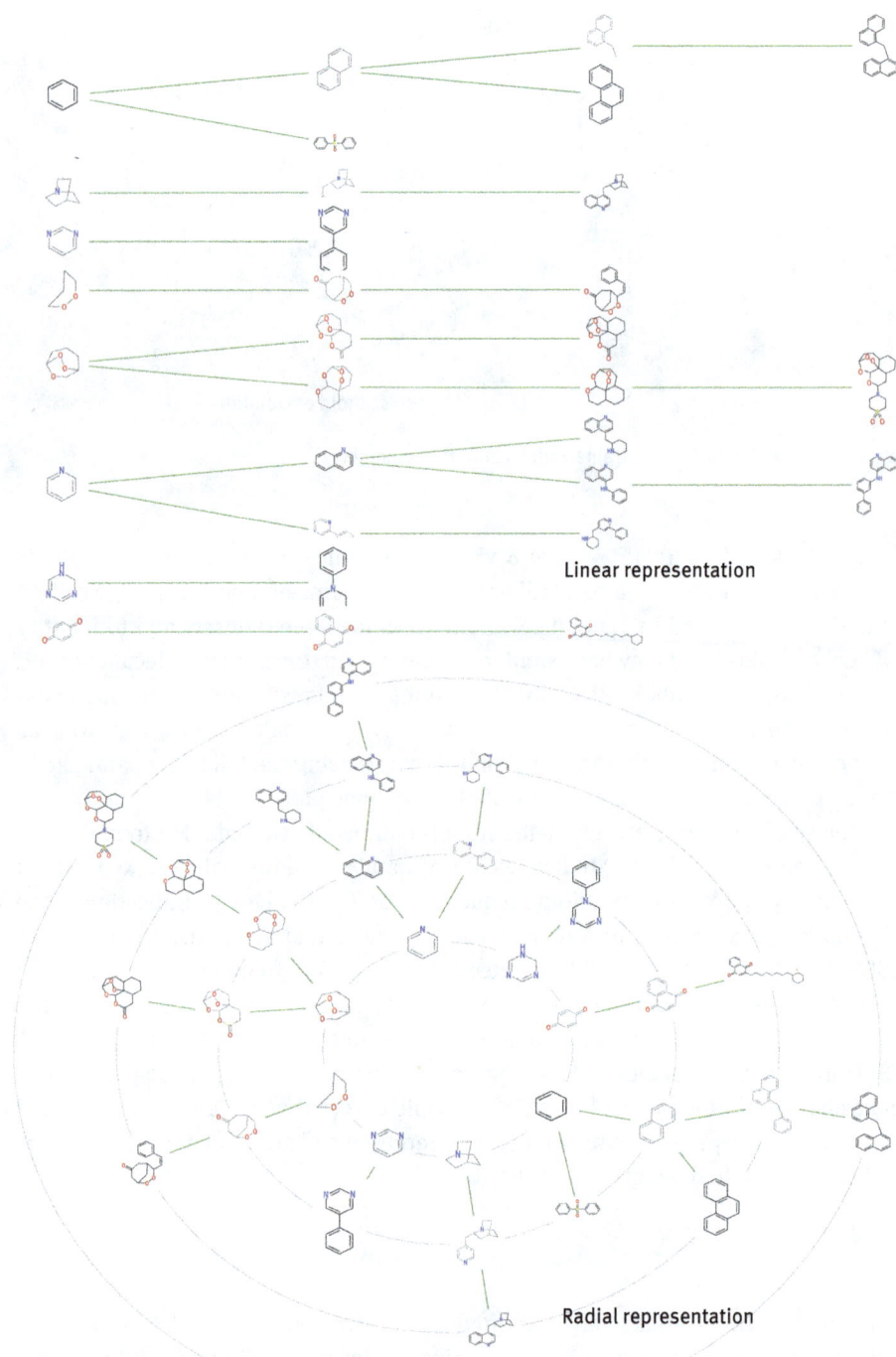

Linear representation

Radial representation

Figure 16.2: A section of a Scaffold Tree generated with Scaffold Hunter for selected compounds in ChEMBL. The technique first prunes all terminal side chains from parent molecules (rightmost molecules in the linear representation and the outermost ring in the radial representation), then iteratively removes ring one by one according to a set of prioritization rules from the parent molecules to get the scaffolds.

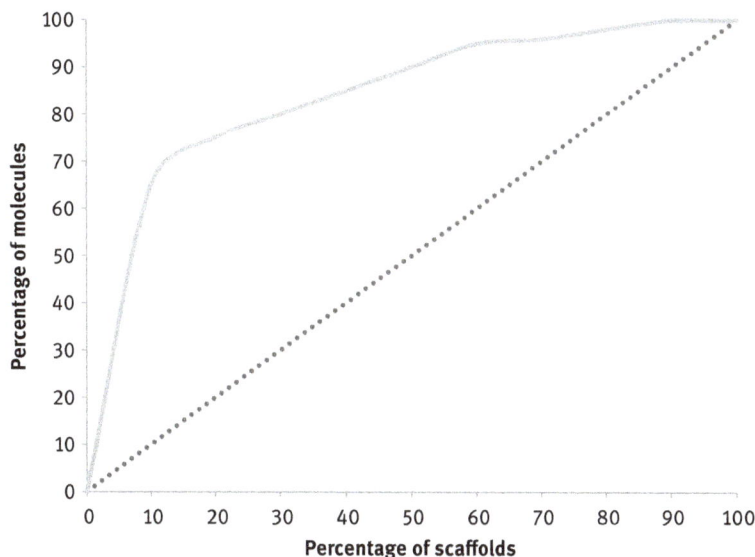

Figure 16.3: A typical Cumulative Scaffold Frequency Plots (CSFP) of scaffolds from an exemplar compound set. The closer the curve is to the diagonal line, the greater the scaffold diversity of the compound set.

based on molecular properties and ML approaches. ML refers to a large set of tools or algorithms for supervised and unsupervised learning of data [47]. In unsupervised statistical learning, there are inputs but no consequent output. Nonetheless, this approach allows us to learn relationships and structure from such input data as well as to cluster the input data into classes based on their properties. Supervised ML involves building a model from an input data and a known output that can then be used to predict the outputs from new inputs. The model encompasses systematic information that the input variable (*x*) provides about the output (*y*) (Equation 16.1).

$$y = f(x) + e \tag{16.1}$$

Where *f* is the systematic information, that *x* provides about *y*; *e* is an error term.

ML algorithms that learn by regression analysis are most suitable when the input and output variables are numeric (e. g. activity values and molecular descriptors). However, when the output variable is categorical (e. g. active/inactive) then classification algorithms are most appropriate to learn from such data.

In cheminformatics, supervised ML methods have been widely used to predict molecular properties and biological activities. Generally, compounds are labeled based on their bioactivity (i. e. active/inactive) and the machine-learning algorithm comes up with a binary classification model based on a set of molecular descriptors. The model may then be used to predict the activity of a set of compounds using their molecular descriptors or molecular fingerprints. Various predictive models have been developed to predict anti-tubercular [48, 49], anti-malarial [50, 51] and anti-HIV [52] bioactivity as well as for drug repurposing [53]. The increasing number of natural products with desirable *in vitro* and/or *in vivo* antiplasmodial activities [40, 54–59] present a platform to build predictive models that may be used to screen other natural products and predict their potential antiplasmodial activities.

16.3.1 Some machine-learning algorithms

Various algorithms and implementations (techniques) have been previously used in supervised ML including M5P [M5 model trees] [60], Sequential Minimization Optimization [61, 62], Multilayer Perceptron [63, 64], Naïve Bayesian [65, 66], Random forest [RF] [67, 68] and Voted perceptron [VP] [69, 70]. Brief descriptions of these algorithms are given in Table 16.1.

Table 16.1: A brief description of some machine learning algorithms.

ML algorithm	Brief description
M5P [M5 model trees]	M5P combines a conventional decision tree with the possibility of linear regression functions at the nodes of the tree.
Sequential Minimization Optimization [SMO]	SMO is a variant of support vector machine (SVM) that simplifies the process of solving the complex quadratic problem inherent in SVM.
Multilayer Perceptron	Multilayer perceptron (MLP) is deep learning, an artificial neural network composed of more than one perceptron (a simple model of a neuron in an artificial neural network). It usually consists of an input layer to receive a signal, an output layer that makes a prediction and in between those two, an arbitrary number of hidden layers.
Voted Perceptron	Voted perceptron is a simple modification over the perceptron algorithm that uses multiple weighted perceptrons. The algorithm initiates a new perceptron every time an object is wrongly classified, resetting the weights vector with the final weights of the last perceptron. The final output will be determined by a weighted vote on all the perceptrons.
Naïve Bayesian	Naïve Bayesian is based on Bayes theorem. It estimates the probability of occurrence of an event (**X**) given that another event (**Y**) has occurred.
Random Forest	Random forest performs both regression and classification tasks with the use of multiple decision trees and a technique called **"Bootstrap Aggregation"** (also known as **bagging**: training each decision tree on a different data sample where sampling is done with replacement) to determine the final output.

16.3.2 Platforms and steps for building predictive models via ML

Various open-source platforms (software) have been developed in the field of ML from the academic and industrial communities. This includes, but not limited to, H2O.ai, KNIME Analytics Platform, Scitkit-Learn, Jupyter, RapidMiner, Weka, Orange, and Octave [71]. In addition to the open-source software outlined, there are also commercial software packages for ML approaches. The following steps are typically used to build predictive models: data preprocessing, training and testing of the model and evaluation of the performance of the models.

16.3.2.1 Data preprocessing
The goal of this step is to clean the input data (e. g. add missing hydrogen assigned to a molecule) and calculate the required attributes for each object (molecule). Numerical data are also normalized using appropriate normalization model e. g. minimum-maximum.

16.3.2.2 Selection of descriptors or features
Descriptors or features selection is a vital step in ML particularly for datasets with tens or hundreds of thousands of variables e. g. gene expression array and combinatorial chemistry. The selection of features, prior to building a model, will ensure an improved prediction performance of the predictive model, faster and more economical predictive models, and a better understanding of the underlying process that generated the predictions [72].

16.3.2.3 Training and testing of predictive models
ML algorithms need to effectively learn from appropriate training data to build a robust model (e. g. regression and classification predictive models) that can accurately predict the class of or variable for a test data. The regression predictive models determine optimal parameters for the prediction of biological activity ($pIC_{50} = -Log\ (IC_{50})$) from the molecular descriptors and/or molecular fingerprints. The classification predictive models assign class (e. g. active/inactive) to molecules defined by a set of attributes (e. g. molecular descriptors or molecular fingerprints). Figure 16.4 shows a typical KNIME workflow for training and testing a linear regression model.

16.3.2.4 Evaluation of the performance of the predictive model
The correlation coefficient (R^2) is used to quantify the extent or degree of fit between the predicted bioactivity (e. g. pIC_{50}) by a regression model and the observed bioactivity (e. g. pIC_{50}). The R^2 values lie between 0 and 1, with 1 corresponding to an ideal fit. The root mean squared error (RMSE) estimates how much error there is between the observed bioactivity (pIC_{50}) and the predicted bioactivity (pIC_{50}) by the regression models. It represents the sample standard deviation of the differences between

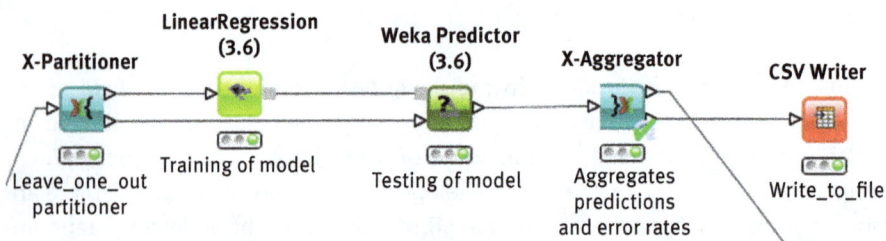

Figure 16.4: A typical simple KNIME workflow designed to build a linear regression model. The X-Partitioner node was used to split the data into the training set and test set. The former was piped into the linear regression node (to train the model) while the later was passed to the predictor node (to test the trained model). The X-Aggregator collates the prediction and the error rates from the predictions. The X-Partitioner and X-Aggregator nodes were used to implement cross-validation or out-of-sample testing (various techniques for assessing how the predictions by the trained model will generalize to an independent data set.). For example, in a leave-one-out cross-validation, the X-Partitioner node partitions the training data such that one sample of the data is left out as the test data and the remaining ones used as the training set. This is iterated until each sample of the data is used as a test data [51].

predicted pIC_{50} values and observed pIC_{50} values. Hence, lower RSME will suggest minimal deviation between the observed and predicted pIC_{50} values and good performance by the model.

The performance of a classifier model is usually represented by a confusion matrix (Table 16.2) and a number of evaluation statistics. Some of the evaluation statistics generated comprise sensitivity, specificity, and accuracy. Sensitivity relates to the test's ability to identify positive results whereas specificity relates to the test's ability to identify negative results. A test with high sensitivity and specificity will have a low error rate. Accuracy may be defined, specifically in drug discovery, as the proportion of compounds that were correctly classified as active and inactive (i. e. the number of compounds correctly classified divided by the total number of compounds classified multiply by 100).

Table 16.2: An example of a confusion matrix used to describe the performance of a classification model on a set of test data for which the true values are known.

N = 165	Predicted: Inactive	Predicted: Active	
Actual/Observed: Inactive	True Negative: 50	False Positive: 10	60
Actual/Observed: Active	False Negative: 5	True Positive: 100	105
	55	110	

Accuracy indicates the proximity of measurement of results to the true value. These evaluation statistics can be mathematically expressed as:

$$Sensitivity = \frac{TP}{TP + FN} *100 \tag{16.2}$$

$$Specificity = \frac{TP}{TN + FN} *100 \tag{16.3}$$

$$Accuracy = \frac{TP + TN}{TP + FP + TN + FN} *100 \tag{16.4}$$

Where TP is the number of true positives; FP is the number of false positives; TN is the number of true negatives and FN is the number of false negatives.

Other evaluation statistics include Receiver Operating Characteristic (ROC) curve and area under the ROC curve. ROC curve is a graphical plot that shows the performance of a binary classifier model as its discrimination threshold is varied (Figure 16.5). It is a plot of the true positive rate (sensitivity) against the false positive rate (1 – specificity) for a binary classification system (e. g. active and inactive) at various threshold settings.

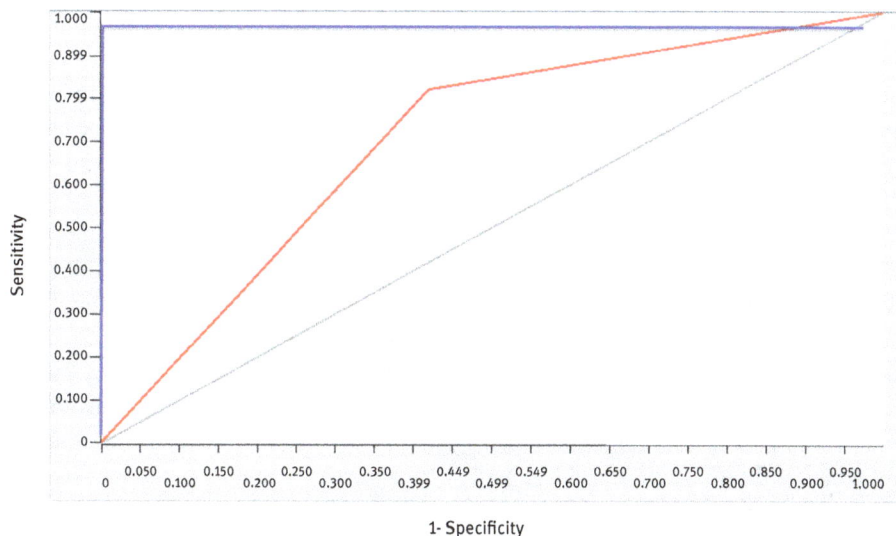

Figure 16.5: A typical Receiver Operating Curve (ROC) for a classifier model. The area under the ROC curve (AUC) is a measure of bioactivity class discriminatory power of a classifier model. A classifier model that randomly assigns bioactivity class to compounds will have an AUC of 0.5 (diagonal grey line), while a model that perfectly predicts bioactivity class of compounds will have an AUC of 1 (blue line). The red line is the ROC plot for a classifier model with an AUC of 0.70 [51].

The area under the ROC curve (AUC) is a measure of how well a model can discriminate between two classes in a dataset (e. g. active and inactive compounds) [73] (Figure 16.5). If the bioactivity class prediction of a compound set by a model is purely random, the AUC will be equal to 0.5 (i. e. the ROC curve will coincide with the diagonal line) (Figure 16.5). A perfect separation of the bioactivity class of the compounds by a model gives an area under the ROC curve of one.

16.3.3 Applicability domain (AD) of a predictive model

Generally, ML models methods are more likely to show good predictive performance for objects in the test set that share similar properties to objects in the training set. Thus, it is necessary to define the "applicability domain" i. e. the boundary defined by the property space (or chemical space) of the objects in the training set of the model, and to check if any new object falls within such domain. With regards to drug discovery, new chemical compounds located within or at the boundary of the chemical space of a training data are most likely to be reliably predicted from models built from such training data [74]. Numerous simple and complex methods are used to define AD; based on range, distance, geometric and density distribution [75–77]. One of the simplest and commonly applied methods is the AD based on range-based definition with a preliminary Principal Components (PC) rotation [76, 78].

16.4 Conclusions

In our previous chapter [22] we described four cheminformatics techniques and parameters (molecular descriptors, ligand efficiency metrics, molecular diversity, and chemical space exploration as well as molecular structural-activity relationships) that can add value to and advance the quest for antimalarial drug development from natural products. Here, we have highlighted two chemoinformatics techniques that have made and can make huge contributions to drug discovery/design. Molecular scaffold exploration enables the identification of novel scaffolds and chemotypes that may result in the design of new compound libraries and the development of drug candidates to combat malaria. Bioactivity predictive models via ML will enable medicinal chemists to prescreen compounds prior to the expensive step of synthesis and *in vitro* assay. Thus these techniques can improve decision-making processes in antimalarial drug design and development to achieve better and cost-effective outcomes (i. e. drug candidate for malaria). Overall, these techniques could significantly contribute to and accelerate the on-going efforts for antimalarial drug discovery, especially from natural products, by identifying novel scaffolds that may be used as starting point for the design of new antimalarial drugs and generate antimalarial bioactivity predictive models that can be used to predict potential antimalarial bioactivity of newly discovered natural products.

16.5 Case study: Cheminformatic profiling of selected natural product with in vitro antiplasmodial activities (NAA)

16.5.1 Dataset

The dataset used in this case study consists of 1040 natural products with antiplasmodial activity (IC_{50}) (NAA). The selected NAA were sub-divided into four categories based on IC_{50} (pIC_{50}): highly active (HA) with IC_{50} less than 1 µM, active (A) with IC_{50} between 1 µM and 5 µM, moderately active (MA) with IC_{50} between 5 µM and 10 µM and low active (N) with IC_{50} equal or greater than 10 µM. Currently available antimalarial drugs (CAD) were also included in the dataset. Three-dimensional (3D) structures were generated, corrected and minimized with Molecular Operating Environment (MOE) 2013 software [79].

16.5.2 Analysis of molecular descriptors of NAA and CAD

Key molecular descriptors and physicochemical properties of NAA and CAD were calculated with MOE and visualized with Datawarrior [80]. The mean of the molecular descriptors and physicochemical properties for NAA and CAD were compared and statistical differences were assessed with analysis of variance (ANOVA) (significance set at $p < 0.05$). A typical boxplot is shown for the number of hydrogen bond acceptors, the number of hydrogen bond donors, total molecular weight and cLogP (Figure 16.6). This analysis allows the comparison of key molecular descriptors of NAA to CAD, which may guide the optimization strategies for NAA towards the next step of antimalarial drug development.

16.5.3 Molecular similarity and exploration of chemical space of NAA and CAD

The extent of molecular similarity within the dataset (consisting of NAA and CAD) was evaluated using "Fragment fingerprint" in Datawarrior. The result was visualized as a 3D chemical space (Figure 16.7) plotted from the first three PCs computed from the "Fragment fingerprint". Most of the NAA (A, HA, MA and N) are structurally diverse from CAD and may be regarded as hit compounds with a potential novel mechanism of antimalarial actions. Such structurally diverse active antiplasmodial compounds create a road into uncharted antimalarial chemical space.

16.5.4 Exploration of molecular scaffolds from NAA

For this analysis, a subset of NAA (20 compounds) was selected and Scaffold Tree was generated using Scaffold Hunter [81]. The Scaffold Tree (Figure 16.8) showed the

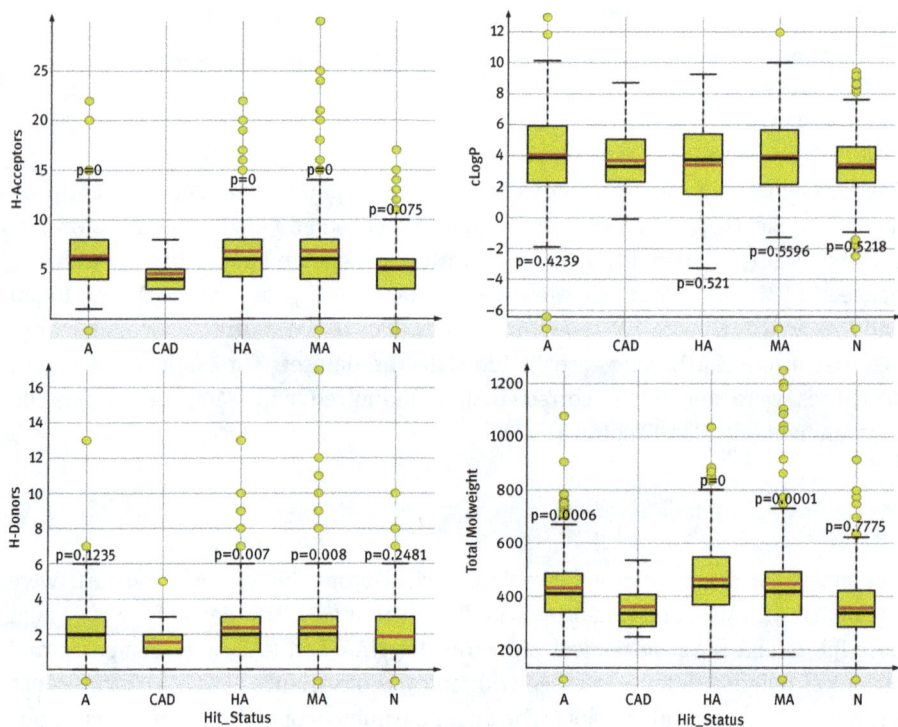

Figure 16.6: Boxplots showing distribution and summary statistics of key molecular descriptors for compound sets: the number of hydrogen bond acceptors, number of hydrogen bond donors, cLogP and total molecular weight for currently available antimalarial drugs (CAD) and subgroups of natural products with antiplasmodial activity (NAA). The red and black lines represent the mean and median respectively for each distribution. The p-values from the statistical analysis of the difference between CAD and subgroups of NAA are indicated on the graph. (Highly active (HA) with IC_{50} less than 1 µM, active (A) with IC_{50} equal or greater than 1 µM but less than 5 µM, moderately active (MA) with IC_{50} equal or greater than 5 µM but less than 10 µM and Low active (N) with IC_{50} equal or greater than 10 µM.

preponderance of ring systems in NAA and identified virtual scaffolds, which may be potential bioactive compounds (Figure 16.8). Murcko scaffolds were also generated for the subset of NAA using Datawarrior. Figure 16.9 shows the frequency of the Murcko scaffolds generated. These scaffolds possess desirable drug-like properties (e. g. cLogP < 5 and molecular weight < 500) making them ideal starting points for antimalarial drug design and scaffold hopping.

Scaffold hopping is an essential task of modern medicinal chemistry that substitutes the central scaffold of a molecule with another one while retaining the pharmacophoric interaction points of the molecule. The aim of scaffold hopping is to discover structurally novel compounds from a known active compound, address the issue around Intellectual property (IP), toxicity, selectivity and Pharmacokinetic (PK) properties among other lead optimization strategy. The scaffolds from NAA are

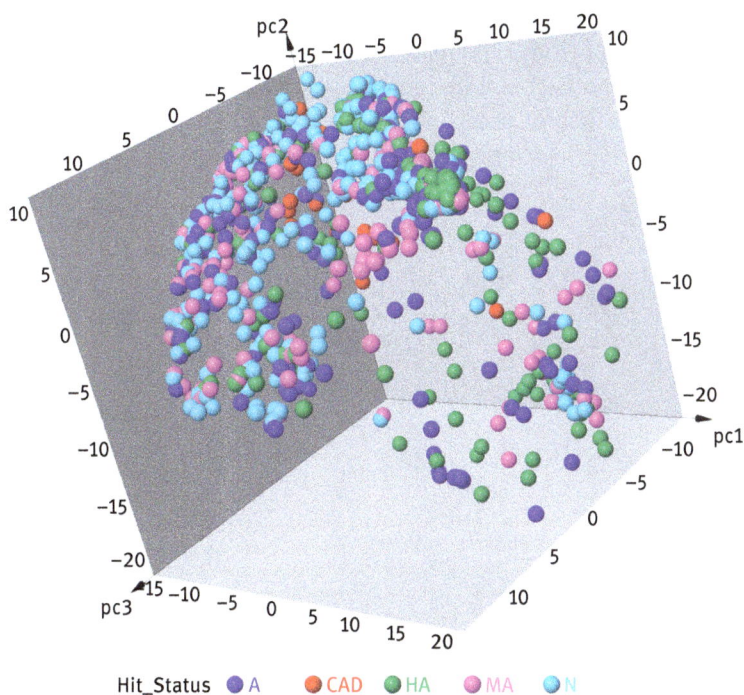

Figure 16.7: A three-dimensional (3D) visualization of the antimalarial chemical space of the dataset. Most subgroups of natural products with antiplasmodial activity (NAA) occupy distinct positions >in the chemical space in relative to CAD. (Highly active (HA) with IC_{50} less than 1 µM, active (A) with IC_{50} equal or greater than 1 µM but less than 5 µM, moderately active (MA) with IC_{50} equal or greater than 5 µM but less than 10 µM and Low active (N) with IC_{50} equal or greater than 10 µM).

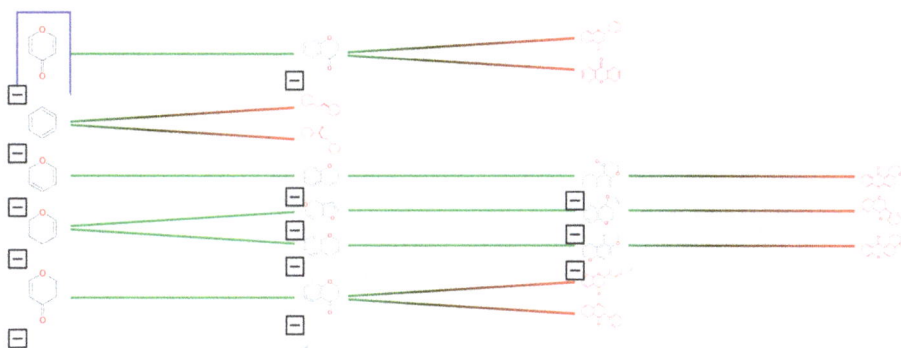

Figure 16.8: Linear representation of Scaffold Tree generated from a subset of NAA.

amenable for scaffold hopping because of their unique three-dimensionality and their desirable drug-like properties (Figure 16.9), which may allow varied spatial orientation within binding pockets of drug targets and gives "room" to add appropriate side chains for the desired pharmacophore.

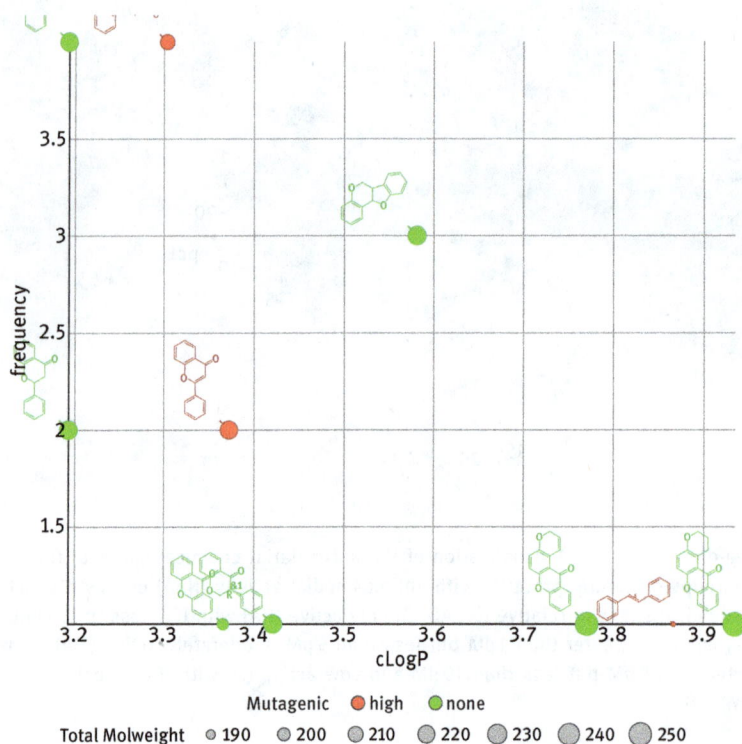

Figure 16.9: Frequency of the Murcko scaffolds generated from the subset of NAA. The cLogP, total molecular weight, and tendency for mutagenicity were estimated for these Murcko scaffolds. More physicochemical properties may be estimated which will allow rational selection of Murcko scaffolds that may be used for scaffold hopping.

16.5.5 Predictive classifier models built from NAA using ML approach

Classical ML approaches were used to build four classifier models (Naïve Bayesian, Voted Perceptron, Random Forest and Sequence Minimization Optimization of Support Vector Machines) from bioactivity data of natural products with *in-vitro* antiplasmodial activity (NAA) using a combination of the molecular descriptors and two-dimensional molecular fingerprints of the compounds. Models were evaluated with an independent test dataset. Figure 16.10, shows the KNIME workflow used to build the predictive models.

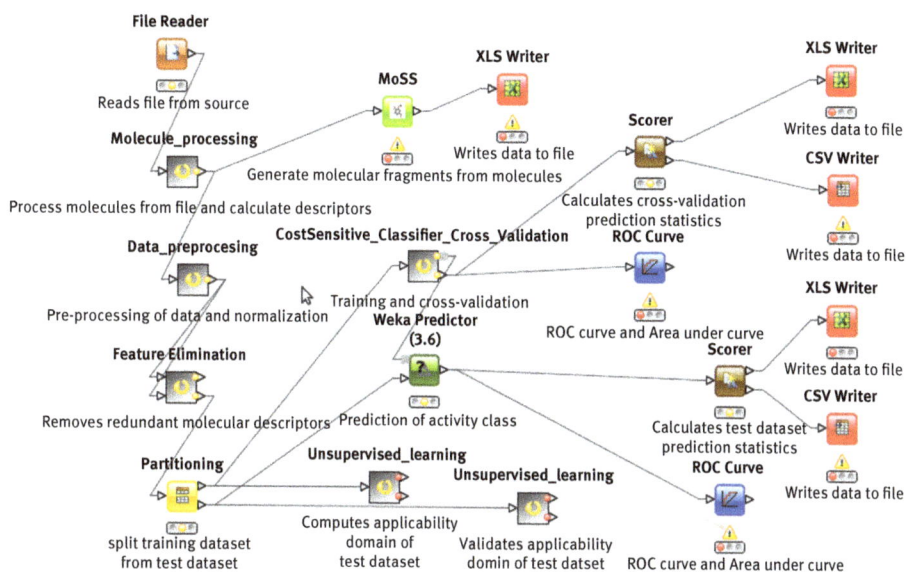

Figure 16.10: Screenshot of the KNIME workflow that was used to build the classifier and regression machine learning models [51].

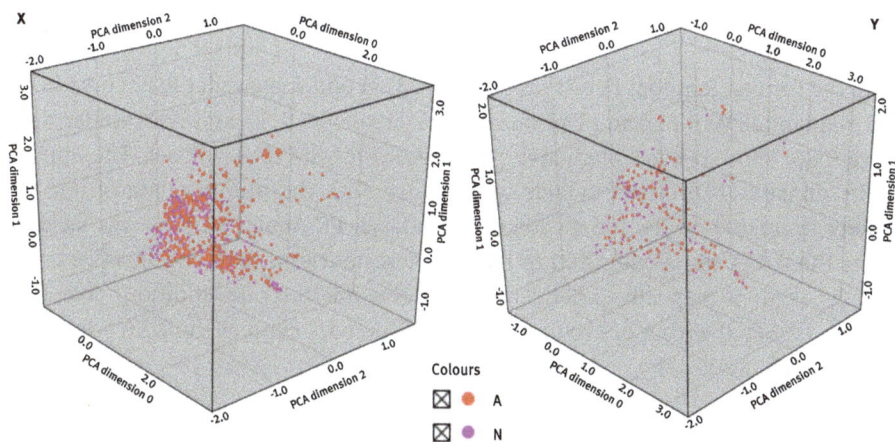

Figure 16.11: Visualization of classifier models' applicability domain (chemical space). Active compounds (red dots) and inactive compounds (purple dots) are represented using the first three Principal Components. Panel X depicts the range of Principal Components of compounds in the training set that define the applicability domain (AD). Panel Y shows that almost all compounds in the test set fell within the AD of the defined by the training set. Therefore, classifier models generated in this study can reliably predict the bioactivity class of new compounds that fall within this AD. NAA: natural products with in-vitro antiplasmodial activity [51].

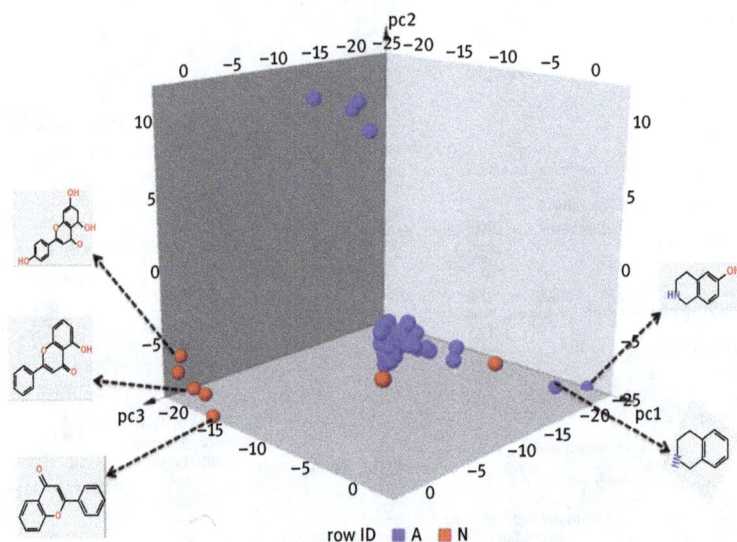

Figure 16.12: Chemical features from active and inactive compounds from NAA dataset. The blue markers represent the most common substructures from active compounds ($IC_{50} \leq 10$ µM) while the red markers represent the most common substructures from inactive compounds ($IC_{50} > 10$ µM). The most common substructures were projected in a three-dimension (3D) space based on molecular similarity. Some of the most common substructures that are peculiar to the active and inactive compounds are highlighted. This may guide the rational selection and design of active antiplasmodial compounds. NAA: natural products with in-vitro antiplasmodial activity [51].

From the results, Random Forest (accuracy 82.81 % and Area under ROC curve 0.91) and Sequential Minimization Optimization (accuracy 85.93 % and Area under ROC curve 0.86) showed good predictive performance for the NAA dataset. The applicability domain (AD) of the models was defined with the training dataset and its validity evaluated on the independent test dataset. PC Analysis (PCA) was used to define the AD of the models and to map the test dataset (active and inactive compounds) in their respective chemical spaces (Figure 16.11). In addition, chemical features enriched within active and inactive bioactivity class of the NAA were also extracted (Figure 16.12).

References

[1] World Health Organization. World malaria report 2011. Accessed at https://www.who.int/malaria/publications/atoz/9789241564403/en/ Accessed: 09 Apr 2019.

[2] Bahia AC, Dong Y, Blumberg BJ, Mlambo G, Tripathi A, BenMarzouk-Hidalgo OJ, et al. Exploring *Anopheles* gut bacteria for *Plasmodium* blocking activity. Environ Microbiol. 2014;16:2980–94.

[3] Garnham PCC. Malaria parasites and other haemosporidia. Oxford: Blackwell Scientific, London, 1966.

[4]   World Malaria Report 2018. Geneva: World Health Organization, 2018. Licence: CC BY-NC-SA
       3.0 IGO. Accessed at: https://apps.who.int/iris/bitstream/handle/10665/275867/
       9789241565653-eng.pdf?ua=1. Accessed: 09 Apr 2019.

[5]   Hviid L, Jensen ATR. PfEMP1–A parasite protein family of key importance in *Plasmodium
       falciparum* Malaria immunity and pathogenesis. Adv Parasitol. 2015;88:51–84.

[6]   Beeson JG, Chu CS, Richards JS, Nosten F, Fowkes FJI. *Plasmodium vivax* Malaria: challenges in
       diagnosis, treatment and elimination. Pediatr Infect Dis J. 2015;34:529–31.

[7]   Gill CJ. *Plasmodium vivax* malaria in the UK. BMJ. 2015;350:h1840.

[8]   Das A. The distinctive features of Indian malaria parasites. Trends Parasitol. 2015;31:83–6.

[9]   Tomar LR, Giri S, Bauddh NK, Jhamb R. Complicated malaria: a rare presentation of *Plasmodium
       ovale*. Trop Doct. 2015;45:140–2.

[10]   ElSohly HN, Croom EM, Jr, El-Feraly FS, El-Sherei MM. A large-scale extraction technique of
       artemisinin from *Artemisia annua*. J Nat Prod. 1990;53:1560–4.

[11]   De Vries PJ, Dien TK. Clinical pharmacology and therapeutic potential of artemisinin and its
       derivatives in the treatment of malaria. Drugs. 1996;52:818–36.

[12]   Price RN, Nosten F, Luxemburger C, Ter Kuile F, Paiphun L, Chongsuphajaisiddhi T, et al. Effects
       of artemisinin derivatives on malaria transmissibility. The Lancet. 1996;347:1654–8.

[13]   World Health Organization. Guidelines for the treatment of malaria. World Health Organization,
       2006. Available at: http://archives.who.int/publications/2006/9241546948_eng.pdf.
       Accessed: 09 Apr 2019.

[14]   World Health Organization. Guidelines for the treatment of malaria. World Health Organization,
       2015. Avialable at: https://apps.who.int/iris/bitstream/handle/10665/162441/
       9789241549127_eng.pdf?sequence=1. Accessed: 09 Apr 2019.

[15]   World Health Organization. Artemisinin resistance and artemisinin-based combination therapy
       efficacy: status report. World Health Organization, 2018. Available at: https://apps.who.int/
       iris/handle/10665/274362. Accessed: 09 Apr 2019.

[16]   Laurent D, Pietra F. Antiplasmodial marine natural products in the perspective of current
       chemotherapy and prevention of malaria. A review. Mar Biotechnol. 2006;8:433–47.

[17]   Kaur K, Jain M, Kaur T, Jain R. Antimalarials from nature. Bioorg Med Chem. 2009;17:3229–56.

[18]   Batista R, De Jesus Silva Júnior A, De Oliveira AB. Plant-derived antimalarial agents: new leads
       and efficient phytomedicines. Part II. Non-alkaloidal natural products. Molecules.
       2009;14:3037–72.

[19]   Mojab F. Antimalarial natural products: a review. Avicenna J Phytomed. 2012;2:52–62.

[20]   Nondo RSO, Zofou D, Moshi MJ, Erasto P, Wanji S, Ngemenya MN, et al. Ethnobotanical survey
       and in vitro antiplasmodial activity of medicinal plants used to treat malaria in Kagera and Lindi
       regions, Tanzania. J Med Plant Res. 2015;9:179–92.

[21]   Harvey AL, Edrada-Ebel R, Quinn RJ. The re-emergence of natural products for drug discovery in
       the genomics era. Nat Rev Drug Discov. 2015;14:111–29.

[22]   Egieyeh S, Malan SF, Christoffels A. Cheminformatics techniques in antimalarial drug
       discovery and development from natural products 1. Basic concept.Phys Sci Rev. 2019.
       DOI: 20180130.

[23]   Hu Y, Stumpfe D, Bajorath J. Lessons learned from molecular scaffold analysis. J Chem Inf
       Model. 2011;51:1742–53.

[24]   Yongye AB, Waddell J, Medina-Franco JL. Molecular scaffold analysis of natural products
       databases in the public domain. Chem Biol Drug Des. 2012;80:717–24.

[25]   Rodrigues T, Reker D, Schneider P, Schneider G. Counting on natural products for drug design.
       Nat Chem. 2016;8:531–41.

[26] Laraia L, Waldmann H. Natural product inspired compound collections: evolutionary principle, chemical synthesis, phenotypic screening, and target identification. Drug Discov Today: Technol. 2017;23:75–82.

[27] Schneider P, Schneider G. Privileged structures revisited. Angew Chem Int Ed Engl. 2017;56:7971–4.

[28] Ertl P, Jelfs S, Mühlbacher J, Schuffenhauer A, Selzer P. Quest for the rings. In silico exploration of ring universe to identify novel bioactive heteroaromatic scaffolds. J Med Chem. 2006;49:4568–73.

[29] Krier M, Bret G, Rognan D. Assessing the scaffold diversity of screening libraries. J Chem Inf Model. 2006;46:512–24.

[30] Lipkus AH, Yuan Q, Lucas KA, Funk SA, Bartelt III WF, Schenck RJ, et al. Structural diversity of organic chemistry. A scaffold analysis of the CAS Registry. J Org Chem. 2008;73:4443–51.

[31] Dimova D, Bajorath J. Assessing scaffold diversity of kinase inhibitors using alternative scaffold concepts and estimating the scaffold hopping potential for different kinases. Molecules. 2017;22:730.

[32] Laraia L, Robke L, Waldmann H. Bioactive compound collections: from design to target identification. Chem. 2018;4:705–30.

[33] Boufridi A, Quinn RJ. Harnessing the properties of natural products. Annu Rev Pharmacol Toxicol. 2018;58:451–70.

[34] Bemis GW, Murcko MA. The properties of known drugs. 1. Molecular frameworks. J Med Chem. 1996;39:2887–93.

[35] Egieyeh S, Syce J, Christoffels A, Malan SF. Exploration of scaffolds from natural products with antiplasmodial activities, currently registered antimalarial drugs and public malarial screen data. Molecules. 2016;21:104.

[36] Shang J, Sun H, Liu H, Chen F, Tian S, Pan P, et al. Comparative analyses of structural features and scaffold diversity for purchasable compound libraries. J Cheminform. 2017;9:25.

[37] Wetzel S, Schuffenhauer A, Roggo S, Ertl P, Waldmann H. Cheminformatic analysis of natural products and their chemical space. CHIMIA. 2007;61:355–60.

[38] Wetzel S, Klein K, Renner S, Rauh D, Oprea TI, Mutzel P, et al. Interactive exploration of chemical space with Scaffold Hunter. Nat Chem Biol. 2009;5:581–3.

[39] Xu Y-J, Pieters L. Recent developments in antimalarial natural products isolated from medicinal plants. Mini-Rev Med Chem. 2013;13:1056–72.

[40] Beaufay C, Bero J, Quetin-Leclercq J. Antimalarial terpenic compounds isolated from plants used in traditional medicine [2010–July 2016]. In: Mérillon JM, Riviere C, editors. Natural antimicrobial agents. Sustainable development and biodiversity, Vol. 19. Cham: Springer, 2018:247–68.

[41] Mayer AMS, Rodríguez AD, Berlinck RGS, Fusetani N. Marine pharmacology in 2007–8: marine compounds with antibacterial, anticoagulant, antifungal, anti-inflammatory, antimalarial, antiprotozoal, antituberculosis, and antiviral activities; affecting the immune and nervous system, and other miscellaneous mechanisms of action. Comp Biochem Physiol C Toxicol Pharmacol. 2011;153:191–222.

[42] Davis RA, Buchanan MS, Duffy S, Avery VM, Charman SA, Charman WN, et al. Antimalarial activity of pyrroloiminoquinones from the Australian marine sponge *Zyzzya sp*. J Med Chem. 2012;55:5851–8.

[43] Shao C-L, Mou X, Cao F-F, Spadafora C, Glukhov E, Gerwick L, et al. Bastimolide B, an Antimalarial 24-membered marine macrolide possessing a tert-butyl group. J Nat Prod. 2018;81:211–15.

[44] Wang Y, Xiao J, Suzek TO, Zhang J, Wang J, Bryant SH. PubChem: a public information system for analyzing bioactivities of small molecules. Nucleic Acids Res. 2009;37:W623–33.

[45] Spangenberg T, Burrows JN, Kowalczyk P, McDonald S, Wells TNC, Willis P. The open access malaria box: a drug discovery catalyst for neglected diseases. PLoS One. 2013;8:e62906.

[46] Bathurst I, Hentschel C. Medicines for Malaria Venture: sustaining antimalarial drug development. Trends Parasitol. 2006;22:301–7.

[47] Deo RC. Machine learning in medicine. Circulation. 2015;132:1920–30.

[48] Periwal V, Rajappan JK, Open Source Drug Discovery Consortium, Jaleel AUC, Scaria V. Predictive models for anti-tubercular molecules using machine learning on high-throughput biological screening datasets. BMC Res Notes. 2011;4:504.

[49] Kovalishyn V, Grouleff J, Semenyuta I, Sinenko VO, Slivchuk SR, Hodyna D, et al. Rational design of isonicotinic acid hydrazide derivatives with antitubercular activity: machine learning, molecular docking, synthesis and biological testing. Chem Biol Drug Des. 2018;92:1272–8.

[50] Jamal S, Periwal V, Open Source Drug Discovery Consortium, Scaria V. Predictive modeling of anti-malarial molecules inhibiting apicoplast formation. BMC Bioinformatics. 2013;14:55.

[51] Egieyeh S, Syce J, Malan SF, Christoffels A. Predictive classifier models built from natural products with antimalarial bioactivity using machine learning approach. PLoS One. 2018;13: e0204644.

[52] Kaiser TM, Burger PB, Butch CJ, Pelly SC, Liotta DC. A machine learning approach for predicting HIV reverse transcriptase mutation susceptibility of biologically active compounds. J Chem Inf Model. 2018;58:1544–52.

[53] Ekins S, Puhl AC, Zorn KM, Lane TR, Russo DP, Klein JJ, et al. Exploiting machine learning for end-to-end drug discovery and development. Nature Mater. 2019;18:435.

[54] Frederich M, Tits M, Angenot L. Potential antimalarial activity of indole alkaloids. Trans R Soc Trop Med Hyg. 2008;102:11–19.

[55] Adebayo JO, Krettli AU. Potential antimalarials from Nigerian plants: a review. J Ethnopharmacol. 2011 Jan 27;133:289–302.

[56] Hertweck C. Natural products as source of therapeutics against parasitic diseases. Angew Chem Int Ed. 2015;54:14622–4.

[57] Nogueira CR, Lopes LMX. Antiplasmodial natural products. Molecules. 2011;16:2146–90.

[58] Lawal B, Shittu OK, Kabiru AY, Jigam AA, Umar MB, Berinyuy EB, et al. Potential antimalarials from African natural products: a reviw. J Intercultural Ethnopharmacol. 2015;4:318–43.

[59] Gabriel HB, Sussmann RAC, Kimura EA, Rodriguez AAM, Verdaguer IB, Leite GCF, et al. Terpenes as potential antimalarial drugs. In: Perveen S, Al-Taweel A, editors. Terpenes and terpenoids. London: IntechOpen, 2018:45.

[60] Barros RC, Basgalupp MP, Ruiz DD, de Carvalho ACPLF, Freitas AA Evolutionary model tree induction. In: Proceedings of the 2010 ACM Symposium on Applied Computing 2010;1131–7.

[61] Chu W, Keerthi SS. Support vector ordinal regression. Neural Comput. 2007;19:792–815.

[62] Chang -C-C, Lin C-J. LIBSVM: A library for support vector machines. Acm T Intel Syst Tec [TIST]. 2011;2:27.

[63] Gardner M, Dorling S. Artificial neural networks [the multilayer perceptron]—a review of applications in the atmospheric sciences. Atmos Environ. 1998;32:2627–36.

[64] Vianna GK, Cruz SMS. Using multilayer perceptron networks in early detection of late blight disease in tomato leaves. In: Proc. ICAI 2014;14:158–64.

[65] Nidhi, Glick M, Davies JW, Jenkins JL. Prediction of biological targets for compounds using multiple-category Bayesian models trained on chemogenomics databases. J Chem Inf Model. 2006;46:1124–33.

[66] Zhang H, Yu P, Xiang M-L, Li X-B, Kong W-B, Ma J-Y, et al. Prediction of drug-induced eosino-philia adverse effect by using SVM and naïve Bayesian approaches. Med Biol Eng Comput. 2016;54:361–9.

[67] Sheridan RP. Three useful dimensions for domain applicability in QSAR models using random forest. J Chem Inf Model. 2012;52:814–23.

[68] Singh H, Singh S, Singla D, Agarwal SM, Raghava GP. QSAR based model for discriminating EGFR inhibitors and non-inhibitors using random forest. Biol Direct. 2015;10:10.

[69] Martišius I, Šidlauskas K, Damaševičius R. Real-time training of voted perceptron for classifi-cation of EEG data. Int J Artif Intell. 2013;10:41–50.

[70] Loukeris N, Eleftheriadis I. Further higher moments in portfolio selection and *a priori* detection of bankruptcy, under multi-layer perceptron neural networks, hybrid neuro-genetic MLPs, and the voted perceptron. Int J Fin Econ. 2015;20:341–61.

[71] Nguyen G, Dlugolinsky S, Bobák M, Tran V, García ÁL, Heredia I, et al. Machine learning and deep learning frameworks and libraries for large-scale data mining: a survey. Artif Intell Rev. 2019;52:1–48.

[72] Guyon I, Elisseeff A. An introduction to variable and feature selection. J Mach Learn Res. 2003;3:1157–82.

[73] Jiménez-Valverde A. Insights into the area under the receiver operating characteristic curve (AUC) as a discrimination measure in species distribution modelling. Global Ecol Biogeogr. 2012;21:498–507.

[74] Sahigara F, Mansouri K, Ballabio D, Mauri A, Consonni V, Todeschini R. Comparison of different approaches to define the applicability domain of QSAR models. Molecules. 2012;17:4791–810.

[75] Netzeva TI, Worth AP, Aldenberg T, Benigni R, Cronin MTD, Gramatica P, et al. Current status of methods for defining the applicability domain of (quantitative) structure-activity relationships. ATLA Altern Lab Anim. 2005;33:155–73.

[76] Jaworska J, Nikolova-Jeliazkova N, Aldenberg T. QSAR applicability domain estimation by projection of the training set in descriptor space: a review. ATLA Altern Lab Anim. 2005;33:445.

[77] Roy K, Kar S. Importance of applicability domain of QSAR models. Pharmaceutical Sciences: Breakthroughs in Research and Practice. 2017;1012–43.

[78] Aït-Sahalia Y, Xiu D. Using principal component analysis to estimate a high dimensional factor model with high-frequency data. J Econ. 2017;201:384–99.

[79] Molecular Operating Environment (MOE). Montreal. QC, Canada: Chemical Computing Group ULC, 2016.

[80] López-López E, Naveja JJ, Medina-Franco JL. DataWarrior: an evaluation of the open-source drug discovery tool. Expert Opin Drug Discov. 2019;14:335–41.

[81] Schäfer T, Kriege N, Humbeck L, Klein K, Koch O, Mutzel P. Scaffold Hunter: a comprehensive visual analytics framework for drug discovery. J Cheminf. 2017;9:28.

Eleni Koulouridi, Sergi Herve Akone, Marilia Valli, Vanderlan da
Silva Bolzani, Fernanda I. Saldívar-González, Angélica Pilón-
Jiménez, José L. Medina-Franco, Berhanu M. Abegaz, Hanok Kinfe,
Aurélien F. A. Moumbock, Mohd Athar, Gabin T.M. Bitchagno, Serge
A.T. Fobofou, David Newman, Rita C. Guedes, Tiago Rodrigues,
Maruca Annalisa, Bagetta Donatella, Lupia Antonio, Ricardo
B. Hernández-Alvarado, Abraham Madariaga-Mazón,
Karina Martinez-Mayorga, Stefan Günther and Fidele Ntie-Kang

Glossary of terms used in chemoinformatics of natural products: fundamental principles

Activity and property landscape modeling: A strategy to describe systematically structure-property relationships of datasets.

ADMET: Abbreviation for the following pharmacokinetic and pharmacological properties; absorption, distribution, metabolism, excretion (or elimination) and toxicity.

Affinity: Affinity is a measure of the tightness with which a drug binds to the receptor. Intrinsic activity is a measure of the ability of a drug that is bound to the receptor to generate an activating stimulus and produce a change in cellular activity. Both agonists and antagonists can bind to a receptor.

Aflatoxins: Mycotoxins mainly produced by the fungi *Aspergillus flavus* and *Aspergillus parasiticus*, biosynthesized by the polyketide route.

AfroCancer: The African anticancer natural products library. Available at: *http://african-compounds.org/about/afrocancer/*.

AfroDB: A database of natural products with at least one known biological activity, from African medicinal plants. Available at: *http://african-compounds.org/about/afrodb/*.

AfroMalariaDB: The African antimalarial natural products library. Available at: *http://african-compounds.org/about/afromalariadb/*.

Agonist: A compound (endogenous ligand or drug) that causes a physiological or a pharmacological action when in complex with a receptor.

AK Scientific natural product library: The company's catalogue includes various compounds. Among them are natural products. Available at: *www.aksci.com*.

Algorithm: A logical procedure for answering scientific queries, usually performed by the use of a computer.

https://doi.org/10.1515/9783110579352-018

Alkaloid: A class of naturally occurring organic compounds that contain (mostly heterocyclic) nitrogen atoms.

Alkamid®: A database focused on *N*-alkylamides from plants. Available at: *http://alkamid.ugent.be/*.

Ambinter collection: Compounds provided by the Ambinter company *via* greenpharma libraries. Available at: *www.ambinter.com*.

Amino acids: The building blocks of proteins.

AnalytiCon Discovery Natural Product Libraries: Various commercial collections of natural products provided by the company AnalytiCon. Available at: *www.ac-discovery.com*.

Antagonist: A compound or drug that competes with the endogenous ligand in order to create a complex with a receptor.

AntiBase database: A database of natural products stemming from microorganisms and higher fungi. Available at: *https://application.wiley-vch.de/stmdata/antibase.php*.

Antibody-Drug-Conjugates (ADC): Using monoclonal antibodies that are directed against a specific "epitope" on the cell, very potent modified natural products are carried to the desired cell, and on enzyme cleavage, the "warhead" is released to poison the specific cell. Four are currently approved as antitumor drugs with over 100 in clinical trials.

AOP: Abbreviation of Adverse Outcome Pathway. The AOP is a way to link existing knowledge to one or more series of causally connected key events (KE) between two points – a molecular initiating event (MIE) and an adverse outcome (AO) that occur at a level of biological organization.

Area under the curve (AUC): A two-dimensional area under the entire ROC curve, which is a measurement of an algorithm's ability to classify compounds into specific categories such as active and inactive. See, also "*ROC curve* and *ROC analysis*".

Artificial neural network (or connectionist systems): Computing systems vaguely inspired by the biological neural networks that constitute animal brains. The neural network itself is not an algorithm, but rather a framework for many different machine learning algorithms to work together and process complex data inputs. Such systems "learn" to perform tasks by considering examples, generally without being programmed with any task-specific rules.

AUC: See "*Area under the curve*".

Bayesian: A branch of statistical methods, including "Bayesian inference", "Bayesian probability", Bayesian classifier, Bayesian regression, etc. Bayesian regression, for example, could be thought of as a way of performing a range of QSAR analyses between two extremes. There exists prior information in the form of a previous regression on the relevant set of previous data – the training set of complexes used to fit the original scoring function.

Bayesian classifier: A supervised learning algorithm used for classification tasks.

BEDROC (Boltzmann-enhanced discrimination of ROC): It contains the discrimination power of the RIE metric but incorporates the statistical significance from ROC and its well-behaved boundaries.

Binding Affinity: How strongly two things bind (e.g. a protein-ligand interaction).

BindingDB: A free database containing binding affinities of protein targets with small molecules. Available at: *https://www.bindingdb.org/bind/index.jsp*.

Binding site: This is a region on a macromolecule such as a protein that binds to another molecule with specificity.

BioAustralis: The compound catalogue from the company BioAustralis Fine Chemicals. Available at: *http://www.bioaustralis.com/*.

Biological activity: Biological activity or pharmacological activity describes the beneficial or adverse effects of a drug on living matter. When a drug is a complex chemical mixture, this activity is exerted by the substance's active ingredient or pharmacophore but can be modified by the other constituents. Among the various properties of chemical compounds, pharmacological/biological activity plays a crucial role since it suggests uses of the compounds in the medical applications. However, chemical compounds may show some adverse and toxic effects which may prevent their use in medical practice (see "***Toxicity***").

Biosynthesis (biosynthesize): The production of a chemical compound by a living organism.

Building blocks: In the chemical sense, building blocks are small molecules that are used in the enzymatic process of assembling larger molecules like proteins, polysaccharides, polyterpenes, etc.

CamMedNP: The Cameroonian 3D structural natural products database. A database which can be freely used for virtual screening purposes. Available at: *http://african-compounds.org/about/cammednp/*.

Carotenoids Database: A database containing chemical structures and data on carotenoids from various organisms. Available at: *www.carotenoiddb.jp*.

CAS: Acronym for Chemical Abstracts Services. The CAS Registry Number® is used to identify your substance of interest. This is universally used to provide a unique, unmistakable identifier for chemical substances. The CAS registry information includes daily updated information on literature references to the substance experimental and predicted property data (e.g. boiling and melting points, etc.), CA Index Names and synonyms, commercial availability of compounds/substances, preparative methods, spectra, regulatory information from international sources, etc. See also *"Chemical Abstracts Services (CAS) Number"*.

CASE: Computer-assisted structure elucidation, a tool created to provide suggestions for the molecular structures of compounds based on spectroscopic data, database information and computational programs.

CASMI: Acronym for Critical Assessment of Small Molecule Identification, an open contest initiated in 2012 for the determination of the molecular formula and structure of unknown metabolites, with automated software tools from either GC-MS or LC-MS/MS data only.

Catalyst: A substance that accelerates a chemical reaction without itself being affected.

CATS2: Chemically advanced template search. Topological pharmacophore descriptors designed for scaffold hopping, de novo design, and machine learning.

ChEBI: An acronym for Chemical Entities of Biological Interest. A free database source focused on small molecules. Available at: *http://www.ebi.ac.uk/chebi/*.

ChemBank: A free, online database of small molecules. Available at: *http://chem bank.broadinstitute.org/*.

ChEMBL: A curated database of compounds accompanied by information on biological activities and ADMET properties. Available at: *https://www.ebi.ac.uk/chembl/*.

Chemical Abstracts Services (CAS) Number: A unique code for identifying a compound in the CAS Directory. See *"CAS"*.

Chemical scaffold: See *"Scaffold"*.

Chemical space: A collection of molecules meaningfully related by a mathematical object. For practical applications, chemical space can be used as a "tool" that helps to find associations in complex data and rapidly exploit the increasing information available for the discovery of drugs and other research areas such as food science.

Chemistry Development Kit (CDK): A computer software, a library in the programming language Java, for chemoinformatics and bioinformatics. It is available for Windows, Linux, Unix, and macOS. It is free and open-source software distributed under the GNU Lesser General Public License (LGPL) 2.0.

Chemoinformatics (cheminformatics or chemical informatics): A scientific branch which involves searching and managing information stemming from chemical structures. This is the science of handling, indexing, archiving, searching, and evaluating information that is specific to chemical structures and is used in *data mining*, information retrieval, information extraction, and *machine learning*.

Chemoproteomics: A method for discovering mechanisms of action based on small molecule chemistry.

ChemSpider: A free database of chemical structures published by the Royal Chemical Society, UK. Available at: *http://www.chemspider.com/Default.aspx.*

Chem-TCM: The chemical database of traditional Chinese medicine. A virtual collection of molecules from plants in Chinese traditional medicine. Available at: *http://www.chemtcm.com/.*

CID: Collision-induced dissociation, dissociation of an ion after collision excitation. In this process, the internal energy of the ion, increased by the collision with an inert gas, is dissipated by means of its fragmentation.

cLogP (see also logP): Calculated log of the octan-1-ol/water partition coefficient.

Cluster of chemical compounds: A group of chemical compounds sharing similar properties such as structural or physicochemical properties.

Clustering of compounds: See "***Cluster of chemical compounds***".

Clustering algorithm: See "***Cluster of chemical compounds***".

CN collection: French Acronym for "Chimiothèque Nationale" chemical collection. Available at: *http://chimiotheque-nationale.cn.cnrs.fr/?Presentation,18.*

CoCoCo: Acronym for Commercial Compound Collection. A free chemical database for *in silico* experiments. Available at: *http://cococo.isof.cnr.it/cococo.*

Compound library (or chemical database): A collection of chemical compounds in electronic format and/or with samples.

Computational toxicology: A scientific discipline that focuses on the development of mathematical and computer-based models for understanding and predicting adverse health effects caused by chemicals, such as environmental pollutants and pharmaceuticals.

Consensus diversity plot (CDPlot): An intuitive two-dimensional graph that represents in low dimensions the diversity of chemical libraries considering simultaneously multiple molecular representations.

COSY: Acronym for correlation spectroscopy. It is a homonuclear two-dimension NMR technique used to identify spins, which are coupled to each other. It shows the frequencies for a single isotope, most commonly hydrogen (^{1}H).

Cytotoxicity: The possession of such destructive action, particularly in reference to lysis of cells by immune phenomena and to antineoplastic agents that selectively kill dividing cells.

Database: A collection of objects organized according to specific characteristics and rules.

Data mining: Useful information which can be extracted from large sets of molecules using various methods.

Dataset (or data set): A collection of data. Most commonly a data set corresponds to the contents of a single database table, or a single statistical data matrix, where every column of the table represents a particular variable (e.g. chemical structure file or input in the form of SMILES entry or InChI key), and each row corresponds to a given member of the data set in question.

DEcRyPT: Drug–Target Relationship Predictor. A machine learning method built with CATS2 descriptors.

Dereplication: A strategy used for the previous identification of known compounds in a complex sample, based on comparing spectroscopic and spectrometric data with the previously reported data from known compouds.

Derivatives of Natural Products: Chemical modification of natural product structures. Examples would be all of the beta-lactams ($\square$30,000) synthesized from the base penicillin G and cephalosporin C.

DFT: Density functional theory, a computational quantum mechanical modeling method used to investigate the electronic or nuclear structure of many-body systems, in particular atoms and molecules.

DiaNat-DB: A database of natural products with antidiabetic properties, collected by the Martinez & Madariaga group.

Diversity of a chemical database: Variety of chemical scaffolds, property space, and functional group space within a chemical database. The degree of the variety can be measured by various properties such as physical properties or descriptors generated by the use of scientific software.

3DMET: The three-dimensional structure database of natural metabolites. Available at: *http://www.3dmet.dna.affrc.go.jp/*.

DMNP: Dictionary of Marine Natural Products, a large commercially available collection of natural products data for compounds from marine sources. Available at: *http://dmnp.chemnetbase.com*.

DNA methyltransferases: Families of enzymes that attach the methyl group to DNA.

DNP: Dictionary of Natural Products, a large commercially available collection of natural products data. Available at: *http://dnp.chemnetbase.com*.

Docking: Computational technique used to find the binding mode of a ligand in a ligand-receptor complex by searching the conformational and orientational space of a ligand and receptor for geometry with favorable binding energy. It is useful for predicting the preferred orientation of one molecule to a second when bound to each other to form a stable complex.

Docking experiments: *In silico* experiments in order to investigate if there is a fit between a ligand and a binding site and the stability of the complex created.

Docking Score: See "*Scoring function*".

Drug: According to IUPAC's definitions, a drug is a compound against a human or an animal disease.

Drug discovery: Process of identifying a new drug and bringing it to market. It involves different scientific disciplines including biology, chemistry, and pharmacology.

DrugBank: A freely available online database describing drugs and their targets. Available at: *https://www.drugbank.ca/*.

Druglike (or Drug-like): A comparison between various chemical compounds and known drugs in order to estimate the degree of similarity based on different properties. A drug-like compound is defined as a compound with sufficiently acceptable ADME properties and sufficiently acceptable toxicity properties.

2D/3D structure: Structure of a chemical compound in two-dimensional space/structure of a chemical compound in three-dimensional space.

Dynophore: Dynamic pharmacophore that takes into account the conformational flexibility of both the ligands and the targets derived from MD simulations.

EF (Enrichment Factor): Concentration of the annotated ligands among the top-scoring docking hits compared to their concentration throughout the entire database.

EI (Electron Ionization): Also known as electron impact ionization and electron bombardment ionization, a high fragmentation ionization method in which highly energetic electrons interact with solid or gas-phase atoms or molecules to produce ions.

eMolecules Plus Database Download: The free version of eMolecules Plus Database with structural data. Available at: *https://www.emolecules.com/info/plus/download-database*.

Enzyme: A protein catalyst that regulates the rate at which chemical reactions proceed in living organisms without itself being altered in the process. Enzymes help speed up chemical reactions in the body and regulate several functions, e.g. breathing and digestion.

Erythromycins: Macrolide compounds with antibiotic activity isolated from *Saccharopolyspora erythraea*.

ESI: Electrospray ionization, spray ionization process in which either cations or anions in solution are transferred to the gas phase via formation and desolvation at atmospheric pressure of a stream of highly charged droplets that result from applying a potential difference between the tip of the electrospray needle containing the solution and a counter electrode.

ETD: Electron transfer dissociation, a process in which molecules with multiple protonations receive an electron from an ion with relatively low electron affinity. Electron capture frees energy and reduces its charge to form the corresponding electron-odd ion $[M + nH] (n-1) +$, which dissociates readily.

Extended connectivity fingerprints (ECFP fingerprint): Hashed *circular fingerprints* that encode each atom according to its atomic symbol and hybridization for path length 2, ECFP2; path length 4, ECFP4; and path length 6, ECFP6.

File formats: Types of files for storing chemical information that can be read and interpreted by computers.

Filters: A set of criteria used to characterize or exclude chemical compounds of a database according to specific rules.

Fingerprint: See **"Molecular Fingerprints"**.

Flavonoid: A class of (often bitter) plant and fungus secondary metabolites, whose name is derived from the Latin word *flavus* meaning yellow, their color in nature. Flavonoids in plants are generally responsible for the color of leaves and flowers. They consist of a diverse group of polyphenolic compounds commonly found in the human diet. See also **"Phenolics"**.

Focused library: A set of molecules with bioactivities against a drug target or family of targets.

Foodinformatics: Application of chemical information to food chemicals.

FooDB *(http://foodb.ca/):* A free database for chemical compounds found in foods.

Forcefield (or force field): A mathematical expression that describes the dependence of the energy of a molecule on the coordinates of the atoms in the molecule.

Fragmentation trees: Computationally generated spectral trees used to predict the CID MS(n) fragmentation pathway of a molecule.

GC: Gas chromatography (more accurately gas-liquid chromatography), an analytical separation technique for the qualitative and quantitative determination of volatile compounds that utilizes an inert gaseous mobile phase and a liquid stationary phase.

General-purpose database: A database containing, usually, molecules with drug-like properties.

Genotoxicity: The ability of a chemical compound to induce changes (damage) to genetic material.

GVKBIO: A commercially available library of natural compounds by GVKBIO company. Available at: *https://www.gvkbio.com/*.

Hepatotoxicity: The ability to poison liver cells.

Heteronuclear spectra: Spectra processed between nuclei of two different types. Frequently the nuclei are protons and another nucleus (heteronucleus).

High-throughput screening (HTS): Method of scientific experimentation that comprises the screening of large compound libraries for activity against biological targets via the use of automation, miniaturized assays, and large-scale data analysis.

HIM: The herbal ingredients *in vivo* metabolism database. A free database to academic researchers concerning active herbal ingredients combined with *in vivo* metabolism data. Available at: *http://www.bioinformatics.org.cn/*.

HIT: Herbal ingredients' targets database. This includes data on herbal ingredients, combined with data on their protein targets. Available at: *http://lifecenter.sgst.cn/hit/*.

Hit identification: The first step of a drug discovery project to obtain a molecule which binds to its target.

HMBC: Acronym for heteronuclear multiple-bond correlation spectroscopy. This two-dimension NMR spectroscopy reveals correlations between carbons and protons that are separated by two, three, and, sometimes in conjugated systems, four bonds.

Homonuclear spectra: Spectra processed between nuclei of two same types.

HOSE code: An acronym for Hierarchically Ordered Spherical Description of Environment.

HSQC: Acronym for heteronuclear single-quantum correlation spectroscopy. A type of two-dimension NMR spectroscopy with one axis for proton (^{1}H) and the other for a heteronucleus (an atomic nucleus other than a proton), which is usually ^{13}C or ^{15}N.

InChI: Acronym for IUPAC's "International Chemical Identifier", a text-based identifier for chemical substances which ensures the uniqueness of the structure represented.

InChI key: The compressed version of InChI, which is 27-characters long.

Indofine collection: The INDOFINE Chemical Company offers various libraries for HTS, including natural products. Available at: *www.indofinechemical.com*.

In silico: Experiments with the use of computers.

Inte:Ligand's libraries: Free, sample, virtual compound libraries provided by the company Inte:Ligand. Available at: *http://www.inteligand.com/ilibdiverse/samplelibs.shtml*.

Interbioscreen collections: Libraries of chemical compounds for screening purposes including natural products via Interbioscreen Ltd. Available at: *www.ibscreen.com*.

Interconversion: Chemical data exchange between different file formats.

In vitro: Experimental assay conducted in the wet lab within a glass apparatus.

In vivo: Experimental assay conducted in the wet lab within a living organism, e.g. mice.

Iridoid: A type of monoterpenoid in the general form of cyclopentanopyran, found in a wide variety of plants and some animals. They are biosynthetically derived from 8-oxogeranial. Iridoids are typically found in plants as glycosides, most often bound to glucose.

Isoflavonoid: Natural product derived from 3-phenylchromen-4-one, often regarded as a sub-class of flavonoids. Also, see "*Flavonoid*".

k-Nearest neighbors: Used for classification of chemical compounds or prediction of their properties, such as biological activities, *via* machine learning. Based on similarity defined by the property of k, a number which is used for separation of various compounds into groups.

KDD: Acronym for Knowledge Discovery in Databases.

KNIME: The Konstanz Information Miner (*www.knime.org*), a free and open-source data analytics, reporting, and integration platform. KNIME integrates various components for machine learning and data mining through its modular data pipelining concept.

Kohonen map: See "*Self-organizing map*".

LBVS (Ligand-based Virtual Screening): Most popular approach for drug discovery and lead optimization in the absence of the 3D structure of potential drug target.

LCMS/MS: Liquid chromatography/tandem mass spectrometry, a technique in which a mixture of compounds is separated into individual components by liquid chromatography. The components are ionized at the ion source of the mass spectrometer and further analyzed. In the first stage, ions are separated by mass-to-charge ratio, after which the ions are selected for fragmentation.

Lead: An active chemical compound which can be optimized in order to be considered a potential drug.

Lead Generation: The second step generally coming after hit identification (maybe called "hit to lead" or "H2L"), which refers to eliminating some unfavoured molecular properties to obtain developable compound.

Lead-likeness: It is characterized by the similarity with the structural and physico-chemical properties of a "lead" compound.

LELP index: Acronym for ligand efficiency dependent lipophilicity, a binding efficiency metric that combines lipophilicity, molecular size, and potency.

Library design: The overall procedure which takes place for the construction of a virtual chemical database. For example, it can include the choice of tools for filtering and management tools.

Ligand-based drug design: A rational drug design method based on a ligand presenting considerable biological activity.

Ligand efficiency (LE): Measure of the binding of a molecule in terms of free energy. LE values are often used to rank fragments and to monitor the progress of the optimization.

Ligand lipophilicity efficiency (LLE): Measure of a ligand-binding taking into account lipophilicity's beneficial effect. This is a metric used to monitor the lipophilicity with respect to an *in vitro* potency of a molecule.

Limonoid: A class of phytochemicals (sub-class of terpenoids), found in citrus fruit and certain other plants, believed to have various therapeutic effects.

Lipinski Rule of Five: An ensemble of rules useful to evaluate drug-likeness. It predicts a high probability of success or failure due to drug-likeness for molecules complying with two or more of the following rules. Lipinski's rule establishes that, in general, an orally active drug has no more than one violation of the following criteria: a molecular mass less than 500 Da; no more than five hydrogen bond donors; no more than ten hydrogen bond acceptors; an octanol-water partition coefficient logP not greater than 5. Also, see **"Rule of five"**.

Lipophilicity: The affinity of a molecule for lipophilic environments.

logP: A logarithm which defines the capability of a chemical compound to being dissolved in *n*-octanol (a measurement of its hydrophobicity) compared to being able to dissolve in water. Therefore, the log P value is the ratio of the concentration of the uncharged form of a compound in a nonpolar phase, traditionally water-saturated octan-1-ol, to that in water. Note: Common algorithms to calculate log P are CLOGP and ALOGP. Also, see **"clog P"**.

Machine learning (ML): An algorithm used in computers which generates models for predictions after being trained with a set of carefully chosen objects. Generally speaking, ML refers to a subset of artificial intelligence which uses algorithms to build statistical models based on training data in order to make predictions or decisions without being explicitly programmed to perform the task.

Machine learning techniques: The use of various algorithms in order to classify molecules or predict properties based on their structure.

Macrolides: A class of natural products that contain a macrocyclic ring lactone.

MALDI: Matrix-assisted laser desorption/ionization, the formation of gas-phase ions from molecules present in a solid or liquid matrix that is irradiated with a laser. The matrix is a material that absorbs the laser energy and promotes ionization.

MarinLit database: A database focused on marine natural products. Available at: *http://pubs.rsc.org/marinlit/.*

Masked Compounds: These are often based on natural products that when seen in 2D do not resemble a known NP structure but on (usually) ring-opening by enzymes, the active molecule is released. Seen with HCV inhibitors in particular.

Mass spectrometry: An analytical technique that measures the mass/charge ratio of the ions formed when a molecule or atom is ionized, vaporized and introduced into a vacuum. Mass spectrometry may also involve breaking molecules into fragments, thus enabling its structure to be determined.

MedChem Express collection: MedChem Express company offers various screening libraries including a natural products library. Available at: *http://www.medchemexpress.com/.*

Metabolism: For the field of medicinal chemistry, this term refers to the biotransformation of drugs.

Metabolite: A product obtained due to metabolism.

Metabolomics: A branch of "omics" science dealing with the comprehensive profiling of metabolites within an animal, plant, and microbial metabolome.

Metadynamics: Computational method aimed at enhancing the sampling of the configuration of space adding a time-dependent repulsive bias potential function of coarse-grained variables, called collective variables.

Mimics of Natural Products: A variation on the pharmacophore/active NP that binds at the active site of the targeted protein. People forget that chemical structures are 3-dimensional and a molecule that in 2D does not look like the NP, in 3D has groups that bind to the active site. Examples would be the HIV protease inhibitors which use peptidomimetics to imitate the native 6-peptide substrate.

MOAD: Acronym for Mother of All Databases, a database including high-quality assemblies.

Mode of action: The mechanism by which a chemical compound exerts a biological activity, e.g. by inhibition of an enzyme, by blocking an ion channel, etc.

Molecular Connectivity Diagram: Human-like computation method of building the structure of a compound from 2D NMR (HMBC, HSQC, COSY, NOESY) interpretation.

Molecular descriptors: Properties calculated from 2D/3D structure of molecules and expressed as arithmetical values in chemical databases. Molecular descriptors are terms that characterize a specific aspect of a molecule, e.g. it's molecular weight.

Molecular diversity: A measure of the spread of various properties or chemotypes within a set of compounds.

Molecular dynamics (MD): The study of the physical movement of collections of atoms and molecules as they interact. The goal is to attempt to identify the exact positions and velocities of each atom in the system by solving Newton's Second Law of motion ($F = ma$). MD has come to represent a major computational tool for the calculation of the time-dependent behavior of a molecular system.

Molecular editor/Structure sketcher: Scientific software used for the digital representation of chemical compounds.

Molecular fingerprints: Descriptions of the structure of a molecule calculated from the properties of each of its atoms and bonds that are then usually condensed into a fixed-length string by a *hashing algorithm*. A molecular fingerprint could be simply regarded as a binary theoretical descriptor encoding certain 2D or 3D structural features of a molecule. It is used to identify the presence or absence of a (sub) structure.

Molecular Interaction: Is used to refer to physicochemical contacts occurring between a biological target (generally protein) and ligand.

Molecular modeling: A method using computers and visualization techniques in order to obtain reliable 3D representations of molecules.

Molecular Networking: Computational tool used to assist dereplication procedures, in which is possible to visualize similarities in compound fragmentation (LC-MS/MS).

Molecular similarity: The degree to which two molecules resemble one another as calculated from their respective *2D or 3D properties*, *molecular fingerprints*, *fragment keys*, or superimposed *3D structures* that usually ranges from 1 (identical) to 0 (dissimilar). **Note:** Examples include *Tanimoto* or *Tversky similarities* for *2D structures* and *Carbo* or *Hodgkin* for *3D structures*.

MOL file: A file format which uses text-based connection tables in order to encode a chemical structure, substructure, and conformations.

Monte Carlo: Stochastic method that propagates the positions of atoms or groups of atoms in a molecule or collection of molecules through conformational space using a Boltzmann sampling of phase space.

MPD3: Acronym for Medicinal Plant Database for Drug Designing. A free database of medicinal plants. Available at: *http://bioinform.info/*. (*http://bioinformation.info/*, new site)

MS(n): Sequential multistage mass spectrometry, a technique in which the product ions are trapped after the second stage (MS2), allowing the repeated isolation and fragmentation of the product ions, thereby resulting in higher order {MS(n)} spectra, where "n" represents the number of times the isolation-fragmentation-measurement cycle has been carried out.

Multiplet: An NMR signal that is split, but is too complex to interpret easily. This might arise from non-first-order splitting or two or more overlapping signals.

Murcko scaffold: This refers to the most central ring system, plain ring systems, etc. The *Murcko scaffold* contains all plain ring systems of the given molecule plus all direct connections between them. Substituents, which don't contain ring systems are removed from rings and ring connecting chains. *Note 1:* Other terms, for example, "framework", "substructure", "cyclic system" or "fragment" are often synonymously used to refer to scaffolds. *Note 2:* see also **"Scaffold"** and **"Chemical scaffold"**.

Naïve Bayesian classifier: An algorithm for supervised learning classifications. See also **"Bayesian classifier"** and **"Bayesian"**.

NANPDB: Northern African natural products database, developed by Ntie-Kang and co-workers. Available at: *http://african-compounds.org/nanpdb/*.

NAPRALERT: Natural Products Alert database, a database based on information from literature sources. Available at: *https://www.napralert.org/*.

Natural products (NPs): Chemical substances that are produced by living organisms, used by them for specific purposes, such as protection, defense against predators, attracting mating partners, etc. NPs are pure materials isolated from plants, microbes or marine invertebrates usually by "Bioactivity Driven Isolation Techniques".

Natural product-likeness: Similarity with structural and physicochemical properties of a natural product.

Natural product-likeness score: A mathematical Bayesian equation that measures the NP-like score to ensure proximity to NP.

NCI: Abbreviation for National Cancer Institute, USA.

NCI Natural Products Repository: A collection of natural products provided by NCI. Available at: *https://dtp.cancer.gov/organization/npb/introduction.htm*.

Neural network (plural neural networks): A real or virtual computer system designed to emulate the brain in its ability to "learn" to assess imprecise data. By doing so, they attempt to mimic the functioning of the network of neurons that form the brain, i.e. that function together to achieve a common purpose.

Neuron (in an Artificial Neural Network): The virtual analogous of a neuron in the human brain. It resembles the neuron's acting in the brain by receiving inputs, processing information and extracting results, and finally giving the output of the previous procedure. This is used in machine learning approaches.

NMR: See "**Nuclear magnetic resonance**".

NOESY: Acronym for nuclear Overhauser effect spectroscopy. It is a homonuclear two-dimension NMR technique utilizing the nuclear Overhauser effect. The NOESY spectrum reveals which protons are close enough (usually no more than about 3–4 Angstroms apart) to transfer energy this way.

Non-ribosomal peptide synthases (NRPSs): A family of multimodular enzymes, consisting of repeated modules, using regiospecific and stereospecific reactions for peptides biosynthesis.

NPACT: Naturally occurring plant-based anticancerous compound-activity-target database. Available at: *http://crdd.osdd.net/raghava/npact/*.

NPCARE: Database of natural products for cancer gene regulation, natural products with anticancer activity.Available at: *http://silver.sejong.ac.kr/npcare/*.

NPDI collection: Natural Products Discovery Institute collection of natural products. Available at: *http://www.npdi-us.org/collection/*.

NPLI: Acronym for The Natural Products Library Initiative. The Natural Products Library Initiative at The Scripps Research Institute. Available at: *http://www. scripps.edu/shen/NPLI/npliattsri.html*.

NPs: See "**Natural Products**".

NuBBE database (also **NUBBE_{DB}):** A natural product database from Brazilian biodiversity (including the Bolzani Lab). Available at: *http://nubbe.iq.unesp.br/portal/ nubbedb.html*.

Nuclear magnetic resonance (NMR): Nuclear magnetic resonance, a physical phenomenon in which atom nuclei in a strong static magnetic field are perturbed by a weak oscillating magnetic field and respond by producing an electromagnetic signal with a frequency characteristic of the magnetic field at the nucleus.

Nuclear Overhauser effect (NOE): When a nucleus (usually a proton) relaxes from an excited nuclear spin state to its ground nuclear spin state the energy may be transferred to another nucleus in the molecule.

Open access (OA): Refers to free, unrestricted online access to research outputs such as journal articles and books. OA content is open to all, with no access fees.

Open source: The practice of providing open-source code for a product, e.g. software in general.

Orthosteric site: (the word orthosteric means "not comparable"). This describes the primary, unmodulated binding site (on a receptor) of a ligand.

p-ANAPL: The pan-African natural products library. A physical collection of molecules from African medicinal plants. A virtual version of this collection has been constructed in order to be used for in silico experiments.

PAINS (pan-assay interference compounds): Compounds presenting false positive biological activity during various tests. Compounds presenting PAINS alerts are compounds influencing the identification of new bioactive compounds since they can origin false activity signals.

PBPK modeling: Abbreviation of Pharmacologically based pharmacokinetics modeling. A mathematical modeling method for the prediction of ADME properties of a chemical compound in humans or animals.

PDB (protein databank/RCSB PDB): A database of 3D structures of macromolecules and macromolecules' complex assemblies with ligands, presenting biological interest. Available at: *http://www.rcsb.org/*.

Pestimep: Acronym for Pesticide Multiple Endpoint Database, collected by Madariaga & Martinez group.

Pharmacophore (ph4): The ensemble of necessary steric and electronic features which result in the optimal supramolecular interaction between a specific biological target and a molecule by triggering or blocking the relative biological response. A ph4 could be a proposal for the ensemble of steric and electronic features that define the optimal supermolecular intermolecular interaction of a ligand with a specific biological target structure with the result that it triggers or blocks its biological response.

Pharmacophore-based virtual screening: Screening virtual molecules using pharmacophore models as a query.

Pharmacophores of Natural Products: The base structure within a natural product that is the part of the molecule that binds to the target enzyme or protein.

Phenolics: Naturally occurring compounds containing hydroxyl group directly linked to a phenyl ring.

Phenylpropanoids: Secondary metabolites synthesized by plants and formed by the amino acids phenylalanine and tyrosine.

Pheromones: Chemical substances produced by one species for the purpose of making other members in the same species to respond for a specific purpose. In the case of insects, pheromones may be used to attract sexual partners (sex pheromones) or causing other members to come together for collective action (aggregating pheromones), etc.

Physical NP libraries: Physical collections of natural products, i.e. with available physical samples ready for testing.

Phytochemica: A database of plant-derived compounds from the Himalayas. Available at: *faculty.iiitd.ac.in/~bagler/webservers/Phytochemica/index.php*.

Phytochemical: A chemical substance of plant origin, i.e. naturally biosynthesized by a plant in the course of surviving, adapting to the environment or protecting itself from disease or predators.

PI Chemicals Collection: A chemical library provided by the company serving as compound suppliers. Available at: *www.pipharm.com*.

Polyketides: This refers to a large group of secondary metabolites built by condensation of acetate or propionate units deriving in poly-β-keto chains. Natural compounds which are composed of alternating carbonyl and methylene groups. The term also refers to further condensation, reduction, dehydration products of the polymethylene/keto compounds.

Polyketide synthases (PKSs): A family of enzymes consisting of some modules for the corresponding building-block of polyketides in fungi, bacteria, and plants.

Polypharmacology: Design or use of pharmaceutical agents that act on multiple targets or disease pathways.

PPDB: Acronym for Pesticide Property Database, developed by the Agriculture and Environment Research Unit (AERU) at the University of Hertfordshire. The information is available at *https://sitem.herts.ac.uk/aeru/ppdb/*.

Principal components: Non-correlated variables. Also, see "**Principal component analysis**".

Principal component analysis (PCA): A technique for machine learning methods used for the reduction of the variables needed to apply a classifier or an algorithm for property predictions. PCA is, therefore, variable reduction method that operates on the correlation matrix of the variables to construct a small set of new orthogonal, i.e. non-correlated, variables (principal components) derived from linear combinations of the original variables.

Private database: A commercial collection of molecules or an in-house library not open to the public.

Problematic functional groups: Parts of a chemical compound which contain the toxicophore or trigger off the reactions responsible for the observed adverse side effects or toxicity.

Protein target: Proteins are biochemical catalysts, i.e. chemical substances that regulate the speed of biochemical reactions. A protein that is a drug target is one for which the reaction involved is connected to a disease or biochemical function within the body. The inhibition of such a protein would affect the function of the body, hence "target" the disease.

PubChem: A free chemical database including data on chemical structures, biological activities and more, published by the National Center for Biotechnology Information (NCBI), USA. Available at: *https://pubchem.ncbi.nlm.nih.gov/.*

Public database: A collection of molecules freely downloadable and free to use.

PubMed: A free database including abstracts and citations from the literature, along with links to the original publications, published by the National Center for Biotechnology Information (NCBI), USA. Available at: *https://www.ncbi.nlm.nih. gov/pubmed/.*

Purchasable chemical space: A combination of diversity and purchasability of compounds or simply the compounds in literature which can be purchased for screening purposes.

Quality Phytochemicals collection: A commercially available library of natural compounds. A collection of phytochemicals provided by the company Quality Phytochemicals. Available at: *http://www.qualityphytochemicals.com/.*

Quantitative structure-activity relationships (QSAR): A method aiming at the construction of models which describe the relationship between structures and biological activities. Quantitative structure-activity relationships (QSAR) are mathematical relationships linking chemical structure and pharmacological activity in a quantitative manner for a series of compounds. Methods which can be used in QSAR include various regression and pattern recognition techniques. QSAR is often taken to be equivalent to chemometrics or multivariate statistical data analysis. It is sometimes used in a more limited sense as equivalent to Hansch analysis. QSAR is a subset of the more general term "structure-property correlations" (SPC).

Racemic: Used to describe a mixture of two substances that are identical in all respects but having opposite effects in terms of interaction with plane-polarized light. Equimolar amounts of dextrorotatory and levorotatory substances lead to a racemic mixture.

Random forest: An algorithm for the classification of compounds and for the prediction of their biological activities (machine learning technique). An ensemble learning approach that constructs a large number of decisions trees, and outputs predictions that are a collection of the votes of the individual trees. A subset of the training dataset is chosen to grow individual trees, with the remaining samples used to estimate the optimal fit. Trees are grown by splitting the training set (subset) at each node according to the value of the random variable sampled independently from a subset of variables.

Rational drug design: Drug design, often referred to as rational drug design or simply rational design, is the inventive process of finding new compounds as medications based on the knowledge of a biological target. The drug is most commonly an organic small molecule that activates or inhibits the function of a biological molecule such as a protein, which in turn results in a therapeutic benefit to the patient. In the most basic sense, drug design involves the design of molecules that are complementary in shape and charge to the biomolecular target with which they interact and therefore will bind to it. Drug design frequently but not necessarily relies on computer modeling techniques. This type of modeling is sometimes referred to as computer-aided drug design.

RDKit tool: An open-source cheminformatics software (*https://www.rdkit.org/*), also useful for computational chemistry, and predictive modeling.

Read-across: A technique for predicting endpoint information for one substance (target substance), by using data from the same endpoint from (an)other substance(s), (source substance(s)).

Reaxys: This is a vast commercial database that includes about 220,000 NP annotations. Apart from supporting chemistry research, including pharmaceutical development, the chemicals industry, and academic research, Reaxys provides integrated access to the eMolecules database containing more than 8 million unique molecules, including screening compounds and building blocks, from over 150 commercial suppliers, which is updated weekly. More information about Reaxys at: *https://www.elsevier.com/solutions/reaxys*.

Receiver Operating Characteristic (ROC) curve: A metric used as an objective way to evaluate the ability of a given test to discriminate between two populations.

Receptor: Macromolecules, usually proteins, found in cells. They recognize and make complex assemblies with specific compounds. A receptor is a protein or a protein complex in or on a cell that specifically recognizes and binds to a compound acting as a molecular messenger (neurotransmitter, hormone, lymphokine, lectin, drug, etc). In a broader sense, the term receptor is often used as a synonym for any specific (as opposed to non-specific such as binding to plasma proteins) drug binding site, also including nucleic acids such as DNA.

Regression: A method for statistical analysis. See "*Regression analysis*".

Regression analysis: The use of statistical methods for modeling a set of dependent variables, Y, in terms of combinations of predictors, X. It includes methods such as multiple linear regression (MLR) and partial least squares (PLS).

RIE (Robust Initial Enhancement): Metric using a continuously decreasing exponential weight as a function of rank.

ROC analysis: A method for decision making. The ability of the method to predict accurately is shown with a graphical plot known as the ROC curve. see "**Receiver Operating Characteristic (ROC) curve**".

ROESY: Acronym for rotating frame nuclear Overhauser effect spectroscopy. A type of two-dimension NMR spectroscopy similar to NOESY, but the initial state is different. Spin-locking is employed to ensure that correlation between spins, which are close in space but have zero NOE, are seen. This technique is used for molecules whose tumbling regime lies at the border of positive and negative NOE responses.

Rotatable Bond: The chemical bond type which can turn around itself, generally a single bond which is not in a ring system or not done with a terminal heavy atom.

Rule of Five: The rule of five states that molecules that violate two or more of the following rules are likely to have permeability problems: (1) CLOGP calculated octan-1-ol/water $log P$ greater than 5.0; (2) molecular weight greater than 500; (3) more than five hydrogen bond donors; and (4) the sum of oxygen and nitrogen atoms is greater than 10. *Note 1:* Natural products, peptides and other substrates for biological transporters are exceptions. *Note 2:* The original authors also noted that one can calculate the octan-1-ol/water $log P$ with the program MLOGP for which the cut-off is 4.15. Users often use $log P$'s calculated with other programs or measured values.

Rule of Three: Assemble of rules useful to evaluate drug-likeness. It predicts high probability of success or failure due to drug-likeness for molecules complying with 2 or more of the following rules. These rules establish that, in general, an orally active drug has no more than one violation of the following criteria: a molecular mass less than 300 Daltons; an n-octanol/water partition coefficient $log P$ not greater than 3; no more than 3 hydrogen bond donors and acceptors; no more than 3 rotatable bonds.

SANCDB: South African natural compound database. A free database of compounds from South Africa. Available at: *https://sancdb.rubi.ru.ac.za/*.

Scaffold: A term used in medicinal chemistry or computational chemistry to refer to core structures of natural products which may contribute to some form of desirable biological activity. This refers to the molecular core to which functional groups are attached. The method used to locate the core structure(s) depends on the chosen scaffold type. See also "**Murcko scaffold**".

Scaffold hopping: Process for the discovery of bioactive compounds that are structurally distinct from their reference molecules.

Scaffold Hunter: A tool that helps to navigate chemical and biological spaces interactively and provide new synthetic directions based on the scaffold hierarchy.

Scaled Shannon Entropy (SSE): SE values normalized. In chemical diversity analysis, the values of SSE range between 0, where all P compounds are contained in one cyclic system, and 1.0, where each cyclic system contains an equal number of compounds. Therefore, SSE values closer to 1.0 indicate large scaffold diversity within the *n* most populated cyclic systems. Also, see *"Shannon Entropy"*.

Scoring function: A mathematical function used to approximately predict the binding affinity between two molecules after they have been docked.

SDF file: Structure-Data File, a file format readable by most computational chemistry software. It uses text-based connection tables in order to represent a chemical structure.

Secondary metabolism: The process resulting in the production of metabolites by routes other than the normal pathways.

Secondary metabolites: Molecules which are not essential for the normal growth, development, or reproduction of the producer organisms but specifically modulate health-maintaining processes. They are chemical compounds produced by organisms in biochemical processes different from those life processes that are common in all living organism.

Selectivity or sensitivity (S_e): The preference of a ligand to perform a stable complex with a specific receptor. It describes the ratio of the number of the active molecules found by a virtual screening method to the number of all active database compounds.

Self-organizing map (SOM): A type of artificial neural network that uses unsupervised *machine learning* to project high-dimensional data into two dimensions that are usually presented as a contour plot. *Note:* This is sometimes called a Kohonen map.

Selleckchem library: A commercially available library of natural compounds. Various screening libraries including Natural Product Library provided via company Selleckchem (Selleck Chemicals). Available at: *http://www.selleckchem.com*.

Sequence alignment: In bioinformatics, a sequence alignment is a way of arranging the sequences of DNA, RNA, or protein to identify regions of similarity that may be a consequence of functional, structural, or evolutionary relationships between the sequences. Aligned sequences of nucleotide or amino acid residues are typically represented as rows within a matrix. Gaps are inserted between the residues so that identical or similar characters are aligned in successive columns.

Shannon Entropy (SE): The Shannon entropy equation provides a way to estimate the average minimum number of bits needed to encode a string of symbols, based on the frequency of the symbols. In chemical diversity analysis, the Shannon entropy (SE) of a population of P compounds distributed in n systems is defined as:

$$SE = -\sum_{i=1}^{n} p_i \log_2 p_i \qquad p_i = \frac{c_i}{P}$$

where p_i is the estimated probability of the occurrence of a specific chemotype i in a population of P compounds containing a total of n acyclic and cyclic systems, and c_i is the number of molecules that contain a particular chemotype c.

Shikimate pathway: Metabolic route found in microorganisms and plants (absent in animals) for the biosynthesis of folates and aromatic amino acids.

Similarity (Molecular Similarity): To what extent two compounds resemble based on various criteria such as 2D/3D properties or molecular fingerprints. Tanimoto similarity is commonly used.

Similarity Ensemble Approach: *In silico* method for the discovery of drug targets.

Similarity search: Screening of a database for molecules similar to a query molecule.

SMILES: Acronym for Simplified Molecular-Input Line-entry System, a specification in the form of a line notation for describing the structure of chemical species. SMILES is a string notation used to describe the nature and topology of molecular structures.

Specificity (S_p): The ratio between the number of inactive compounds not selected by the VS protocol, and the number of all inactive molecules included in the chemical database. Specificity ranges from 0 to 1 and denotes the percentage of truly inactive compounds. $S_p = 0$ defines the worst-case scenario where all inactive compounds are selected by error as actives, whereas $S_p = 1$ means that all inactive compounds have been correctly rejected during the screening process.

Specs Natural Products: A commercially available library of natural compounds provided by Specs. Available at: *http://www.specs.net/page.php?pageid= 2004111115353984&smenu=2008111411133023.*

Spectral trees: The sequential stages and relationships of CID MS(n) fragmentations, representing precursor and product ions as nodes and neutral losses as edges. It is used to (sub)structure relationships in a hierarchical order.

SPiDER: Self-Organizing Maps (SOM)-based prediction of drug equivalence with CATS2 and physicochemical descriptors.

Stilbenes: Secondary metabolites structurally characterized by a diarylethene moiety. A widely known stilbene derivative is resveratrol.

StreptomeDB: A database focused on natural products produced by *Streptomycetes* species, developed by the Günther Lab. Available at: *http://132.230.56.4/strepto medb2/.*

Structural alert/expert rule/toxicophore: Molecular substructures associated with a particularly adverse outcome, used to inform about the toxicity of pharmaceuticals or agrochemicals.

Structure-activity landscape: See *"Activity and property landscape modeling"*.

Structure-activity landscape index (SALI): A metric used to identify activity cliffs, defined as pairs of structurally similar compounds with large differences in bioactivity/potency.

Structure-based drug design: A method to design a potential drug based on the 3D structure of the macromolecule target.

Structure-based virtual screening (SBVS): A computational approach used in the early-stage drug discovery to search a chemical compound library for novel bioactive molecules against a drug target for which the 3D structure is known. An example is docking various compounds against a macromolecule-target in order to investigate the interactions between them and the stability of the resulted complex.

Substructure search: Virtual screening of compound libraries in order to identify those containing the substructure used as the query.

SuperNatural II: A free to use database of natural products, developed by the Preissner Lab. A database of natural products with various details such as 2D structures and physicochemical properties. Available at: *http://bioinf-applied.char ite.de/supernatural_new.*

Support vector machine (VSM): A technique used in machine learning approaches for the tasks of classification and prediction (of properties).

Sweetener: A substance that imparts a sense of sweetness in the mouth like sugar. Sweeteners may be natural or synthetic; they may also be calorific or not. Examples of non-calorific sweeteners include stevioside (a diterpene glycoside), or some plant-derived proteins like brazzein and miraculin which are taste modifying sweeteners. Synthetic sweeteners include sucralose, cyclamate, etc.

Tandem MS (MS/MS): A technique encompassing the formation of precursor and product ions in the first stage (MS1), and the separation and detection of product ions in a second stage (MS2).

Tannins (or tannoids): A class of astringent, polyphenolic biomolecules, widely distributed in many species of plants, where they play a role in protection from predation (including as pesticides) and might help in regulating plant growth. They bind to and precipitate proteins and various other organic compounds including amino acids and alkaloids.

Target fishing: Discovery of the interactions between bioactive, chemical molecules and different targets.

Target-focused library: A collection of molecules presenting bioactivity against a target.

TargetMol's Natural Compound Library: Various collections of natural compounds (e.g. antitumor natural products library) provided by the company TargetMol. Available at: *www.targetmol.com*.

TCM: Abbreviation for Traditional Chinese Medicine.

TCM Database@Taiwan: Traditional Chinese medicine database@Taiwan. A database regarding the traditional Chinese medicine appropriate for in silico screening. Available at: *http://tcm.cmu.edu.tw/*.

TCMID: Traditional Chinese medicine integrated database. A database with data on traditional Chinese medicine combined with data on drugs and diseases. Available at: *www.megabionet.org/tcmid*.

Terpenoid (sometimes called isoprenoids): A large and diverse class of natural products derived from isoprene (about 60% of known natural products are terpenoids). Although sometimes used interchangeably with "terpenes", terpenoids contain additional functional groups, usually O-containing.

TIGER: Target Inference Generator. Method inspired in SPiDER for target prediction.

TimTec NPL: TimTec's Natural Product Library. The collection of natural products by the company TimTec LLC. Available at: *http://www.timtec.net/natural-compound-library.html*.

TIPdb: The Taiwan indigenous plant database. A database of phytochemicals from plants of Taiwan. Available at: *http://cwtung.kmu.edu.tw/tipdb/*.

TOCSY: Acronym for total correlation spectroscopy. In this experiment, cross peaks are observed not only for nuclei that are directly coupled, but also between nuclei which are connected by a chain of couplings.

Topological polar surface area (tPSA): An approximation to the *polar surface area* that is calculated from the 2D structure of a molecule.

Toxicity: The degree to which a chemical substance or a particular mixture of substances can damage an organism. This could refer to the effect on a whole organism, such as an animal, bacterium, or plant, as well as part of the organism, e.g. a cell (cytotoxicity) or an organ like the liver (hepatotoxicity). A central concept of toxicology is that the effects of a toxicant (or toxin) are dose-dependent.

Toxicophore: A functional group (structural group) which can probably cause toxic effects during metabolic activation.

Toxin: A substance (mixture or pure compound) exhibiting toxicity.

Transcript: The result copying of part of a DNA sequence into a transfer RNA (codon).

Transcription: The process by which the information provided by a specific area of a gene is copied into a transfer RNA.

UEFS Natural Products Database: A database provided by Universidade Estadual De Feira De Santana *via* the ZINC database. Available at: *http://zinc.docking.org/cata logs/uefsnp.*

UNPD: Acronym for Universal Natural Products Database. A freely available database of natural products from plants, animals, and microorganisms.

Virtual chemical database: A virtual collection of chemical compounds accompanied by various data (generated or derived from the literature), usually biological activities and physicochemical properties.

Virtual screening: Computational method that ranks the molecules in a database by their forecast continuous or categorical biological or chemical properties. It is often used to predict new ligands on the basis of tridimensional biological structure, with the goal to filter enormous virtual chemical databases of small organic molecules.

ZINC (an acronym for "ZINC is not commercial"): The largest virtual collection of compounds, whose 3D structures are readily available for docking purposes. A free database of compounds which are commercially available. Available at: *http://zinc15. docking.org/.*

References

Martin YC, Abagyan R, Ferenczy GG, Gillet VJ, Oprea TI, et al. Glossary of terms used in computational drug design, part II (IUPAC Recommendations 2015), Pure Appl Chem. 2016;88:239–64.
Wermuth CG, Ganellin CR, Lindberg P, Mitscher LA. Glossary of terms used in medicinal chemistry. Pure Appl Chem. 1998;70:1129–43.
Nagel B, Dellweg H, Gierasch LM. Glossary for chemists of terms used in biotechnology (IUPAC Recommendations 1992). Pure Appl Chem. 1992;64:143–68.
Duffus JH. Glossary for chemists of terms used in toxicology (IUPAC Recommendations 1993). Pure Appl Chem. 1993;65:2003–122.
Nordberg M, Duffus JH, Templeton DM. Glossary of terms used in toxicokinetics (IUPAC Recommendations 2003). Pure Appl Chem. 2004;76:1033–82.
Moss GP, Smith PA & Tavernier D. Glossary of class names of organic compounds and reactive intermediates based on structure (IUPAC Recommendations 1995), Pure Appl Chem. 1995;67:1307–75.
Buckle DR, Erhardt PW, Ganellin CR, Kobayashi T, Perun TJ, et al. Glossary of terms used in medicinal chemistry. Part II (IUPAC Recommendations 2013), Pure Appl Chem. 2013;85:1725–58.
Bajorath J. Integration of virtual and high-throughput screening. Nat Rev Drug Discovery. 2002;1:882–94.
Gu J, Gui Y, Chen L, Yuan G, Lu H-Z, Xu X. Use of natural products as chemical library for drug discovery and network pharmacology. PLoS ONE. 2013;8:e62839.

Index

https://doi.org/10.1515/9783110579352-019